Crystal Engineering: From Molecules and Crystals to Materials

T0137469

NATO Science Series

A Series presenting the results of activities sponsored by the NATO Science Committee. The Series is published by IOS Press and Kluwer Academic Publishers, in conjunction with the NATO Scientific Affairs Division.

A.	Life Sciences	IOS Press
B.	Physics	Kluwer Academic Publishers
C.	Mathematical and Physical Sciences	Kluwer Academic Publishers
D.	Behavioural and Social Sciences	Kluwer Academic Publishers
E.	Applied Sciences	Kluwer Academic Publishers
F.	Computer and Systems Sciences	IOS Press
1.	Disarmament Technologies	Kluwer Academic Publishers
2.	Environmental Security	Kluwer Academic Publishers
3.	High Technology	Kluwer Academic Publishers
4.	Science and Technology Policy	IOS Press
5.	Computer Networking	IOS Press

NATO-PCO-DATA BASE

The NATO Science Series continues the series of books published formerly in the NATO ASI Series. An electronic index to the NATO ASI Series provides full bibliographical references (with keywords and/or abstracts) to more than 50000 contributions from internatonal scientists published in all sections of the NATO ASI Series.
Access to the NATO-PCO-DATA BASE is possible via CD-ROM "NATO-PCO-DATA BASE" with user-friendly retrieval software in English, French and German (WTV GmbH and DATAWARE Technologies Inc. 1989).

The CD-ROM of the NATO ASI Series can be ordered from: PCO, Overijse, Belgium

Series C: Mathematical and Physical Sciences – Vol. 538

Crystal Engineering: From Molecules and Crystals to Materials

edited by

Dario Braga
Dipartimento di Chimica,
Università di Bologna,
Bologna, Italy

Fabrizia Grepioni
Dipartimento di Chimica,
Università di Sassari,
Sassari, Italy

and

A. Guy Orpen
School of Chemistry,
University of Bristol,
Bristol, United Kingdom

Kluwer Academic Publishers

Dordrecht / Boston / London

Published in cooperation with NATO Scientific Affairs Division

Proceedings of the NATO Advanced Study Institute on
Crystal Engineering: From Molecules and Crystals to Materials
Erice, Italy
12–23 May 1999

A C.I.P. Catalogue record for this book is available from the Library of Congress.

ISBN 0-7923-5898-8 (HB)
ISBN 0-7923-5899-6 (PB)

Published by Kluwer Academic Publishers,
P.O. Box 17, 3300 AA Dordrecht, The Netherlands.

Sold and distributed in North, Central and South America
by Kluwer Academic Publishers,
101 Philip Drive, Norwell, MA 02061, U.S.A.

In all other countries, sold and distributed
by Kluwer Academic Publishers,
P.O. Box 322, 3300 AH Dordrecht, The Netherlands.

Printed on acid-free paper

TABLE OF CONTENTS

vi

PREFACE

Modern crystal engineering is the planning and utilization of crystal-oriented syntheses and the evaluation of the physical and chemical properties of the resulting crystalline materials. Hence crystal engineering is an interdisciplinary area of research that cuts horizontally across traditional subdivisions of chemistry, e.g. organic, inorganic, organometallic, materials chemistry and biochemistry. The spectacular growth in the chemistry and properties of complex supramolecular systems has fueled further interest in controlling and exploiting the aggregation of molecules in crystals. The interest in the potential of this research field is increasing rapidly. This has sprung from a diverse set of research fields such as topochemical reactions of crystalline organic species, interest in new organometallic and inorganic materials, biomineralization and biomimetics, the packing of organic molecules in crystals and the effort to predict crystal structures theoretically. It has been said that crystal engineering is at the intersection of supramolecular chemistry and materials chemistry. There is a synergistic relationship between design and synthetic methods of supramolecular chemistry and those exploited to build crystalline aggregates. In both cases the collective properties of the aggregate depend on the choice of inter- molecular and inter-ion interactions between components. Materials for applications in optoelectronics, conductivity, superconductivity, magnetism as well as in catalysis, molecular sieves, solid state reactivity, and mechanics are being sought.

The Erice 1999 NATO Advanced Study Institute is intended to foster close interaction between emerging scientists and those who are established leaders in the areas of crystal engineering. We hope that this book will provide a state-of-the-art description of the field and offer new ideas to young scientists so that the multidisciplinary field will emerge strengthened. The topics discussed in the 28 contributions deal with many of the sources of scientific input to the discipline, e.g. supramolecular chemistry, crystal-directed synthetic strategies, preparation of coordination polymers, hydrogen bonded frameworks, networks, clathrates, inclusion compounds, techniques for the characterization of solid supramolecular aggregates, hydrogen bonding, non-covalent interactions between closed shell atoms, theoretical calculations of crystal structures and properties, applications in optoelectronics, second harmonic generation for optical devices, conductivity and superconductivity, charge-transfer and magnetism, nano- and meso-materials, kinetics and thermodynamics of molecular crystals and inclusion compounds, and crystal nucleation. Though long, this list can has no ambitions to be exhaustive, but we trust that the reader will also find fresh and current entries in the relevant literature on the various topics.

We expect that the dissemination of the principles and methods of crystal engineering will stimulate new research activity and, to some extent, also lead to conversion of some obsolete lines of research into new ones. The ambitious objective of the 1999 Erice School on Crystal Engineering and of this book is that of assisting the establishment and growth of a synergistic interaction (and a common scientific language) between those who are expert in modelling and theoretical evaluations, those who have developed reproducible strategies for crystal synthesis, and those who can characterize the products structurally and measure their physical properties. Collaboration with those expert in structure simulation, datamining, theoretical calculations and in the utilization of the tools of molecular chemistry to modify molecular architecture and prepare new building blocks is also essential.

We thank the NATO Science Committee for financially supporting the 28th Course of the International School of Crystallography. We also thank the EC Commission, DG XII, TMR Programme, Euroconferences, Bruxelles, and the International Union of Crystallography, Chester, for sponsorship and the "Ettore Majorana" Centre for Scientific Culture in Erice for hosting the School. We are particularly grateful to Professor Lodovico Riva di Sanseverino and to Professor Paola Spadon for their relentless assistance and invaluable comments and suggestions.

Finally, we wish to thank the Authors for their work. The interest in participating in the Erice School, the number of papers quoted and published in the best scientific journals and the high quality of the contributions collected in this Book demonstrate that crystal engineering, on the verge of the new millennium, has become a fully-fledged discipline.

Dario Braga
Fabrizia Grepioni
A. Guy Orpen

ON THE CONCEPT OF A SINGLE CRYSTAL IN BIOMINERALIZATION

L. ADDADI, J. AIZENBERG, E. BENIASH, S. WEINER
Dept of Structural Biology, Weizmann Institute of Science
76100 Rehovot, Israel

1. Introduction

Organisms produce a myriad of different mineral forms and types [1,2]. Some of these are amorphous, but the vast majority are crystalline. In some cases, such as bone or mollusk shell, the material is composed of many small single crystals. In other cases, such as the spines and plates of sea urchins, the whole skeletal element is one enormous single crystal. The single crystals themselves are often most unusual. For example, the crystals of bone (Fig. 1a) present a series of surprises. Bone crystals are composed of a form of calcium phosphate, called carbonated apatite. They are probably the smallest crystals known to be formed biologically. They are just 20 to 30 Å thick, that is about 5 unit cells, and a few hundred Ångstroms long and wide [3]. Their surface area is around 250 m^2/g [4]. Even more perplexing is the fact that their platy shapes do not reflect the hexagonal symmetry of the apatite crystal lattice.

The same carbonated apatite mineral is also present in the outer enamel layer of vertebrate teeth. In enamel, however, the crystals have a shape that is completely different from those of bone. They are extremely long (maybe up to 100 μm or so), about 450 nm wide and 17 nm thick (Fig 1b)[5]. They are arranged in bundles, and the bundles in turn are oriented in 3 different directions [6] to form a highly complex and very stiff material.

The magnetic iron oxide mineral, magnetite, forms only under very extreme conditions of pH and Eh at atmospheric temperatures and pressures [7]. Organisms, from bacteria to mammals, form single crystals of magnetite. Many of these crystals are the size of single domain magnets, as organisms use them for navigation purposes [8]. Figure 1c shows an example of biologically produced magnetite crystals, whose morphologies are rounded and thus quite different from inorganically formed magnetite, which normally expresses well defined crystal faces.

Sea urchins have five continuously forming teeth, which they use for scraping the rocky substrate to extract food. The whole tooth is composed of calcite with small amounts of organic material [9]. The rocks that they scrape are often also composed of calcite, or sometimes of materials that are even harder than calcite. One of several "tricks" used by the sea urchin to harden the tooth scraping surface, is to incorporate needle-shaped single crystals of calcite, that taper in size from about 20 μm to less than

1

D. Braga et al. (eds.), Crystal Engineering: From Molecules and Crystals to Materials, 1–22.
© *1999 Kluwer Academic Publishers. Printed in the Netherlands.*

1 μm in diameter at the scraping tip [10](Fig. 1c). The decrease in size significantly reduces the chance of a defect being present in the crystal lattice, and hence increases its hardness. Thus the world of biomineralization challenges us to understand the manners in which these single crystals form and function. It also forces us to carefully examine the concept of a single crystal.

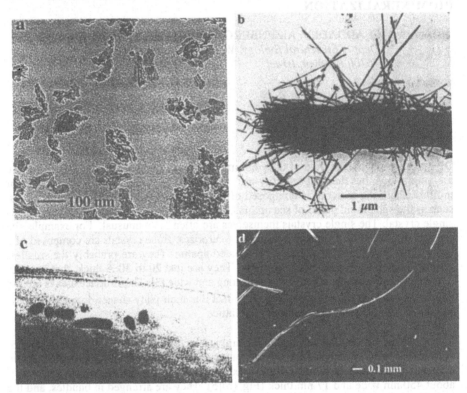

Figure 1, Examples of biologically produced single crystals. a) Plate shaped crystals of carbonated apatite extracted from a 14 months old rat bone (tibia) [4]. b) Carbonated apatite crystals from the forming tooth enamel of a rat. The individual highly elongated crystals are seen spalling off a prism composed of thousands of such crystals all aligned along their *c* axes. c) Rounded to cone-shaped single crystals of magnetite inside a magnetotactic bacterium as viewed in the transmission electron microscope. The species of bacterium is unknown. The crystals are 50-150 nm long. The picture is from the collection of the late H.A. Lowenstam, given to him by H. Vali. d) Scanning electron micrograph of a fiber from the tooth of the sea urchin, *Paracentrotus lividus.* The whole fiber is one single crystal of calcite *[10].*

What is a single crystal? A minimalist definition is based on a unit cell, namely "a parallelepiped containing the smallest atomic group of the crystal which repeats itself identically throughout space .... by translations equal to the fundamental lattice vectors" [11]. This in turn implies that, "To define the local electron density", on which diffraction of X-rays by crystals is based, "we require two quantities: .....the *form factor*, which defines the exterior shape of the object, and..... the distribution of matter within the crystal, assumed to be *homogeneous and infinite*." Other assumptions, tacit or explicit, are that the crystal contains a large number of unit cells (dimensions of the order of one micron), and that it has an external shape with a symmetry higher or equal to the internal symmetry of the lattice. Based on these assumptions, one can then proceed to define and study the parameters specific to crystalline states, such as structure, perfection, long and short range order, superlattices, morphology, etc.

Within the realm of biomineralization, it is evident that many of the postulates of crystallography break down. Organisms, of course, are unaware of our definitions, but appear to be so well "acquainted" with the principles of crystallography, so as to be able to overrule, control and exploit them at will. The concepts of "single crystal", "amorphous phase" or "morphology", are very difficult to define in the biological world, and the boundaries become confused.

The following notes are aimed at illustrating some issues arising around the often peculiar single crystals formed by organisms. They try to explain how these crystals are formed in a few cases, and attempt to derive a better understanding of the concepts involved. For simplicity we exclusively use examples derived from the study of biogenic single crystals of calcite. We do not even attempt to review all known examples, but focus on some that we have studied in some detail.

2. The spicule of *Clathrina*: a single crystal that is not exactly what it appears to be.

Sponges belong to the animal kingdom. They first evolved more than one billion years ago. The *Calcarea* are marine sponges which form calcium carbonate skeletons. They however developed much later, in the Lower Cambrian, around 540 million years ago [12]. The *Calcarea* deposit spicules of various shapes and sizes, whose main function is to provide mechanical support to the soft tissues [13]. Individual spicules may protrude from the surface of the sponge body, or be distributed inside the tissue. They are closely associated with the cells and the fibers of the organic skeleton.

The processes by which calcareous sponge spicules form have been studied by biologists using light microscopy since the last century [14,15,16,17]. Spicule structures, composition and development were documented by these biologists in a series of extraordinarily accurate and perceptive monographs, which to this day form the basis of our understanding of mineral deposition in this class of animals.

One well studied example, that is often referred to as a classical prototype of spicule formation, is the triradiate spicule of *Clathrina* (Fig 2a). These spicules assume, as the name implies, the characteristic shape of three radii emerging from a center at angles of exactly 120°. They are crystalline, birefringent, and extinguish simultaneously under polarized light [15,17,18]. They therefore behave under light and

4

also in X-ray diffraction as single crystals of calcite. They are described by Haeckel as "... the absolutely regular triradiate, which can be considered as a hemiaxon form of the hexagonal crystalline system, in which calcium carbonate crystallizes as calcite" [14]. The plane of the triradiate is perpendicular to the optical axis (crystallographic axis c), while each ray follows the a^* crystallographic axes of calcite [19].

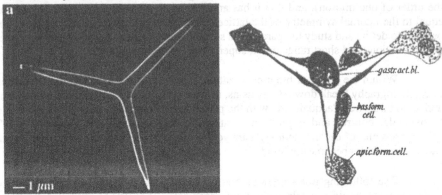

Figure 2, a) Triradiate spicule from the calcareous sponge *Clathrina contorta*. The whole spicule is a single crystal of calcite, containing amorphous calcium carbonate. The crystallographic c axis of calcite is perpendicular to the plane of the spicule. Each radius is elongated along the a^* direction. b) A young spicule from *Clathrina sp*, surrounded by the six formative cells of the basal system. Drawing taken from Minchin's monograph dated 1898[17].

The development of a spicule is described as occurring inside a membrane (syncytium) associated with a sextet of cells: one ray is secreted between each of the three pairs of cells (Fig 2b). One cell per pair moves outwards lengthening the ray, while the other remains at the central junction. Surprisingly, Minchin [17], and many scientists citing him to the present day, report the initial deposition of three separate rays, that join together at a later stage to form a triradiate spicule. Minchin did recognize the inherent problem: "A great objection, at first sight, to regarding the many rayed calcareous spicules as compound bodies, is the well established fact that they behave optically as single crystals. ......It is difficult to comprehend why the united products of a number of independent cells should behave as a single crystal."

We measured individual triradiate spicules by X-ray diffraction using highly collimated synchrotron radiation. The diffraction peaks are single, and there is no evidence whatsoever of three crystals even slightly misaligned. Thus the suggestion that three crystals fuse together is simply unacceptable. Scattered in the literature, however, are observations that point to the solution of the dilemma. The specific gravity of the spicule is lower than that of calcite, heating results in loss of weight accompanied by formation of vacuoles and eventually disintegration[16,19]. The mineral (unlike calcite) contains water, and is etched by acids and alkalis, especially in the central part.

The first suspicion that an additional amorphous mineral phase might be present within the spicules arose when we quantitatively evaluated the intensities of the X-ray reflections of the spicules relative to the volume of mineral, and compared them to that of pure calcite. The integrated diffraction intensities accounted for only 20% of the expected value! It was only when we eventually consulted the addendum to the original 120 page monograph of Minchin [17], that the suspicion was confirmed. We also learned that little evidence eluded this pioneering scientist, notwithstanding the limited means at his disposal. Minchin noted (apologetically) (page 574):"The chief point of interest which I have discovered is that the spicule rays are not crystalline when first laid down. So long as the rays are quite separate they will not light up when rotated between cross nicols. After union of the rays they gradually become crystalline, the change appearing to start from the centre, and in fact from the secondary deposit of calcite by which the rays are united together. This central portion seems to be in many cases the part that first lights up between crossed prisms. On the other hand, spicules may be found which have the rays completely joined, but which fail to show any crystalline properties; others, again, light up so feebly.... that it is very difficult to detect....". He concludes:"The rays are non crystalline so long as they are distinct from one another."

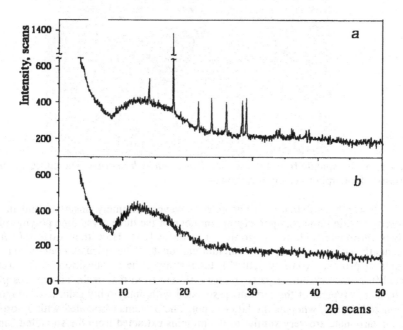

Figure 3, Synchrotron powder X-ray diffraction spectra of biologically deposited CaCO₃: the sharp reflections correspond to calcite. The broad peaks centered around 2θ= 15° are characteristic of amorphous material. a) Triradiate spicules from *Clathrina*. b) Antler spicules from *Pyura pachydermatina* [20].

From here on progress was easy. A high resolution powder X-ray diffraction pattern of the spicules using synchrotron radiation revealed, in addition to the sharp calcite reflections, an intense broad peak around $2\theta=15°$, indicative of the presence of an amorphous phase (Fig 3a). This was further confirmed by infrared spectroscopy. A series of additional etching and partial dissolution experiments allowed us to differentiate the amorphous phase from the crystalline one, leading to the following conclusions. The spicules of *Clathrina* are composed of separate domains of amorphous and crystalline calcium carbonate. The amorphous phase is stable, and does not transform with time into calcite. It is located in the core of each ray, and probably corresponds to the first deposits observed by Minchin. The crystalline part is located in the central junction of the rays and also forms a thin sheath enveloping all the spicule, and is responsible for the single crystal optical behavior of the whole triradiate. It can be assumed that the presence of amorphous material inside the spicule somehow endows the material with advantageous properties, and was thus preserved during evolution [20].

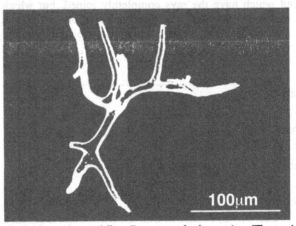

Figure 4, Antler spicule from the ascidian *Pyura pachydermatina*. The entire spicule is composed of amorphous calcium carbonate.

Amorphous calcium carbonate is however extremely unstable, and in vitro transforms within minutes, upon drying, into one of the three crystalline polymorphs of calcium carbonate, calcite, aragonite or vaterite [21]. How then does *Clathrina* permanently stabilize amorphous calcium carbonate? The spicules contain very low amounts of proteins (0.02 weight %). Interestingly, the composition of the proteins isolated from the crystalline parts is very different from that of the amorphous phase. The former is typical of the proteins associated with many other calcitic or aragonitic mineralized tissues, whereas the latter is not. The proteins associated with amorphous calcium carbonate are very similar to the proteins extracted from the so-called "antler" spicules of the ascidian, *Pyura pachydermatina* (Fig 4), which are composed entirely of stable amorphous calcium carbonate (Fig 3b). Furthermore, the introduction of low amounts of these proteins in vitro into supersaturated solutions of calcium carbonate, resulted in the formation of amorphous deposits, that persisted in the amorphous state even when dried, for as long as we cared to check them (months)[20]. Specialized glycoproteins are thus used by organisms to permanently stabilize unstable amorphous phases in selective, spatially defined, compartments.

3. Sea urchin larval spicules. Single crystals that grow from an amorphous transient phase.

Sea urchins belong to the phylum *Echinodermata*. Interestingly, the body form of many of these organisms has pentagonal symmetry. Echinoderms are among the most active mineralizers in biology. They form hard parts with diverse functions, including protective shells or tests, ossicles, spicules, spines and teeth, all composed of magnesium-bearing calcite. Many skeletal parts have a characteristically fenestrated or foam-like ultrastructure (stereom), with interconnected labyrinthic cavities filled by cells[22]. The surfaces of the mineral are smooth and curved, even when observed at the highest magnification. Whole spines, individual test plates and spicules, reaching up to several centimeters in length, diffract X-rays as single crystals of calcite [23,24].

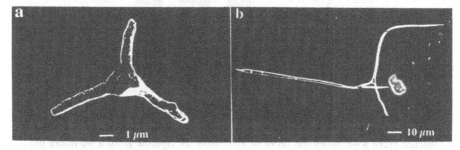

Figure 5, Larval spicules from the sea urchin *Paracentrotus lividus* at two stages of development. a) Triradiate spicule stage (25 hours in culture). The initially deposited calcite rhombohedron is still visible in the center of the spicule. The radii grow along the *a* axes of the calcite crystal. b) Fully developed pluteus spicule (48 hours in culture). The rods that lie approximatively in the plane of the picture are parallel to the *c* axis of the calcite crystal. The triradiate portion is visible in the center, almost edge-on to the plane of the picture.

The sea urchin larva forms an endoskeleton composed of two spicules, each a single crystal of calcite, based on their behavior in polarized light and when diffracting X-rays[25]. Sea urchin larval development, and with it also spicule development, has been the subject of many investigations [26,27,28]. As in the sponge spicules described above, spiculogenesis takes place in a membrane-delineated compartment inside a syncytium formed by fused cell membranes. At the onset of mineral deposition, a single crystal of calcite with the characteristic rhombohedral shape, forms [25,29]. Three rays develop within the next few hours along the directions of the *a* axes of the calcite crystal (not *a**, as above for *Clathrina*), to yield a triradiate spicule with a size of approximately 10 μm [30] (Fig 5a). Additional rods form during the next stage of development. They emanate from the extremities of two of the radii, and grow in the direction perpendicular to the triradiate (ie along the *c* axis of calcite), while the triradiate part thickens. In the

8

Mediterranean species *Paracentrotus lividus*, spicule development is complete after 48 hours in culture without feeding (Fig 5b). From this stage on, the spicules do not grow any more in volume or weight, although the larvae may continue to live for several days.

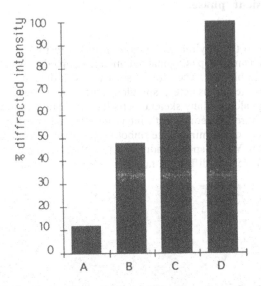

Figure 6, Relative intensities of X-ray diffraction from the {104} planes of calcite, normalized to a unit volume of synthetic calcite crystals. **A**. Individual spicule from a 48 hours old larva of *P. lividus*. **B**. Spicule from a 96 hours old larva. **C**. Spicule from a 96 hours old larva, stored dry for 4 months. **D**. Synthetic calcite crystal[32].

In 1992, we measured freshly grown spicules of *P. lividus* larvae by X-ray diffraction with synchrotron radiation [31]. Although each spicule produced, as expected, a single crystal diffraction pattern, the intensities of the peaks were surprisingly low. In 1996, having been alerted by the demonstrated presence of amorphous calcium carbonate in the *Clathrina* sponge spicules and by the peculiar behavior under etching, we suspected its presence in the sea urchin larval spicules as well. Indeed amorphous calcium carbonate is present in the larval spicules, but its location and fate are quite different from that of the *Clathrina* spicules. The X-ray diffraction intensities per unit volume of larval spicules grown for 48 hours are 12% of the expected value for a calcite crystal of the same volume (Fig 6). Those of spicules grown for 96 hours in culture (of the same size and volume as the 48 hour spicules) are about 50% of the expected value. The diffraction intensities reach 60% of the value expected for a calcite crystal of the same volume after preserving the spicules for 4 months in the dry state [32]. From these and other observations, we concluded that the amorphous calcium carbonate phase is transient. It is present everywhere in the spicule, and eventually transforms into calcite by accretion of the single crystal deposited initially in the center of the triradiate spicule.

How and where is the amorphous phase first deposited, what is its composition, how is it transiently stabilized and how does it eventually transform into calcite? Those are still, by and large, questions under study [33]. Mechanistically, sea urchin spicule development illustrates an important new concept: the growth of macroscopic biogenic single crystals may be achieved through the deposition of an amorphous phase under accurately controlled and well timed conditions. This may be the key to an understanding of many surprising properties of these peculiar single crystals, such as their complicated morphologies and their smooth curved surfaces. This same strategy may well be used in the construction of other skeletal elements within the echinoderms, and possibly even by species of other taxa. The phenomenon may have escaped observation, because amorphous phases present transiently in small volumes are very difficult to detect, especially when only the mature final product is available for study.

As more examples are found in which metastable amorphous phases are exploited by organisms, either transiently or permanently, stimulating new questions arise. What is the mechanism of stabilization? Whatever it is, it has to effectively prevent localized and instantaneous formation of critical nuclei of crystallization, that may not be larger than a few tens of ions. What is the structure of amorphous calcium carbonate? The question appears at first sight to be a contradiction. Even so-called amorphous phases may, however, have nascent short range order. We cannot exclude the possibility that the amorphous calcium carbonate phase of the *Clathrina* sponge spicules may be different from that of the sea urchin larval spicules. This may provide one additional variable that directs the transformation of the latter into calcite, while the former is permanently stabilized. No answers to these questions are yet available. Additional tools are needed to detect the presence of short range structural order that escapes detection by diffraction. When this knowledge becomes available, it may well provide fruitful ideas for new strategies of producing artificial materials under better controlled conditions and with advantageous properties.

4. The adult sea urchin spine. Protein modulation of crystal texture and mechanical properties.

Some adult sea urchins grow spines up to several tens of centimeters in length. These are also single crystals of calcite (Fig 7a), with their long axes parallel to the calcite c axis [22]. In mature sea urchin spines, the foam-like fenestrated stereom spaces (Fig 7b,c) are secondarily filled in part, to form areas of apparently solid mineral (Fig 7d). These are separated in radial sectors, joined by regions of unfilled stereom.

The spines are much more resistant to fracture than single crystals of calcite, which are characteristically brittle because of easy cleavage along the {104} planes of the hexagonal structure [34,35,36]. Part of the fracture resistance is attributable to the fenestrated structure itself, because the presence of cavities increases the elasticity of the material and deviates the progression of cracks [37]. This phenomenon occurs, however,

10

at length scales of several microns, and is not sufficient to account for all the superior properties of the crystalline material itself.

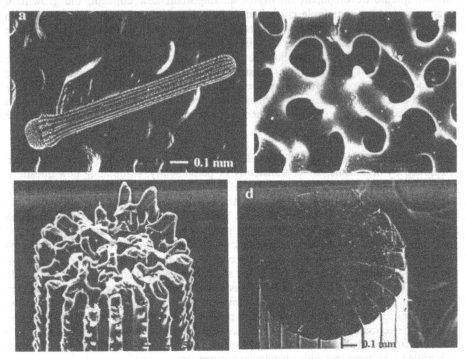

Figure 7, Spines from the sea urchin *P. lividus.* a) Intact small spine. The spines can reach up to several cm in length. They are each composed of one single crystal of calcite. b) Characteristic 'foam-like' structure of the fenestrated stereom. c) Fractured young spine. The fracture surface does not follow the cleavage planes of calcite. It rather has the appearance of a glass, with a curved fracture surface. d) Fractured mature spine, showing the development of the sectors that fill the stereom. Note that the central unfilled portion in (d) has the diameter of the young spine in (c).

Sea urchin spines contain a family of glycoproteins located inside the single crystal, in amounts not exceeding 0.05 weight%[38,39]. The hypothesis that the proteins are actively involved in modeling of crystal morphology during their formation, and modulating their mechanical properties is substantiated by the observation of calcite crystals artificially overgrown on cleaned spine surfaces [40]. Overgrowth is a technique that is used to determine the orientation of the crystallographic axes of biogenic calcite crystals, as the characteristic morphology of calcite rhombohedra epitaxially grown on the biogenic substrates may be easily related to the substrate crystal orientation. We, however, obtained some additional unexpected information when overgrowth was induced very slowly (over days) on the sea urchin spine surfaces. The overgrown crystals displayed, in addition to the stable {104} faces of calcite, a set of rough faces {10l}, almost parallel to the c axis (Fig 8). Extraction of the proteins from the spines followed by growth of calcite in vitro in the presence of the proteins, results in crystals displaying the same set of faces. During the slow overgrowth experiments, the proteins apparently leak out from the spine calcite and are readsorbed on the growing crystals along specific planes. Their adsorption causes a selective decrease in the rate of growth of the crystal in the directions perpendicular to these planes, resulting in the appearance of the new faces. The proteins thus display, at least in vitro, recognition for specific calcite crystal planes. Furthermore the crystals grown in vitro contain protein occluded within the crystal bulk. These crystals do not fracture along the {104} cleavage planes, but exhibit a curved fracture surface similar to that of glass. The sea urchin spine fractures in the same manner. This evidence led to the formulation of a model in which proteins, intercalated inside the calcite crystals along the {100} planes, modulate crystal growth and morphology [41]. They also strengthen the crystalline material against fracture, by continuously interfering with the propagation of cracks at length scales of nanometers.

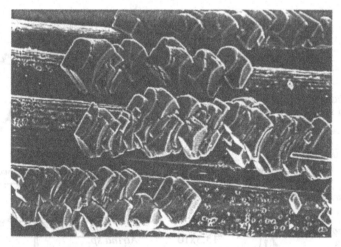

Figure 8, Calcite crystals overgrown epitaxially on the lateral surface of a young spine. The ridges lie in the direction of the spine axis, and thus also of the c axis of the crystals. The overgrown crystals develop a family of faces almost parallel to the c axis. The holes on the spine surface indicate partial dissolution.

In order to determine the distribution and location of the proteins inside the biogenic crystals, we initiated a series of measurements of crystal texture, ie of the distribution of imperfections that limit domains of perfect structure [42]. The rationale for these experiments relates to the size of the proteins. They are at least one order of magnitude larger than the unit cell components of the crystal. Their intercalation inside the crystal lattice is therefore not compatible with the formation of a solid solution inside the calcium carbonate lattice. Wherever a protein macromolecule is adsorbed, an imperfection must be generated. If the distribution of the proteins inside the crystal is anisotropic, there should be a detectable correspondence between the protein locations and the location of imperfections.

The distribution of imperfections inside single crystals can be mapped by measuring the coherence length (the size of domains of perfect structure) and the angular spread (or mosaic spread), which is the extent of misalignment between the domains of the crystal, in the different crystallographic directions. The coherence length is derived from the width of the corresponding diffraction peaks measured in the ω/2θ scan mode, while the angular spread is derived from the width of the diffraction peaks measured in the ω scan mode[43]. For a description of the parameters of crystal texture and their measurement, see the Appendix. Suffice it to note here that the higher the coherence length and the lower the angular spread, the sharper are the diffraction peaks (Fig 9 a,b). In the case of calcite, which forms in quasi-perfect crystals with high coherence length and low mosaic spread, measurements must be performed using the highly collimated beam of synchrotron radiation, in order to achieve sufficient resolution (a few thousandths of a degree) [31,44].

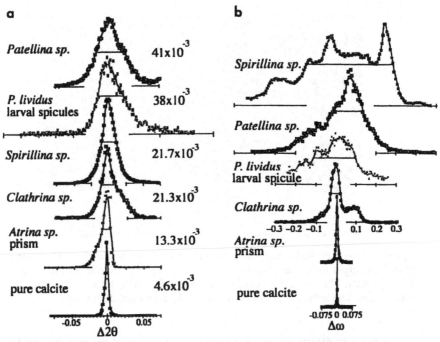

Figure 9, Representative peak profiles from {104} reflections of various biologically produced single crystals, and from a synthetic calcite crystal. The profiles were measured at very high resolution with synchrotron radiation, at line X7B, SNLS, Brookhaven National Laboratory. a) Profiles measured in the ω/2θ mode. The numbers on the right side of the profiles correspond to the full width at half maximum of the peaks, in degrees. b) profiles measured in the ω mode[31].

Texture measurements may provide information on the control mechanisms whereby crystals are grown by the organism, only if the protein locations and orientations are related to the imperfections. This was demonstrated when textural data collected on synthetic calcite crystals grown with and without the proteins extracted from the sea urchin spines, were correlated with the morphological modifications resulting from protein adsorption during crystal growth [45]. Crystals that developed {10l} faces in the presence of proteins, showed a pronounced reduction (60%) in coherence length in the a^* direction (corresponding to [100]), relative to pure calcite crystals. The measured coherence lengths progressively increased towards the c direction (corresponding to [001]), reaching 90% of the values in pure calcite along the c direction (Fig 10a). In the ab plane, the ratio of the reduction in coherence lengths in the [100] direction relative to [110], was close to the maximum reduction that can be theoretically expected, considering the 30° angle between the two directions (Fig 10b). The angular spreads mirror the coherence lengths in inverse proportion, indicating that increased misalignment is directly associated with the reduction in coherence length. We therefore deduced that the proteins are adsorbed on the {100} planes during growth, with high selectivity. They remain occluded inside the crystal bulk, and leave a permanent imprint of their adsorption site on the crystal texture. Mapping the texture can therefore provide information on the distribution of proteins within the biogenic crystals.

The distribution of defects was studied in sea urchin spines at different levels of maturation, ie with increasing proportions of filled wedges relative to stereom. The most mature spines show distinct anisotropy, similar to the synthetic crystals (Fig 10c,d). In contrast, in the youngest spines, devoid of filled sectors, the distribution of imperfections is substantially more isotropic. We therefore concluded that cells can override the intrinsic tendency of the macromolecules to direct themselves to specific crystal planes.

The formation of the convoluted labyrinthic stereom structure requires continuous interference and control of the cells. If the intracrystalline macromolecules are involved in limiting and modulating growth at each stage, they need to be secreted and directed to specific sites in a continuous process. Possible support for this hypothesis is derived from the observation that in sea urchin spines, the protein content in the stereom is more than 3 times larger than in the secondarily filled wedges, which require far less modeling and interference during growth. The stereom architecture is widely used by echinoderms as a building material, and is modeled into many different macroscopic shapes. It is conceivable that the stereom only evolved into an all-purpose material, after the basic difficulty related to the intrinsic brittleness and anisotropy of the building material, calcite, was overcome [46].

The concept of modulating crystal growth, morphology and texture through targeted adsorption of specialized proteins is exploited by organisms from different phyla, and is clearly important in shaping biogenic crystals and altering their mechanical properties [44]. It was demonstrated in the triradiate spicules of the sponge *Clathrina*, in single fibers isolated from sea urchin teeth, in the sea urchin larval spicules, and in the foraminifer *Spirillina*, whose entire shell is composed of one single crystal of calcite [47] (Fig 9). In all these examples the anisotropy of the crystal texture corresponds to the macroscopic directions of growth: wherever the rate of growth has been decreased relative to other growth directions, the corresponding coherence lengths are reduced. This

14

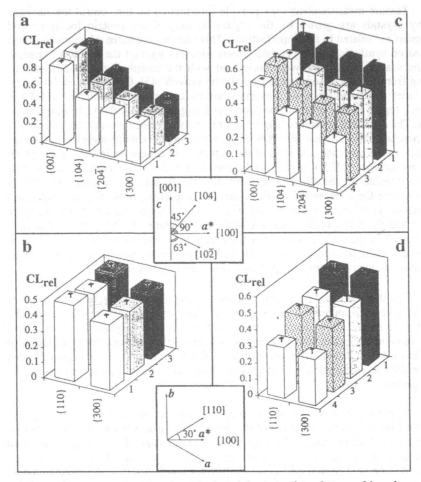

Figure 10, Coherence lengths of synthetic calcite crystals and sea urchin spines. Measurements were performed from sets of planes in different orientations, as indicated in the inserts. The values were normalized to the corresponding coherence lengths of control crystals of pure synthetic calcite. a,b) Synthetic crystals 1,2,3 with occluded proteins from sea urchin spines: a) normalized coherence lengths measured in the *ca** plane in directions progressively oblique to the c axis. The coherence lengths along the [100] directions are significantly reduced relative to the [001] direction; b) normalized coherence lengths measured in the *ab* plane. The coherence lengths along the [100] directions are reduced relative to the [110] directions. The reduction (~13%) is the maximum expected, considering that the dihedral angle between the planes is 30° (see insert). c,d) Sea urchin spines at different stages of maturation, measured and represented as in a,b) respectively. In the youngest spine (1) the domains are almost isotropic. In the most mature (4), there is a substantial decrease in coherence length from the [001] to the [100] direction[45].

implies that more proteins were introduced on crystal planes in those specific directions to limit growth. Thus, for the *Clathrina* sponge spicules and the shell of the foraminifera *Spirillina*, which develop mainly in the *ab* plane, lower coherence lengths were measured along the *a* axes, relative to *c*. The opposite occurs for the sea urchin spines, larval spicules and tooth fibers, all elongated along *c*. Exceptions are the prisms isolated from the mollusk shell prismatic layer and the shell of the foraminifer, *Patellina*. The former are elongated along *c*, but proteins are intercalated mainly perpendicular to *c*, on the *ab* plane. It is interesting to note that the growth of the prisms does not need to be limited in the *ab* plane, because growth in these directions stops when adjacent prisms come into contact with each other. They eventually form a compact palisade of crystals all aligned with their *c* axes parallel to each other. In *Patellina*, the measured texture is by and large isotropic. The *c* axis lies in a direction tangential to the shell, and thus the morphology does not correspond to any main crystallographic direction[48].

Despite the complex and sophisticated processes described above, we know of one case of an organism exerting such exquisite control over the formation of a single crystal, that it boggles the mind. The single crystals produced by the sponge *Sycon*. These are discussed below.

5. *Sycon* curved monaxons. Modeling shape and breaking symmetry at will.

The calcareous sponge *Sycon* forms four types of spicules of different shapes, each composed of a single crystal of calcite [49]. The slender monaxon spicules are found around the sponge osculum, the opening through which sea water is filtered. They are up to a few millimeters in length and not more than 5 micrometers thick. They do not contain detectable amounts of protein. Interestingly, their textural parameters are practically identical to that of pure synthetic calcite, ie isotropic, with large coherence lengths and relatively small omega spreads [50,51]. The slender monaxons further strengthen the evidence that textural anisotropy is directly associated with protein location, rather than, for example, with macroscopic anisotropy of shape or rate of crystal growth.

In contrast to the slender monaxons, the curved monaxons are located all over the sponge tissue, protruding slightly from the surface. They are about 1-3 mm long and 50 μm thick, and have a characteristic curved shape, which gives the spicule its name (Fig 11a). All the spicules have the same crystallographic orientation, with the *c* axis forming an angle of approximately 70° to the direction of elongation of the spicule, which is roughly along the [012] direction [52]. Consequently, the macroscopic morphology of the spicule does not respect the hexagonal (or rhombohedral) symmetry of calcite. In particular, the growth of the spicule is much more restricted in the directions [$\bar{1}$02] and [1$\bar{1}$2], rather than in the symmetry related [012] direction ,which is the direction of elongation of the spicule. The widths of the reflections from the three families of planes {012} strikingly reflect this macroscopic reduction in symmetry.

16

While the (012) reflection (both in the ω and ω/2θ modes) is as sharp as the corresponding reflection in pure synthetic calcite, the ($\bar{1}$02) and ($1\bar{1}$2) reflections are much broader. The evaluated coherence lengths are 8000-9000Å for the former, and 1500-2000Å for the ($\bar{1}$02) and ($1\bar{1}$2) reflections. The same phenomenon occurs for other families of symmetry related planes, when one member of the family is more aligned with the morphological axis than the others. The smallest coherence lengths are observed in the ac plane, which is roughly perpendicular to the longitudinal direction of the spicule. In this plane, the coherence lengths are six-fold smaller than those of pure calcite crystals, and are isotropic, independent of the specific crystallographic orientation within the calcite structure. A reconstruction of the envelopes of the measured coherence lengths in space reproduces, with astounding fidelity, the macroscopic shape of the spicules (Fig 11b).

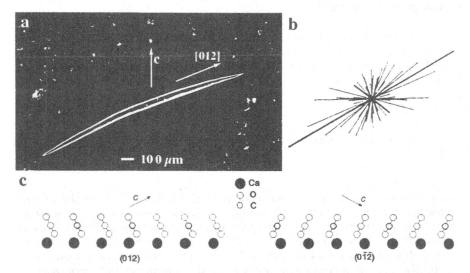

Figure 11, a) Curved monaxon spicule from the calcareous sponge *Sycon sp.* The crystallographic orientation of the c axis of calcite and the direction of the morphological axis of the spicule, [012], are indicated. b) Representation of the coherence lengths measured in different orientations of the curved monaxon spicule in a). Each line represents a vector of the same orientation as in a), with length proportional to the coherence length measured in that direction. The thicker lines correspond to the vectors of the symmetry-related [012] reflections.

c) Comparison of the structure of calcite in the (012) plane (left) and ($0\bar{1}\bar{2}$) plane (right) [52].

In these crystals, the modulation of texture is strictly associated with the macroscopic morphology of the object under construction, and is independent of lattice symmetry. Another amazing observation is that the positive direction of the crystallographic axis c is always oriented out of the convex curve of the spicule, forming an angle of approximately 30° with the perpendicular to it. Thus the combination of the morphology of the spicule with the orientation of the axes is constant and chiral. This implies that not only is there a distinction between three symmetry related $\{hkl\}$ planes, but also between the positive and negative surfaces of the same plane ($\{hkl\}$ as opposed to $\{\bar{h}\bar{k}\bar{l}\}$) (Fig 11c)[52]. There is no simple explanation of how such an astounding level of recognition can be achieved even by accurate control of the microenvironment. We can only speculate that it may result from the combination of controlled oriented nucleation and controlled growth, modulated by specialized proteins.

This is not the only case in which crystallographic/morphological symmetry is broken in biologically formed single crystals. *Coccolithophoridae*, monocellular marine organisms, excrete and assemble complex architectures out of calcite crystals. These coccoliths were shown to have chiral morphologies [53]. Some species of magnetotactic bacteria also display a chiral morphology of the magnetite crystals [54,55].

6. Concluding Comments

Organisms are able to control almost every facet of single crystal growth, as well as the textural properties of the bulk of the crystal. Following the Darwinian perspective on evolution, we can assume that in so doing they benefit in one way or another. It is difficult and presumptuous to try to second guess what these benefits might be. Even disregarding the benefit to the organism, it is still most perplexing to try and contemplate hypothetical reasons why, for example, the sea urchin larval spicule architecture follows so closely the crystallographic axes of the calcite crystal, especially when its mode of growth is through a transient amorphous precursor phase. Is the correlation between the shape of the nanometer sized mosaic block and the sub-millimeter sized macroscopic shape of the single crystal related only to the manner in which the crystal grows? The mere fact that organisms from many completely different taxonomic groups control the distribution of dislocations in the crystal, is an indication in itself that they must derive important benefits. One benefit certainly relates to the change in mechanical properties – a feature that to our knowledge has not been utilized in the production of improved synthetic crystal containing materials.

There is no doubt that our knowledge of this subject is still in its infancy. For us, it is fascinating in its own right. We suspect, however, that some of the insights gained will, sooner or later, benefit mankind.

Acknowledgments

This work was supported by a United States-Israel Binational Foundation grant.

18

7. References

1. Lowenstam H.A. and Weiner S. (1989) *On Biomineralization*, Oxford University Press. New York; Oxford.
2. Simkiss K. and K. W. (1989) *Biomineralization. Cell Biology and Mineral Deposition.*, Academic Press, Inc. San Diego.
3. Robinson R. (1952) An electron mocroscope study of the crystalline inorganic component of bone and its relationship to the organic matrix. *J. Bone Joint Surg.* **34A**, 389-434.
4. Weiner S. and Price P. (1986) Disaggregation of bone into crystals. *Calcif Tissue Int.* **39**, 365-375.
5. Ronnholm E. (1963) The amelogenesis of human teeth as revealed by electron microscopy. II. The development of enamel crystallites. *J. Ultrastructure Res.* **6**, 249-303.
6. Boyde A. (1965) The structure of developing mammalian enamel, in Stack, MV Fearnhead, RW (ed.), *Tooth Enamel*, J. Wright & Sons. Bristol.
7. Curtis C. , and Spears D. (1968) The formation of sedimentary iron minerals. *Economic Geology* **63**, 257-270.
8. Kirschvink J. , and Gould J. (1981) Biogenic magnetite as a basis for magnetic field detection in animals. *Biosystems* **13**, 181-201.
9. Markel K., Gorny P. , and Abraham K. (1977) Microarchitecture of sea urchin teeth. *Fortschr. Zool* **24**, 103-114.
10. Wang R.Z., Addadi L. , and Weiner S. (1997) Design Strategies of Sea-Urchin Teeth - Structure, Composition and Micromechanical Relations to Function. *Phil. Trans. Roy. Soc. London B* **352**, 469-480.
11. Guinier A. (1963) *X-Ray Diffraction in Crystals, Imperfect Crystals, and amorphous Bodies*, W.H. Freeman & Co. San Francisco.
12. Bengston S., Conway Morris S., BJ C., Jell P. , and Runnegar B. (1990) *Early Cambrian Fossils from South Australia*, Association of Australasian Paleontologists. Brisbane.
13. Koehl M.A.R. (1982) Mechanical design of spicule-reinforced connective tissue: stiffness. *J. Exp. Biol.* **98**, 239-267.
14. Haeckel E. (1872) *Die Kalkschwamme* 1, Berlin.
15. Ebner V.von (1887) Uber den fieneren bau der skelettheile der kalkschwämme nebst bemerkung uber Kalkskelet überhaupt Sber. *Akad. Wiss. Wien (Abt. I)* **95**, 55-149.
16. Minchin E.A. (1909) Sponge-spicules. Summary of present knowledge. *Ergebn. Fortschr. Zool.* **2**, 171-274.
17. Minchin E.A. (1898) Materials for a monograph of the ascons. I. On the origin and growth if the triradiate and quadriradiate spicules in the family Clathrinidae. *Q. J. Microsc. Sci.* **40**, 469-587.
18. Jones W.C. , and Jenkins D.A. (1970) Calcareous sponge spicules: A study of magnesian calcites. *Calc. Tiss. Res.* **4**, 314-329.
19. Jones W.C. (1970) The composition, development, form and orientation of calcareous sponge spicules. *Symp. Zool. Soc. Lond.* **25**, 91-123.
20. Aizenberg J., Lambert G., Addadi L. , and Weiner S. (1996) Stabilization of amorphous calcium carbonate by specialized macromolecules in biological and synthetic precipitates. *Adv. Mat.* **8**, 222-225.

21. Lippmann F. (1973) *Sedimentary Carbonate Minerals*, Springer Verlag. Berlin. .
22. Raup D. (1966) *The endoskeleton*, J. Wiley & Sons, Interscience. New York.
23. Donnay G. and Pawson D.L. (1969) X-Ray diffraction studies of echinoderm plates. *Science* **166**, 1147.
24. Dubois P. and Chen C. (1989) Calcification in echinoderms, in Lawrence (ed.), *Echinoderm Studies*, A.A. Balkema. Rotterdam, pp. 109-178 .
25. Okazaki K. and Inoue S. (1976) Crystal property of the larval sea urchin spicule. *Dev. Growth Diff.* **18**, 413-434.
26. Davidson E.H. (1993) Later embryogenesis: regulatory circuitry in morphogenetic fields. *Development* **118**, 665-690.
27. Decker G.L. and Lennarz W.J. (1988) Skeletogenesis in the sea urchin embryo. *Development* **103**, 231-247.
28. Wilt F. and Benson S. (1988) Development of the endoskeletal spicule of the sea urchin embryo, in (ed.), *Self-assembling architecture*, Alan R. Liss, Inc. pp. 203-227 .
29. Theel H. (1892) On the development of *Ehinocyamus pusillus*. *Nova Acta Res. Soc. Sci. Upsala* **15**, 1-57.
30. Wolpert L. and Gustafson T. (1961) Studies of the cellular basis of morphogenesis of the sea urchin embryo (Development of the skeletal pattern). *Exp. Cell Res.* **25**, 311-325.
31. Berman A., Hanson J., Leiserowitz L., Koetzle T.F., Weiner S. and Addadi L. (1993) Biological control of crystal texture: A widespread strategy for adapting crystal properties to function. *Science* **259**, 776-779.
32. Beniash E., Aizenberg J., Addadi L. , and Weiner S. (1997) Amorphous calcium carbonate transforms into calcite during sea-urchin larval spicule growth. *Proc. Roy. Soc. London B* **264**, 461-465.
33. Beniash E., Addadi L. , and Weiner S. (1999) Cellular control over spicule formation in sea urchin embryos: a structural approach. *J. Struct. Biol.* in press.
34. Nichols D. and Currey J. (1968). The secretion, Structure and Strength of Echinoderm Calcite. in McGee-Russel, S.M., Ross, K.F.A. (ed.), *Cell Structure and its Interpretation.*, Edward Arnold London, pp. 251-261. .
35. Weber J., Greer R., Voight B., White E. and Roy R. (1969) Unusual strength properties in echinoderm calcite related to structure. *J Ultrastr Res* **26**, 355-366.
36. Burkhardt A., Hansmann W., Markel K. and Niemann H.J. (1983) Mechanical design in Spines of Diadematoid Echinoids. *Zoomorphology* **1023**, 189-203.
37. Gibson L.J. and Ashby M.F. (1988) *Cellular Solids. Structure & Properties*, Pergamon Press. Oxford, UK.
38. Weiner S. (1985) Organic matrix like macromolecules associated with the mineral phase of sea urchin skeletal plates and teeth. *J.. Exp. Zool.* **234**, 7-15.
39. Albeck S., Weiner S. and Addadi L. (1995) Polysaccharides of Intracrystalline glycoproteins modulate calcite crystal growth in vitro. *Chem. Eur. J.* **2**, 278-284.
40. Aizenberg J., Albeck S., Weiner S. and Addadi L. (1994) Crystal-protein interactions studied by overgrowth of calcite on biogenic skeletal elements. *J. Cryst. Growth* **142**, 156-164.
41. Berman A., Addadi L. and Weiner S. (1988) Interactions of sea urchin skeleton macromolecules with growing calcite crystals-A study of intracrystalline proteins. *Nature (London)* **331**, 546-548.

42. Berman A., Addadi L., Kvick Å., Leiserowitz L., Nelson M. and Weiner S. (1990) Intercalation of sea urchin protein in calcite: Study of a crystalline composite material. *Science* **250**, 664-667.

43. Alexander L.E. and Smith G.S. (1962) Single-Crystal Intensity Measurements with Three-circle Counter Diffractometer. *Acta Crystallogr.* **15**, 983-1004.

44. Berman A., Hanson J., Leiserowitz L., Koetzle T., Weiner S. and Addadi L. (1993) Crystal-protein interactions: controlled anisotropic changes in crystal microtexture. *J. Phys. Chem.* **97**, 5162-70.

45. Aizenberg J., Hanson J., Koetzle T.F., Weiner S. , and Addadi L. (1997) Control of Macromolecule Distribution Within Synthetic and Biogenic Single Calcite Crystals. *J. Amer.Chem.Soc.* **119**, 881-886.

46. Weiner S. and Addadi L. (1997) Design strategies in mineralized biological materials. *J. Mater. Chem.* **7**, 689-702.

47. Towe K.M. and Cifelli R. (1967) Wall Ultrastructure in the Calcareous Foraminifera: Crystallographic Aspects and a model for Calcification. *J. Paleontol.* **41**, 742-762.

48. Towe K.M., Berthold W.-U. , and Appleman D.E. (1977) The crystallography of *Patellina corrugata* Williamson: a-axis preferred orientation. *J. Foram. Res.* **7**, 58-61.

49. Ledger P.W. and Jones W.C. (1977) Spicule formation in the calcareous sponge *Sycon ciliatum*. *Cell Tiss. Res.* **181**, 553-567.

50. Aizenberg J., Hanson J., Ilan M., Leiserowitz L., Koetzle T.F., Addadi L. and Weiner S. (1995) Morphogenesis of calcitic sponge spicules - A role for specialized proteins interacting with growing crystals. *FASEB J.* **9**, 262-268.

51. Aizenberg J., Ilan M., Weiner S. and Addadi L. (1996) Intracrystalline Macromolecules Are Involved in the Morphogenesis of Calcitic Sponge Spicules. *Conn. Tiss. Res.* **35**, 17-23.

52. Aizenberg J., Hanson J., Koetzle T., Leiserowitz L., Weiner S. and Addadi L. (1995) Biologically induced reduction in simmetry: a study of crystal texture of calcitic sponge spicules. *Chem.Eur. J.* **1**, 414-422.

53. Mann S. and Sparks N. H. C. (1988) Single crystalline nature of coccolith elements of the marine alga *Emiliania huxleyi* as determined by electron diffraction and high-resolution transmission electron microscopy. *Proc. R. Soc. Lond.* B **234**, 441-453.

54. Mann S., Sparks N.H.C. , and Blakemore R.P. (1987) Structure, morphology and crystal growth of anisotropic magnetite crystals in magnetotactic bacteria. *Proc. R. Soc. Lond.* B **231**, 477-487.

55. Mann S. and Frankel R.B. (1989) Magnetite Biomineralization in Unicellular Microorganisms, in Mann, Webb and Williams (ed.), *Biomineralization. Chemical and Biochemical Perspectives*, VCH. Weinheim.

56. Dunitz J.D. (1979) *X-ray Analysis and the Structure of Organic Molecules*, Cornell University Press. Ithaca.

57. Klug H.P., and Alexander L.E. (1974) *X-ray Diffraction Procedures*, Wiley Interscience. New York.

58. Stokes A.R. (1960) *X-ray Diffraction by Crystalline Materials*, Chapman &Hall. London, Ch 17 .

59. Hoche H., Schulz H. and Weber H.P. (1986) Measurement and Correction of Secondary extinction in CaF2 by Means of Synchrotron X-ray Diffraction Data. *Acta Crystallogr.* **A42**, 106-110.

60. Mathieson A.M. (1988) Small Single Crystal Diffractometry with Monochromated Synchrotron Radiation- the Wavelength Dispersion Minimum Condition for Bragg Reflection Profile Measurement. *Acta Crystallogr.* **A44,** 239-243.

61. Mathieson A.M. (1982) Anatomy of a Bragg Reflection and an Improved Prescription for Integrated Intensity. *Acta Crystallogr.* **A38,** 378-387.

62. Wilson A.J.C. (1962) On Variance as a Measure of Line Broadening in Diffractometry. General Theory and Small Particle Size. *Proc. Phys. Soc.* **80,** 286-294.

63. Warren B.E. (1941) *Phys Rev.* **59,** 693-699.

64. Stokes A.R. , and Wilson A.J.C. (1942) A Method of Calculating the Integral Breadths of Debye-Scherrer Lines. *Proc. Camb. Phil. Soc.* **38,** 313-322.

65. Warren B.E. , and Bodenstein P. (1965) The diffraction pattern of Fine Particles of Carbon Blacks. *Acta Crystallogr.* **18,** 282-286.

8. Appendix. Texture parameters. Coherence length and angular spread

The conditions for diffraction to occur from a set of crystal planes is based on constructive interference of the diffracted X-rays. An ideally perfect crystal would give no diffraction because of extinction, ie X-rays would be diffracted infinitely within the perfect lattice and never emerge. In contrast, a real crystal was described initially as an 'ideally imperfect' crystal, composed of discontinuous small perfect domains (mosaic blocks), slightly misaligned one relative to the other [56,57]. The concept of mosaic blocks evolved later into that of 'coherence length'. This description does not imply a crystal composed of distinct blocks; it rather reflects the presence of dislocations or imperfections within the crystal, separated on the average by a distance defined as the coherence length. The latter thus corresponds to the average size of domains of perfect lattice structure. For any real crystal of finite size, the small deviations from the Bragg condition caused by the imperfections yield an appreciable diffracted intensity over an angular spread around the theoretical value θ [58].

The width of the diffraction peak results from the combined contributions of instrumental departure from ideal conditions and crystal imperfections. The instrumental contributions are the wavelength spread of the X-ray source ($\Delta\lambda/\lambda$) and its angular divergence, or lack of perfect collimation (β) [59]. The instrumental parameters may be minimized by using a highly monochromatic, highly collimated beam from synchrotron radiation (in Brookhaven beam-line X7B, $\Delta\lambda/\lambda = 2\times10^{-4}$ and $\beta = 0.003$).

The intrinsic parameters of the crystal that cause peak broadening are the domain size (coherence length), and the misalignment between domains (angular spread) [43,60,61]. The latter may be caused by lattice strain, inducing distortions within the perfect lattice.

Deconvolution of the contributions of misalignment and lattice strain from coherence length can be achieved by scanning the diffracted beam in two orthogonal directions, performing an ω-scan (so-called rocking curve) and an $\omega/2\theta$ scan respectively. An analyzer crystal is introduced in the detector arm of the diffractometer.The angular spread is derived directly from the width of the diffraction peak in the ω mode. The coherence length is derived by application of the Scherrer formula to the width of the diffraction peak in the $\omega/2\theta$ scan. Modifications have been introduced to the Scherrer formula during time, but the basic parameters and formulation is still essentially the original one [62].

$$CL = K\lambda/cos\ \theta(B^2-\beta^2)^{1/2}$$

CL=coherence length
K= Scherrer constant
λ=X-ray wavelength
θ=Bragg angle
B= peak width (measured from the integrated intensity/peak height)
β= angular divergence of the beam

The constant K is crystal shape dependent, and may assume values between 0.5 and 2. K=1 for cubic crystals [63,64,65]. It assumes values >1 for needles measured in the direction along the needle axis, and was calculated to be K=2 for reflections from thin plates measured in the plane of the plate.

In the analyses of biogenic crystals, K was taken to be 1, although the crystal morphology is not cubic or spherically symmetric. We note that, while it would be extremely difficult to exactly evaluate K for these complicated morphologies, the application of exact values of K would, in most cases, further exaggerate the observed anisotropies (see, as an example, the CL in the direction of elongation of the *Sycon* curved monaxon, which was measured to be five times larger relative to the other symmetry related directions).

The Scherrer formula tends to provide progressively lower CL values for higher order reflections, ie for higher θ. In order to neutralize the effect of θ on the CL, the data may be normalized each to the correspondent values in reference (pure calcite) crystals. This procedure avoids artifacts that may ensue from comparison of CL in the different directions, measured from reflections at different Bragg angles.

INTRODUCTION TO PACKING PATTERNS AND PACKING ENERGETICS OF CRYSTALLINE SELF-ASSEMBLED STRUCTURES

Predicting Crystal Structures using Kitaigorodskii's Aufbau Principle

JERRY PERSLTEIN
Center for Photoinduced Charge Transfer
Department of Chemistry
University of Rochester
Rochester, NY 14627
perlstein@chem.chem.rochester.edu

"The following is the most useful way for dealing with packing in molecular crystals. We must first of all deduce all possible methods of constructing chains of molecules (formations extending in one dimension) and then demonstrate what layers are possible (formations extending in two dimensions), followed finally by considering layer stacking in the crystal (a formation extending in three dimensions)"

A. I. Kitaigorodskii[1]

ABSTRACT. To understand how molecules assemble themselves in crystalline forms requires knowledge of space group symmetry. The complexity of the symmetry relationships between molecules, although well understood by crystallographers, makes it very difficult for others to understand the chemical significance of the close packing interactions present in the solid state. With the use of Kitaigorodskii's Aufbau Principle (KAP) and a few simple ideas concerning symmetry and the energetics of packing, complex crystal structures can be analyzed in terms of the substructures that make them up. KAP analysis along with some simple molecular simulation methods can be used to predict crystal structures of arbitrary shaped molecules in specific space groups, even those that have numerous internal rotational degrees of freedom.

1. The Substructures of KAP

1.1 ONE-DIMENSIONAL PERIODIC STRUCTURES: CHAINS

A one-dimensional chain of molecules consists of an array of molecules packed along one spatial direction. Examples of this abound. They appear in molecular conductors[2](see also Geiser, this Volume), charge-transfer complexes[3], molecular magnets [4](See also Miller, Kahn, and Gatteschi, this Volume), organometallics[5] (See also Braga, this Volume) and 1-dimensional molecular networks[6](see also

23

D. Braga et al. (eds.), Crystal Engineering: From Molecules and Crystals to Materials, 23–42.

Hosseini, this Volume). These structures and many others like them have a common feature; they consist of a periodic stack of molecules along one axis and are usually a substructure of a more complex arrangement. As we will show below, one-dimensional chains occur as substructural units in almost all molecular crystalline solids. Such structures belong to the general class of objects called "rods" which consists of objects belonging to one of the 75 space-group symmetries of 3-dimensional objects arranged periodically in one-dimensional space.[7] While in principle any of these rod symmetries may occur, in practice only 4 simple ones occur most often in molecular crystals as exemplified by Scaringe and Perez.[8]

These four chain space groups, easy to visualize and comprehend, are shown schematically for 1-acetyl-4-bromo-pyrazole[9] in Figure 1. In each case there is a single molecule (the packing unit) which repeats either itself or a symmetry related equivalent by translation along the chain axis. The four chains are:

  a) the translation chain: - an identical molecule repeats
  b) the glide chain-a mirror image of the molecule repeats itself
  c) the 2_1 screw chain- a 180° rotation of the molecule repeats itself
  d) the inversion chain-an inversion molecule repeats itself

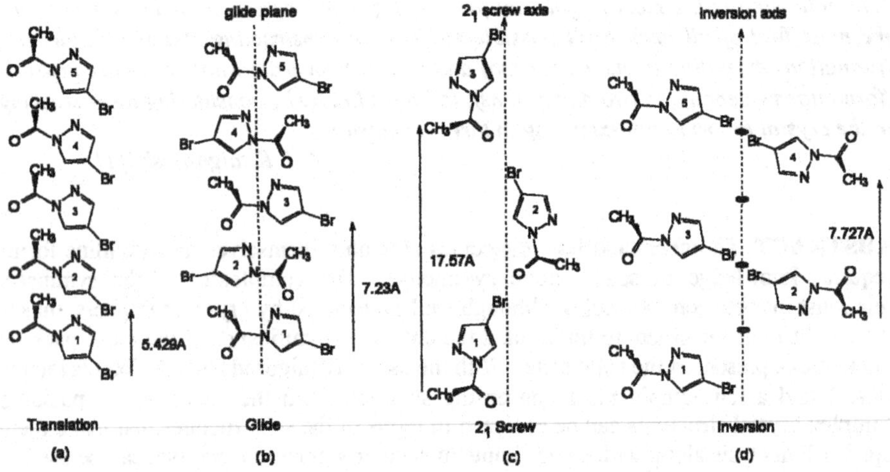

Figure 1. Schematic of the four most common chain types as they occur in the crystal structure of 1-acetyl-4-bromo pyrazole. a) translation chain-all molecules are identical b) glide chain-mirror image of molecules are translated along the repeat direction c) 2_1 screw chain-180° rotated molecules are translated along the repeat direction d) inversion chain- molecules related by inversion through the inversion point(ellipse).

Of course more complex chains can be envisioned, but these four are sufficient to understand the packing in 90% of crystalline molecular solids[10].

1.2 TWO-DIMENSIONAL PERIODIC STRUCTURES: LAYERS

Layer structures are well known. They occur in Langmuir and Langmuir-Blodgett films[11] and as epitaxial films grown on crystal substrates.[12, 13] Not well recognized is the observation that most organic crystals have a low energy layer as a substructure. The presence of these layers in organic molecular crystals prompted Kitaigorodskii to raise their existence to that of a crystallographic law:

"A crystal of an organic compound can be considered as a system of very closely packed layers... This rule should be considered as a generalization from all existing experimental data; it can with complete justice be taken as the fundamental law of organic chemical crystallography".[1]

The packing pattern observed in the vast majority of layer structures can be understood by combining the four one-dimensional chain structures of Figure 1 in various ways. Scaringe has outlined in detail how to do this at a previous Erice conference.[14] He showed that of the 80 possible layer types [15, 16] only seven types of layers occur most frequently.

Figure 2 shows schematic examples of the seven layer types using a stick figure as the packing unit. Five of them are considered simple in that they are constructed by simple translation of the 4 chain types. Their space group designations are as follows:

a) P1 translation layer: a translation chain repeats itself

b) Pc glide∥ layer: a glide chain repeats itself with the glide plane parallel to the layer

c) Pc11 glide⊥ layer: a glide chain repeats itself with the glide plane perpendicular to the layer

d) P112$_1$ screw layer: a 2$_1$-screw chain repeats itself

e) P-1 inversion layer: an inversion chain repeats itself

The two remaining layers are complex. The chains that make them up are not related by simple translation but by another symmetry element. These layers are constructed by combining a screw chain with its mirror image perpendicular to the screw axis (P112$_1$/a) or parallel to the screw axis (Pca2$_1$).

More than 90% of all crystalline organic molecular solids contain one of these seven layers.

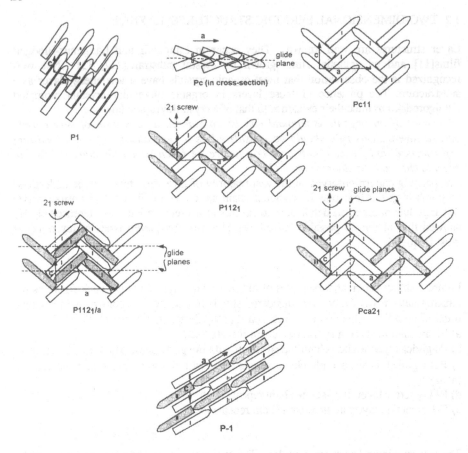

Figure 2. Schematic of the seven most common KAP layer types constructed from the assembly of KAP chains and designated by their space group notation. The packing unit is a stick figure (front side white, back side shaded). Within a chain the repeat vector is designated c and between chains it is designated a.

1.3 THREE-DIMENSIONAL PERIODIC STRUCTURES-CRYSTALS.

When the seven layers are combined, one obtains the crystals in the most frequently occurring space groups. These are triclinic (P1, P-1), monoclinic (P2₁, P2₁/c, C2/c), and orthorhombic (P2₁2₁2₁, Pca2₁, Pna2₁, Pbca). Kitaigorodskii reached these conclusions based on simple geometric arguments of close packing of arbitrary shaped molecules. These nine space groups account for more than 90% of all crystal structures in the Cambridge Structural Database.[17] To understand the packing pattern in these frequently occurring space groups in terms of the chains and layers which make them up, a method for computing the intermolecular packing energy is needed. This is described in the next section.

2. The Energetics of Intermolecular Packing

2.1 FORCE FIELDS

We use a classical force field description to compute the packing energy of a crystal. A force field is simply a potential energy equation with associated parameters. The equation used by most workers in crystal engineering is an atom-atom potential equation consisting of three terms

$$E_{packing\ energy} = E_{non-bonded\ energy} + E_{electrostatic\ energy} + E_{intramolecular\ energy} \qquad (1)$$

The first term, $E_{non-bonded\ energy}$, consists of two parts, a van der Waals attractive component (a negative term) which varies as $1/r_{ij}^6$, where r_{ij} is the distance between any two atoms i and j in different molecules and a repulsive component (a positive term). Some typical examples of this force field term are shown in equations (2)-(4) along with the names of the force field from which they come.

MM2[18]

$$E_{non-bonded-energy} = \sum_{i,j} A_{ij} \left[2.90x10^5 \exp\left(\frac{-12.50r_{ij}}{B_{ij}} \right) - 2.25 \left(\frac{B_{ij}}{r_{ij}} \right)^6 \right] \qquad (2)$$

CFF91[19]

$$E_{non-bonded-energy} = \sum_{i,j} A_{ij} \left[2 \left(\frac{B_{ij}}{r_{ij}} \right)^9 - 3 \left(\frac{B_{ij}}{r_{ij}} \right)^6 \right] \qquad (3)$$

AMBER[20]

$$E_{non-bonded-energy} = \sum_{i,j} \left[\left(\frac{A_{ij}}{r_{ij}} \right)^{12} - \left(\frac{B_{ij}}{r_{ij}} \right)^6 \right] \qquad (4)$$

In each equation, the parameters A_{ij} and B_{ij} are the force field specific parameters for each atom pair. The force fields above differ only in the form of the repulsive term.

The electrostatic term in the force field represents the interaction between atoms in different molecules due to the presence of a static charge distribution associated with the fact that the molecules are polar. This is usually simplified to associating a partial charge with each atom in the molecule and then summing the Coulomb interaction over all atoms in different molecules as in eq. (5).

$$E_{electrostatic-energy} = \sum_{i,j} \frac{q_i q_j}{\varepsilon_o r_{ij}} \qquad (5)$$

Here the q_i's are the partial charges on each atom and ε_o is the static dielectric constant.

There are other ways to determine the electrostatic contribution. For example, MM2 uses a sum over dipoles at each atom site.[18] More recently methods have been developed using a multipole expansion of the electrostatic term.[21] In addition if the space group is polar, then corrections to the surface-polarization-energy contribution have to be made to the sum[22] or and Ewald summation[23] has to be done for this term (See discussion by Williams, this volume).

Finally lumped into the intra-molecular contribution are all those terms representing, bond stretching, bond bending, and bond rotation energy. For the most part these are usually neglected in crystal energy computations. They are nevertheless very important contributors to the crystal packing and cannot be neglected when comparing a molecule in different polymorphic structures. We will come back to this when we discuss methods for predicting KAP structures.

2.2 PACKING ENERGY COMPUTATIONS

It would seem from the previous discussion that to determine the packing energy of a crystal, one would have to compute the interaction of every molecule with every other molecule out to infinity. Fortunately, this is not necessary for a periodic structure. As shown by Busing,[24] one need only compute the interaction of a **single** molecule with every other molecule and then divide the result by 2 to get the packing energy for the entire crystal.

$$E_{packing\ energy} = \tfrac{1}{2}(E_{non-bonded\ energy} + E_{electrostatic\ energy}) + E_{intramolecular\ energy} \qquad (6)$$

Equation 6 says the following, pick any molecule in the packed crystal structure and compute the energy necessary to remove it from the crystal. Divide the result by 2 and add the intramolecular energy necessary to bring the molecular conformation into some standard state. The result is the energy necessary to separate all the molecules to infinity in the standard state (-$E_{packing\ energy}$).

Since the non-bonded term falls off rapidly with distance, neglecting molecules in unit-cells beyond say the third is a reasonable approximation for the non-bonded interactions. The electrostatic term however is not so easy. For non-polar crystals, the same approximation can be used as for the non-bonded term. For polar lattices, the problem is more acute. Many workers use a linear distance dependent dielectric constant which decreases the electrostatic contribution so that the same cutoff can be used as for the non-bonded term and then add a surface polarization correction[22] or do and Ewald summation which ignores the surface polarization

3. Finding KAP substructures

3.1 KAP CHAINS

On a computer, the packed crystal structure is easily displayed, knowing the coordinates for the packing unit, the space group and the unit cell parameters. Many programs can now do this. I use CHEMX/CHEMLIB from the Oxford Molecular Group for this purpose since it readily displays the symmetry related molecules in different colors for distinction as well as allowing me to add my own software subroutines to the program.[25] *As* an example, consider the chains that occur in the crystal structure of 1-acetyl-4-bromo-pyrazole shown in figure 1 where each molecule has been labeled. The crystal of this molecule belongs to space group $P2_1/c$ (a short hand notation for $P1,2_1/c,1$ designating a primitive lattice, P, with a 2_1-screw along the b-unit-cell axis and a glide layer perpendicular to it). In terms of the KAP chains that we have described, this space group has at least a 2_1-screw chain along b and a glide chain along c as shown in Figure 1. However, there are other chains in this structure as well. An inversion chain (fig. 1d) along the a unit-cell direction and a translation chain (fig. 1a) along the a-c face diagonal. Of interest is to ask which of these chains is the lowest energy chain in the structure? The screw chain or the glide chain or perhaps some other chain. The space-group information doesn't tell you. At most it presents the minimum number of symmetry elements to describe the structure but provides no detail as to the interactions between the molecules that produce the lowest energy KAP structures. To determine the KAP structures we do a packing energy computation using eq. 3, and 5 in eq. 1, neglecting the intramolecular term. An arbitrary molecule is chosen (called the test molecule) and the interaction of this molecule is computed with each molecule that surrounds it (its nearest neighbors) and molecules further away (its next nearest neighbors and so forth). The results are shown in Table 1 for each chain type.

TABLE 1. Packing Energy (in Kcal/mol) for the four chain types occurring in 1-acetyl-4-bromo-pyrazole (Figure 1) using the CFF91 Force Field (eq. 3 and 5). Molecule numbering is as in Figure 1.

	Non-bonded-energy	Electrostatic-energy	Total-energy
Translation Chain			
Molecule 3 with			
1	-0.07	-0.01	-0.08
5	-0.07	-0.01	-0.08
2	-2.75	-0.19	-2.94
4	-2.75	-0.19	-2.94
Totals	**-5.64**	**-0.4**	**-6.04**
Glide Chain			
Molecule 3 with			
1	-0.45	+0.06	-0.39
5	-0.45	+0.06	-0.39
2	-5.94	+0.31	-5.63
4	-5.94	+0.31	-5.63
Totals	**-12.78**	**+0.74**	**-12.04**
Screw Chain			
Molecule 2 with			
1	-0.16	+0.16	0
3	-0.16	+0.16	0
Totals	**-0.32**	**+0.32**	**0**
Inversion Chain			
Molecule 2 with			
5	-0.04	+0.01	-0.03
1	-0.12	0	-0.12
3	-0.82	-0.02	-0.84
4	-2.75	-0.19	-2.94
Totals	**-3.73**	**-0.20**	**-3.93**

Notice the following: a) for each chain type, the intermolecular potential energy occurs in pairs except for the inversion chain. This will always be true for the translation, screw and glide chains because the molecules are equally spaced along the chain axis, but not true for the inversion chain since the molecules are not usually equally spaced. b) The lowest energy chain is a glide chain with a repeat vector of 7.23 Å along the $\underline{c}$ unit-cell direction.

3.2 KAP LAYERS

Given the KAP chain, the lowest energy KAP layer is readily found. Looking down the KAP chain, one will see other chains surrounding the test molecule. The interaction of the test molecule with all the molecules in each of these chains is computed using equations 3 and 5. The chains that interact with the test molecule with the lowest energy make a layer. For 1-acetyl-4-bromo-pyrazole this turns out to be a Pc glide‖ layer shown in Figure 3.

Figure 3. Three glide chains of 1-acetyl-4-bromo-pyrazole packed together to form a glide‖ layer. The glide plane is parallel to the paper.

3.3 KAP CRYSTALS

The crystal is composed of stacked KAP layers shown in Figure 4. The interaction among the layers is readily computed by subtracting the layer energy from the crystal energy. Table 2 displays the complete KAP energy components (Note that the glide chain energies in this Table computed from eq. 6 are one-half those shown in Table 1 computed from eq. 1). Also shown is the anisotropy ratio for each KAP structure. This is the ratio of the substructure energy to the total energy. These numbers give some idea as to how strongly the molecules interact in each of the substructures relative to say

32

the packing of spheres in an isotropic lattice (ideally 0.16 for chains, 0.50 for layers). The larger these numbers the more anisotropic the crystal. 1-acetyl-4-bromo-pyrazole is quite anisotropic with more than 75% of the crystal energy tied up in the layer.

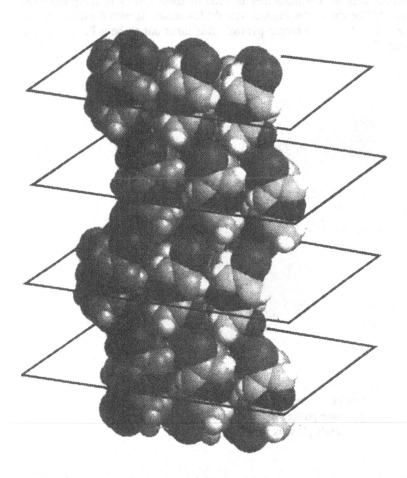

Figure 4. The stacking of four glide‖ layers of 1-acetyl-4-bromo-pyrazole which completes the crystal structure. Each layer is related to the one above or below it by a 180° rotation (the 2_1-screw) perpendicular to the layers.

TABLE 2. Packing energy (in Kcal/mole) for each of the KAP substructures of 1-acetyl-4-bromo-pyrazole as computed using the CFF91 force field (eqs. 3 and 5 in 6).

Structure type	Non-bonded-energy	Electrostatic-energy	Total energy	Anisotropy ratio
Glide chain	-6.39	+0.37	-6.02	0.36
Pc glide‖ layer	-11.23	-1.50	-12.73	0.77
Crystal energy	-15.08	-1.45	-16.53	1.00

Using the method outlined above, any crystal structure can be analyzed for low energy KAP structures. My colleagues and I have presented a number of examples of this type of analysis.[26], [27] This analysis should be of interest to anyone interested in understanding the specific types of interaction in crystals that produce the packing patterns observed and for predicting new ones. Moreover the method computes the total non-bonded and electrostatic energy as separate terms and shows how each contributes to the packing energy. In Table 1 for instance we see that the electrostatic contribution is only a small fraction of the total. This should not be surprising. The non-bonded term is always attractive whereas the electrostatic term has both attractive (negative) and repulsive (positive) terms in the sum. Any other specific interactions among atoms can be extracted from the sum at one's discretion to determine its significance to the sum. (See Taylor, Orpen, this Volume). Examples include strong hydrogen-bond interactions that we have discussed in some detail.[28](See also Aakeroy, Katrusiak, this Volume). Desiraju calls these specific interactions, synthons, since they may act as vectors for the crystal packing pattern.[29] Other types of synthons may be treated in the same way. I am presently looking at weak hydrogen bond synthons[30] to determine their significance to the crystal packing patterns observed.

4. Predicting KAP Structures From First Principles

4.1 THE NATURE OF THE PROBLEM

If the atom-atom force field described above contains all the information for obtaining the intermolecular interaction energy, then it should be possible to predict the molecular packing pattern of a crystal simply from a knowledge of the molecular structure and the force field without any additional assumptions. However, simply placing a bunch of molecules in a box and minimizing the energy will not yield a crystalline arrangement. The nature of this problem is equivalent to that associated with the packing of amino acids in proteins, the protein-folding problem, for which there presently is no general solution.[31] Like the protein-folding problem, the crystal-packing problem suffers from the local minimum explosion. The potential energy surface is bumpy, not smooth. The more degrees of freedom the bumpier it gets. Energy minimization simply finds the nearest local minimum which maybe far from the observed crystal minimum.

A brute force conformational analysis of the problem also fails. The number of conformations that has to be examined is explosively large. The nature of this problem can best be appreciated by looking at Figure 5 where an arbitrarily shaped molecule is displayed in a unit-cell.

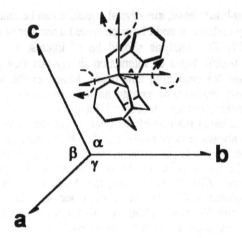

Figure 5. An organic molecule of unknown shape with N internal
unknown torsion angles in a unit-cell of unknown repeat dimensions a, b, c
and angles α, β, γ. The molecule is positioned in an unknown location
within the cell in an unknown orientation.

There are 6 degrees of freedom associated with the cell dimensions (a, b, c, α, β, γ), and
six degrees of freedom associated with the molecule's rotation and translation for a total
of 12. In addition if the molecule has N internal bond rotational degrees of freedom, the
total number of variables is 12+N. Each variable is independent, so if there were 10
possible values for each, the total number of possible crystal "conformations" that
would have to be examined is 10^{12+N}, a very large number. Alternatively, one can do a
random search for the global energy minimum and nearby local minima. This is the
Monte Carlo technique used by many workers. Verwer and Leusen have presented a
recent review of ten different methods.[32]

4.2 USING KAP TO PREDICT PACKING PATTERNS

The approach I have taken uses the KAP philosophy in a Monte Carlo simulated
annealing process to predict packing structures. It has some advantages and
disadvantages. The advantages are that it builds up the structure by first predicting
chains for which there are at most 6+ N degrees of freedom. Then uses these chains to
predict layers for which there are only 3 degrees of freedom and finally uses the layers
to predict crystal packing for which there are only 3 more degrees of freedom. The idea
has some similarities to one described by Gavezzotti[33] but differs in methodology.
By using KAP, a smaller number of degrees of freedom are needed at each stage, thus

allowing the introduction of internal degrees of freedom (bond rotation, bond bending) into the problem early on. In fact we have shown that up to 18 internal bond rotations can be incorporated into the analysis[26] and we have used this effectively to predict layer structures of a large number of Langmuir and Langmuir monolayers.[34]. The disadvantages of the method are a) the space group has to be known a-priori (this is a requirement of all the other methods except those of the Williams'[35] and the Scheraga [36] groups) and b) the number of substructures that would have to be computed for a space group like P21/c is very large.

4.3 MONTE CARLO SIMULATED ANNEALING

As its name implies, Monte Carlo methods are random sampling techniques.[37] In the crystal packing problem, the energy, E_i, of an arbitrary initial configuration of the molecules is computed using equation 6. Then, using a random number generator, the value of one of the degrees of freedom is changed by a small amount and the energy, E_f, of the new structure computed. The new structure is accepted or rejected based on a comparison of the Boltzmann probability, p(E), equation 7, with a random number R chosen between 0 and 1

$$p(E) = \exp-\frac{\left(E_f - E_i\right)}{kT} \tag{7}$$

If p(E) < R then the new state is rejected. If p(E) > R, then the new state is accepted. Clearly if E_f < E_i, the new state is always accepted. The process is then repeated over and over again with the new state as the starting state (or the old state if the new state was rejected). Two problems arise with this method, a) when do you stop it and b) what temperature, T, do you use. The acceptance or rejection of a new state clearly depends on the temperature T. For large T, new states will be accepted more often than at low T. To simulate the effect of temperature, a simulated annealing method has been developed where T is gradually lowered during the simulation.[38] In our method we start at T = 4000K and change T in 10% increments until T reaches 300K and then stop.[39] The lowest energy structure is then saved and the annealing process is repeated until 700 structures have been saved. Application of the method to the various stages of KAP has been described in detail to KAP chains[40],[41] and KAP layers.[26] Here we give a few examples of its application to 3-dimensinal crystal structure prediction for molecules with rotating single bonds.

4.4 PREDICTING KAP CHAINS

Given the chain type, KAP chain packing patterns are easily predicted using a variety of techniques. Scaringe and Perez used systematic conformational analysis to find the low energy glide chain structures of thiapyrlium dyes.[8] Perlstein has used Monte Carlo simulated annealing methods to find the low energy packing patterns of KAP chains in crystals.[39] The results of these analyses reveal the following:

36

a) The experimentally observed chain is a local energy minimum (more exactly a stationary point) of the simulation
b) The experimentally observed chain is usually not the global minimum
c) The electrostatic contribution to the total energy is usually much smaller than the non-bonded contribution

Point a) is partly a result of the fact that molecular crystals are usually made up of anisotropically shaped molecules (they are not spheres). There is usually one direction in which they pack better than in any other. But the other more important conclusion is that the chain as an isolated entity has its own existence within the constraint of the chain space group. Put another way, the chain is a thermodynamically stable entity. If we remove the lowest energy chain from a crystal then except for some floppy bonds at the edges of the chain, the molecules will not rearrange themselves within the chain space group. Point b) on the other hand emphasizes the fact that the environment of the chain (that is the other molecules around it) contributes to the total packing energy of the crystal. The global minimum chain may not be the best chain for packing into a layer. Point c) is not very remarkable. As molecules get larger the non-bonded energy term gets larger (it is always attractive) but the electrostatic term doesn't change much (it has both attractive and repulsive terms).

4.5 PREDICTING KAP LAYERS FROM CHAINS

KAP layers occur not only as substructures of most molecular crystals but also as monolayers on water surfaces (Langmuir films), as epitaxial grown films on ordered substrates, and as self-assembled monolayers covalently attached to substrates. Monte Carlo simulated annealing software, PACK,[42] is available for predicting layer structures including intramolecular conformational details due to exocyclic bond rotations. The method relies on packing the lowest energy chains into layers of specific symmetry. The most common layers are the KAP layers shown in figure 2 although other layers are possible and are known.[12],[13, 43] The structures of a large number of Langmuir monolayers have been predicted using the PACK routines.[34] Most of these are either simple translation layers or glide layers. As with KAP chains, KAP layers are not usually the global energy minimum structure, but a local minimum, and the electrostatic contribution to the layer energy is usually small compared to the non-bonded energy. An interesting observation made with PACK is that the intramolecular conformational details of a molecule in a three-dimensional crystal is almost completely resolved in the layer substructure, and only a few rotating bonds at the surface of the layer are affected by layer-layer interactions.[26]

4.6 PREDICTING CRYSTAL STRUCTURES FROM KAP LAYERS.

Given the structure of the global and nearby local energy minima of a KAP layer it is a relatively straightforward procedure to pack the layers into a crystal using Monte

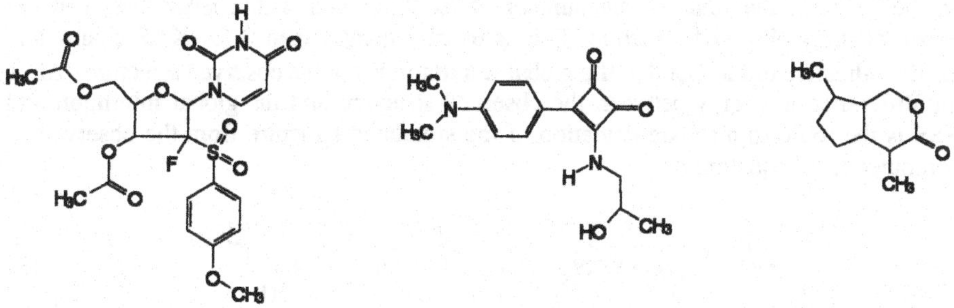

VOYVEK   JIHREX   ISIRIN

Figure 6. Molecular structures of 3 chiral molecules used in PACK crystal structure predictions. Reference codes are from the Cambridge Structural Database[Allen, 1993 #10]: VOYVEK-space group P1, with 10 torsion bonds, JIHREX space-group P1 with 6 torsion bonds and ISIRIN space-group $P2_1$ with 2 torsion bonds.

Carlo simulated annealing techniques. PACK software for the most common chiral space-groups (P1, $P2_1$, $P2_12_12_1$) is available for this type of prediction.[42] Results are shown in Table 3 for the three molecules displayed in figure 6.

TABLE 3. Results of crystal structure predictions for some chiral space-groups

	Rank	ΔE(Kcal/mole)	RSS
VOYVEK-P1			
Translation chain	4	+3.16	6.01
Translation layer	6	+2.86	8.17
Crystal	0	0	13.3
JIHREX-P1			
Translation chain	21	+2.56	8.03
Translation layer	0	0	4.01
Crystal	0	0	11.56
ISIRIN-P2$_1$			
2_1 Screw Chain	80	3.23	15.0
2_1 Screw Layer	40	2.35	9.05
Crystal	0(22)[a]	0(2.45)[a]	15.93

[a]results in parenthesis are without polarization correction.

In the Table 3, the Rank is the number of local minima with energy less than the experimentally observed structure. This is usually greater than 0 for KAP chains and layers indicative of the fact that the global minimum is not the observed structure. ΔE is the difference in energy between the observed structure and the global minimum and RSS is the route sum square deviation of the simulated structure from the observed as computed from equation 8.

$$RSS = \sqrt{\sum_i [18(q_{r_i} - q_{r_0})^2 + (q_{\theta_i} - q_{\theta_0})^2}$$

(8)

In equation 8, the dq_r's are the differences between the translational degrees of freedom of the simulation and the observed and the dq_θ's are the differences between the angular degrees of freedom for the simulation and the observed crystal structure. The factor of 18 is introduced as a weight for the translational degrees of freedom so that and error of 0.3 Å in any q_r is equivalent to a 5.4° error in q_θ. For a crystal with 12 external degrees of freedom, and RSS of 20 would correspond to and average translational error of 0.32 Å and an average angular error of 5.8°.

Note that there are as many as 10 torsion bond-angles that need to be resolved. PACK makes no assumptions about what these torsion bond angles should be and includes them as variables in the simulated annealing runs starting with the chain packing. Also note that for ISIRIN a surface-polarization-energy correction[22] needs to be introduced into the energy computation for the three-dimensional crystal structure in order for the crystal energy minima to be in the correct order. This needs to be done only at the end of the simulation and saves considerable computational time.

5. Space-Group Predictions and Polymorphism

The same molecule can have different packing patterns in the same space group or in different space groups.[44-46] This polymorphism is an extremely important topic especially for drug design where polymorphic variants have different pharmaceutical characteristics. Except for two preliminary reports on predicting space group for some simple molecules of high symmetry,[35, 36] none of the available techniques can predict space-group a-priori. Moreover, none of them predict crystal entropy and thus none of them can predict the phase relationship between the polymorphic variants that inevitably occur. Complicating the problem further is the possibility of solvent incorporation into the crystal structure. A review of the complexity of polymorphism is available. .[47]

6. Packing Patterns Beyond Crystals

The 230 space groups for the packing of 3-dimensional periodic space was established well before the dawn of X-ray crystallography. It represents but one description of how molecules can assemble themselves. There are others. The major premise of space group analysis is that there are units that translationally repeat themselves in a periodic lattice. This however is only one way molecules may assemble (albeit an important one). Molecules may assemble in many other ways, which are not necessarily 3-dimensionally periodic. The beautiful structures associated with micelles, vesicles, the folding patterns of proteins, the structure of viruses and the quasiperiodic lattices of alloys are but a few other examples. In this short introduction we cannot consider all these other assemblies, but the reader is referred to other references for more descriptions[48]. (See also Lehn, this Volume).

7. Acknowledgments

The author thanks the National Science Foundation Center for Photoinduced Charge Transfer (CHE-9120001) for partial support of this research.

8. References

1.   Kitaigorodskii, A.I. (1961) *Organic Chemical Crystallography*, Consultants Bureau, NY, p. 67.
2.   Dahl, T. (1994) The Nature of Stacking Interactions between Organic Molecules Elucidated by Analysis of Crystal Structures, *Acta Chem. Scand.* **48**, 95-106.
3.   Sharma, C.C.K. and Rogers, R.D. (1998) CH---X (X=N,O) Hydrogen Bond-Mediated Assembly of Donors and Acceptors: The Crystal Structures of Phenazine Complexes with 1,4-Dinitrobenzene and TCNQ, *Crystal Engineering* **1**, 139-145.
4.   Brandon, E.J., Arif, A.M., Miller, J.S., Suguira, K.-i., and Burkhart, B.M. (1998) The Structure of Several Supramolecular Meso-Tetraaryl porphinatomanganese(III) Tetracyanoethanide Magnets, *Crystal Engineering* **1**, 97-107.
5.   Braga, D., Grepioni, F., and Desiraju, G.R. (1998) Crystal Engineering and Organometallic Architecture, *Chemical Reviews* **98**, 1375-1405.
6.   Hajek, F., Graf, E., Hosseini, M.W., De Cian, A., and Fisher, J. (1998) Crystal Engineering: Formation of 1D-Networks Based on the Self-Assembly of Self-Complimentary Hollow Molecular Modules in the Solid State, *Crystal Engineering* **1**, 79-85.
7.   Shubnikov, A.V. and Koptsik, V.A. (1974) *Symmetry in Science and Art*, Plenum Press, NY, pp. 103-127.
8.   Scaringe, R.P. and Perez, S. (1987) A Novel Method for Calculating the Structure of Small-Molecule Chains on Polymeric Templates, *J. Phys. Chem.* **91**, 2394-2403.

40

9.   Lapasset, J., Escande, A., and Falgueirettes, J. (1972) Structure Cristalline et Moléculaire de l'Acétyl-1-Bromo-4-Pyrazole, *Acta Crystallogr.* **B28**, 3316-3321.

10.   Brock, C.P. and Dunitz, J.D. (1994) Towards a Grammar of Crystal Packing, *Chem. Mater.* **6**, 1118-1127.

11.   Ulman, A. (1991) *An Introduction to Ultra Thin Organic Films*, Academic Press, Boston.

12.   Rabe, J.P. (1992) *Nanostructures Based on Molecular Materials,*, W. Gopel and C. Ziegler (eds.) VCH, Weinheim, pp 313-327.

13.   Patrick, D.L. and Beebe, T.P., Jr. (1994) Substrate Defects and Variations in Interfacial Ordering of Monolayer Molecular Films on Graphite, *Langmuir* **10**, 298-302.

14.   Scaringe, R.P. (1990) A Theoretical Technique for Layer Structure Prediction, in J.R. Fryer and D.L. Dorset (eds.), *Electron Crystallography of Organic Molecules*, Kluwer Academic Publishers, Dordrecht, pp. 85-113.

15.   Wood, E.A. (1964) The 80 Diperiodic Groups in Three Dimensions, *Bell Sys. Tech. J.* **43**, 541-559.

16.   Wood, E.A. (1964) The 80 Diperiodic Groups in Three Dimensions, *Monograph 4680-Bell Telephone System Technical Publication*, AT&T Archives, Murray Hill.

17.   Allen, F.H. and Kennard, O. (1993) 3D Search and Research Using the Cambridge Structural Database, *Chem. Design Automat. News* **8**, 31-37.

18.   Allinger, N.L. (1977) Conformational Analysis 130. MM2. A Hydrocarbon Force Field Utilizing V1 and V2 Torsional Terms, *J. Am. Chem. Soc.* **99**, 8127-8134.

19.   (1992) *Discover User Guide, Version 2.8 Part 2*, Biosym Technologies(now Molecular Simulations Inc.), San Diego.

20.   Weiner, S.J., Kollman, P.A., Nguyen, D.T., and Case, D.A. (1986) An All Atom Force Field for Simulations of Proteins and Nucleic Acids, *J. Comp. Chem.* **7**, 230-252.

21.   Coombes, D.S., Price, S.L., Willock, D.J., and Leslie, M. (1996) Role of Electrostatic Interactions in Determining the Crystal Structures of Polar Organic Molecules. A Distributed Multipole Study, *J. Phys. Chem.* **100**, 7352-7360.

22.   Smith, E.R. (1981) Electrostatic Energy in Ionic Crystals, *Proc. R. Soc. Lond. A* **375**, 475-505.

23.   Catti, M. (1978) Electrostatic Lattice Energy in Ionic Crystals: Optimization of the Convergence of Ewald Series, *Acta Crystallogr.* **A34**, 974-979.

24.   Busing, W.R. (1978) An Error in the Calculation of the Rotational Barrier in Molecular Crystals, *J. Phys. Chem. Solids* **39**, 691.

25.   CHEMX/CHEMLIB is a molecular modeling program distributed by the Oxford Molecular Group at http://www.oxmol.com/chemdesign/products.html

26.   Perlstein, J. (1994) Molecular Self-Assemblies. 4. Using Kitaigorodskii's Aufbau Principle for Quantitatively Predicting the Packing Geometry of Semiflexible Organic Molecules in Translation Monolayer Aggregates, *J. Am. Chem. Soc.* **116**, 11420-11432.

27. Abdallah, D., Bachman, R.E., Perlstein, J., and Weiss, R. (1999) Crystal Structures of Symmetrical teta-n-Alkyl Ammonium and Phosphonium Halides. Dissection of Competing Interactions Leading to 'Biradial' and 'Tetraradial' Shapes, *J. Phys. Chem to be published*.

28. Perlstein, J., Steppe, K., Vaday, S., and Ndip, E.M.N. (1996) Molecular Self-Assemblies. 5. Analysis of the Vector Properties of Hydrogen Bonding in Crystal Engineering, *J. Am. Chem. Soc.* **118**, 8433-8443.

29. Nangia, A. and Desiraju, G.R. (1998) Supramolecular Synthons and Pattern Recognition, *Topics in Current Chemistry* **198**, 57-95.

30. Steiner, T. (1998) Donor and Acceptro Strengths in C-H---O Hydrogen Bonds Quantified from Crystallographic Data of Small Solvent Molecules, *New J. Chem.* **22**, 1099-1103.

31. Doye, J.P.K. and Wales, D.J. (1996) On Potential Energy Surfaces and Relaxation to the Global Minimum, *J. Chem. Phys.* **105**, 8428-8445.

32. Verwer, P. and Leusen, F.J.J. (1998) Computer Simulation to Predict Possible Crystal Polymorphs, in K.B. Lipkowitz and D.B. Boyd (eds.), *Reviews in Computational Chemistry*, Wiley-VCH, NY, pp. 327-365.

33. Gavezzotti, A. (1991) Generation of Possible Crystal Structures from the Molecular Structure for Low-Polarity Organic Compounds, *J. Am. Chem. Soc.* **113**, 4622-4629.

34. Whitten, D.G., Chen, L., Geiger, H.C., Perlstein, J., and Song, X. (1998) Self-Assembly of Aromatic-Functionalized Amphiphiles: The Role and Consequences of Aromatic-Aromatic Noncovalent Interactions in Building Supramolecular Aggregates and Novel Assemblies, *J. Phys. Chem. B* **102**, 10098-10111 and references therein.

35. Williams, D.E. (1996) Ab Initio Molecular Packing Analysis, *Acta Crystallogr.* **A52**, 326-328.

36. Wawak, R.J., Pillardy, J., Liwo, A., Gibson, K.D., and Scheraga, H.A. (1998) Diffusion Equation and Distance Scaling Methods of Global Optimization: Applications to Crystal Structure Prediction, *J. Phys. Chem.* **102**, 2904-2918.

37. Binder, K. (1997) Applications of Monte Carlo Methods to Statistical Physics, *Rep. Prog. Phys.* **60**, 487-559.

38. Kirkpatrick, S., Gelatt, C.D., and Vecchi, M.P. (1983) Optimization by Simulated Annealing, *Science* **220**, 671-680.

39. Perlstein, J. (1992) Molecular Self-Assemblies: Monte Carlo Predictions for the Structure of the One-Dimensional Translation Aggregate, *J. Am. Chem. Soc.* **114**, 1955-1963.

40. Perlstein, J. (1994) Molecular Self-Assemblies. 2. A Computational Method for the Prediction of the Structure of One-Dimensional Screw, Glide, and Inversion Molecular Aggregates and Implications for the Packing of Molecules in Monolayers and Crystals, *J. Am. Chem. Soc.* **116**, 455-470.

41. Perlstein, J. (1994) Molecular Self-Assemblies. 3. Quantitative Predictions for the Packing Geometry of Perylenedicarboximide Translation Aggregates and the Effects of Flexible End Groups. Implications for Monolayers and Three-Dimensional Crystal Structure Predictions, *Chem. Mat.* **6**, 319-326.

42

42.   PACK is a series of Monte Carlo routines for packing chains, layers, and crystals. It is available as part of the CHEMX/CHEMLIB molecular modeling package. Contact the author at perlstein@chem.chem.rochester.edu.

43.   Swanson, D.R., Hardy, R.J., and Eckhardt, C.J. (1994) A Cross-Section Potential for Calculation of Close Packing Geometries of Monolayer Films, *Thin Solid Films* **244**, 824-826.

44.   Anthony, A., Desiraju, G.R., Jetti, R.K.R., Kuduva, S.S., Madhavi, N.N.L., Nangia, A., Thaimattam, R., and Thalladi, V.R. (1998) Crystal Engineering: Some Further Strategies, *Crystal Engineering* **1**, 1-18.

45.   Bernstein, J. and Henck, J.-O. (1998) Disappearing and Reappearing Polymorphs- An Anathema To Crystal Engineering, *Crystal Engineering* **1**, 119-128.

46.   Gavezzotti, A. and Filippini, G. (1995) Polymorphic Forms of Organic Crystals at Room Conditions: Thermodynamic and Structural Implications, *J. Am. Chem. Soc* **117**, 12299-12305.

47.   Caira, M.R. (1998) Crystalline Polymorphism of Organic Compounds, *Topics in Current Chemistry* **198**, 163-208.

48.   Hargittai, I. and Hargittai, M. (1995) *Symmetry Through the Eyes of a Chemist*, Plenum Press, NY, pp. 446-458.

MOLECULE-BASED MAGNETS

JOEL S. MILLER[a] and ARTHUR J. EPSTEIN[b]
[a] *Department of Chemistry, University of Utah, Salt Lake City, UT 84112-0850 U. S. A.* [b] *Department of Physics and Department of Chemistry, The Ohio State University, Columbus, OH 43210-1106 U. S. A.*

The employment of molecules, not atoms, as a basic building block to construct solids has lead to the development of new classes of materials exhibiting commercially useful properties.[1,2]   These include electrical conductivity, ferroelectricity as well as magnetic ordering.   Molecules, in contrast to atoms, enables the modulation of the commercially useful properties by low-temperature organic-synthesis methodologies, that can lead to the improvement of the properties, and lead to the development of materials with a combination of properties that will expand their desirability. Herein, we focus solely upon work related to molecule-based magnets.   Molecule-based magnets are defined as substances prepared from molecules (or molecular ions) that maintain aspects of the parent molecular framework, and magnetically order.

Harden H. McConnell in 1963 suggested a mechanism for ferromagnetic coupling between radicals that required a specific spatial arrangement.[3]   In 1967 McConnell discussed another approach for stabilizing ferromagnetic coupling between radicals that involved the admixing of a change transfer excited state into the ground state.[4] Although these models only discussed ferromagnetic coupling between a pair of radicals, not magnetic ordering, experimental research focused toward the testing these models began to appear more than a decade latter.[5, 6]

H. Hollis Wickman, Anthony M. Trozollo *et al.* reported in 1967 that S = 3/2 $ClFe^{III}(S_2CNEt)_2$ 1, was a ferromagnet with a critical or magnetic ordering temperature, T_c, of 2.46 K.[7]   R. L. Martin and co-

D. Braga et al. (eds.), Crystal Engineering: From Molecules and Crystals to Materials, 43–53.

workers at the University of Melbourne in 1970 reported that another S = 3/2 complex, manganese phthalocyanine, **2**, was a ferromagnet.[8] However, in 1983 William E. Hatfield's group showed that it was a canted-ferromagnet with 8.3 K T_c.[9] Claudine Veyret and co-workers in 1973 reported that bis(2,2,4,4-tetramethyl-4-piperidinol-1-oxyl), **3**, tanol suberate, was a ferromagnet with a T_c of 0.38 K.[10] Additional data led to the characterization of tanol suberate as being a metamagnet, *i. e.*, it had an antiferromagnetic ground state; however, above a critical applied magnetic field of 100 Oe and below a T_c of 0.38 K it had a high moment, ferromagnetic-like state.[11] In 1979 [Fe^{III}(C$_5$Me$_5$)$_2$][TCNQ] [TCNQ = 7,7,8,8-tetracyano-*p*-quinodimethane, **4**], [**5**][**4**], was characterized to be a metamagnet below the T_c of 2.55 K with a critical applied magnetic field of 1600 Oe.[12]

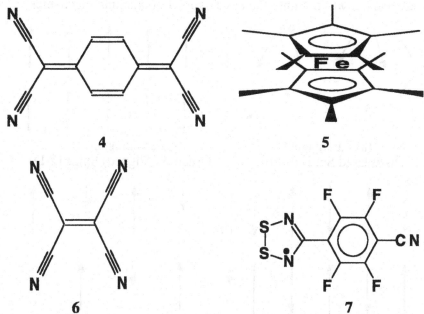

4   5

6   7

Unlike classical magnets these magnets were prepared from organic preparative methodologies, and lack extended ionic, covalent, or metallic bonding in the solid state and as a consequence are readily soluble in conventional organic solvents, but do not exhibit magnetic hysteresis characteristic of hard ferromagnets. Hysteresis was first established for [$Fe^{III}(C_5Me_5)_2$][TCNE] [TCNE = tetracyanoethylene, **6**], [**5**][**6**]. In accord with [TCNE]·⁻'s smaller size with respect to [TCNQ]·⁻ and expected greater spin density, [**5**][**6**] was expected to have stronger ferromagnetic coupling and order as ferromagnet as subsequently observed.[13]

Magnets comprised on molecular components can be grouped by the orbitals in which the spins reside. The organic moiety may be *active* with its spin sites contributing to both the magnetic moment and the spin coupling. The organic moiety also may be *passive* by only providing a structural framework to position the spins that solely reside on metal ions and hence enabling spin coupling. For these reasons, molecule-based magnets differ from conventional magnets studied for millennia.

Magnetic Behavior - An Overview

Magnetism is a materials' response (repulsive or attractive) to a magnet. Although aspects of this phenomenon were studied for several thousand years, it was not until the development of quantum mechanics

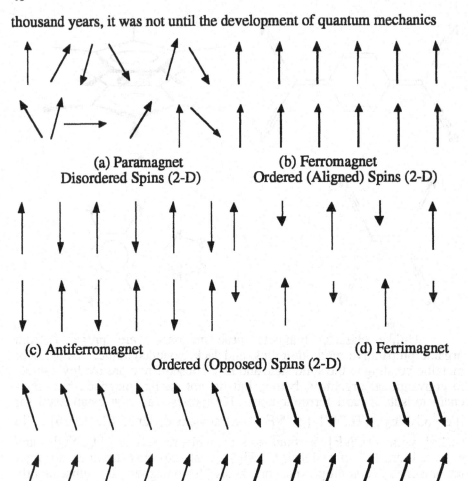

(a) Paramagnet
Disordered Spins (2-D)

(b) Ferromagnet
Ordered (Aligned) Spins (2-D)

(c) Antiferromagnet

(d) Ferrimagnet

Ordered (Opposed) Spins (2-D)

(e) Canted Ferromagnets (2-D)

Figure 1. Schematic illustration of spin coupling behaviors.

that a satisfactory explanation evolved. Magnetism is due to the spin associated with an unpaired electron ($S = \pm 1/2$; '$\uparrow$' or '$\downarrow$') and how the electron spins interact among themselves. Distant spins interact weakly and molecules, which are typically large and far apart, have weak spin-spin interactions (coupling) and the coupling energy, J (as deduced from the Hamiltonian $H = -2JS_a \bullet S_b$) is small compared to the thermal energy which randomize the spins. When spins do not couple, $i.\ e.$, they have random spins, Figure 1a. These materials are termed paramagnetic. When the spins get closer together the coupling energy increases and J can become sufficiently large to enable an effective parallel/ferromagnetic or antiparallel/antiferromagnetic alignment. This increases or decreases the measured susceptibility, χ. For uncoupled or weakly coupled spins the susceptibility can be modeled by the Curie-Weiss expression, eqn. (1), where μ_B is the Bohr magneton, k_B is the Boltzmann constant, N is Avogadro's number, g is the Landé g value, and S is the spin quantum number. When the spins couple ferromagnetically, χ is enhanced, i. e., θ > 0, and when spins couple antiferromagnetically, χ is suppressed, i. e., θ < 0. The magnetization, M, (χ/H) (H is the applied magnetic field) as a function of H can be calculated from the Brillouin function, eqns. (2).

The magnetization, M, increases with H prior to reaching an asymptotic value, the saturation magnetization, Figure 2. The Curie Law, eq. (1), is valid in the region where M linearly increases with H as illustrated for a paramagnet, Figure 2, and for temperatures much larger then θ. The saturation magnetization, M_s, and is $Ng\mu_B S$, eqn. (3). Antiferromagnetic coupling occurs when the initial slope of $M(H)$ is less than expected from eqn. (2), while ferromagnetic coupling occurs when the initial slope of $M(H)$ exceeds the expectation from eqns. (2). A ferromagnet exhibits a spontaneous magnetization, Figure 2. Hence, the shape of the $M(H)$ curve, Figure 2, as well as the determination of θ reveals the dominant magnetic coupling.

θ-values reflect pairwise, not long-range spin coupling. Although rare, pairwise ferromagnetic coupling ('↑↑'), can lead to long range ferromagnetic order, Figure 1b. Antiferromagnetic order can result from pairwise antiferromagnetic coupling ('↑↓'), Figure 1c. Ferrimagnets, as well as canted ferromagnets are a consequence of antiferromagnetic coupling which does not lead to complete cancellation of the spin and thus has a net magnetic moment, Figures 1d and e. Ferro-, antiferro-, or ferrimagnetic ordering only occurs below a critical or magnetic ordering temperature, T_c. Magnetic ordering, albeit ferro-, antiferro-, or ferrimagnetic requires spin coupling three dimensions, or in a rare case of a two-dimensional layered system, but never from coupling solely in a chain, i. e., 1-D. Hence, magnetic ordering is not a property of a molecule or ion; it is a cooperative solid state (bulk) property.

$$\chi = \frac{Ng^2\mu_B^2(S)(S+1)}{3k_B(T-\theta)} \qquad (1)$$

$$M = \mu_B NSgB \qquad (2)$$

where
$$B = \frac{2S+1}{2S}\coth\left(\frac{2S+1}{2S}x\right) - \frac{1}{2S}\coth\left(\frac{x}{2S}\right)$$

and
$$x = \frac{gS\mu_B H}{k_B T}$$

$$M_S = Ng\mu_B S \qquad (3)$$

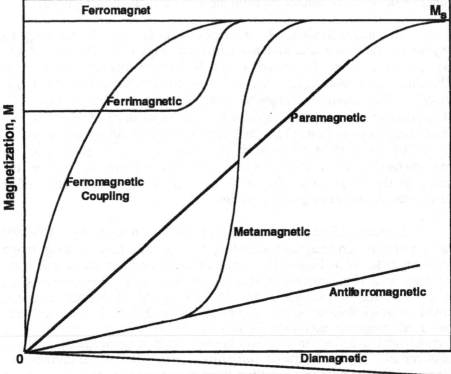

Figure 2. Field, H, dependencies of the magnetization, M, for different types of magnetic coupling. The saturation magnetization, M_S, and spontaneous magnetization (magnetization at zero applied field) only depends upon the number of spins per repeat unit.

Below T_c the magnetic moments for ferri-, ferromagnets, and canted antiferromagnets (not antiferromagnets) align in small regions termed domains, e. g., Figure 3. The direction of the magnetic moments of adjacent domains differ, but can be aligned by a application of a minimal magnetic field, H_c (coercive field).   This leads to history dependent magnetic behavior (hysteresis), $e.$ $g.$, Figure 4, characteristic of ferri- and ferromagnets. Large values (hundreds of Oe) of H_c are necessary for magnetic storage of data, while low values (mOe) are necessary for ac motors and magnetic shielding.   Examples of molecule-based magnets with both large and small values of H_c have been reported.   Since, the coercive field reflects the magnetic anisotropy, low values are expected for organic compounds due to their inherent isotropy. Magnetic materials that are subdomain in size exhibit superparamagnetic (high-spin), but not extended ordering, behavior. The T_c, M_s, and H_c etc. are key parameters in ascertaining the commercial utility of a magnet.

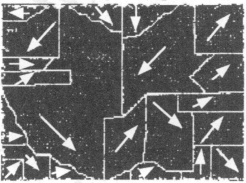

Figure 3. Thin slab or iron showing the different magnetic domains, each with an arrow showing the direction of its magnetization. When aligned in an external magnetic field, all the arrows will align in the direction of the applied magnetic field, H.

Additional types of magnetic ordering phenomena, $e.$ $g.$, metamagnetism and spin-glass behavior have been reported for molecule-based magnets. The transformation from an antiferromagnetic state to a high moment state, $i.$ $e.$, the spin alignment depicted in Figure 1c is transformed to that depicted in Figure 1b, by an applied magnetic field, is metamagnetism, Figure 2. α-Decamethylferrocenium 7,7,8,8-tetracyano-p-quinodimethanide, α-[Fe(C$_5$Me$_5$)$_2$][TCNQ] {[5][4]},[12] and tanol suberate (3)[11] are molecule-based metamagnets. A canted (or 'weak') ferromagnet results from the relative canting of antiferromagnetically coupled spins that reduces the moment, Figure 1e.   Manganese phthalocyanine (2)[8] and 4'-cyanotetrafluorophenyldithiadiazolyl[15] (7) are examples of molecule-based canted ferromagnets. A spin-glass occurs when local spatial correlations in the directions of neighboring spins exist, but long-range order is lacking. The spin alignment for a spin glass is identical to that of a paramagnet, Figure 1a; but, while the spin directions vary with time for a paramagnet, the spin orientations for a spin-glass remain fixed or change slowly with time. $V[TCNE]_x \cdot zMeCN$ at low temperature is an example of a correlated spin-glass.[17a,18]

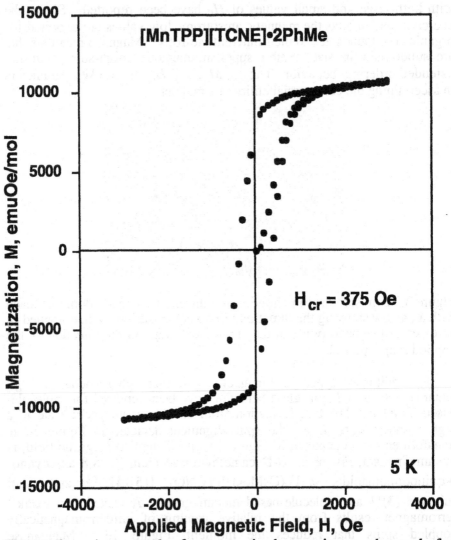

Figure 4. An example ferromagnetic hysteresis as observed for [Mn^{III}TPP][TCNE]·2PhMe.[14]

As noted earlier, magnetic ordering requires spin alignment throughout the solid. The spin coupling energy, J, reflects the type [ferro- ($J > 0$) or antiferromagnetic ($J > 0$)] and magnitude of coupling between nearest neighbor (pairwise) spins. J, unlike T_c, is not directly determinable, but is determined from a fit to a mathematical model used to describe the magnetic data and is not discussed herein. Three distinct mechanisms for spin coupling, Table 1,[17a,f] have been identified. Additionally, the antiferromagnetic spin coupling of spins on adjacent sites possessing a differing number of spins per site can lead to ferrimagnet ordering.

TABLE 1. Mechanisms for Achieving Ferro- or Antiferromagnetic Spin Coupling

Mechanism	Spin Coupling[a]	Spin Interaction
1 Spins in Orthogonal Orbitals (no CI) - Hund's rule	FO	Intramolecular
2 Spin Coupling via Configuration Interactions (CI)	FO or AF	Intra- or intermolecular
3 Dipole-dipole (through-space) interactions	FO or AF	Intra- or intermolecular

[a] AF = antiferromagnetic ($J < 0$); FO = ferromagnetic ($J > 0$)

Which spin coupling mechanism dominates for a particular system, may not be clear and more than one mechanism may contribute. Furthermore, as a consequence of the many levels of complexity of the CI model, this mechanism can be described in many ways. As it is always important to note that magnetic ordering is not a property of an isolated molecule; it is a cooperative solid state (bulk) property, thus, to achieve bulk magnetic ordering for a molecular system, the spin coupling interactions must be present in all three directions. This is described in more detail in ref. 17a,f.

Unpaired electron spins are essential for the formation of any magnet. The number spins and type and strength of spin coupling dictate the magnetic behavior, e. g., the type of coupling as well as magnetic ordering. In the past decade magnets have been made from molecular precursors and these magnets maintain aspects of the molecules in the structure.[17] Hence, molecule-based magnets enable organic species to contribute spins or magnetization as well as providing a means to couple the spins. Hence, like proteins, the primary, secondary, and tertiary structures of the molecule-based solid are crucial for achieving the desired cooperative magnetic properties. Presently the development of new molecular-based magnets is thwarted by the rational design of solid state structures which remains an art. This is due to the formation of numerous polymorphs, isomers,[19] as well as complex and solvated compositions,[20] and unwanted structure types. Crystals growth is also essential for determining the single crystal structure and single crystal magnetic properties and is also limitation. The advent of CCD-based single crystal diffractometers has, however, enabled the study of substantially smaller crystals. Hence, this problem is reduced in importance. Nonetheless, new spin-bearing radical anions and cations, not to mention new structure types are needed for this area of research to develop. Due to the rapid development of the field of molecule-based magnets, significant advances during the next decade are expected for this embryonic multidisciplinary arena of solid state science.

Acknowledgment

The authors gratefully acknowledge the continued partial support by the Department of Energy Division of Materials Science (Grant Nos. DE-FG02-86ER45271, DE-FG03-93ER45504, DE-FG02-96ER12198) as well as the National Science Foundation (Grant No. CHE9320478) and the ACS-PRF (Grant No. 30722-AC5).

References

1   J. S. Miller, *Adv. Mater.* **1990**, *2*, 98.
2   A. P. Alivisatos, P. F. Barbara, A. W. Castleman, J. Chang, D. A. Dixon, M. L. Klein, G. L. McLendon, J. S. Miller, M. A. Ratner, P. J. Rossky, S. I. Stupp, M. E. Thompson, *Adv. Mater.* **1998**, *10*, 1297\.
3   H. M. McConnell, *J. Chem. Phys.* **1963**, *39*, 1910.
4   H. M. McConnell, H. M., *Proc. R. A. Welch Found. Chem. Res.* **1967**, *11*, 144.
5.   A. Izoka, S. Murata, T.Sugawara, H. Iwamura, *J. Am. Chem. Soc.* **1985**, *107*, 1786; *J. Am. Chem. Soc.* **1987**, *109*, 2631.
6.   R. Breslow, B. Juan, R. Q. Kluttz, C.-Z. Xia, *Tetrahedron* **1982**, *38*, 863. Breslow, R., *Pure App. Chem.* **1982**, *54*, 927.
7.   H. H. Wickman, A. M. Trozzolo, H. J. Williams, G. W. Hull, F. R. Merritt, *Phys. Rev.* **1967**, *155*, 563.
8.   C. G. Barraclough, R. L. Martin, S. Mitra, R. C. Sherwood, *J. Chem. Phys.* **1970**, *53*, 1638.

9. S. Mitra, A. Gregson, W. E. Hatfield, R. R. Weller, *Inorg. Chem.* **1983**, *22*, 1729.

10. M. Saint Paul, C. Veyret, *Phys. Lett.* **1973**, *45A*, 362.

11. G. Chouteau, C. Veyret-Jeandey, *J. Physique* **1981**, *42*, 1441.

12. G. A. Candela, L. J. Swartzendruber, J. S. Miller, M. J. Rice, *J. Am. Chem. Soc.* **1979**, *101*, 2755.

13 J. S. Miller, J. C. Calabrese, D. A. Dixon, A. J. Epstein, R. W. Bigelow, J. H. Zhang, W. M. Reiff, *J. Am. Chem. Soc.* **1987**, *109*, 769. J. S. Miller, J. C. Calabrese, A. J. Epstein, R. W. Bigelow, J. H. Zhang, W. M. Reiff, *J. Chem. Soc., Chem. Commun.* **1986**, *1026*. S. Chittipeddi, K. R. Cromack, J. S. Miller, A. J. Epstein, *Phys. Rev. Lett.* **1987**, *22* , 2695.

14. J. S. Miller, J. C. Calabrese, R. S. McLean, A. J. Epstein, *Adv. Mater.* **1992**, *4*, 498.

15. A. J. Banister, N. Bricklebank, I. Lavender, J. M. Rawson, C. I. Gregory, B. K. Tanner, W. Clegg, M. R. J. Elsegood, F. Palacio, *Angew. Chem. internat. Ed.* **1996**, *35*, 2533. A. J. Banister, N. Bricklebank, W. Clegg, M. R. J. Elsegood, C. I. Gregory, I. Lavender, J. M. Rawson, B. K. Tanner, *J. Chem. Soc., Chem. Commun.* **1995**, 679.

17. (a) J. S. Miller, A. J. Epstein, *Angew. Chem. internat. Ed.* **1994**, 106, 399; *Angew. Chem. int. Ed.* **1994**, *33*, 385. (b) O. Kahn, *Molecular Magnetism*, VCH Publishers, Inc. **1993**. (c) D. Gatteschi, *Adv. Mat.* **1994**, *6*, 635. (d) M. Kinoshita, *Jap. J. Appl. Phys.* **1994**, 33, 5718. (e) A. L. Buchachenko, *Russ. Chem. Rev.* **1990**, *59*, 307. (f) J. S. Miller, A. J. Epstein, *Chem. & Engg. News*, **1995**, *73#40*, 30 .

18. P. Zhou, B. G. Morin, J. S. Miller, A. J. Epstein, *Phys. Rev. B* **1993**, *48*, 1325.

19. J. S. Miller, *Adv. Mater.* **1998**, *10*, 1553.

20. *e. g.*, J. S. Miller, A. J. Epstein, *J. Chem. Soc., Chem. Commun.* **1998**, 1319.

STRUCTURE DETERMINATION BY NEUTRON AND SYNCHROTRON X-RAY SCATTERING

H. FUESS
Department of Materials Science
Darmstadt University of Technology
Petersenstraße 23
64287 Darmstadt/Germany

Abstract

The basis for modelling crystals is the enormous wealth of crystal structures determined by diffraction methods. Neutron diffraction has added complementary precision to structural data and inelastic neutron scattering gives valuable information on interatomic and intermolecular forces. The tunability and high brilliance of third generation synchrotron sources opens new fields in structure determination of large molecules by using the anomalous dispersion of X-rays near to the absorption edge. Tremendous progress in computer facilities, mono-chromators and detection devices allow experiments at extreme conditions (temperature, pressure) and at very short measuring times. The precision of deformation density distribution is considerably enhanced by the use of short wavelengths with synchrotron radiation.

1. Introduction

Crystallography is beyond any doubt the method which contributed most to our knowledge of the three-dimensional periodic arrangement of atoms in matter. Generally applied as a laboratory tool is X-ray diffraction. X-rays do not only provide the idealized structure of molecules and crystals but they elucidate at the same time all kinds of defects of a real crystal structure. As X-ray diffraction is well established since the first structure determinations by the Braggs [1] in 1913 it has been complemented by neutron scattering after the construction of research reactors about 1948. The advent of Synchrotron Radiation (SR) sources with

D. Braga et al. (eds.), Crystal Engineering: From Molecules and Crystals to Materials, 55–78.

extreme brilliance opened new fields of application to X-ray crystallography. Whereas the use of neutrons in molecular structures is determined by elastic and inelastic interactions, synchrotron radiation is useful for diffraction and spectroscopic applications. Inelastic X-ray scattering has been very succesful on third generation synchrotron sources where even phonon measurements in amorphous solids have been carried out. The present contribution shall give some fundamentals of both neutron scattering and synchrotron X-ray diffraction including some principles of inelastic scattering and deals with some examples of structure determination, and precise electron density measurements.

2. Principles of Scattering Processes

2.1 PRODUCTION AND MODERATION OF NEUTRONS

The neutron is a unique particle as it matches simultaneously typical energy and wavelength scales in condensed matter. The neutron wavelengths (1-10Å) are ideal for the study of interatomic distances. The energy range from 1-1000 meV corresponds to the vibrational energy in a molecule or crystal. The energy transfer is thus comparable to the incident energy of neutrons and easily detectable. An energy transfer occurs of course during an X-ray

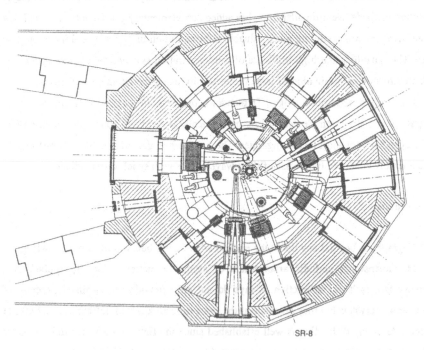

SR-8

Figure 1. Arrangement of beam-tubes around the core of the nuclear research reactor at Munich

diffraction process also. But X-rays of 1Å have energies of about 12 keV, an energy transfer of several meV is difficult to detect and the observation of inelastic processes has only been achieved at third generation synchrotron sources. Neutrons are produced either in a fission reactor from ^{235}U or by spallation when protons strike a heavy element target. The nuclei are excited and "spallate" particles (protons, neutrons). Whereas the production of neutrons in a reactor is a steady process, spallation sources (Rutherford Lab., UK) reach a higher concentration of neutrons in the pulse. In both cases high energy neutrons are obtained which have to be slowed down by inelastic collision with light elements (water, heavy water, graphite) to "thermal" energies, i.e. E=300 K. The range of energy may be extended by collision with liquid hydrogen (T=22K) or a graphite bloc (T~2000K).

Many energy units are used in neutron scattering. The more common are: $E= 0.08617T = 5.227 v^2 = 81.81/\lambda^2 = 2.072 k^2$ (1) where E is in meV, T in K, v in kmsec^{-1}, λ in Å and k in 10^{10} m^{-1}. For a neutron moderated at T=300 K (25.85 meV) the de Broglie equation gives a wavelength of $\lambda = 1.78$ Å. Whatever the moderator is neutrons always show a continuous distribution of energy similar to synchrotron radiation. The "intensity" is mainly determined as flux, the highest flux in a research reactor is available at the HFBR at the Institute Laue-Langevin in Grenoble with 2.10^{15} neutrons cm^{-2}sec^{-1} in the core. Most of the neutrons, however, are lost before they reach the sample. Under best conditions a flux of about 10^8 ncm^{-2}sec^{-1} is registered at the sample. First the neutrons are extracted out of the reactor by collimators. Fig. 1 gives the arrangement of the beam-tubes for the new reactor FRM II at Munich. Most experiments use monochromatic neutrons which are selected out of the continuous distribution either by monochromators (silicon, copper, germanium) or by velocity selectors (chopper). According to the experimental conditions the process is optimized for high resolution ($\Delta\lambda/\lambda \sim 10^{-3}$) or high intensity. Neutrons scattered by the sample are registered in detectors by nuclear reactions, usually in ^3He detectors.

2.2 INTERACTION OF NEUTRONS WITH MATTER

As neutrons have no charge they penetrate deeply into the atom and into matter. Thus they interact with nuclear forces in the atom and are only weakly absorbed. The low absorption means that neutrons can penetrate furnaces, cryostats, pressure cells etc. to view samples under extreme conditions. For example cryostat walls are made out of aluminium which has a small absorption cross section. The linear absorption coefficient for neutrons (x-rays) of 1.5 Å is 0.008 cm^{-1} (131 cm^{-1}) which means that 50 % transmission will take place through the thickness of 86 cm (5x10^{-3} cm) for neutrons (X-rays). This is quite a difference! Neutron scattering is a bulk method, laboratory X-ray experiments are mainly surface sensitive. High energy X-rays from synchrotrons (above 50 KeV) penetrate also and are used for bulk analysis.

The interaction of the neutron with the nuclei is governed by the scattering potential. As no complete theory of nuclear forces exists which should allow to calculate the potential from first principles the cross sections for elements and different isotopes of an element are tabulated from experiments. Nuclear forces have a range of 10^{-14} to 10^{-15} m. The nucleus acts as a point scatterer and the nuclear potential is characterized by a single number, called scattering length (symbol b) and tabulated either in units of 10^{-12} cm or 1 Fermi $=10^{-15}$ m. The total cross-section σ_T is given by the total solid angle (4π) times the square of the scattering potential

$$\sigma_T = 4 \pi b^2 \tag{1}$$

Since the interaction is nuclear it is dependent on the isotopic state and different isotopes have different scattering potentials. The most prominent examples are hydrogen ($b_H = -3.7$ Fermi) and deuterium $b_D = 6.8$ Fermi). Scattering power is therefore fairly easily modified by isotopic substitution, whereas a linear increase with the number of electrons prevails in electromagnetic waves The tunability of synchroton radiation offers, however, the opportunity to change the scattering power near to resonance energies at an absorption edge. Furthermore by the neutron spin of $\pm 1/2$ a compound nucleus with a nucleus of spin I with the spin states $I \pm 1/2$ can be formed and these have different cross sections. Normally all these effects are present and average quantities are used. The canonical average over all isotopes and possible spin states is defined as coherent cross section.

$$b_{coh} = \sum_i f_i \, b_i \tag{2}$$

The b_i are the individual scattering lengths of isotopes or compound nuclei which occur with relative frequency f_i. The measurements in crystallography are concerned with the space correlation of the different average nuclei with respect to one another. This information is given by elastic coherent scattering. The collective motion of atoms in a solid gives rise to energy transfer, hence inelastic coherent scattering. Coherent scattering may be described by amplitude coupling of the scattered waves. The elastic part produces Bragg reflections, the inelastic one phonons.

Additionally the random distribution of the scattering lengths from their mean value

$$b^2_{inc} = (\overline{b^2_i} - b^2_{coh}) \tag{3}$$

gives no interference effect, but intensities arising from the correlation of the same nucleus at different times. In elastic scattering (all times equal zero) the incoherent scattering contributes to the background. A detailed analysis of the background reveals information on point defects in crystals. The inelastic incoherent part observes vibrations of individual nuclei or their diffusion through the lattice. For most elements the cross-section σ_{inc} for incoherent scattering is rather small compared with σ_{coh}. The most prominent exception from that rule are the two isotopes of hydrogen. Therefore inelastic incoherent scattering is a valuable method for spectroscopic studies of hydrogeneous materials.

The spin of the neutron was already briefly mentioned. By the interaction of the spin with the moments of unpaired electrons surrounding a nucleus magnetic scattering is produced. Again elastic or inelastic, coherent and incoherent processes may occur. For unpolarized neutrons scattering of magnetic and nuclear origin does not interfere, hence the scattered intensities add up. The possible range of interactions is summarized in Table 1. It should be noted that basically the same interactions exists for electromagnetic waves but the relative importance of the various interactions is different.

TABLE 1: Interaction of neutrons with matter

elastic (momentum transfer)		inelastic (energy transfer)	
coherent	incoherent	coherent	incoherent
three-dimensional order	defects	collect. motions	vibrations diffusion
crystal structure	clusters	phonons	spectroscopy
cation order	vacancies	soft modes	rotations
H-positions	interstitials		
liquids			
Magnetic structures	spin glasses	magnons	magn. excitations
Magn. density			
Magn. form factors			

2.3 PROPERTIES OF SYNCHROTRON RADIATION

Synchrotron radiation was first discovered in 1947 as a by-product in high-energy physics installations. During the 1960s the potential of synchrotron radiation for the study of matter has been realized but as only parasitic use was possible then the results were rather limited. A real break-through has only been achieved when the first dedicated sources were built about 20 years ago. Approximately 70 synchrotrons are either planned or in operation now. SR is produced by accelerators. The accelerated particles are electrons or positrons and the radiation

60

emanates when these particles follow a curve in their path. The dipole radiation emitted by a charge that moves at a speed v close to the speed c of light is sharply contracted into the forward direction. This results in an opening angle

$$\psi_{nat} \sim \frac{1}{\gamma} \qquad (4)$$

where $\gamma = E/mc^2 = [1-(v/c)^2]^{-1/2}$ is the ratio of the kinetic energy E to the energy-equivalent of the rest mass m of the accelerated particle. For electrons or positrons the natural opening angle is $\psi_{nat} < 0.1$ mrad for energies E in the range of several GeV.

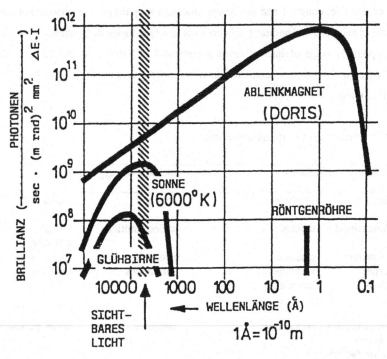

Figure 2: Wavelength distribution of synchrotron radiation compared with X-ray tube or sun. Only a very short flash is seen from each of the circulating electrons. Deviations of the individual electrons from both ideal orbit and ideal energy result in a continuous electromagnetic spectrum which extends from the infrared into the X-ray region. This range is eight orders of magnitude larger than the sensibility of the human eye. The spectrum is characterized by the so-called critical energy

$$[keV] \; \varepsilon_c = 2.218 \; \frac{E^3 \, [GeV]^3/[m]}{\rho} \qquad (5)$$

which is the photon energy above and below which half of the total radiation power is emitted. For the European Synchrotron Radiation Facility (ESRF) at Grenoble (E=6 GeV, γ= 20 m) ε_c is about 24 keV, i.e. well in the hard X-ray region. The hard range above 1 KeV (shorter wavelengths than 10 Å) is used for structural investigations, whereas the low energy range is suited for spectroscopic investigations related to electronic structure, collective excitations or vibrational modes of the system under investigation (Fig. 2).

The production of the electromagnetic wave occurs in the plane of the accelerator, thus an extended range is covered horizontally but the beam is only 0.2 mm high. The radiation is extremely polarized (in plane polarization, and out of plane elliptical polarization). Storage rings show a twofold time dependence. Firstly, the electrons (or positrons) are not distributed uniformly around their orbit in a storage ring, rather they are packed into short bunches with a length in the ps-ns range, circulating with a frequency of about 1 MHz. Secondly, the stored electron current decays with time, requiring refills ("injections") at regular time intervals. Only little use is made at present of that time structure of bunches by the experimental set-up. Almost all instruments regard the beam as continuous. The "intensity" of SR is defined as flux (photons/sec mrad 0.1% $\Delta\lambda/\lambda$) or brilliance (photons/(sec mrad^{2}mm^{2}0.1% $\Delta\lambda/\lambda$)).

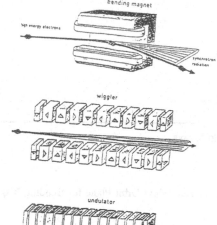

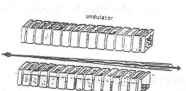

Figure 3: Optical elements for the production of SR

SR radiation extracted by a bending magnet (Fig. 3 above) is by several orders of magnitude more intense than that of an X-ray tube and may considerably be increased by wigglers (Fig. 3 middle) and undulators (Fig. 3 below). These optical devices produce radiation at each individual bending. The individual beams then interfere coherently and thus enhance the intensity considerably. Rays from synchrotrons are strictly parallel with a very small divergence. By a suitable monochromator (e.g. Si(111)) a highly resolved wavelength

62

$(\Delta\lambda/\lambda\sim5.10^{-4})$ is obtained and highly resolved X-ray powder patterns may be realized. Some

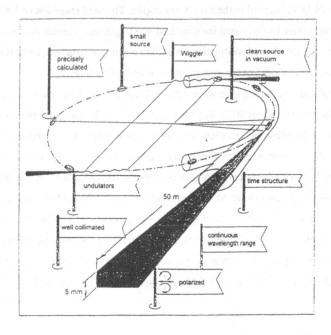

Figure 4: Properties of synchrotron radiation

Table 2: Properties of synchroton radiation

1)	Continuum from the Infrared Range to X-rays
(2)	High Intensities
(3)	High Degree of Collimation
(4)	Linear Polarization in Orbit Plane
(5)	Elliptical or Circular Polarization above and below Orbit Plane for Bending-Magnet-Radiation
(6)	Small Source Size
(7)	Well defined Time Structure
(8)	Quantitatively known Characteristics
(9)	Clean Environment: Light is generated inside an Ultrahigh Vacuum System
(10)	High Stability of Intensity and Source Position

3. Experimental

The general problems in neutron scattering and SR-experiments are rather similar. In both cases a monochromatic beam has to be extracted from a continuous wavelength distribution. Typical experimental set-ups for powder diffraction are despicted in Fig. 5 and 6. In Fig. 5 a high-resolution powder diffractometer at beam-line SR 8 of the reactor FRM II (Fig. 1) is shown. A neutron guide of 13 m length is conducting neutrons by total reflection to a monochromator which takes a small band-width out of the range of thermal neutrons. The wavelength is changed by the variation of the take-off angle θ_M (in our example $\theta_{M=}$ 155° or $\theta_{M=}$ 90°). A high take-off angle is chosen to obtain a good resolution (small values of the half-width of individual reflections). As intensity is lost by that experimental set-up the instrument will be equipped with a large detector bank in order to collect simultaneously a complete diffraction pattern.

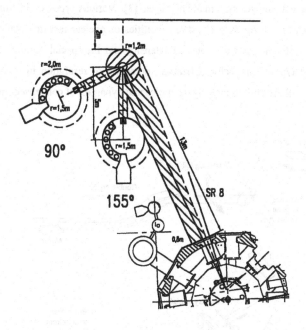

Figure 5: High resolution powder diffractometer at the research reactor FRM II (under construction)

Single crystal diffractometers do not need extensive collimation of the beam. However to compensate for low intensity they are normally equipped with two-dimensional detectors. Traditionally gas-filled (BF_3 or ^3He) detectors are in use. Recently progress has been made to develop image plate detectors similar to those in use for X-ray experiments. For the application in neutron experiments an element which absorbs thermal neutrons has to be incorporated. Essentially two Gd isotopes are used. The general use of image plates is still hampered by their high sensitivity for γ-rays.

64

A set-up for synchrotron powder diffraction is shown in Fig. 6. The first optical device in the beam is a mirror which may cut a part of the spectrum by total reflection. The instrument may operate with or without mirror. The device B2 at HASYLAB [2] may be used for high angular resolution or extreme high intensity to allow for time-resolved studies. Whereas the instrumentation of Fig. 6a is optimized for low divergence and well resolved pattern in 2θ, the arrangement in Fig. 6b is conceived for time-resolved experiments in connection with a position sensitive detector. Measuring times in the range of milliseconds were realized at low angular resolution but not yet systematically used [3]. Various types of instruments were constructed at beamlines arround synchrotron facilities or nuclear research plants. They are optimized to use different parts of scattered intensity and for special scattering processes. They differ in the type of monochromatization, detection or resolution in momentum and energy transfer but all are built to investigate statics and dynamics of condensed matter.

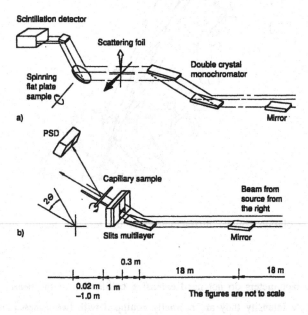

Figure 6: The two types of monochromators and their arrangement in the beamline B2 at Hasylab a) Double crystal monochromator and scattering foil to monitor intensity and polarization, b) multilayer monochromator.

4. Results

4.1 CRYSTAL STRUCTURE DETERMINATION AND REFINEMENT FROM POWDERS

Powder diffraction is the most powerful technique for the structural characterization of solids. Routinely applied to phase analysis and the refinement of crystal structures the advent of high brilliance synchrotron sources and improvements of detectors and computers brought a new quality to the field which reflected back to laboratory use. By the introduction of established methods for the solution of structures (Patterson, Fourier and Direct Methods) many crystal structures were determined from powder patterns. Furthermore time and temperature resolved studies gave information on phase transition and kinetics. Especially successful were studies at elevated temperatures and pressures. Neutron powder diffraction was extremely successful in the determination of the structures of high temperature superconductors and GMR-materials. In these cases the location of oxygen, especially the deviation of stoichiometry was performed. Thus neutron powder diffraction is ideally adapted to refine details of an otherwise known structure. A typical example is given in Fig. 7. The powder pattern of the well-known frame-work structure of the microporous zeolite faujasite loaded with aromatic molecules is shown. In this particular case a charge-transfer complex of TCNQ and TTF is formed. The difference in scattered intensity due to the organic molecule is despicted in the

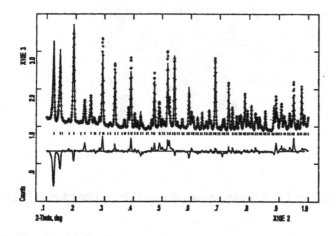

Figure 7: Neutron powder pattern of tetrathiafulvalene in zeolite NaY. The difference is due to the organic molecule.

66

lower part of the figure, thus indicating that the molecule is located at well-defined positions within the voids and produces coherent X-ray scattering. The positional parameters of the framework atoms are well established and are used as starting values for the calculation of the diffraction pattern. The difference between the two patterns serves to locate cations and guest-molecules by difference Fourier techniques. Fig. 8 presents the position of the tetracyanochinodimethane (TCNQ) molecule in the framework of zeolite NaX [4]. Supplementary inelastic and quasielastic neutron scattering exeriments are in progress to elucidate the dynamics of the motion and diffusion of the molecule. Molecular mechanics calculations gave for other systems evidence fo the minimum of potential energy and showed good agreement between the calculated positions and those determined by experiment [5].

Figure 8: The location of 7, 7, 8, 8-tetracyanodimethane (TCNQ) in the frame of faujasite

Due to the extreme resolution powder synchrotron data are well-suited for the refinement of low symmetry structures. The data quality of an organic substance with pharmaceutical applications is shown in Fig. 9.

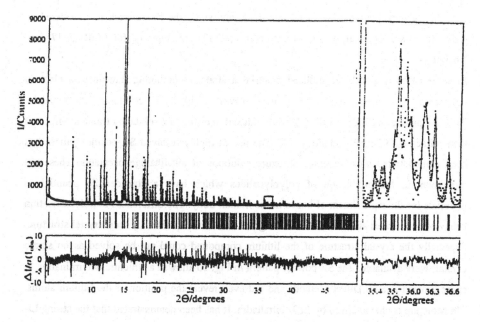

Figure 9: Rietveld refinement of marmesin. The pattern at the right is the marked part at left.

Still at higher angles a good profile peak/background (see right side of Figure) is obtained. The full powder pattern was successfully refined in the monoclinic system.

4.1.1 *Structure determination*

In principle a powder pattern contains much less information than a single crystal measurement as the three-dimensional location of each reciprocal lattice point is lost by projection on to a single dimension. Thus observed diffraction lines are superimposed. Based on high-resolved synchrotron powder patterns, methods were worked out to solve crystal structures from powder data. Several discrete steps are required from a careful sample preparation over powder pattern indexing, a solution of the structure and finally the refinement. High resolution synchrotron data are preferable but structure solution from laboratory equipment has been reported as well. The Patterson method is relatively robust against errors in collected datasets and has been a first choice for use in powder diffraction. Rius and Miravittles [6] have introduced a strategy for structure solution based on known molecular fragments. These rigid fragments are oriented and positioned in the unit cell. The method has successfully been used for molecular crystals which contain well known structural parts but no heavy atoms and to zeolitic materials where the frame-work is considered as rigid. Successful applications of the Patterson method were reported for $MnReO_4$ [7] and Yb_6ReO_{12}.

In some cases of poorly crystallized material a strategy combining transmission electron microscopy and X-ray powder data proved to work. With lattice constants determined by TEM the pattern of Si_2CN_4 or SiC_2N_4 were indexed, the structure solved by trial and error and then refined by Rietveld-techniques [8]. The use of highly resolved SR powder pattern has been demonstrated by a series of structure solutions of alkalihalogenoacetates. These are monomers in the production of polyglycolides which have a considerable poential as materials for the replacement of bones in the human body. The differences in polymerisation behaviour of various alkali metals and halogenes could be explained by their crystal structure. Especially the crystal structure of the lithium compound could not be solved as no single crystals were available. The SR powder pattern in Fig. 10 allowed a structure determination in space group $P2_1/a$ by Direct Methods and did even reveal the position of the lithium atoms. The structure is characterized by LiO_4-tetrahedra. It has been demonstrated that the strong Li-O bonds polarize the structure and prevent a polymerization for that particular compound [9]. By choosing an appropriate alkalihalogenoacetate a variety of compounds with different porosities for potential applications as biomaterial is produced.

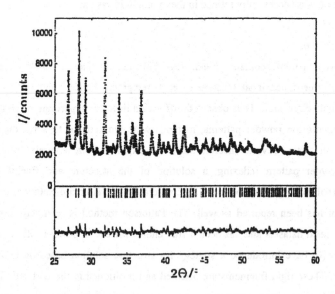

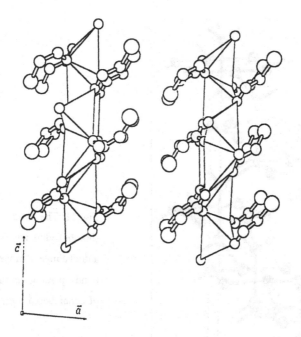

Figure 10: previous page: High angle part of the powder pattern of Li-chloroacetate; above: Crystal structure showing the LiO_4-tetrahedron and the acetate groups

4.2 SINGLE CRISTAL STUDIES

As stated earlier the location of "light"elements was the main concern of neutron diffraction. Whereas in organic compounds and molecules the "light" element is hydrogen, carbon and oxygens are comparably light in inorganic materials. As ice is the most important material containing hydrogens a large amount of work has been devoted to the extremely rich phase diagram of ice comprising eleven confirmed crystalline phases in which the water molecules link through hydrogen bonds to form tetrahedral frameworks. These results were obtained both by single crystal and powder experiments under pressure at the Institute Laue-Langevin in Grenoble. Among those phases was a totally new structure, that of phase XII with a topology unlike any of the known ice phases. It contains a mixture of 5 and 7 membered rings. The fivemembered rings are organized to form channels along the unique c-axis [10]. Ice XII is the densest known phase of water Hydrogen bonding is of special importance in the ice phases. The H-bonded framework of the rhombohedral ice IV is depicted in Fig. 11.

70

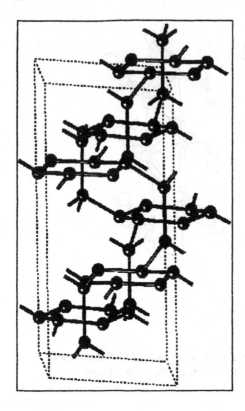

Fig. 11: The H-bond framework of rhombohedral ice IV showing the autoclathrate arrangement with H-bonds passing through the centre of 6 membered rings.

A huge amount of knowledge has been collected on the hydrogen bond. This information is presented in diagrams connecting O-H bonds to O-....O. distances. Expecially the question of the existence of symmetrical hydrogen bonds has extensively been discussed. Of course metal hydrides, organometallic compounds (Fig. 12) and zeolites are among the more popular crystals studied.

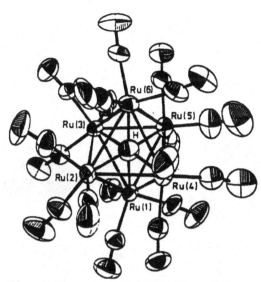

Figure 12: The structure of the anion $[HRu_6(CO)_{18}]$ representing the hydrogen atom in the middle of the Ru_6-cage [11]

These studies were clearly limited by the requirement for crystal sizes. The location of one hydrogen atom in the middle of a cage of ruthenium ions showed the possibilities of single crystal neutron diffraction with classical counters [11]. With the development of new detector systems, however, smaller crystals are now studied and important results on the structure of solvent molecules in biomolecular crystals may be obtained. Results on a test case are cited. The image plate detector LADI at the ILL on a cold neutron beam allows the data collection time and/or required crystal volume to be decreased by up to 2 orders of magnitude for medium-resolution studies. On the same crystal specimen of a coenzyme B12 grown from D_2O/D_6-acetone solution data were collected by neutron diffraction and SR. In the neutron case data were obtained with the LADI detector and with the system on the high-resolution single-crystal diffractometer D 19. The LADI data extend to a medium resolution of 1.43Å while both D 19 and SR data extend to high resolution of less than 1 Å. As seen from Figure 13 the high resolution data show considerable details of deuterium positions. The comparison of LADI and SR results reveal, however, that enough information for useful interpretation of results may already be obtained from medium resolution data. Thus a new route to solvent - molecule interaction may be opened [12]. Due to the large difference in scattering power

72

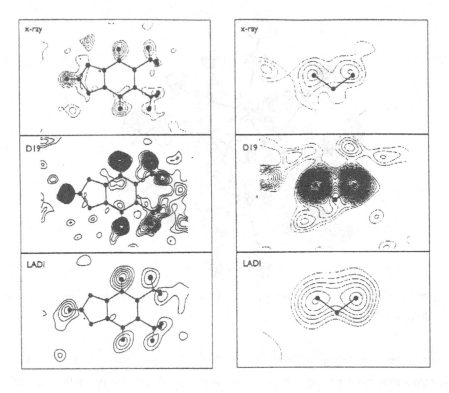

Figure 13: Difference density maps for the determination of hydrogens. Synchrotron data (above), two dimensional neutron detector (middle) and neutron image plate (below)

between deuterium and hydrogen the scattering contrast may easily be varied by solutions having different ratios of H_2O/D_2O. This methods has extensively been used in small angle studies of proteins in solution. Recently the method has been applied to proteins embedded in cellular membrans. The membrane protein porin allows the passage of small molecules. Crystals of about 0.1 mm^3 in volume containing detergent solubilized porin and water are soaked in different H_2O/D_2O mixtures. The scattering of parts of the crystal may be matched by the D_2O/H_2O ratio. For example, a neutron density map calculated for a D_2O content of 40 % will render the protein invisible and reveal only the detergent. Conversely at around 20% D_2O only the protein is visible. Experiments were carried out for two different detergent molecules. The results revealed that the two detergents (β-octyl-glycoxide and C8E4) bind the protein in an identical way. The cohesion of the crystal in a three-dimensional structure depends on the contacts between detergents belts attached to protein molecules [13].

Synchroton radiation has become the most powerful method for crystal structure determination due to high brilliance and the tunability of the radiation. High brilliance together with two-dimensional detectors allowed the study of specimens with volumes in the micrometer region. Grains of that size are normally classified as "powder". Therefore that traditional classification depends on the method applied. Not only the intensity but also the pure parallel beam contributes enormously to better data quality by an increased peak/background ratio. Thus many otherwise (e.g. in a rotating anode laboratory experiment) "unoberserved" reflections are observed. An example is the study of dimethylindigo. Not only was the structure solved from a poorly crystalline specimen but even anisotropic thermal displacement factors could successfully be refined [Fig. 14]. Tremendous success has been

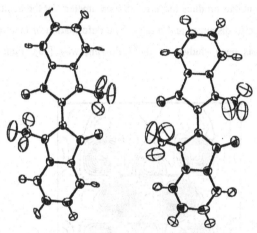

Figure 14: Two molecules of dimethylindigo with anisotropic thermal parameters refined from synchrotron data [14]

obtained in the field of protein crystallography with SR. Exposure time for a single simultaneous data collection of several thousand reflections has been reduced to seconds. Thus not only structure determination but also dynamical processes of protein inhibitor interaction are feasible. Structures are solved by the isomorphous replacement technique and more and more by multiple anomalous diffraction (MAD). Changing the wavelength close to an absorption edge of an element in the protein crystal modifies the scattering power due to anomalous dispersion. With MAD no heavy metal derivatives are needed and the isomorphism in the isomorphous replacement method is no longer a crucial point. Many succesful structure determinations are reported. Selenium replacing sulfur in amino acids of

proteins proved to be extremely valuable for MAD. Experiments on the absorption edge of
sulfur are not yet conclusive.

4.3 ELECTRON DENSITIES

Though the ability to obtain charge densities experimentally from accurate X-ray data has
only become widely accepted during the last three decades, the possibility to probe electronic
structure was recognized almost immediately after the discovery of X-ray diffraction. First
electron density studies on ionic crystals revealed deviations of the electron distribution from
sphericity but more quantitative results were only obtained from so-called deformation or
X-N densities. As positional atomic parameters are generally biased by electron distribution
the centre of the electron charge around an atom is determined by X-ray diffraction. Neutrons
interacting with the nucleus produce more reliable parameters for the position of an atom in a
crystal and more precise distances and angles. The deformation of electron densities due to
chemical bonding has been studied in the following way [15]. First the total electron

Figure 15: Deformation density in the plane of the cyclobutadiene ring, showing bond-
bending.Contours at 0.05 Å^{-3}

distribution is determined by an X-ray diffraction experiment, in general at low temperatures
to avoid thermal motion. Secondly precise atomic positional parameters are obtained from
neutron diffraction. The third step consists in a calculation of a theoretical density by
assuming spherical atoms at the position determined by the neutron experiment. This
calculated density is then deduced from the experimental X-ray density. The remaining
density is then due to the deformation from chemical bonds. An early example of this kind of

work is given in Fig. 15 which shows that the bonds in the cyclobutadiene ring are not equivalent [16]. The early studies concerned mainly molecular crystals with light elements. During recent years many metal-complexes were examined revealing not only valence electrons but also the distribution of electrons in transition elements. An example of the deformation density in copper complexes [17] is shown in Fig. 16 for the trans-bis (cyanamidonitrato-N:O) bis (imidazole-N^3)copper (II) complex $[Cu(CN_3O_2)_2(C_3H_4N_2)_2]$. Static deformation density around the copper atom (Fig. 16a) and for the imidazole ring (Fig. 16b) is shown. By multipole refinement parameters for the population of the d-electrons on the copper were derived. The copper atom is about neutral, oxygen atoms are slightly negative and the biggest negative charge is on one of the nitrogens. The most surprising feature is the relative important charge on carbon atom C (8) (Fig. 16) of about +0.35 (4) electrons which is clear evidence for possible nucleophilic addition reactions.

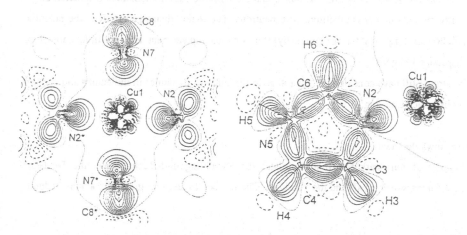

Figure 16: Deformation electron density in a copper atom complex. (a) lone pair density pointing in the direction of unoccupied d-orbitals. (b) imidazole ring

As pointed out deformation densities determined by X-ray and neutron measurements (X-N densities) are based on two independent measurements, hence two sources of experimental error. Another experimental approach has been the X-X-technique. Since the atomic form factor at high diffraction angles is mainly determined by electrons in inner atomic shells, high angle data give reliable parameters for atomic positions. Another main source of experimental errors is the extinction which is difficult to correct for. As the extinction effect is diminishing with wavelength experiments at short wavelength and small samples should produce more precise data. All these requirements point to the use of synchrotron radiation. Pioneering experiments at $\lambda = 0.3\text{Å}$ on $Cr(NH_3)_6Cr(CN)_6$ were reported by Coppens and coworkers [18] and by Kirfel and Eichhorn [19] on Al_2O_3 and Cu_2. The whole field of Charge Density research and Drug Design has been covered in a recent volume of NATO ASI series [20].

5. Summary

Neutron scattering and synchroton radiation are complementary methods for the investigation of structure and dynamics of condensed matter. The traditional strongholds of neutrons are the determination of light elements, isotopic substitution and the distinction between neighbouring elements in the periodic table. Third generation synchrotron radiation facilities with tunable wavelength, high brilliance and the possible use of anomalous dispersion are in many respects highly competitive with neutrons. The main advantage of SR is the potential solution of large crystal structures by isomorphous replacement and multiple anomalous dispersion methods.

Neutrons still have considerable advantage in the study of dynamics in molecules and crystals by inelastic scattering.

Acknowledgement

Support of our work by Bundesminister für Forschung und Technologie, the Deutsche Forschungsgemeinschaft and the Fonds der Chemischen Industrie is gratefully acknowledged.

References

1. Bragg, W.H., and Bragg, W.L. (1913) The reflection of X-rays by crystals, *Proc. R. Soc. London* A **88**, 428-438

2. Löchner, U., Pennartz, P.U., Miehe, G., and Fuess, H. (1993) Synchrotron powder diffractometry at Hasylab/DORIS reviewed, *Zeitschr. f. Kristallogr.* **204**, 1-41.

3. Pennartz, P.U., Loechner, U., Fuess, H., and Wroblewski, T. (1992) Powder diffraction in the millisecond range, *J. Appl. Cryst.* **25**, 571-577.

4. Bähtz, C., and Fuess, H. (1999)
    Structure and dynamics of charge transfer complexes in zeolite faujasite
    To be published.

5. Klein, H., Kirschhock, C., and Fuess, H. (1994) Adsorption sites of aromatic hydrocarbons in zeolite Y by molecular mechanics calculation and X-ray powder diffraction. *J. Chem. Phys.* **98**, 12345-12360.

6   Rius, J., and Miravittles, C. (1988)
    Determination of crystal structures with large known fragment directly from measured X-ray powder diffraction intensities, *J. Appl. Cryst.* **21**, 224-227.

7. Butz, A. Miehe, G., Paulus, H. Strauss, P. and Fuess H. (1998) The crystal structures of Mn $(ReO_4)_2$. $2H_2O$ and of the anhydrous perrhenates $M(ReO_4)_2$ of divalent manganese, cobalt, nickel and zinc, *J. Sol. State Chem.* **138**, 232-237.

8. Riedel, R., Greiner, A., Miehe, G., Dreßler, W., Fuess, H., and Bill, J. (1997) The first crystalline solids in the ternary Si-C-N-system. *Angew. Chem., Int. Ed. Eng.* **36**, 603-606.

9. Ehrenberg, H. and Epple, M. (1999) Structure solution of Li-halogenoacetate from powder synchrotron data. *Acta Crystallogr.* B, in press

10. Lobban, C., Finney, J.L., and Kuhs, W.F. (1998) The structure of a new phase of ice. *Nature* **391**, 268-270.

11. Jackson, P.F. Johnson, B.F.G., Lewis, J. Raithby, P.R. Maparthin, M., Nelson, W.J.H., Rouse, K.D., Allibon, J., and Mason, S.A. (1980) Direct location of interstitial hydride ligand in $[HRu_6(CO)_8]^-$ by X-ray and neutron analyses of $[Ph_6As][HR_6(CO)_{18}]$, *J. Chem. Soc. Chem. Comm.*, 296-297.

12. Langan, P., Lehmann, M., Mason, S. Wilkinson, C., Jogl, G. and
    Kratky, C. (1997) Crystallographic analysis of solvent regions in biomolecular
    crystals: Coenzyme B12r as a test case, *Institut Laue-Langevin, Annual Report* P. 93

13. Pebay-Peyroula, E., Garavito, R.M., Rosenbusch, J.P., Zulauf M.,
    Timmins, P.A. (1995) *Structure* 3, 1051-

14. Miehe, G., Süsse, P., Kupcik, V., Egert, E., Nieger, M., Kunz, G., Gerke, R.,
    Knieriem, B., Niemeyer, M., and Lüttke, W. (1991) Light absorption as well as crystal
    and molecular structure of N,N'-dimethyl-indigo: An example of the use of
    synchrotron radiation. *Angew. Chem., Intern. Ed. Engl.* 30, 964-967.

15. Coppens, P., and Hall, M.B., eds. (1982) *Electron distributions and the chemical bond,*
    Plenum, New York.

16. Irngartinger, H. (1982) Electron distribution in the bonds of cumulenes and
    small ring compounds, in P. Coppens and M.B. Hall (eds.), *Electron distribution
    and the chemical bond,* Plenum, New York.

17. Kozisek, J., Dvorsky, A., and Fuess, H. (1999) Charge density studies of some
    copper complexes. Conf. coord. chem., Smolinice, Slovakia

18. Nielsen, F.S., Lee, P., and Coppens, P. (1986) Crystallography at 0.3Å:
    Single crystal study at the Cornell High Energy Synchrotron Source, *Acta
    Crystallogr.* B42, 359.

19. Kirfel, A., and Eichhorn, K. (1990) Accurate structure analysis with synchrotron
    radiation. The electron density in Al_2O_3 and Cu_2O, *Acta Crystallogr.* A46, 271-284.

20. Jeffrey, G.A. and Piniella, J.F., eds (1991) *The application of charge density research
    to chemistry and drug design.* Plenum New York

THEORY OF INTERMOLECULAR INTERACTIONS

PEKKA PYYKKÖ
Department of Chemistry, University of Helsinki,
P.O.B. 55 (A. I. Virtasen aukio 1),
FIN-00014 Helsinki, Finland

1. Introduction

We discuss in this lecture the theory of intermolecular interactions. The discussion will start from the case of two monomers in free space. Certain further aspects concerning crystals will then be added. Finally a few additions to the author's recent review [1] are made.

The discussion draws heavily from the reviews of Buckingham [2, 3]. Other general references are Maitland et al. [4] and Stone [5] (see also those quoted in ref. [1]).

2. Two monomers in free space

The interaction potential between two closed-shell monomers in vacuum can be split in several ways [2, 3]. One way is the partitioning into short-range, electrostatic, induction, and dispersion parts:

$$V^{int} = V^{short} + V^{elec} + V^{ind} + V^{disp} \qquad (1)$$

2.1. SHORT-RANGE REPULSION

The short-range repulsion can be fitted to an exponential form,

$$V^{short} = A \exp(-bR), \qquad (2)$$

79

D. Braga et al. (eds.), Crystal Engineering: From Molecules and Crystals to Materials, 79–88.

or to a series of them, including angular dependent terms if one or both monomers are weakly anisotropic. An example is the rare-gas–hydrogen complex, Rg-H_2. If the interaction is strongly anisotropic, a situation already occurring for Rg–CO_2, alternative approximations must be found [3], for instance in terms of atom-atom pair potentials, assumed to be additive:

$$V^{\text{short}}(R,\theta) = \sum_{ij} V_{ij}^{\text{short}}(R_{ij}). \tag{3}$$

2.2. ELECTROSTATIC INTERACTIONS AND MULTIPOLE EXPANSIONS

The electrostatic potential between two undeformed monomers a and b can be expressed in terms of their charge densities ρ by summing the Coulomb law over their volume elements:

$$V^{\text{elec}} = \frac{1}{4\pi\epsilon_0} \int \rho_a(\mathbf{r}_a) \frac{1}{r_{ab}} \rho_b(\mathbf{r}_b) d\mathbf{r}_a d\mathbf{r}_b. \tag{4}$$

As long as the two charge distributions ρ remain 'stiff', this is exact. At large distances we can describe this interaction by a multipole expansion in the Cartesian coordinates $\alpha, \beta, \gamma... = (x, y, z)$. We shall use the notation of Buckingham [2]. The leading multipoles are the charge

$$\xi^{(0)} = q = \sum_{j}^{\text{charges}} e_j, \tag{5}$$

the dipole moment

$$\xi_\alpha^{(1)} = \mu_\alpha = \sum_{j}^{\text{charges}} e_j r_{j\alpha}, \tag{6}$$

the quadrupole moment

$$\xi_{\alpha\beta}^{(2)} = \Theta_{\alpha\beta} = \frac{1}{2} \sum_{j}^{\text{charges}} e_j(3r_{j\alpha}r_{j\beta} - r_j^2\delta_{\alpha\beta}), \tag{7}$$

and so on. Expressed in scalar form, the quadrupole moment Θ equals its largest component Θ_{zz}. The octupole moment is

$$\xi^{(3)}_{xyz} = \Omega = \Omega_{xyz} = \frac{5}{2} \sum_{j}^{\text{charges}} e_j x_j y_j z_j, \tag{8}$$

and the hexadecapole moment is

$$\xi^{(4)}_{xxxx} = \Phi = \Phi_{xxxx} = \frac{1}{8} \sum_{j}^{\text{charges}} e_j(35x_j^4 - 30x_j^2 r_j^2 + 3r_j^4). \tag{9}$$

All these moments occur as the lowest one in some common molecules, see Table I.

TABLE I.   The lowest multipole moments for certain symmetries. For a comprehensive list, see Buckingham [2].

Molecule	Symmetry	Lowest multipole	L
HCl	C_∞	Dipole	1
CO_2	$D_{\infty h}$	Quadrupole	2
C_6H_6	D_{6h}	"	2
CH_4	T_d	Octupole	3
SF_6	O_h	Hexadecapole	4
Ar		none	

These multipoles yield the following electrostatic energy

$$V^{\text{elec}} = Tq^{(a)}q^{(b)} + T_\alpha(q^{(a)}\mu_\alpha^{(b)} - q^{(b)}\mu_\alpha^{(a)}) +$$
$$T_{\alpha\beta}\left(\frac{1}{3}q^{(a)}\Theta_{\alpha\beta}^{(b)} + \frac{1}{3}q^{(b)}\Theta_{\alpha\beta}^{(a)} - \mu_\alpha^{(a)}\mu_\beta^{(b)}\right) +$$
$$\frac{(-1)^{n'}}{(2n-1)!!(2n'-1)!!}T_{\alpha\beta...\nu\alpha'\beta'...\nu'}\xi_{\alpha\beta...\nu}^{(n)(a)}\xi_{\alpha'\beta'...\nu'}^{(n')(b)}. \tag{10}$$

Here the T tensors are the successive derivatives

$$T = (4\pi\epsilon_0)^{-1}R^{-1}, \tag{11}$$
$$T_\alpha = (4\pi\epsilon_0)^{-1}\nabla_\alpha R^{-1} = -(4\pi\epsilon_0)^{-1}R_\alpha R^{-3}, \tag{12}$$
$$T_{\alpha\beta} = (4\pi\epsilon_0)^{-1}\nabla_\alpha\nabla_\beta R^{-1} = (4\pi\epsilon_0)^{-1}(3R_\alpha R_\beta - R^2\delta_{\alpha\beta})R^{-5}, \tag{13}$$
$$T_{\alpha\beta...\nu} = (4\pi\epsilon_0)^{-1}\nabla_\alpha\nabla_\beta...\nabla_\nu R^{-1}, \tag{14}$$

and R is the intermolecular distance, a vector from an origin in molecule a to an origin in molecule b. Repeated Greek subscripts stand for summation and $n!! \equiv 1 \cdot 3 \cdot 5 \cdot \ldots n$. If the expansion (10) is truncated, the energy V^{elec} will depend on the choice of origin.

A faster convergence can be reached by a distributed multipole analysis, such as that of Stone (see [3], p. 966). Anyway, the idea of dominant multipole-multipole interactions is qualitatively useful. All listeners know what two stationary dipoles will do. For two static, linear, polar molecules the answer is

$$V^{dipole-dipole} = (4\pi\epsilon_0)^{-1}\mu^{(a)}\mu^{(b)}R^{-3}[-2\cos\theta_a\cos\theta_b + \sin\theta_a\sin\theta_b\cos\phi]. \quad (15)$$

Here θ is the angle between each dipole moment vector and the $+\mathbf{R}$ intermolecular axis while ϕ is the relative rotational angle around the axis. The dipole-quadrupole and quadrupole-quadrupole terms become

$$
\begin{aligned}
V^{dq+qq} = (4\pi\epsilon_0)^{-1}\Big[&\frac{3}{2}\mu^{(a)}\Theta^{(b)}R^{-4}\big[\cos\theta_a(3\cos^2\theta_b - 1) - 2\sin\theta_a\sin\theta_b\cos\theta_b\cos\phi\big] \\
&+\frac{3}{2}\mu^{(b)}\Theta^{(a)}R^{-4}\big[-\cos\theta_b(3\cos^2\theta_b - 1) + 2\sin\theta_b\sin\theta_a\cos\theta_a\cos\phi\big] \\
&+\frac{3}{4}\Theta^{(a)}\Theta^{(b)}R^{-5}\big[1 - 5\cos^2\theta_a - 5\cos^2\theta_b + 17\cos^2\theta_a\cos^2\theta_b \\
&+2\sin^2\theta_a\sin^2\theta_b\cos^2\phi - 16\sin\theta_a\cos\theta_a\sin\theta_b\cos\theta_b\cos\phi\big]\Big]. \quad (16)
\end{aligned}
$$

Example: Consider two molecules with zero dipole moment but non-zero quadrupole moment, such as benzene. It follows from eq. (16) that the $\Theta - \Theta$ interaction is repulsive for two stacked molecules and attractive for a "T". Between two monomers having Θ:s of opposite sign, such as C_6H_6 and C_6F_6, the opposite is true. The equimolar $C_6H_6 \cdot C_6F_6$ crystallises in a stacked, columnar structure, see ref. [6].

2.3. INDUCTION EFFECTS

The next, induction term, V^{ind}, arises from the interaction of the induced electric moments on one side with the static electric moments on the other side.

If the monomer a carries a charge, $q^{(a)}$, the leading term is the charge-polarizability interaction

$$V^{cp} = -\frac{1}{2}(4\pi\epsilon_0)^{-2}[q^{(a)}]^2R^{-4}[\alpha^{(b)} + \frac{1}{3}(\alpha_{||} - \alpha_\perp)^{(b)}(3\cos^2\theta - 1)], \quad (17)$$

where z is the intermolecular axis. The induction terms are *not* additive. Rather, one first has to sum the electric fields F_α, the field gradients $F_{\alpha\beta}$ etc. from all sources at the site (b) and then let them interact with the static polarizability tensor α and the higher polarizabilities $A, C, \ldots$ of the molecule b:

$$V^{(b)}_{induction} = -\frac{1}{2}\alpha^{(b)}_{\alpha\beta}F_\alpha F_\beta - \frac{1}{3}A^{(b)}_{\alpha,\beta\gamma}F_\alpha F_{\beta\gamma} - \frac{1}{6}C^{(b)}_{\alpha\beta,\gamma\delta}F_{\alpha\beta}F_{\gamma\delta} - \ldots, \quad (18)$$

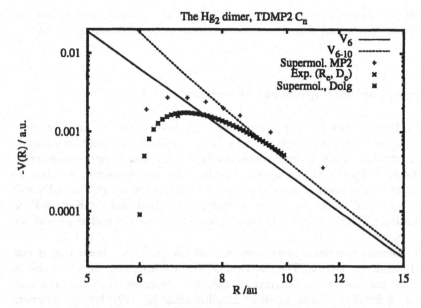

The Hg$_2$ dimer, TDMP2 C$_n$

Figure 1. Potential curves for the Hg$_2$ dimer. The experimental minimum is given by ×. Note the unusually large effect of the higher coefficients $C_{8,10}$. Reproduced from ref. [1], Fig. 31.

2.4. DISPERSION EFFECTS

The leading, R^{-6}, term of the dispersion interaction between two isotropic monomers a and b can be written as

$$V_{\text{dispersion}}(R^{-6}) = -(4\pi\epsilon_0)^{-2}\frac{3\hbar}{\pi}R^{-6}\int_0^\infty \alpha^{(a)}(iu)\alpha^{(b)}(iu)du = -C_6 R^{-6}, \quad (19)$$

where the dipole polarizabilities of imaginary argument for the state m of monomer a are defined as

$$\alpha_{\alpha\gamma}^{(a)}(iu) = \sum_{p\neq m}\frac{\omega_{pm}\left[\langle m|\mu_\alpha|p\rangle\langle p|\mu_\gamma|m\rangle + \langle m|\mu_\gamma|p\rangle\langle p|\mu_\alpha|m\rangle\right]}{\hbar(\omega_{pm}^2 + u^2)}. \quad (20)$$

These $\alpha(iu)$ functions are monotonically decreasing functions of u. Eq. (19) is often known as the Casimir-Polder equation. Actually it seems to first appear in ref. [7]. The combination of one dipole operator μ and one quadrupole operator Θ will lead to a C_8 term and two quadrupole operators will lead to a C_{10} term. Such terms can be extracted from supermolecular calculations [8] or from the analogues of eq. (19-20).

Example: For the rare gas dimers of Xe or Rn, the higher, C_8 and C_{10} contributions are small, compared with the leading, C_6 one [8]. For the Hg$_2$ dimer they are relatively important, see Figure 1.

The dispersion terms as such are additive. The first non-additive terms are the Axilrod-Teller ones arising from three dipole operators in third-order perturbation theory.

3. Connection with quantum chemistry

SCF. The Hartree-Fock (HF) approximation includes, obviously without dynamic and static correlation effects, the short-range, electrostatic and induction contributions. It does not include the dispersion terms. They would require correlation contributions with at least one virtual excitation at each monomer. At large R one can start from properties of the monomers, calculate the $\alpha(iu)$ of eq. (20) and do the integral (19) to get a C_6. This is called the time-dependent Hartree-Fock approximation (TDHF). A supermolecular treatment with HF is useless for van der Waals systems.

DFT. Density-functional theory behaves like the SCF one, except that it can include dynamic correlation. The short-range, electrostatic, and induction effects are there but the van der Waals terms are not, in a supermolecular treatment near R_e. The long-distance C_n can again be handled using Eq. (19), but *the listeners are strongly warned against touching DFT methods if dispersion contributions are important within the system considered.*

This situation is slightly paradoxical because in principle the Kohn-Sham theorem covers all cases. For some recent attempts to include dispersion effects at DFT level, see [9, 10, 11, 12] and ref. [1], p. 600.

MP2. The lowest level of approximation that does include dispersion effects at the supermolecular level is the 2nd-order Møller-Plesset one. In all cases so far studied, it actually exaggerates the strength of the interaction (see ref. [1], Fig. 33). When the reasons were analyzed [8, 14, 15] it was found that the MP2 treatment of the combined system a+b corresponds to a treatment of each monomer (a or b) at the uncoupled Hartree-Fock level. This corresponds to energy denominators of the type $E_i - E_j$.

MP3. At this level the interaction comes out too weak and it is not recommended for this purpose.

MP4. The fourth-order Møller-Plesset treatment is the lowest supermolecular one that gives a clear correlation improvement on the dispersion interaction.

CCSD and CCSD(T). The coupled-cluster approximation with single and double excitations, with possible perturbative treatment of the triple excitations is the best commonly available approximation. The way the convergence approaches experimental values can be seen from the intramolecular example of $S(AuL)_2$ in Figure 2.

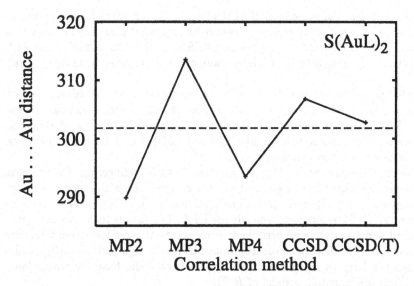

Figure 2. Convergence of the Au ··· Au distance in the A-frame molecule S(AuL)₂, L = PH₃ or PPh₃. The horizontal line gives the experimental value [25].

SAPT. A different and perhaps more intelligent approach to intermolecular interactions is the 'symmetry-adapted perturbation theory' (for recent references, see [5], pp. 90-94, or [18]). In this case the interaction energy is calculated directly, as a sum of physically distinct polarization and exchange contributions

$$E_{\text{int}} = E_{\text{pol}}^{(1)} + E_{\text{exch}}^{(1)} + E_{\text{pol}}^{(2)} + E_{\text{exch}}^{(2)} + \cdots \tag{21}$$

The correction $E_{\text{pol}}^{(1)}$ is the classical electrostatic energy. The second-order corrections are split into induction and dispersion components

$$E_{\text{pol}}^{(2)} = E_{\text{ind}}^{(2)} + E_{\text{disp}}^{(2)}, \tag{22}$$

$$E_{\text{exch}}^{(2)} = E_{\text{exch-ind}}^{(2)} + E_{\text{exch-disp}}^{(2)}. \tag{23}$$

The exchange corrections $E_{\text{exch}}^{(n)}$, $n = 1,2$, result from the antisymmetrization (symmetry adaptation) of the wave function.

Basis sets. Especially in a supermolecular treatment of the dimer, rather large basis sets are necessary. Between closed-shell main-group compounds with np valence shells, such as the $(H_2Te_2)_2$ dimer, the secondary bond length appears to stabilize at the $[4s4p2d2f]$ level of contracted functions but the energy still goes down when a third f-function is added [19].

For the intra- or intermolecular attraction between two Au(I) ions in compounds, a 19-valence-electron (19-VE) pseudopotential is preferable over a 11-VE one, in order to obtain the correct nodal structure for the $6s6p$ shells. Two f functions should be used if possible [20]. Adding a g function will still influence the result [21].

Spin-orbit effects are expected to invariably increase the attraction between closed-shell monomers (ref. [1], p. 628). As noticed by Han et al. [22] for Rn₂, they should be treated at the same level of correlation as the rest, the non-relativistic or quasirelativistic (= scalar relativistic = spin-orbit averaged) part. For instance it is not a good idea to combine SO effects from CISD (configuration interaction with single excitations) level with high-level quasirelativistic treatments, as done in ref. [8].

Concluding, although valuable insight can be obtained at a relatively modest level, such as MP2 with a basis having a couple of polarization functions, if 'the right result for the right reason' is desired from a supermolecular calculation, one should use very big basis sets, levels like MP4 or CCSD(T) and, for heavy elements, large-valence-space pseudopotentials.

Solid-state approaches. This is a school on Crystal Engineering. The ultimate goal is to predict entire crystal structures. That can be done with success, even for quite complicated crystal structures, using methods like LAPW (linear augmented plane waves) [13]. Such methods typically use DFT. This means that covalent, ionic, or metallic bonding can be well described. The electrostatic and induction terms are in the model, but the dispersion contributions can vary in sign and have magnitudes from much too large to much too small. Besides they would have the wrong long-distance limit (exponential, instead of R^{-6}).

4. Some Applications

4.1. THE METALLOPHILIC ATTRACTION

In inorganic and organometallic systems it is often found that d^{10}, s^2, or d^8 metal cations attract each other. In the case of Au(I) Schmidbaur called this the *aurophilic* attraction. A more general adjective would be *metallophilic*. Likewise, attractions between closed-shell molecules, such as Bunsen's cacodyl, $(CH_3)_2As\text{-}As(CH_3)_2$, are found. Solid cacodyl contains chains of the type ... As-As ... As-As ... [19]. Similar interactions are ubiquitous in tellurium chemistry [1].

In all ab initio studies so far we find no attraction at SCF level, precluding the idea that the interaction would be due to hybridization (in the case of Au(I), $5d - 6s - 6p$ hybridization). When correlation effects are included, at MP2 level or higher levels, the attraction appears. Reasonable secondary bond lengths, R_e, and interaction energies, $V(R_e)$, are found. At large distances the interaction goes over to an R^{-6} one [23].

An analysis in terms of localized orbitals verifies the dispersion origin, from double virtual excitations (A → A', B → B') [24]. In addition to this term, Runeberg et al. [21] discovered an *ionic term* of the type (A → A', B → A'). At large R it decays exponentially. Near R_e it approaches in importance the traditional one.

Apart from the review [1], our own papers in this area can be traced back from the latest ones [25, 26].

4.2. MIXED INDUCTION AND COVALENT INTERACTIONS

The first chemical bonds between noble metals and noble gases were proposed in 1995 [27] and the predicted species $AuXe^+$ and $XeAuXe^+$ were observed using mass spectroscopy in ref. [28]. This case shows how fluid the borders are between the various types of bonding: At large R the interaction between the closed-shell species Au^+ ($5d^{10}$) and Xe ($5p^6$) is dominated by the R^{-4} one of eq. (17). At shorter distances a substantial charge transfer takes place from the $5p\sigma$ of Xe to the originally empty $6s$ orbital of Au. The best calculated D_e is 30.3 kcal/mol at an R_e of 257.4 pm at the CCSD(T) level with ($8s6p5d5f1g$) and ($8s8p6d6f$) basis sets for Au and Xe, respectively. This suggests an honest covalent bond.

Furthermore, to show how the various views overlap, Read and Buckingham [29] have considered the analogous system $AuAr^+$ and explain its properties using the expansions quoted in Section 2.

Further examples are the neutral isoelectronic analogues to $AuXe^+$, viz. PdXe and PtXe, that again could be characterized as covalent bonds near the equilibrium. PdXe dissociates into the closed-shell systems Pd ($4d^{10}$) and Xe. For PtXe the molecular ground-state goes over to the $5d^{10}$ excited state of Pt [30].

References

1. Pyykkö, P. (1997) Strong closed-shell interactions in inorganic chemistry, *Chem. Rev.*, **97**, 597–636

2. Buckingham, A. D. (1978) Basic theory of intermolecular forces: Applications to small molecules, in *Intermolecular Interactions: From Diatomics to Biopolymers*, ed. B. Pullman, Wiley, Chichester, pp. 1–67

3. Buckingham, A. D., Fowler, P. W., and Hutson, J. M. (1988) Theoretical studies of van der Waals molecules and intermolecular forces, *Chem. Rev.*, **88**, 963–988

4. Maitland G. C., Rigby, M., Smith, E. B., and Wakeham, W. A. (1981) *Intermolecular Forces. Their Origin and Determination*, Clarendon Press, Oxford, 616 p.

5. Stone, A. J. (1996) *The Theory of Intermolecular Forces*, Clarendon Press, Oxford, 264 p.

6. Heaton, N. J., Bello, P., Herradón, B., del Campo, A., and Jiménez-Barbero, J. (1998) NMR study of intramolecular interactions between aromatic groups: van der Waals, charge-transfer or quadrupolar interactions?, *J. Am. Chem. Soc.*, **120**, 12371-12384

7. Mavroyannis, C. and Stephen, M. J. (1962) Dispersion forces, *Mol. Phys.*, **5**, 629-638

8. Runeberg, N., and Pyykkö, P. (1998) Relativistic pseudopotential calculations on Xe_2, RnXe and Rn_2: The van der Waals properties of radon, *Int. J. Quantum Chem.*, **66**, 131-140

9. Kohn, W., Meir, Y., and Makarov, D. E. (1998) van der Waals energies in density functional theory, *Phys. Rev. Lett.*, **80**, 4153-4156

10. Hult, E., Rydberg, H., Lundqvist, B. I., and Langreth, D. C. (1999) Unified treatment of asymptotic van der Waals forces, *Phys. Rev. B*, **59**, 4708-4713.

11. Engel, E. and Dreizler, R.M. (1999) From explicit to implicit density functionals, *J. Comp. Chem.*, **20**, 31-50

12. Lein, M., Dobson, J.F., and Gross, E.K.U. (1999) Towards the description of van der Waals interactions within density functional theory, *J. Comp. Chem.*, **20**, 12-22.

88

13. Petrilli, H. M., Blöchl, P. E., Blaha, P., and Schwarz, K. (1998) Electric-field-gradient calculations using the projector augmented wave method, *Phys. Rev. B*, **57**, 14690-14697

14. Chałasiński, G., Szczęśniak, M. M., and Cybulski, S. M. (1990) Calculation of nonadditive effects by means of supermolecular Møller-Plesset perturbation theory approach: Ar_3 and Ar_4, *J. Chem. Phys.*, **92**, 2481-2487

15. Chałasiński, G., Szczęśniak, M. M., and Kendall R. A. (1994) Supermolecular approach to many-body dispersion interactions in weak van der Waals complexes: He, Ne, and Ar trimers *J. Chem. Phys.*, **101**, 8860-8869

16. Chałasiński, G. and Szczęśniak, M. M. (1994) Origins of structure and energetics of van der Waals clusters from ab initio calculations, *Chem. Rev.*, **94**, 1723-1765

17. Woon, D. E. (1994) Benchmark calculations with correlated molecular wave functions. V. The determination of accurate *ab initio* intermolecular potentials for He_2, Ne_2, and Ar_2, *J. Chem. Phys.*, **100**, 2838-2850

18. Bukowski, R., Sadlej, J., Jeziorski, B., Jankowski, P., Szalewicz, K., Kucharski, S. A., Williams, H. L., and Rice, B. M. (1999) Intermolecular potential of carbon dioxide dimer from symmetry-adapted perturbation theory, *J. Chem. Phys.*, **110**, 3785-3803

19. Klinkhammer, K. W. and Pyykkö, P. (1995) Ab initio interpretation of the closed-shell, intermolecular E...E attraction in dipnicogen (H_2E-EH_2)$_2$ and dichalcogen (HE-EH)$_2$ hydride model dimers, *Inorg. Chem.*, **34**, 4134-4138

20. Pyykkö, P., Runeberg, N., and Mendizabal, F. (1997) Theory of the d^{10}-d^{10} closed-shell attraction. I. Dimers near equilibrium, *Chem. Eur. J.*, **3**, 1451-1457

21. Runeberg, N., Schuetz, M., and Werner, H.-J. (1999) The aurophilic attraction as interpreted by local correlation methods, *J. Chem. Phys.*, **110**, 7210-7215

22. Han, Y.-K., Bae, C., and Lee, Y. S. (1999) Two-component calculations of spin-orbit effects for a van der Waals molecule Rn_2, *Int. J. Quantum Chem.* **72**, 139-143

23. Pyykkö, P. and Mendizabal, F. (1997) Theory of the d^{10}-d^{10} closed-shell attraction. II. Long-distance behaviour and non-additive effects in dimers and trimers of type (X-Au-L)$_{n=2,3}$ (X=Cl, I, H; L=-PH_3, PMe_3, -NCH), *Chem. Eur. J.*, **3**, 1458-1465

24. Doll, K., Pyykkö, P., and Stoll, H. (1998) Closed-shell interaction in silver and gold chlorides, *J. Chem. Phys.*, **109**, 2339-2345

25. Pyykkö, P. and Tamm, T. (1998) Theory of the d^{10}-d^{10} closed-shell attraction. IV. $X(AuL)_n^{m+}$ centered systems, *Organometallics*, **17**, 4842-4852

26. Tamm, T. and Pyykkö, P. (1999) Structure and stability of gold-substituted diborane, boranes and borohydride ions, *Theor. Chem. Accounts*, accepted

27. Pyykkö, P. (1995) Predicted chemical bonds between rare gases and Au^+, *J. Am. Chem. Soc.*, **117**, 2067-2070

28. Schröder, D., Schwarz, H., Hrušák, J. and Pyykkö, P. (1998) Cationic gold(I) complexes of xenon and of ligands containing the donor atoms oxygen, nitrogen, phosphorus, and sulfur, *Inorg. Chem.*, **37**, 624-632

29. Read, J. P. and Buckingham, A.D. (1997) Covalency in $ArAu^+$ and related species?, *J. Am. Chem. Soc.*, **119**, 9010-9013

30. Burda, J. V., Runeberg, N. and Pyykkö, P. (1998) Chemical bonds between noble metals and noble gases. Ab initio study of the neutral diatomics NiXe, PdXe, and PtXe, *Chem. Phys. Lett.*, **288**, 635-641

HYDROGEN-BOND ASSISTED ASSEMBLY OF ORGANIC AND ORGANIC-INORGANIC SOLIDS

C.B. AAKERÖY and D.S. LEINEN
Department of Chemistry, Kansas State University
Manhattan, Kansas 66506, USA.

Abstract

The ability to predict and control the assembly of molecular and ionic species into ordered networks has become an important target area in chemistry, biochemistry and materials science. In this context, the strength, selectivity and directionality of the hydrogen bond have been instrumental in the preparation of distinctive and predictable structural aggregates, and the use of hydrogen bonding as a steering force has emerged as the most important strategy in crystal engineering.

An improved insight into the structural consequences of intermolecular interactions is crucial when considering that most physical properties, *e.g.* melting point, solubility, mechanical strength, *etc.*, are a function of the relative orientation of molecules in a 3-D solid. At this stage, however, we do not have the necessary understanding of non-covalent forces and their influence over solid-state structures. This point is amply illustrated by the fact that many fundamental properties of solids can not be simulated or calculated with any great degree of confidence. Furthermore, we are still lacking sufficient information about crystallization processes, structural phase changes and other events where thermodynamic and kinetic factors are precariously balanced. Nevertheless, considerable efforts and progress have been made in recent years, and this Chapter provides an overview of some of the strategies, structures, successes and pitfalls (notably involving hydrogen bonding) that have been presented in the crystal engineering of organic and organic/inorganic materials.

1.   Introduction

Crystal engineering, *"the understanding of intermolecular interactions in the context of crystal packing and in the utilisation of such understanding in the design of new solids with desirable physical and chemical properties."*[1], is concerned with challenges and opportunities associated with supramolecular synthesis and molecular recognition [2,3,4,5]. Unfortunately, the ability to modify reactivity and properties of a crystalline solid through precise control over its structure presents a formidable task, since a small change in molecular structure may lead to dramatic changes in the crystal structure.

Covalent bonds are the primary design elements in organic chemistry and these bonds have been extensively studied to the point where even very complex biomolecules can be prepared through organic synthesis. The next step should take us beyond the molecule to organized polymolecular systems held together by non-covalent interactions; this is the field of "supramolecular chemistry" [2]. Intermolecular interactions form the basis of the highly specific recognition, reaction and regulation processes that occur in life, *e.g.* antigen-antibody association, gene expression, cellular recognition, *etc.* These processes display astoundingly high selectivity based upon efficient manipulation of the energetic and stereochemical features of non-covalent intermolecular forces. Any attempt to predict or design solid-state assembly must take these forces into account.

Despite recent advances in supramolecular synthesis, detailed control over the assembly of molecules into ordered solids remains an elusive goal, and we are rarely able to predict the structure of a simple crystalline material. Much more work is required to increase the choice of reliable building blocks of low-dimensional architectures and, more importantly, to allow us to rationalize and predict structural arrangements of

D. Braga et al. (eds.), Crystal Engineering: From Molecules and Crystals to Materials 89 –106.
© 1999 *Kluwer Academic Publishers. Printed in the Netherlands.*

molecules within new crystalline materials. An enhanced understanding of intermolecular forces may also enable us to better address crucial questions about polymorphism and crystal morphology.

As crystal engineering is still in its infancy, it *is* necessary to assemble structural information, to identify recurring packing motifs and to test the reliability of intermolecular interactions. At the same time, we are at a stage where supramolecular skills should be put into practice by advancing towards functional materials that are cheaper, faster or more efficient that current alternatives.

The term 'crystal engineering' was coined by Schmidt in 1971 as a result of a study of the interrelationship between crystal structure and solid-state reactivity [6]. It was shown that a photodimerizable derivative of cinnamic acid crystallized in three forms, α, β and γ. When irradiated in the solid state, the α and β forms reacted in a 2+2 manner to form cyclobutane dimers, whereas the γ form was photostable. It was further demonstrated that the distance and orientation of double bonds in adjacent molecules determined the solid-state reactivity; a reaction would only occur if the distance between reactive parallel double bonds was less than 4.2 Å. In the case of the γ form, with a separation of 4.9 Å, the distance was simply too large for a topochemical reaction to take place.

Subsequent studies of crystal packing and photoreactivity of anthracene derivatives brought about the notion of 'crystal engineering'. Further work on carboxylic acid derivatives [7] and quinones [8] strengthened the topochemical argument. At the time, Schmidt stated that further development of this concept would "be difficult, if not impossible, until we understand the intermolecular forces responsible for the stability of the crystal lattice" [6]. Nevertheless, many elegant examples of crystal engineering have been reported and a range of intermolecular interactions is being utilized to preferentially orient molecules into organic and inorganic-organic structures with the ultimate intent of preparing new functional solids.

2. Intermolecular forces

The unambiguous identification of a molecule or of some particular feature within the molecular structure represents the ultimate goal in many areas of chemistry. However, for a crystal engineer, this stage is only a starting point. Since the crystal structure represents a system where all covalent and non-covalent forces are poised at an energetic minimum [9], it contains all the information regarding the structural influence of, and balance between, intermolecular forces. If we could effectively decode this information, then the prospects of designing materials with specific properties would be greatly enhanced.

2.1 HYDROGEN BONDING

"Hydrogen-bonding is surely the most frequently encountered and hence most important interaction in crystals. The frequency with which the elements nitrogen and oxygen occur in organic compounds, coupled with the strength and directionality of the hydrogen-bond as compared to other intermolecular forces account for its significance."[1]. It is the most reliable directional force in supramolecular synthesis, and as such, can be considered as the "master key" for molecular recognition [2].

2.1.1 *What is a hydrogen bond?*
Hydrogen-bond interactions have been defined as "short-range site-bounded cohesion forces that considerably weaken a given chemical bond of one of the partners" [10]. A hydrogen bond represents a particular case where the weakened chemical bond involves a hydrogen atom and a region of increased electron density. This region is often provided by a more electronegative atom *e.g.* O, N, S, Cl, *etc.*, but hydrogen-bond acceptors may also include π-electrons in unsaturated organic molecules. Since the hydrogen bond causes a weakening of the covalent D-H bond, it can be identified and quantified using IR and NMR spectroscopy, and X-ray and neutron diffraction. Note that it is important to consider the particular prejudicial quality of the technique that is used to locate the hydrogen atom [11]; neutron and X-ray diffraction detect two non-coincidental areas of the hydrogen atom. As "the nucleus and the electron density of an atom are essential parts, it is therefore impossible to assert rationally that the position of either the one or the other is *the* position of the atom" [12]. Despite this apparent paradox, the hydrogen bond has been extensively studied both experimentally and theoretically primarily due to its importance in life itself.

2.1.2 Hydrogen-bond strength and geometry

The thermodynamic strength of a hydrogen bond depends crucially upon the atoms involved in the bond. For neutral molecules, typical values fall in the range of 10-65 kJmol^{-1}. This is greater than for van der Waals forces, but less than for conventional covalent bonds. However, if one of the species involved in the hydrogen bond is ionic (a charge-assisted hydrogen bond) the strength can increase to 80-120 kJmol^{-1}. This means that the strongest hydrogen bonds are energetically equal to some of the weaker covalent bonds [9]. The characteristic geometry of a hydrogen bond D-H····A-X is given by the lengths A····D and H····D, the angles AHD, (θ), and HAX, (Φ), and the planarity of the system, DHAX, Fig. 2.1, [13].

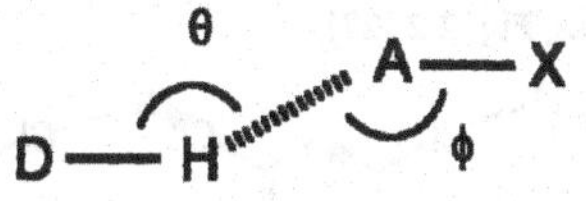

Figure 2.1   Schematic drawing illustrating important geometric parameters of a hydrogen bond.

The most commonly used criteria for identification of a hydrogen bond, are the interatomic distance, H····A, and the angle, θ. If the distance between two atoms is significantly shorter than the sum of their van der Waals radii, [14] and if θ is larger than 110°, then the presence of a hydrogen bond is usually evoked. Some typical values for hydrogen bond distances are 1.80-2.00 Å for an N-H····O bond, and 1.60-1.80 Å for an O-H····O bond [15]. If the hydrogen bond is bifurcated or trifurcated, a more relaxed distance criterion is often applied [16]. When a carbonyl oxygen atom is accepting two hydrogen bonds, the N-H bonds tend to lie close to a 'lone pair' on the oxygen atom (N-H····O=C < 120°), whereas when one bond is formed, the N-H bond is positioned between the formal lone pairs. The directionality of hydrogen bonds and the preference for the N-H or O-H vector to point towards the 'lone pair' direction on the acceptor atom, has been found to depend upon the 'hybridization' of the acceptor atom [17]. This view is further supported by a study of the crystal structures of all ether, epoxide, ketone, and ester structures in the Cambridge Structural Database [18] with a O····X (X=O, N) distance of less than 3.00 Å; most hydrogen-bond donors lie along the lone pair direction, and this angular preference decreases when moving from epoxide to ketone to ether [1].

2.1.3 Weak hydrogen bonds

There is no doubt that hydrogen bonds involving O-H and N-H moieties influence crystal packing but in recent years, several studies have demonstrated that weaker hydrogen bonds, such as C-H····O, C-H····Cl, and C-H····π, can play an important role in determining the crystal structure [19]. The existence of C-H····X hydrogen bonds has been a matter of controversy ever since it was proposed by Glasstone in 1937 [20]. Physical consequences of these interactions were suggested by Pauling who attributed the higher boiling point of acetyl chloride (51°C), compared to that of trifluoroacetyl chloride (<0°C), to hydrogen bonding even though no conventional hydrogen-bonding groups were present [14b]. Even so, the ability of C-H moieties to act as hydrogen-bond donors remained a subject of contention for many years [21], until a survey of 113 organic structures, solved from neutron diffraction, provided conclusive evidence for the existence of C-H····O hydrogen bonding [22]. The study also indicated that the C-H····O bond is essentially electrostatic in nature, which was inferred by the fact that the hydrogen atoms point towards the lone pair of the oxygen atom. As this bond is electrostatic, its strength will decrease slowly with distance, and can therefore be considered an attractive interaction at distances equal to, or larger than the van der Waals limit. The long range orienting nature of a C-H····O interaction underscores its important role in crystal engineering. Despite an accumulating body of evidence supporting the existence of weaker hydrogen bonds, the debate regarding the strength, directionality and relative abundance of such interactions has continued [23]. As a consequence, considerable theoretical and experimental efforts have been made to probe the structural consequences of 'weaker' hydrogen-bond interactions [24]. Typically, the directional preferences of these bonds are satisfied within the geometrical constraints of stronger hydrogen bonds, such as O-H····O and N-H····O interactions. However, the energy of weaker hydrogen bonds (4-8 kJmol^{-1}) is certainly large enough to allow these interactions to compete with van der Waals forces, especially if they are working in concert. Furthermore, C-H····X interactions can tilt the balance between several options of stronger bonded networks, thus acting as an important "steering force" in the solid-state assembly.

92

2.1.4   Halogen-halogen and other non-covalent hetero-atom interactions

Halogen substituents in molecular solids are capable of forming short contacts to other heteroatoms, such as nitrogen, oxygen and sulfur, and they may also participate in hydrogen bonds with N-H, O-H, and C-H moieties. The nature of short halogen-halogen atom distances is still not fully understood; short contacts may be the result of elliptically shaped atoms [25], or due to specific attractive forces [26]. Polarization does play a role since there is a distinction between symmetrical interactions, Cl····Cl and I····I, and unsymmetrical interactions, I····Cl and Br····F [1]. The directional interactions between halogen atoms and oxygen and/or nitrogen atoms are also induced by polarization. Such interactions have been successfully used in creating robust molecular tapes, Fig. 2.2 [27].

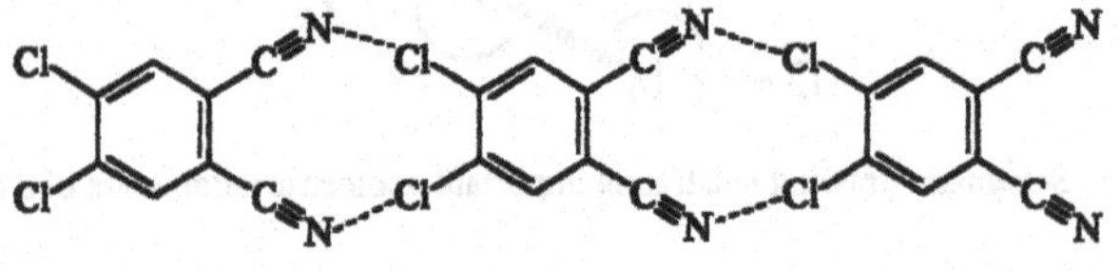

Figure 2.2   One-dimensional tape motif generated by C≡N····Cl interactions

Polarization effects are also important for sulfur atoms, which are known to engage in short intermolecular contacts. Such interactions have been employed to generate "supramolecular carpets" *via* the self-assembly of *bis*-4,4-dihydroxyphenol sulfone sustained by S····O interactions [28].

2.1.5   General rules for hydrogen bonding

After an extensive examination of preferential packing preferences and hydrogen-bond patterns in a vast number of organic crystals, Etter and co-workers put forward the following guidelines to facilitate deliberate design of hydrogen-bonded solids [29];

(i)   All good proton donors and acceptors are used in hydrogen bonding.

(ii)   Six-membered ring intramolecular hydrogen bonds form in preference to intermolecular hydrogen bonds.

(iii)   The best proton donor and acceptor remaining after intramolecular hydrogen bond formation will form intermolecular hydrogen bonds.

These empirical 'rules' are the consequences of correlations between hydrogen-bond patterns and functional groups. The first rule was originally put forth by Donohue [30], and generally as the proton increases in acidity, its donor properties will improve. Good donors are generally found in carboxylic acids, amides, urea derivatives, anilines, imines and phenols, and good proton acceptors are typically acid and amide carbonyl groups, sulfoxides, phosphoryls, nitroxides, and amine nitrogen atoms. The second rule derives from competition studies, which showed that an intramolecular hydrogen bond was more difficult to break than a comparable intermolecular bond formed from similar donors. The third rule can be illustrated by 2-aminopyrimidine succinic acid cocrystals, where the best donors (the acidic protons) are paired with the best acceptors (the ring nitrogen atoms), Fig. 2.3 [31].

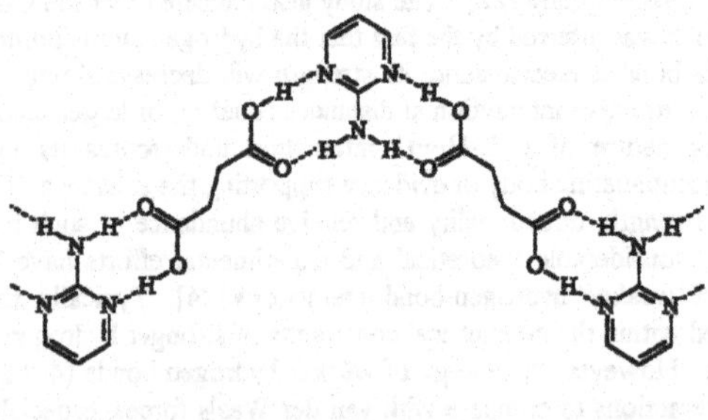

Figure 2.3   Hydrogen-bonding pattern of 2-aminopyrimidine and succinic acid.

2.1.6   Graph sets: classification of hydrogen bonds

Etter and coworkers proposed a systematic method based upon graph sets for describing hydrogen-bond patterns [32]. The process of assigning a graph set begins with the identification of the number of different types of hydrogen bonds present in the structure, and then by defining the bonds by the nature of its donors and acceptors, Fig. 2.4.

$$G_d^a(n)$$

Figure 2.4   A generic graph-set descriptor.

A set of molecules hydrogen bonded in a repeat unit, a motif, is characterized by one of four designators, C (chain), R (ring), D (dimer), or S (self) which denotes an intramolecular hydrogen bond. The number of donors and acceptors used in each motif are assigned as subscripts and superscripts respectively, and the total number of atoms in the repeat unit is denoted in brackets. One of the advantages of using graph sets is that it brings the focus onto the hydrogen-bonded pattern, and not solely on the geometrical constraints of non-covalent interactions. Some examples of hydrogen-bonded motifs and their graph set assignments are given in Fig. 2.5.

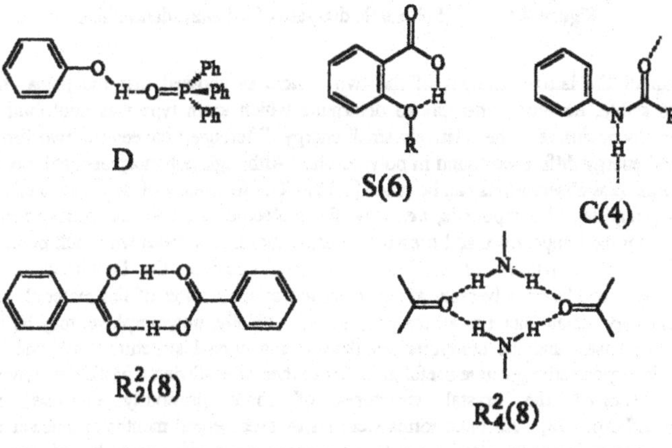

Figure 2.5   Graph set assignments for various hydrogen-bonded motifs.

3.   Polymorphism

Polymorphism, a unique feature of the solid state, has been defined as "a solid crystalline phase of a given compound resulting from the possibility of at least two crystalline arrangements of the molecules of that compound in the solid state." [33]. The extent of polymorphism amongst crystalline materials is difficult to measure, and estimates range from 3% to 30% to virtually every compound [34]. Polymorphism arises because the subtle intermolecular forces that establish the stability in a crystal typically are weak and numerous, and since entropic factors are temperature dependent, the observed crystal structure is often only one of several minima in the free energy surface, not necessarily corresponding to the global minimum. The conflicting demands of entropy and enthalpy, and of different types of intermolecular interactions may yield many local minima in the energy surface. As the free energy difference between these minima starts to merge, it will increase the possibility of polymorphism. As a consequence, it is rare that polymorphs have widely different lattice energies; typically the difference is no greater that 2.5 kJ mol^{-1} at ambient conditions.

Polymorphism in flat rigid molecules can be described as "alternative ways of close packing within the framework of a constant heteroatom pattern," [1] as the directional requirements of heteroatom interactions tend to be satisfied. This is demonstrated by the crystal structure of 1,4-dichlorobenzene [35] which can exist in a monoclinic or a triclinic form, both of which are composed of layers with optimal Cl····Cl interactions, but assembled in different ways. Polymorphism can also be found in compounds that have the same hydrogen-bonding pattern within the layers, but the layers will be stacked differently. This has been observed in the triclinic and tetragonal forms of 2,2-aziridine dicarboxamide [36].

Flexible molecules permit more variation in crystal packing than rigid molecules, since the torsional and conformational behavior of the molecular structure can induce a larger variety of non-covalent interactions between adjacent molecules. Conformational polymorphism occurs when the intermolecular interactions in each polymorph are similar, but each form will contain a specific molecular conformation that can be considered as characteristic of that particular polymorph. These conformations also may exist in solution. An example of this is seen in the chloro derivatives of n-benzylideneaniline [37], Fig. 3.1. In the triclinic form both torsion angles (α and β) are equal to $0°$, and in the orthorhombic form they are equal to $25°$.

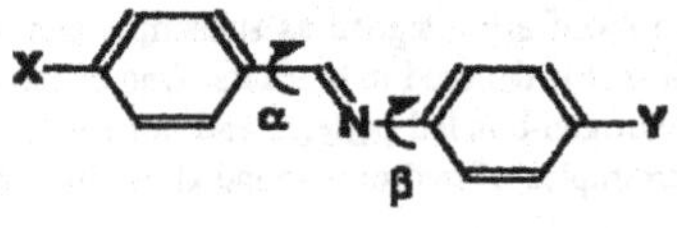

Figure 3.1   Schematic diagram of n-benzylideneaniline.

An examination of the lattice energies of the two structures showed that the planar triclinic form is marginally more stable, however, attempts to determine which atom type was contributing to the extra stabilization were inconclusive. The relatively small energy difference between the two forms is consistent with the expected energy difference found in polymorphs. Although polymorphism still poses many vexing questions, a few general observations can be made [1,13]. The frequency of occurrence of polymorphism is not equal in all categories of compounds, i.e., rigid flat molecules are unlikely to have many polymorphic conformations at normal temperature and pressure. Furthermore, molecules with multipoint recognition are likely to yield a unique pattern of synthons, and thus decrease the likelihood of polymorphism. Polymorphism does not always adversely affect a particular application of the material, as the property-defining fragment may remain intact in all the polymorphs. Finally, polymorphism may be strongly solvent dependent in some cases, and the computer simulation of a crystal structure, with and without solvent molecules being incorporated, may give useful guidelines when crystallizing a particular polymorph.

We recently reported the crystal structures of three previously unknown polymorphs of 2-amino-5-nitropyrimidine [38]. All three forms were stable over several months at ambient temperature and pressure, and no changes in crystal morphology or macroscopic appearance could be detected as a function of time. Each structure is characterized by an extensive network of hydrogen-bond interactions (including C-H····N and C-H····O interactions), and short inter-layer distances. All three polymorphs were also obtained separately through a careful choice of solvent or via sublimation. Although this molecule itself is relatively simple with very limited flexibility (rotation of the amino or nitro groups with respect to the aromatic ring), our crystal structure modeling has shown that this structural system presents a considerable challenge to current models of structure prediction. The best such a prediction can do is to find the minima in the lattice energy (calculated using the assumed model potential), which are closest to the experimental structures, using a rigid molecular structure derived by theoretical calculations. The 2-amino-5-nitropyrimidine polymorphic system has shown the limitations of this crude static thermodynamic criterion as well as the assumed rigidity and intermolecular potential. It is clear that the conformational flexibility of 2-amino-5-nitropyrimidine does play an important role in providing alternative packing arrangements and this is a good example of a system where a small change in molecular geometry allows a qualitatively different and more favourable crystal packing. This observation also serves to underline the pitfalls and difficulties involved with structure predications. A small change in molecular structure may lead to dramatic changes in the crystal structure and we therefore need more refined potential models and more structural and thermodynamic data.

The appearance of polymorphs also raises an important point regarding the reliability of single-crystal X-ray diffraction. When an individual crystal is chosen, it is generally the 'best' in the sample, and therefore, by its very nature, atypical of the majority of the sample. It is thus important to compare the structure of the 'best' crystal with the bulk of the material and this can be done using powder diffraction. By simulating a powder diffraction pattern from the single crystal data and comparing this with the pattern recorded for the bulk sample, it can be easily determined if the chosen crystal is a true representative of the sample; that is, if the sample is structurally homogenous.

4.   Strategies in crystal engineering: Organic solids

4.1 INTRODUCTION

Strategies for organic synthesis are based upon an understanding and utilization of the covalent bond. Supramolecular synthesis, on the other hand, is defined in terms of intermolecular connectivity and interactions. For the rational design of predictable assemblies, hydrogen bonding is the most commonly used supramolecular "glue", due to its strength and directionality. These non-covalent interactions can be refined by judicious placement of specific functional groups onto the molecular skeleton to generate "supramolecular synthons" [13]. Some common examples of supramolecular synthons are given in Fig. 4.1.

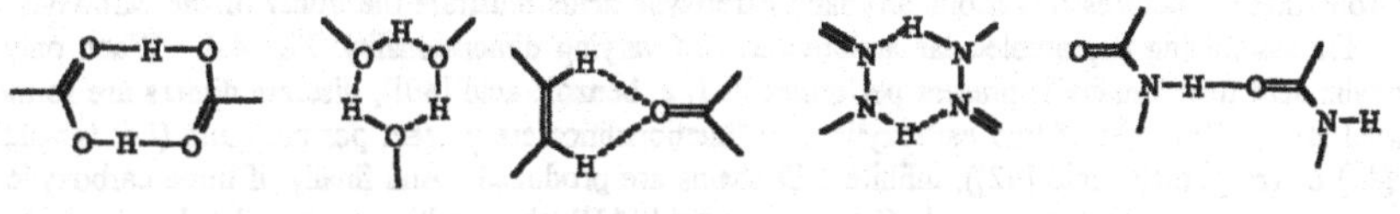

Figure 4.1   Examples of supramolecular synthons.

Supramolecular synthons are structural units within assemblies that can be generated by known or conceivable synthetic operations involving intermolecular interactions. One goal in crystal engineering is to recognize and design synthons that are robust enough to be exchanged from one network to another, thus leading to generality and predictability. This structural predictability can then be used in the construction of one, two, or three-dimensional patterns. In general, with careful molecular design, a 1-D array held together by strong intermolecular forces can be obtained relatively easily. The addition of functional groups or weaker intermolecular forces can then be used to bridge the 1-D networks into 2-D layers or sheets. Finally, these assemblies can be built up to yield a 3-D solid architecture by the same process, or by addition of new molecules/counterions that are capable of connecting the layers. This step-wise process can help to simplify the design process of a new solid, Fig. 4.2.

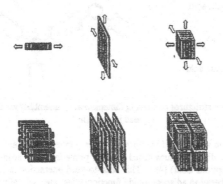

Figure 4.2   Step-wise construction of a 3-D lattice.

4.2   DESIGN OF ORGANIC MOLECULAR SOLIDS

Of the many supramolecular synthons that employ hydrogen bonds as the main intermolecular connectors, the carboxylic acid moiety is the most widely recognized. Leiserowitz [39] has shown that the carboxylic acid functionality is capable of forming two different hydrogen-bonding motifs, Fig. 4.3. They include the head-to-head interaction, which forms cyclic hydrogen-bonded dimers *via* two O-H⋯O hydrogen bonds, and the less common catemer interaction, in which each carboxyl group is linked to two adjacent carboxyl groups *via* one O-H⋯O hydrogen bond.

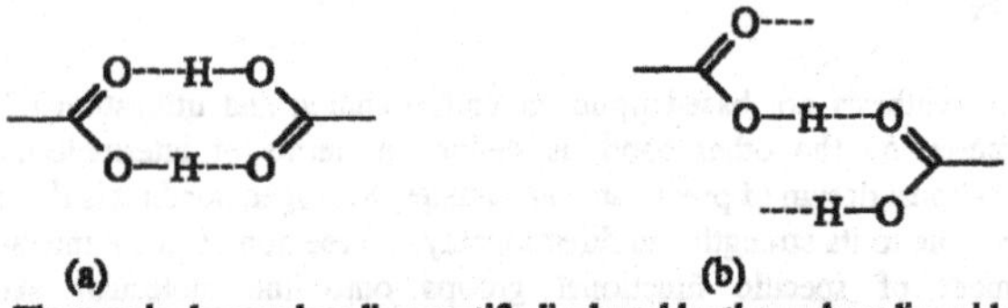

(a)   (b)

Figure 4.3   The two hydrogen-bonding motifs displayed by the carboxylic acid functionality - (a) head-to-head motif, and (b) catemer motif.

The solid-state structures of various aromatic carboxylic acids illustrate the utility of the carboxylic acid moiety for assembling supramolecular architectures of varying dimensionality, Fig. 4.4. When only one carboxylic acid functionality is present per molecule (*i.e.* benzoic acid [40]), discrete dimers are formed in the solid-state. However, if two carboxylic acid functionalities are present per molecule (*i.e. iso*-phthalic acid [41] or terephthalic acid [42]), infinite 1-D chains are produced. And finally, if three carboxylic acid functionalities are present per molecule (*i.e.* trimesic acid [43]), the resulting supramolecular structure is an infinite 2-D sheet.

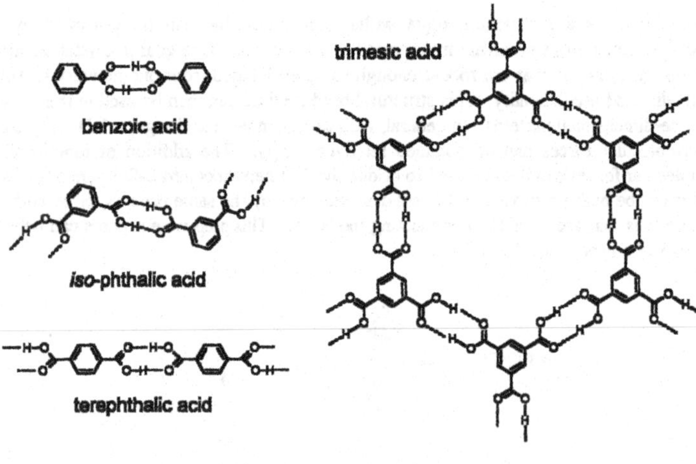

benzoic acid

iso-phthalic acid

terephthalic acid

trimesic acid

Figure 4.4   Solid-state structures of varying dimensionality assembled *via* the carboxylic acid head-to-head interaction.

Another prominent synthon widely used in supramolecular synthesis is the amide functionality. Much like the carboxylic acid cyclic dimer, unsubstituted amides primarily form head-to-head hydrogen-bonding interactions (although exceptions exist) [44]. This head-to-head interaction is composed of complementary N-H⋯O hydrogen bonds between adjacent amide functionalities, Fig. 4.5. However, unlike the carboxylic acid functionality, the second amide hydrogen will often form another hydrogen bond with the carbonyl oxygen of a neighboring amide. This 'extra' hydrogen bond adds to the overall dimensionality of the solid-state structure.

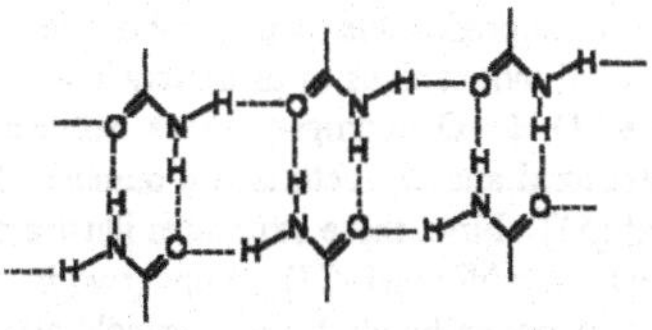

Figure 4.5   Common hydrogen-bonded ribbon in primary amides.

As previously mentioned, the amide functionality may be used to assemble supramolecular architectures in much the same way as the carboxylic acid moiety. This point is illustrated by the similarities of the solid-state structures of various aromatic carboxamides, Fig. 4.6, when compared to their carboxylic acid analogs, Fig. 4.4. When one amide functionality is present per molecule (*i.e.* benzamide [45]), cyclic dimers again are formed. In addition, these dimers are propagated into a 1-D tape through two complementary N-H····O hydrogen bonds between neighboring dimers. When two amide functionalities are present per molecule (i.e. terephthalamide [46]), infinite 1-D chains are produced. These 1-D infinite chains then are propagated into the corresponding 2-D sheets *via* a second N-H····O hydrogen bond involving the 'extra' amide proton and the carbonyl oxygen of a neighboring amide.

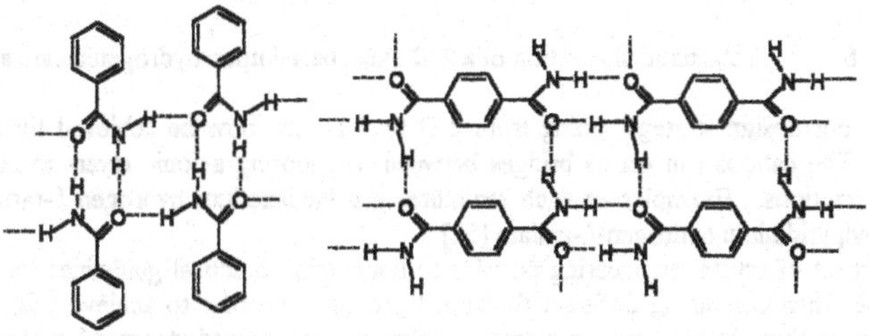

Figure 4.6   Solid-state architectures assembled *via* amide head-to-head interactions.

The oxime moiety represents another synthon capable of generating supramolecular architectures. Although this functionality is nothing new to organic chemists, it remains relatively unexplored as an intermolecular connector. In the absence of other strong hydrogen bond acceptors, the oxime functionality has been shown to form head-to-head cyclic dimers in the solid-state [47]. The cyclic dimer is held together by complementary O-H····N hydrogen bonds between adjacent molecules, Fig. 4.7.

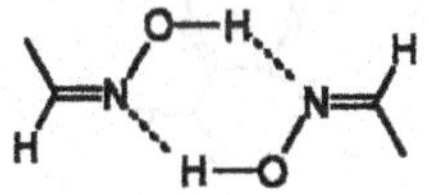

Figure 4.7   Head-to-head hydrogen-bonding interactions between adjacent oxime moieties.

4.3   DESIGN OF IONIC ORGANIC SOLIDS

In recent years, several examples of engineering of ionic organic and organic-inorganic compounds with specific packing motifs have appeared. Ward and co-workers have utilized the complementarity of hydrogen-bond sites on guanidinium cations and substituted sulfonate anions in the preparation of materials containing infinite hydrogen-bonded sheets [48]. The precise topology of these motifs can be modified through appropriate substitution of the anion. Anions of trimesic acid have also been used in the design of porous organic solids [49].

98

In our group, we have developed several predictable hydrogen-bonded architectures generated by anionic species. We have employed a variety of monoanions of dicarboxylic acids in the design of 2-D layers held together primarily by charge-assisted O-H····O hydrogen bonds between the 'head' and 'tail' of adjacent anions. In this way, we have synthesized and characterized a number of salts of tartaric acid [50], malic acid [51] and amino-*iso*phthalic acid [52]. First, these structures illustrate that the expected anionic chains are almost invariable formed. Second, the additional -OH groups (two in the case of tartaric acid, one in the case of malic acid) generate robust hydrogen-bonds between neighboring chains which gives rise to the desired 2-D anionic network, Fig. 4.8.

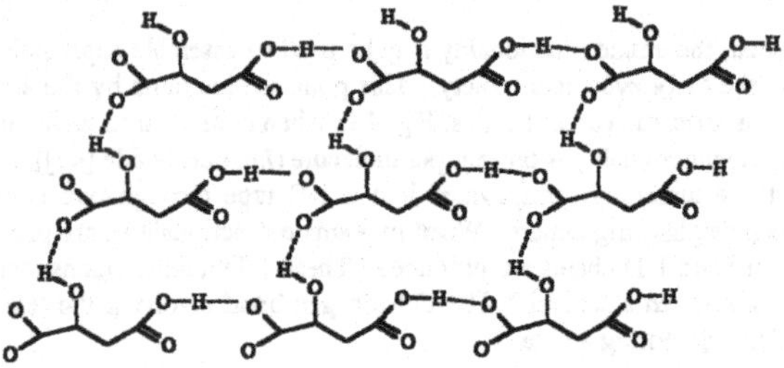

Figure 4.8   Schematic illustration of a 2-D sheet based upon hydrogenmalate anions

The final step in our design strategy, going from 2-D to 3-D, can now be achieved through a suitable choice of cation. The cations can act as bridges between neighboring anionic layers through directional intermolecular interactions. Examples of such structures are imidazolium hydrogen-*L*-tartrate [50b] and 3-hydroxy-6-methylpyridinium hydrogen-*L*-malate [51].

An important aspect of crystal engineering pertains to uncovering practical guidelines for estimating the structural outcome when combining different functional groups. In order to achieve this, it is helpful to study a simple system that can give rise to a small number of well-defined structural motifs. With this in mind we determined the structures of several organic salts based upon the oxamate anion, which contains complementary hydrogen-bond donors and acceptors. *Ab initio* calculations showed three dimeric motifs, Fig. 4.9, with very small energy differences and, in fact, all three features have appeared in crystalline oxamates [53].

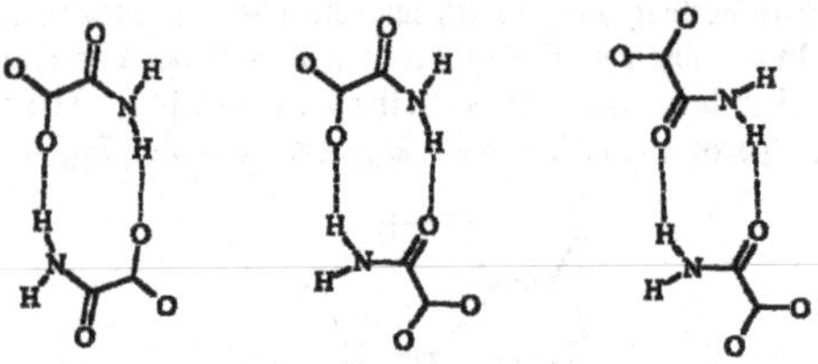

Figure 4.9   Three possible dimeric motif of oxamate anions.

In addition, the anions invariably give rise to either infinite 1-D motifs or 2-D layers and the subsequent detailed arrangement of anions and cations is determined by specific cation-anion interactions. Since certain cations lead to specific anionic motifs, we can propose simple guidelines for predicting oxamate structures. Primary ammonium ions, with three hydrogen-bond donors positioned in close proximity of each other, form hydrogen bonds to at least two different anions and, in doing so, bring anions close together (*via* N-H····O interactions), forcing the anions into a layered arrangement. Secondary or tertiary ammonium ions only bring together a smaller number of anions which is insufficient to cause the formation of an infinite 2-D sheet. In those cases, *e.g.* imidazolium oxamate, the anions instead create 1-D motifs. In addition, cation size may also be important since a small cation will inevitably bring neighboring anions within close proximity of each thereby encouraging anionic layer formation. This is exemplified by the crystal structure of potassium oxamate hydrate [53]. Conversely, a bulky cation will expand the size of the lattice thus increasing the distance between anions resulting in the formation of 1-D anionic motifs. The hydrogen bond is arguably sufficiently strong to link neighboring ions together into infinite architectures, even in the presence of different cations and in competition with other non-covalent interactions, and it is likely that there will be many more journeys into the design of desired structural functionalities in organic salts.

5.   Strategies in crystal engineering: organic-inorganic hybrid materials

5.1   COORDINATION POLYMERS

The most common approach to supramolecular assembly of metal complexes has led to a wide variety of coordination polymers [54]. Several coordination polymers, 1-D, 2-D, and 3-D, have been prepared by coordinating bifunctional ligands such as 4, 4'-bipyridine or pyrazine to a transition metal ion. In these systems, the coordination environment (linear, square planar, trigonal, or octahedral) of the metal ion can dictate the dimensionality of the porous solid, as well as the influencing the nature of the cavity created.

Robson and co-workers have outlined the possibilities of creating infinite 2-D and 3-D metal-containing frameworks based around metal-4,4'-bipyridine derivatives [55]. Interpenetration of 2-D sheets can give rise to 3-D solids with channels and cavities of specific dimensions. Zaworotko and co-workers have employed similar building blocks in the design of porous 'coordination-polymers' with diamondoid structures [56]. Chen, et. al. [57], provide another notable example of this type of design strategy. Coordination of both 4,4'-bipyridine and pyrazine to copper(II) ions produces a 2-D rectangular network with small, rectangular channels, Fig. 5.1. Each copper(II) ion is coordinated by two 4,4'-bpy molecules, two pyrazine molecules, and two water molecules to give an octahedral coordination environment. The bpy molecules and the two pyrazine molecules are mutually *trans* and occupy the equatorial plane. Similar 2-D square networks with open channels have been constructed from Zn(II) ions and 4,4'-bpy [58] and from Cd(II) and 4,4'-bpy [59].

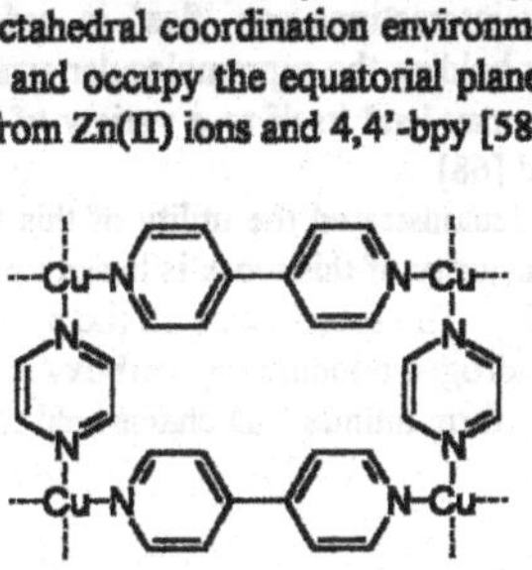

Figure 5.1   Schematic representation of the main structural feature in {[Cu(4,4'-bpy)(pyz)(H_2O)$_2$][PF_6]$_2$}$_n$.

Rogers, et. al. [60], provide another example of a square supramolecular coordination complex. The fundamental difference between this structure and many others lies in how the 'squares' are assembled. In the previous example, the supramolecular squares are assembled by linear ligands connecting metal ions. The coordination geometry of the metal ion ensures that the ligands are orthogonal to each other, and in doing so, forms the 'corners' of the grid-like network. However, in these squares, the 'corners' of the squares are the *ligands* (pyrimidine) and the linear connectors are silver(I) ions, Fig. 5.2.

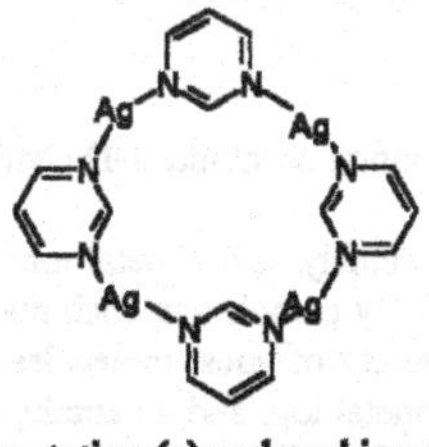

Figure 5.2   Schematic representation (a) and packing diagram of [Ag(pyrimidine)][NO_3].

5.2 INTERMOLECULAR ASSEMBLY OF ORGANIC/INORGANIC MATERIALS

Recently, efforts have been made to combine intermolecular hydrogen-bonding interactions of organic molecules with the strict geometrical constraints imposed by transition metal ions to produce inorganic/supramolecular hybrids [61]. In each of these examples, a coordination complex of a transition metal ion with a coordinating ligand assembles *via* hydrogen bonds into a supramolecular architecture. As described above for coordination polymers, 1-D, 2-D, and 3-D structures are possible from this type of design strategy.

Lehn and co-workers, have generated some extraordinary structures covering silver(I) helicates [62], circular double helicates [63], and supramolecular macrocycles [64]. Many of these motifs are built with a combination of well-established coordination chemistry (coordination number, geometry, *etc.*) and site-specific and self-complementary hydrogen-bonds. Mingos and co-workers [65] have utilized intermolecular connectors (*e.g.* thiourea//carboxylate, diaminopyridine///uracil), *etc.* as a way of linking neighboring metal-complexes into extended 'tapes' and sheets. Munakata *et al* prepared hydrogen bonded copper(I) complexes with large channels occupied by cations, using the well-known dimeric 2-pyridone synthon [66]. Braga, Grepioni and co-workers have been at the forefront of developing an organometallic 'branch' of crystal engineering through a combination of synthesis, database analysis, and packing-energy calculations [67] (and elsewhere in this volume).

Noncovalent assembly of coordination complexes can sometimes provide advantageous physical properties that are not always displayed by coordination polymers. Unlike coordination polymers, these types of assemblies tend to be more readily soluble. Increased solubility is desirable for reasons ranging from ease of synthetic modification and manipulation to crystal growth. The other advantage these types of systems have over coordination polymers is flexibility. Unlike the rigid frameworks of coordination polymers, assemblies generated through hydrogen-bonding interactions can 'flex' in order to accommodate changes in their environment since the hydrogen bonds holding the supramolecular array together are weaker than a covalent bond. Therefore, this type of system may lend itself to a variety of host-guest applications that would be inaccessible if the framework was rigid [68].

Aakeröy and co-workers have also demonstrated the utility of this type of design strategy for assembling low-dimensional architectures. The majority of this work is based upon coordination complexes of silver(I) salts with substituted pyridine rings [69]. The substituents on the pyridine rings are various moieties capable of solid-state self-assembly through hydrogen bonding *e.g.* carboxylic acids, carboxamides, and oximes [70]. The silver(I) complexes predominantly form infinite 1-D chains held together by the head-to-head hydrogen-bonding interactions, Fig. 5.3.

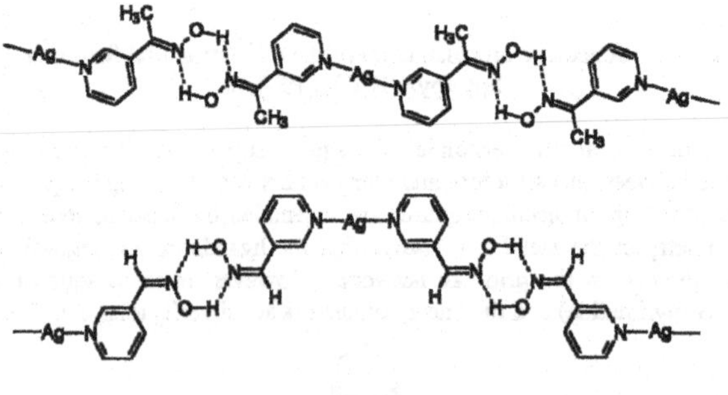

Figure 5.3   Schematic representation of infinite 1-D chains based on a linear silver(I) complex.

To increase the dimensionality of the assembly, other metal ions (*e.g.* nickel(II) and platinum(II)), can be used as the corner stone of the network. By complexing such metal ions with ligands capable of hydrogen bonding, new porous materials with a variety of guest molecules have been prepared [71]. Typically, the pyridine-based ligand coordinates to the metal ion, and an amide, oxime, or carboxylic moiety provides the cross link between adjacent metal-ion complexes, Fig. 5.4.

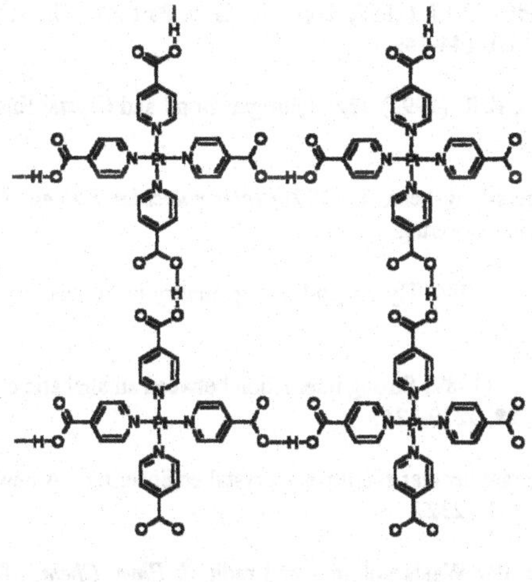

Figure 5.4   Supramolecular square grid based on platinum(II) complexes.

These types of structures demonstrate that metal complexes can be linked through hydrogen bonds to form porous structures with large cavities. The approach is quite versatile, as either octahedral or square planar metal centers can be used. In addition, a variety of ligands can link through hydrogen bonds to assemble porous 3-D networks that are thermally stable (melting points are typically > 300°C).

Crystal engineering, which is built upon an amalgamation of structural techniques, powerful theoretical models and general principles of synthetic methods, is a prime example of a truly interdisciplinary topic. As such, it is in a position to combine resources and expertise that can produce new materials for optical and electronic devices as well as more efficient catalysts, while, at the same time, addressing fundamental issues with broader implications, *e.g.* intermolecular forces and structure-property correlations. Indeed, a rational design of crystalline materials may create opportunities of considerable academic and commercial importance.

References

1. Desiraju, G.R. (1989) *Crystal Engineering: The Design of Organic Solids*, Elsevier, Amsterdam.

2. Lehn, J.-M. (1995) *Supramolecular Chemistry*, VCH, Weinheim.

3. Aakeröy, C.B. (1997) Crystal engineering: strategies and architectures, *Acta Crystallogr.* B53, 569- 586.

4. Desiraju, G.R. (ed.) (1995) *Perspectives in Supramolecular Chemistry, Vol. 2, The Crystal as a Supramolecular Entity*, Wiley, Chichester.

5. Lehn, J.-M. (1990) Perspectives in supramolecular chemistry – From molecular recognition towards molecular information-processing and self-organization, *Angew. Chem. Int. Ed. Engl.* 29, 1304-1319.

6. Schmidt, G.M.J. (1971) Photodimerizations in the solid state, *Pure Appl. Chem.* 27, 647-657.

7. Rabinovich, D. and Schmidt, G.M.J. (1967) Topochemistry. Part XIX. The crystal structure of monomethyl *trans, trans*-muconate, *J. Chem. Soc. (B)*, 286-289.

8. Rabinovich, D. and Schmidt, G.M.J. (1967) Topochemistry. Part XV. The solid-state photochemistry of p-quinones, *J. Chem. Soc. (B)*, 144-149.

9. Aakeröy, C.B. and Seddon, K.R. (1993) The Hydrogen Bond and Crystal Engineering, *Chem. Soc. Rev.* 397-407.

10. Huyskens, P.L. and Zeegers-Huyskens, T. (1991) *Intermolecular Forces: An Introduction to Modern Methods and Results*, Springer-Verlag.

11. Taylor, R. and Kennard, O. (1984) Hydrogen-bond geometry in organic crystals, *Acc. Chem. Res*, **17**, 320-326.

12. Cotton, F.A. and Luck, R.L. (1989) Strong interaction between an aliphatic carbon hydrogen-bond and a metal atom, *Inorg. Chem.* **28**, 3210-3213.

13. Desiraju, G.R. (1995) Supramolecular synthons in crystal engineering – A new organic synthesis, *Angew. Chem. Int. Ed. Engl.* **34**, 2311-2327.

14. (a) Bondi, A. (1964) Van der Waals volumes and radii, *J. Phys. Chem*, **68**, 441-449; (b) Pauling, L. (1940) *The Nature of the Chemical Bond and the Structure of Molecules and Crystals*, Oxford University Press; (c) Nyburg, S.C. and Faerman, C.H. (1985) A revision of van der Waals radii for molecular crystals – N, O, F, S, Cl, Se, Br, and I bonded to carbon, *Acta Crystallogr.* **B41**, 274-279.

15. (a) Jeffrey, G.A. and Saenger, W. (1991) *Hydrogen Bonding In Biological Structures*, Springer; (b) Speakman, J. (1975) *MTP Internationa Review of Science Vol 2*, Butterworths.

16. Allinger, N.L. (1976) Calculation of molecular structure and energy by force-field methods, *Adv. Phys. Org. Chem*, **13**, 1-82.

17. Murray-Rust, P. and Glusker, J.P. (1984) Directional hydrogen bonding to sp^2-hybridized and sp^3-hybridized oxygen atoms and its relevance to ligand macromolecule interactions, *J. Am. Chem. Soc.* **106**, 1018-1025.

18. Allen, F.H. and O. Kennard, (1993) Integrated 3d search facilities for the Cambridge Structural Database (CSD),*Chemical Design and Automation News*, **8**, 3137.

19. Subramanian, S. and Zaworotko, M.J. (1994) Exploitation of the hydrogen bond – Recent developments in the context of crystal engineering, *Coord. Chem. Rev.***137**, 357-401.

20. Glasstone, S. (1937) The structure of some molecular complexes in the liquid state, *Trans. Faraday Soc.* **33**, 200-214.

21. Donohue, J. (1968) *Structural Chemistry and Molecular Biology*, W.H.Freeman.

22. Taylor, R. and Kennard, O. (1982) Crystallographic evidence for the existence of CH····O, C-H····N, and C-H····Cl hydrogen bonds, *J. Am. Chem. Soc.* **104**, 5063-5070.

23. Cotton, F.A., Daniels, L.M., Jordan, G.T., and Murillo, C.A. (1997) The crystal packing of *bis*(2,2'-dipyridylamido)cobalt(II), Co(dpa)$_2$ is stabilized by C-H center dot center dot center dot N bonds: are there any real precedents? *Chem. Commun.* 1673-1674.

24. (a) Allen, F.H., Lommerse, J.P.M., Hoy, V.J., Howard, J.A.K., and Desiraju, G.R. (1996) The hydrogen-bond C-H donor and π-acceptor characteristics of three-membered rings, *Acta Crystallogr.*

B52, 734-735; (b) Reddy, D.S., Goud, B.S., Panneerselvan, K., and Desiraju, G.R. (1993) C-H⋯N mediated hexagonal network in the crystal structure of the 1/1 molecular complex 1,3,5- tricyanobenzene hexamethylbenzene, *J. Chem. Soc. Chem. Commun.* 663-665; (c) Desiraju, G.R. (1992) C-H⋯O hydrogen-bonding and the deliberate design of organic crystal structures, *Mol. Cryst. Liq. Cryst.* 211, 63-74; (d) Desiraju, G.R. (1991) The C-H...O hydrogen bond in crystals – What is it? *Acc. Chem. Res.* 10, 290-296; (e) Steiner, T. and Desiraju, G.R. (1998) Distinction between the weak hydrogen bond and the van der Waals interaction, *Chem. Commun.* 81-892; (f) Mascal, M. (1998) Statistical analysis of C-H⋯N hydrogen bonds in the solid state: there *are* real precedents, *Chem. Commun.* 303-304.

25. Lucas, J., Price, S.L., Rowland, R.S., Stone, A.J., and Thornley, A.E. (1994) The nature of –Cl-center-dot-center-dot-Cl- intermolecular interactions, *J. Am. Chem. Soc.* 116, 4910-4918.

26. Desiraju, G.R. and Parthasarathy, R. (1989) The nature of halogen⋯halogen interactions – Are short halogen contacts due to specific attractive forces or due to close packing of nonspherical atoms? *J. Am. Chem. Soc.* 111, 8725-8726.

27. Desiraju, G.R., Ovchinnikov, Y.E., Reddy, D.S., Shishkin, O.V., and Struchkov, Y.T. (1996) Supramolecular synthons in crystal engineering 3. Solid state architectures and synthon robustness in some 2,3-dicyano-5,6-dichloro-1,4-dialkoxybenzene, *J. Am.Chem. Soc.* 118, 4085-4089.

28. Davies, C., Langler, R.F., Sharma, C.V.K. and Zaworotko, M.J. (1997) A supramolecular carpet formed *via* self-assembly of *bis*(4,4'-dihydroxyphenyl) sulfone, *Chem. Commun*, 567-568.

29. Etter, M.C. (1990) Encoding and decoding hydrogen-bond patterns of organic compounds, *Acc. Chem. Res.* 23, 120-126.

30. Donohue, J. (1952) The hydrogen bond in organic crystals, *J. Phys. Chem.* 56, 502-510.

31 Etter, M.C. and Adsmond, D.A. (1990) The use of cocrystallization as a method of studying hydrogen-bond preferences of 2-aminopyridine, *J. Chem. Soc. Chem. Commun.*, 8, 589-591.

32. (a) Etter, M.C. (1985) Aggregate structures of carboxylic-acids and amides, *Israel. J. Chem*, 25, 312-319; (b) Etter, M.C., Bernstein, J.B. and MacDonald, J.C. (1990) Graph-set analysis of hydrogen-bond patterns in organic crystals, *Acta Crystallogr*, B46, 256-262; (c) Bernstein, J., Davis, R.E., Shimoni, L., and Chang, N. (1995) Patterns in hydrogen bonding – functionality and graph-set analysis in crystals, *Angew. Chem. Int. Ed. Engl.* 34, 1555-1573.

33. McCrone, W.C. (1965) *Physics and Chemistry of the Organic Solid State, Vol. B*, Weissberger.

34. Burger, A. (1982) Thermodynamic and other aspects of the polymorphism of drugs, *Pharm. Int.* 3, 158-163.

35. Mnyukh, Y.V. (1963) The crystal structure of 1,4-dichlorobenzene, *J. Phys. Chem. Solids*, 24, 631-638.

36. Bruckner, S. (1982) An unusual example of packing among molecular layers – The structures of two crystalline forms of 2,2-aziridinecarboxamide, $C_4H_7N_3O_2$, *Acta Crystallogr.* B38, 2405-2408.

37. Bernstein, J. and Hagler, A.T. (1978) Conformational polymorphism. The influence of crystal structure on molecular conformation, *J. Am. Chem. Soc*, 100, 673-681.

38. Aakeröy, C.B., Nieuwenhuyzen, M., Price, S.L. (1998) Three polymorphs of 2-amino-5-nitropyrimidine: Experimental structures and theoretical predictions, *J. Am. Chem. Soc.* 120, 8986-8993.

39. Leiserowitz, L. (1976) Molecular packing modes. Carboxylic acids *Acta Crystallogr.*, *B32*, 775-802

40. Bruno, G. and Randaccio, L. (1980) A refinement of the benzoic acid structure at room temperature, *Acta Crystallogr.* **B36**, 1711-1712.

41. Derissen, J.L. (1974) Isophthalic acid, Acta *Crystallogr.* B30, 2764-2765.

42. Bailey, M. and Brown, C.J. (1967) The crystal structure of terephthalic acid, *Acta Crystallogr.* **22**, 387-391.

43. Duchamp, D. J. and Marsh, R.E. (1969) The crystal structure of trimesic acid, *Acta Crystallogr.* **B25**, 5-19.

44. Leiserowitz, L. and Schmidt, G.M.J. (1969) Molecular packing modes. Part III. Primary amides, *J. Chem. Soc. (A)*, 2372-2382.

45. Penfold, B.R. and White, J.C.B. (1959) The crystal and molecular structure of benzamide, *Acta Crystallogr.* **12**, 130-135.

46. Cobbledick, R. E. and Small, R. W. H. (1972) The crystal structure of terephthalamide, *Acta Crystallogr.* **B28**, 2893-2896.

47. Jerslev, B. and Larsen, S. (1992) Crystallizations and crystal-structure at 122 K of anhydrous (*E*)-4-hydroxybenzaldehyde oxime, *Acta Chem. Scand.* **46**, 1195-1199.

48. (a) Russell, V.A., Evans, C.C., Li, W.J., and Ward, M.D. (1997) Nanoporous molecular sandwiches: Pillared two-dimensional hydrogen-bonded networks with adjustable porosity, *Science* 276 575-579; (b) Russell, V.A. and Ward, M.D. (1996) Molecular crystals with dimensionally controlled hydrogen-bonded nanostructures, *Chem. Mater.* **8**, 1654-1666.

49. Melendez, R.E., Sharma, C.V.K., Zaworotko, M.J., Bauer, C., and Rogers, R. (1996) Toward the design of porous organic solids: modular honeycomb grids sustained by anions of trimesic acid, *Angew. Chem. Int. Ed Engl.* **35**, 2213-2215

50. (a) Aakeröy, C.B., Hitchcock, P.B., and Seddon, K.R. (1992) Organic salts of *L*-tartaric acid: Materials for second harmonic generation with a crystal structure giverned by an anionic hydrogen-bonded *J. Chem. Soc., Chem. Commun.*, 553-555; (b) Aakeröy, C.B. and Hitchcock, P.B. (1993) Hydrogen-bonded layers of hydrogentartrate anions: two-dimensional building blocks for crystal engineering, *J. Mater. Chem.* **3**, 1129-1135.

51. (a) Aakeröy, C.B. and Nieuwenhyzen, M. (1994) Hydrogen-bonded layers of hydrogenmalate anions: A framework for crystal engineering, *J. Am. Chem. Soc.* 116, 10983-10991; (b) Aakeröy, C.B. and Nieuwenhuyzen, M. (1996) Hydrogen Bonding in crystal engineering: 2-D layers of hydrogen-*L*-malate anions, *J. Mol. Struct.*, 374, 223-239.

52. Aakeröy, C.B., Hughes, D.P. McCabe, J.M., and Nieuwenhuyzen, M. (1998) Aromatic dicarboxylic acids as building blocks of extended hydrogen-bonded architectures, *Supramolecular Chemistry*, 9, 127-135.

53. Aakeröy, C.B., Hughes, D.P., and Nieuwenhuyzen, M. (1996) The oxime functionality: A versatile synthon for supramolecular assembly of metal-containing hydrogen-bonded architectures, *J. Am Chem. Soc.* 120, 7383-7384.

54. (a) Janiak, C. (1997) Functional organic analogues of zeolites based on metal-organic coordination frameworks, *Angew. Chem. Int. Ed. Engl.* **36**, 1431-1434; (b) Abrahams, B. F., Batten, S. R., Hamit, H., Hoskins, B. F., and Robson, R. (1996) A cubic (3,4) connected net with large cavities in solvated [Cu₃(tpt)₄](ClO₄)₃, *Angew. Chem. Int. Ed. Engl.* **35**, 1690-1692; (c) Kawata, S., Kitagawa, S., Kumagai, H., Iwabuchi, S., and Katada, M. (1997) The structural characterization of the novel ribbon sheet [Cu₂Cl₂(pyz)]ₙ (pyz=pyrazine), *Inorg. Chim. Acta* **267**, 143-145.

55. Batten, S.R. and Robson, R. (1998) Interpenetrating nets: Ordered, periodic entanglement, *Angew. Chem. Int. Ed. Engl.* **37**, 1460-1494.

56. (a) Copp, S.B., Subramanian, S., and Zaworotko, M.J. (1993) Spontaneous strict self-assembly of distorted super-diamondoid networks that are capable of enclathrating acetonitrile, *J. Chem. Soc., Chem. Commun.* 1078-1079; (b) Robinson, F. and Zaworotko, M.J. (1995) Triple interpenetration in [Ag(4,4'-bipyridine)][NO₃], a cationic polymer with a 3-dimensional motif generated by self-assembly of t-shaped building blocks, *J. Chem. Soc., Chem. Commun.* 2413-2414; (c) Zaworotko, M.J. (1994) Crystal engineering of diamondoid networks, *Chem. Soc. Rev.* **23**, 283-288.

57. Tong, M.-L., Chen, X.-M., Yu, X.-L., and Mak, T. C. W. (1998) A novel two-dimensional rectangular network, *J. Chem. Soc. Dalton Trans.* 5-6.

58. Subramanian, S. and Zaworotko, M. J. (1995) Porous solids by design – [Zn(4,4'bpy)₂ (SiF₆)]ₙ center-dot-xDMF, a single framework octahedral coordination polymer with large square channels, *Angew. Chem. Int. Ed. Engl.* **34**, 2127-2129.

59. Fujita, M., Kwon, Y. J., Washizu, S., and Ogura, K. (1994) Preparation, clathration ability, and catalysis of a 2-dimensional square network material composed of cadmium(II) and 4,4'-bipyridine, *J. Am. Chem. Soc* **116**, 1151-1152.

60. Sharma, C.V.K., Griffin, S.T., and Rogers, R.D. (1998) Simple routes to supramolecular squares with ligand corners: 1:1 Ag(I):pyrimidine cationic tetranuclear assemblies, *J. Chem. Soc., Chem. Commun.* 215-216.

61. (a) Blake, A.J., Hill, S.J., Hubberstey, P. and Li, W.-S. (1997) Rectangular grid two-dimensional sheets of copper(II) bridged by both coordinated and hydrogen bonded 4,4'-bipyridine (4,4'-bipy) in [Cu(μ-4,4'- bipy)(H₂O)₂ (FBF₃)₂] center dot 4,4'-bipy, *J. Chem. Soc., Dalton Trans.* **6**, 913-914; (b) Carlucci, L.,Ciani, G., Proserpio, D. M., and Sironi, A. (1997) Extended networks via hydrogen bond cross-linkages of [M(bipy)] (M=Zn²⁺ or Fe²⁺) linear co-ordination polymers, *J. Chem. Soc., Dalton Trans.* **11**, 1801-1803; (c) Chan, C.- W., Mingos, D.M.P., White, A.J.P., and Williams, D.J. (1996) A crystal engineering study of copper(II) complexes with 2,4-diamino-6-(4-pyridyl)-1,3,5-triazine and 2-amino-4-phenylamino-6-(4-pyridyl)-1,3,5-triazine, *Polyhedron*, **15**, 1753-1767.

62. Garrett, T.M., Koert, U., Lehn, J.-M., Rigault, A., Meyer, D., and Fischer, J. (1990) Self-assembly of silver(I) helicates, *J. Chem. Soc., Chem. Commun.* 557-558.

63. Hasenknopf, B., Lehn, J.-M., Kneisel, B.O., Baum, G., and Fenske, D. (1996) Self-assembly of a circular double helicate, *Angew. Chem. Int. Ed. Engl.* **35**, 1838-1840.

64. Drain, C.M., Russel, K.C., and Lehn, J.-M. (1996) Self-assembly of a multi-porphyrin supramolecular macrocycle by hydrogen bond molecular recognition, *Chem. Commun.* 337-338.

65. (a) Burrows, A.D., Mingos, D.M.P., White, A.J.P., Williams, D.J. (1996) Molecular recognition of 2,6-diaminopyrimidine by platinum orotate complexes, *J. Chem. Soc. Dalton Trans.* 149-151; (b) Chowdhry, M.M., Mingos, D.M.P., White, A.J.P., and Williams, D.J. (1996) Novel supramolecular self-assembly of

a transition-metal organo network based on simultaneous coordinate- and hydrogen-bond interactions, *Chem. Commun.* 899-900; (c) Burrows, A.D., Chan, C.W., Chowdhry, M.M., McGrady, J.E., and Mingos, D.M.P. (1995) Multidimensional crystal engineering of bifunctional metal-complexes containing complementary triple hydrogen bonds, *Chem. Soc. Rev.* **24**, 329-337.

66. Munakata, M., Wu, L-P., Yamamoto, M., Kuroda-Sowa, T., and Maekawa, M. (1996) Construction of three-dimensional supramolecular coordination copper(I) compounds with channel structures hosting a variety of anions by changing the hydrogen-bonding mode and distances, *J. Am. Chem. Soc.* **118**, 3117-3124.

67. (a) Braga, D. and Grepioni, F. (1999) Complementary hydrogen bonds and ionic interactions give access to the engineering of organometallic crystals, *J. Chem. Soc. Dalton Trans.* 1-8; (b) Braga, D., Grepioni, F., and Desiraju, G.R. (1998) Crystal engineering and organometallic architecture, *Chem. Rev.* **98**, 1375-1405.

68. (a) Chen, C.-N., Zhang, H.-X., Yu, K.B., Zheng, K.C., Cai, H., and Kang, B.S. (1998) Extended networks *via* hydrogen bond linkages of zig-zag coordination chains, *J. Chem.Soc. Dalton Trans.* 1133-1136; (b) Aakeröy, C.B., Beatty, A.M., and Helfrich, B.A. (1998) Two-fold interpenetration of 3-D nets assembled *via* three-coordinate silver(I) ions and amide-amide hydrogen bonds, *J. Chem. Soc. Dalton Trans.* 1943-1946; (c) Kumar, R.K., Balasubramanian, S., and Goldberg, I. (1998) Crystal engineering with tetraarylporphyrins, an exceptionally versatile building block for the design of multidimensional supramolecular structures, *Chem. Commun.* 1435-1436; (d) Cameron, B.R., Corrent, S.S., and Loeb, S.J. (1995) Transition metals as both host and guest: Second-sphere coordination between a Pt-azacrown ether host and a Pt-NH$_3$ guest, *Angew. Chem. Int. Ed. Engl.* **34**, 2689-2691; (e) Rivas, J.C.M. and Brammer, L. (1998) Supramolecular assembly of anion-channel and anion-layer structures of [PtL$_4$]X$_2$ (L=nicotinamide; X=Cl, or PF$_6$): surprisingly robust arene ring 'herringbone' motifs and adaptable amide-amide hydrogen bonding, *New J. Chem.* **22**, 1315-1318.

69. (a) Aakeröy, C. B. and Beatty, A. M. (1998) Low-dimensional architectures of silver(I) coordination compounds assembled *via* amide-amide hydrogen bonds, *Crystal Engineering* **1**, 39-49; (b) Aakeröy, C.B. and Beatty, A.M. (1998) Supramolecular assembly of low-dimensional silver(I) architectures *via* amide-amide hydrogen bonds, *Chem. Commun.* 1067-1068.

70. Aakeröy, C.B., Beatty, A.M., and Leinen, D.S. (1998) The oxime functionality: A versatile synthon for supramolecular assembly of metal-containing hydrogen-bonded architectures, *J. Am. Chem. Soc.* **120**, 7383-7384.

71. Aakeröy, C.B., Beatty, A.M., and Leinen, D.S. (1999) A new route to porous solids: Organic/inorganic hybrid materials assembled *via* hydrogen bonds, *Angew. Chem. Int. Ed. Engl.* in press.

SECONDARY BONDING AS A POTENTIAL DESIGN TOOL FOR CRYSTAL ENGINEERING

A. Guy Orpen
School of Chemistry, University of Bristol, Bristol BS8 1TS, United Kingdom
Guy.Orpen@bris.ac.uk

Abstract The concept of secondary bonding in p-block structural chemistry as outlined by Alcock is introduced in the context of its application to crystal engineering. The relationship between secondary and hydrogen bonding is described and the hypothesis advanced that the analogy between these two classes of bond is a deep and close one and that therefore the tools of analysis in the study of hydrogen bonds in crystals may be applied usefully to secondary bonding. In archetypal systems based on Bi(III) and Sb(III) complexes between 70 and 90% of the possible secondary bonds are formed, indicating the robustness of this potential supramolecular synthon. The networks formed by the secondary bonds are also analogous to those formed in hydrogen bonded crystals and are likewise prone to disruption on dissolution. It seems that secondary bonds may be a valuable alternative to conventional directional interactions in the synthesis of novel molecular crystalline solids.

Introduction

The ambition of many crystallographers and chemists as we approach the new millennium is to conquer one of the great challenges of synthesis - the preparation of molecular crystals of designed structure. The aim is to combine the control that synthetic chemists have developed in the past century over the preparation of molecules with desired properties and structure with increasing knowledge of the

107

D. Braga et al. (eds.), Crystal Engineering: From Molecules and Crystals to Materials, 107–127.
© 1999 *Kluwer Academic Publishers. Printed in the Netherlands.*

108

ways in which molecules associate. If the patterns of aggregation can
be exploited so as to arrange molecules in well defined and
advantageous ways then a new era of molecular materials chemistry
beckons in which the properties of the crystalline assemblies can be put
to beneficial use. The dominance of inorganic crystalline solids
(typically based on alloys, metal oxides or similar compounds) and
poorly- or non-crystalline polymers in materials chemistry will then be
challenged by new materials with potential for more easily controlled
and more diverse functionality.

In addressing the goal of controlled crystal structure synthesis over
the past recent past a number of strategies have been adopted. These
fall in to two main camps in which either

(i)   the shape of the molecular building block is used to
        dictate the crystal structure, or,

(ii)   directional interactions between the molecular
        components are used to induce specific intermolecular
        interactions which then control the pattern of
        association leading to the crystal structure.

Thus, for example, in a simple shape directed approach a large
planar molecule might be prepared in the hope of forcing a crystal
structure in which a stacked or other low-dimensional crystal structure
results.[1] In the second strategy, which often takes its inspiration from
biology, multiple hydrogen bonds have frequently been used in order to
enforce a particular pattern of intermolecular association between the
component molecular units forming the crystal.[2]

In this paper we consider the prospects for exploiting a less familiar
class of interaction in order to implement the second strategy. The
approach we will take is that of analysis of a class of crystal structures
to identify and evaluate the interactions and their utility. As is often the
way in this type of study, anecdotal evidence, from one or two key
crystal structures is our starting point and we seek to establish the key
features of these structure, and then evaluate how general these
features may be and develop a design principle that may be applied in

practice and used in synthesis. In this study therefore we use the classical scientific method as expounded by Popper[3] - making conjectures and developing hypotheses for testing and finally establishing a theory that can be exploited.

Directional interactions in crystal synthesis

In crystal engineering, or perhaps more properly crystal synthesis, the second strategy outlined above is the one relevant to this paper.[4] The predominant class of directional interaction that has been used in the synthesis of molecular crystals and in other supramolecular chemistry is the hydrogen bond. In particular those hydrogen bonds of the sort familiar in organic and biological chemistry (NH..O; OH...O; NH...N; OH...N *etc.*) have been heavily studied[5,6] although more exotic hydrogen bonds are increasingly attracting attention.[7-9] Other interactions that have been studied in this way include the dative coordinate bond,[10] CH..π interactions;[11] dipole-dipole interactions,[12] and other interactions such as the C=O...X-C (X = I, Br, Cl) contacts studied by Allen *et al*[13] and related NO_2...I-C interactions exploited by Desiraju and co-workers,[14] a special form of the sort of secondary bond interactions we discuss below at length. Others will address the utility and generality of the other classes of interaction. In this paper we consider the potential for the use of secondary bonding in crystal engineering, a class of intermolecular interaction that is characteristic of the heavier p-block elements.

Secondary bonding

In a classic paper[15] Alcock summarised the evidence available in 1972 for the existence of intermolecular interactions of considerable strength in the molecular complexes of Pb(II), Bi(III), Te(IV), I(III), and other p-block elements. The key characteristic of these species is that the electron configuration at the p-block element E is of the sort $(nd^{10})(n+1s^2)(n+1p^x)$; n = 3, 4 or 5; x = 0-4) *i.e.* that there are one or more lone pairs at E. The result of presence of the lone pair(s) is that the coordination sphere of E is opened so as to allow the approach of ligands (*i.e.* electron pair donors), A, to form E...A contacts. The E centre is typically Lewis acidic, although the degree of Lewis acidity is strongly dependent on the substituents at E, being higher when those substituents are more electronegative. The classic form of secondary bonding occurs when a substituent at E approaches the E atom of a neighbouring and identical molecule forming a contact (in the solid state) which is shorter than the sum of the van der Waals radii of the two atoms involved. In some cases these interactions can be so strong as to be indistinguishable from the "primary" bonds within the molecule. It is worth noting that the Lewis acidity of the E centre in these species allows them to bind electron pair donor ligands, L, through dative coordinate bonds, often in the same site that would otherwise form a secondary bond.

The secondary bond can be classified as a class of intermolecular hypervalent bond in which a molecule forms a supramolecular (often polymeric) association with another identical molecule though a hypervalent interaction involving the p-block element E. The hypervalent bonding capabilities of the heavier p-block elements are exceedingly well known in classic molecular species such as XeF_2, ICl_3, or SF_4 in which the octet rule is violated and the electron count at the p-block element E rises to 10, 12 or more. The modern view of these systems is that they are best understood as involving one or more 3-centre-4-electron bonds in each of which one E p-orbital is used to make two E-X bonds (see Fig. 1 and references 16a-c).

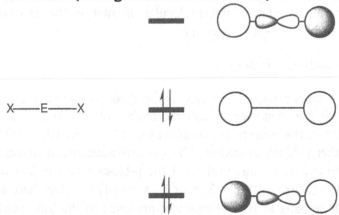

Fig. 1. 3-centre-4-electron bonding in hypervalent p-block compounds

An archetypal hypervalent bond is found in the anion [BiCl$_2$Ph$_2$]$^-$ which has a structure based on a trigonal bipyramid with one equatorial site occupied by a lone pair, a geometry consistent with Nyholm-Gillespie rules (or valence shell electron pair repulsion [VSEPR] theory). The bismuth has oxidation state +3 and its electron configuration is therefore formally 5d^{10}6s^2. In the precursor species BiCl$_2$Ph$_2$ (from which the [BiCl$_2$Ph$_2$]$^-$ anion may be prepared by reaction with Cl$^-$) the bismuth geometry is pyramidal with Cl-Bi-Ph angles not far from 90°. The structure that results from interacting an electron pair donor such as Cl$^-$ with BiClPh$_2$ therefore resembles that one would expect from addition of the Lewis base *trans* to the Bi-Cl bond of BiClPh$_2$ forming a near linear Cl-Bi-Cl moiety by interaction of a Cl- lone pair with the Bi-Cl σ* orbital of BiClPh$_2$ (see Fig. 2).

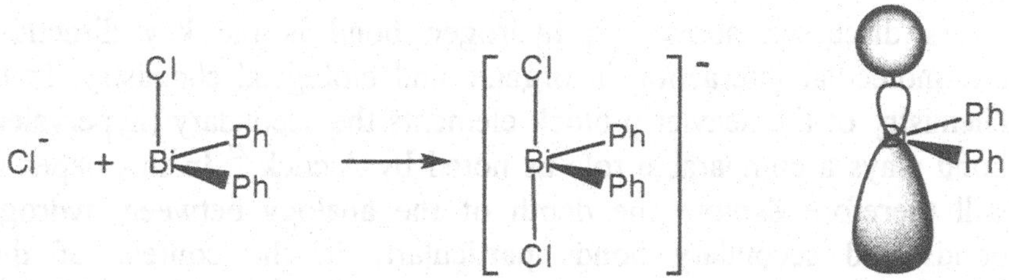

Fig. 2 Formation of [BiCl$_2$Ph$_2$]$^-$ anion from Cl$^-$ and BiClPh$_2$; the Bi-Cl σ* orbital used to form the hypervalent bond with the incoming Cl$^-$.

The bonding in this type of molecule has been carefully studied and widely discussed[16] and can be best summarised as involving the interaction of a Bi-Cl σ* orbital with the lone pair of the incoming Lewis base. The energy of this orbital is lower and it is more localised on Bi when the *trans* substituent (here Cl) is more electronegative. Hence the interaction with the incoming Lewis base is most favoured for highly electronegative *trans* substituents such as halides, alkoxides, *etc.* The [BiCl$_2$Ph$_2$]$^-$ ion provides a good example of the analogy between hypervalent (and secondary bonded) p-block element complexes and hydrogen bonded species. The analogy is here between the [HCl$_2$]$^-$ ion and [BiCl$_2$Ph$_2$]$^-$ in that the linear Cl...H...Cl unit in the former resembles the near-linear Cl-Bi-Cl moiety, with the H$^+$ Lewis acid centre replacing the [BiCl$_2$]$^+$ group.

Secondary bonding has been noted and exploited in a range of inorganic chemistry since the term was introduced by Alcock, although his work has not always been cited. Among the more notable applications of secondary bonding in materials chemistry are the work of Mitzi[17a] and Sobczyk[17b], while polyiodide anion chemistry, in which the secondary bond can be considered to form between I_2 and I^- units, has been explored in the context of crystal engineering by Schroder and co-workers (see reference 18). The influence of secondary bonding in I(III) chemistry has been noted.[19,20] Pyykkö has included secondary bonding in a remarkable and authoritative review of a wide range of closed shell interactions in inorganic chemistry. [21]

Hydrogen bond characteristics

As discussed above the hydrogen bond is the key directional intermolecular interaction in organic and biological chemistry. In the chemistry of the heavier p-block elements the secondary (hypervalent) bond plays a comparable role as noted by Alcock.[22] In this paper we will therefore explore the depth of the analogy between hydrogen bonds and secondary bonds particularly in the context of their application to the control of intermolecular interactions in molecular crystals. The key features of the geometry of hydrogen bonds are well known from some important structures (such as ice, α-helical proteins, DNA etc.) and have been the subject of numerous studies based on the crystallographic literature, often using the accumulated data in the Cambridge Structural Database (CSD).[23] In this respect the CSD plays a unique role in which its both the archive of all molecular crystal structures and the primary source of experimental information on the geometry of intermolecular interactions found in such structures.

The main characteristics of D-H....A hydrogen bonds[24] (D = hydrogen bond donor atom = O, N, halogen, S etc.; A = hydrogen bond acceptor = halide, O, N, S etc.) of relevance to the discussion in this paper are listed below:

(i)   The D-H...A moiety is linear or nearly so.

(ii)   More electronegative D atoms lead to stronger interactions (other things being equal).

(iii)   The DH...A interaction is largely electrostatic in nature, but may have significant covalent character especially when the hydrogen bond is especially short.

(iv)   The H...A distance in "conventional" hydrogen bonds is shorter than the sum of the H and A van der Waals radii.

(v)   D-H...A distances (D-H and H...A) are rather variable and coupled in a hyperbolic manner.

(vi)   Hydrogen bonds form networks in the solid state that may be described using graph theory.[25]

(vii)   These networks of interactions are usually disrupted on dissolution in solvents have hydrogen bonding capability (such as water!).

(viii) Hydrogen bond functionalities have been found to be robust supramolecular synthons.[26]

(ix)   The utility of the hydrogen bond in crystal engineering is well documented and increasingly widely exploited.

The hydrogen bond is in large measure an electrostatic bond in which the major energetic stabilisation arises because of the partial positive charge on the acidic hydrogen is brought into proximity with the partial negative charge of the acceptor atom. However a useful view of the hydrogen bond is shown in Fig 3, in which the covalent character of the hydrogen bond is emphasised in treating it as a 3-centre 4-electron interaction, stabilised by the interaction of the D-H σ^* orbital with the lone pair of the A atom. This view clearly has similarities with that advanced above to describe the hypervalent bonds formed by the heavy p-block elements. While both models lead to the prediction of linearity being favoured for the three centre system, there are significant differences of detail - since for example there is no p orbital contribution to the central atom in the hydrogen bond whereas it is the E atom p-orbital which is the key orbital in forming the hypervalent interaction. For further discussion of the relationships between theoretical treatments of hydrogen and secondary bonding the reader is referred to reference 16a.

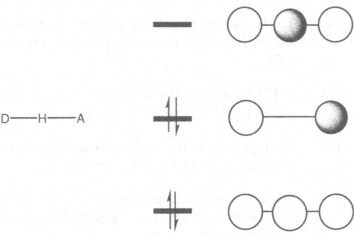

Fig. 3 3-centre-4-electron bonding in hydrogen bonds

The robustness of a supramolecular synthon[4] such as a hydrogen bond can be quantitatively evaluated using the CSD. For example Allen *et al* took a series of popular and well known cyclic bimolecular hydrogen bond motifs and counted the number of occasions on which the expected bond pattern was formed compared with the number of occasions on which the (two) required components were present in crystal structures.[26] The percentages produced were interpreted as a measure of the reliability or robustness of the synthon or motif. This interpretation seems both plausible and intuitively appealing, but is clearly not a rigorous treatment of the problem. However given a reasonably diverse and numerous data set this approach surely tells us much about the usefulness of the synthon under study. Notably the classic carboxylic acid dimer (Fig. 4) which is perhaps one of the most familiar supramolecular synthons is not as robust as one might expect and is formed on only 33% of all possible occasions.

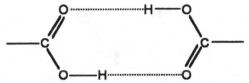

Fig. 4. The carboxylic acid dimer motif

Anecdotes.

In studies of bismuth halide compounds we observed a series of structures in which the analogy between hydrogen bonding and secondary bonding was marked.[27-29] Thus in the structures of $[BiCl\{Mo(CO)_3(\eta\text{-}C_5H_5)\}_2]$[30] and $[BiCl\{Fe(CO)_2(\eta\text{-}C_5H_5)\}_2]$[31] polymeric and ring structures were observed in which secondary bonds with near linear Cl-Bi...Cl moieties formed the links between the molecular units.

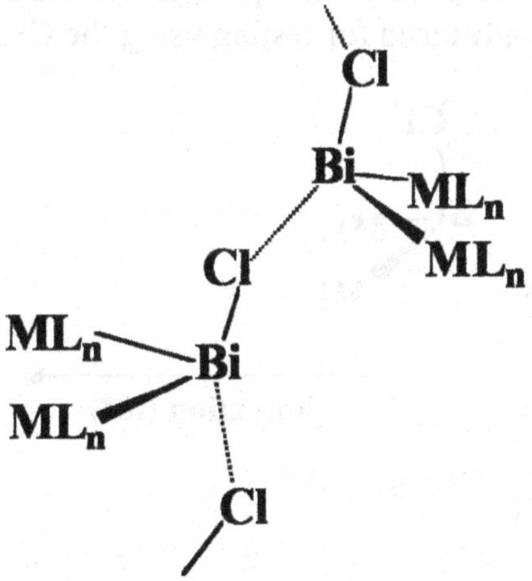

Fig. 5. Polymeric structure of $[BiCl\{Mo(CO)_3(\eta\text{-}C_5H_5)\}_2]$[30] $\{ML_n = \{Mo(CO)_3(\eta\text{-}C_5H_5)\}\}$

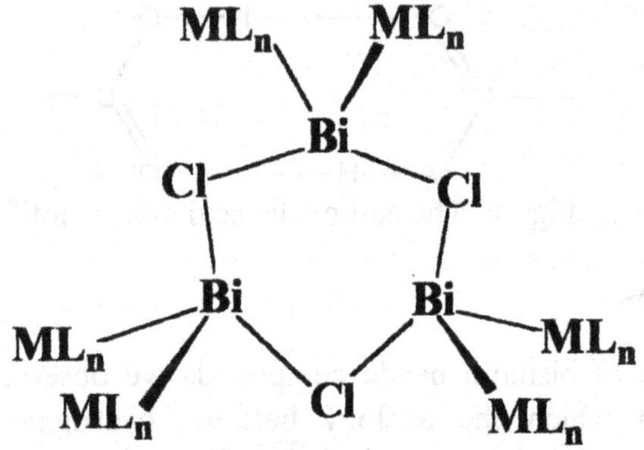

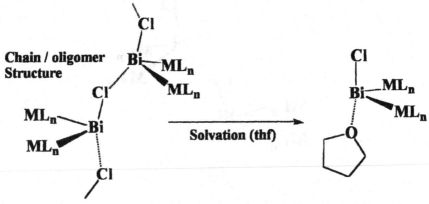

Fig. 6. Cyclic trimer structure of $[BiCl\{Fe(CO)_2(\eta\text{-}C_5H_5)\}_2]$[31]
$\{ML_n = \{Fe(CO)_2(\eta\text{-}C_5H_5)\}$

These networks were shown by EXAFS spectroscopy to be disrupted by dissolution in donor solvents such as thf.[27] Given these observations and the picture of hypervalent bonding outlined above a hypothesis can be advanced for testing using the CSD.

Fig. 7. Schematic illustration of the effects of dissolution in thf on secondary bonds in bismuth(III) chloride species studied by EXAFS spectroscopy.[29]

Hypothesis

The analogy between hydrogen bonding and secondary bonding in crystal structures implies that the tools of analysis developed for the study of hydrogen bonding can be used to explore secondary bonding. For the purposes of this paper, we will restrict our investigation to archetypal secondary bonding bismuth(III) and antimony(III) compounds of the form EX_xR_{3-x} and their complexes with ligands L, $EX_xR_{3-x}L_y$. In these complexes the E atom carries electronegative, formally uninegative, substituents X or relatively electropositive, also formally uninegative, substituents R (*e.g.* X = halide, OR, NR_2, SR *etc.*; R = alkyl, aryl, metal *etc.*) and electron pair donor ligands, typically with electronegative atoms (L = X^-; neutral or anionic O- N- or S- donor ligands). Therefore the coordination number at E, n = 3+y. The number of strong Lewis acid sites, N_s = x - y, which sites lie *trans* to the electronegative substituents X. This assumes octahedral coordination at E, which assumption will fail from time to time but holds widely, and that y ^{2} x, as is usually the case. There should also be N_w = 3-x weak Lewis acid sites at E *trans* to the substituents R.

The hypothesis is therefore that secondary bonds formed between E and a secondary bond acceptor A have the following characteristics analogous to those noted above for hydrogen bonds:

(i)   The secondary bonds are linear or nearly so.

(ii) More electronegative X substituents lead to stronger interactions X-E...A (other things being equal).

(iii) The X-E...A interaction is largely covalent in nature.

(iv) The E...A distance in secondary bonds is shorter than the sum of the Bi and A van der Waals radii.

(v)   The X-E...A distances (X-E and E...A) are rather variable and coupled in a hyperbolic manner.

(vi)   Secondary bonds form networks in the solid state that may be described using graph theory.[25]

(vii)   These networks of interactions are usually disrupted on dissolution in solvents with ligand capability (such as thf).

(viii) The secondary bond is potentially a robust supramolecular synthon

(ix)   The utility of the secondary bond in crystal engineering is yet to be widely exploited.

Testing the hypothesis

To test aspects of this hypothesis we used the CSD and selected from it those structures that contain appropriate bismuth(III) and antimony(III) complexes in order to establish the numbers and geometries of secondary bonds that they form.[32]

The comparable linearity of secondary bonds and hydrogen bonds may be shown by histograms (see Figs. 8 and 9) of the angle at the central atoms in two archetype systems: Cl-Bi...Cl secondary bonds and O-H...O hydrogen bonds (as determined in neutron diffraction studies).[33] In both systems most of the structures have angles greater than 160° and only a few with angles down to 130°. The origin of the preference for linearity in these 3-centre-4-electron systems may differ in detail (as described above) but the outcome is broadly similar.

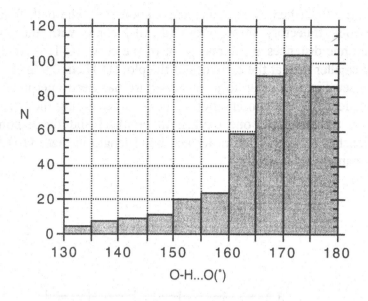

Fig. 8 Histogram of angles in O-HO systems in the CSD

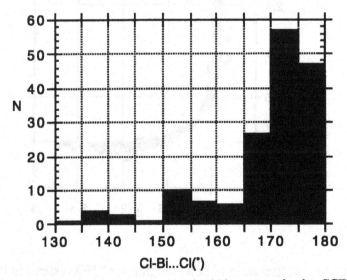

Fig. 9 Histogram of angles in Cl-BiCl systems in the CSD

The coupling of the two O-H distances in O-H...O hydrogen bonds has been well documented and is a classic example of a reaction

120

pathway established by structure correlation.[34,35] The path follows a hyperbolic trajectory through the (d_1, d_2) space with the sum of interatomic distances at the mid point of the curve (O-H = *ca*. 1.2 Å) being smaller than at the extremes of the plot (O-H *ca*. 0.9Å, H...O *ca*. 2.5Å, see Fig. 10). This form of curvature has been interpreted as a consequence of the conservation of bond order during the "reaction" (proton transfer from one oxygen to another) and the exponential relationship between bond order and bond length in these O-H bonds (and many other systems).

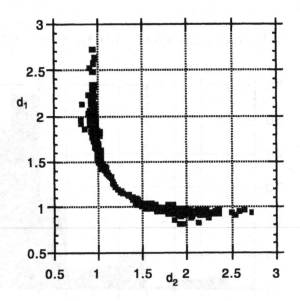

Fig. 10   Scattergram of bond lengths in O-HO systems in the CSD

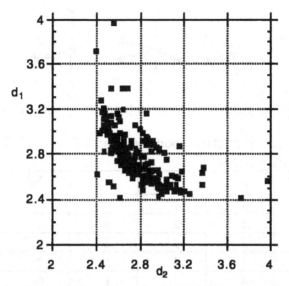

Fig. 11 Scattergram of bond lengths in Cl-BiCl systems in the CSD

The corresponding plot (Fig. 11) for the Cl-Bi...Cl system shows similarity with many points lying close to a hyperbolic curve (or curves). However there is considerably more scatter in the *trans*-BiCl$_2$ dataset with many points lying at some distance from the most obvious hyperbolic path. This difference may be the consequence of much greater variation in the chemical environment of the Bi in these species than is possible for the central hydrogen atom in the O-H...O system. Thus the presence and variation in cis substituents at Bi in the trans Cl-B...Cl moiety surely has a strong influence on the Bi-Cl bond lengths - as we have argued in some specific cases.[29] Despite this difference the general similarity in form of the scatterplots and the presence of considerable variation in the lengths of the secondary/hydrogen bonds is notable.

A set of 5 different systems based on bismuth(III) and antimony(III) compounds was extracted from the CSD, [ER$_2$X, ERX$_2$, EX$_3$L$_2$, EX$_3$L, ER$_2$XL, see Table 1 and Fig. 12]. These cover a range of situations in which there are variations in the number of strong and weak Lewis acid sites and therefore sites at which secondary bonds may be formed.

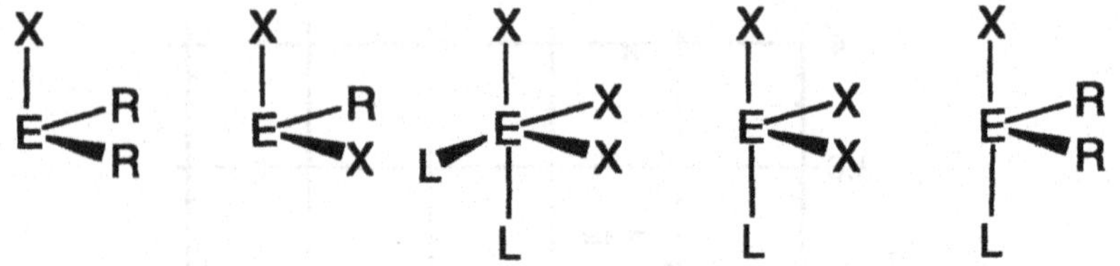

Fig. 12 Bismuth(III) and antimony(III) moieties studied.

Table 1. Secondary bonding capabilities of bismuth and antimony(III) moieties.

Moiety [a]	n [b]	x	y	N_s [c]	N_w
ER_2X	3	1	0	1	2
ERX_2	3	2	0	2	1
EX_3L_2	5	3	2	1	0
EX_3L	4	3	1	2	0
ER_2XL	4	1	1	0	2

[a] moiety formula: E = Sb, Bi; X = halide, alkoxide etc.; L = 2e donor ligand, including cases where L = X⁻; R = alkyl, aryl etc.

[b] n = coordination number of E; x = number of substituents X; y = number of ligands L

[c] number of strong and weak secondary bonding sites at E = N_s, N_w

The frequency with which secondary bonds are formed by these various moieties is given in Table 2. It is striking that the strong Lewis acid sites (*trans* to electronegative substituents) are consistently involved in secondary interactions at a frequency of between 70-90%. This contrasts with frequencies of < 20% for the weaker secondary bonding sites. The marked difference is therefore in accord with the postulate that the electronegativity of the trans substituent is a major determinant in the likelihood and strength of secondary bond formation. The high frequency of secondary bond formation in the "strong" Lewis acid sites implies that such sites have considerable potential as robust supramolecular synthons - they can be relied upon to form secondary bonds.

In addition the graph sets[25] describing the network of secondary bonds formed in each structure were determined and are summarised in outline in Table 2.

Table 2. Secondary bonding interactions in bismuth and antimony(III) structures.

Moiety	n(CSD)[a]	n(frag)[a]	Ns	Fs[c]	Nw	Fw[c]	Network family[d]
ER₂X	18	24	1	71	2	8	44% C(2); 11% R3,3(6)
ERX₂	20	21	2	81	1	19	50% C(2)
EX₃L₂	68	76	1	86	0	-	38% C(2); 38% R2,2(4)
EX₃L	61	89	2	85	0	-	39% C(2); 18% R2,2(4)
ER₂XL	29	30	0	-	2	17	

[a] number of unique crystal structures located from the CSD

[b] number of unique moieties located from the CSD

[c] F = percentage of possible secondary bonds actually formed

[d] family to which graph set[25] of secondary bond network belongs.

The predominant mode of association in these species is that in which chains of molecules are formed - albeit by a variety of networks in which considerable complexity may be present. Rings are also formed in a substantial number of cases and there is some indication that the steric bulk of the substituents at E plays a key role in favouring chain formation (in cases when the substituents are large) or rings (with smaller substituents) or no extended secondary bond network (for very largest substituents). Note that here we have named the secondary bond electron pair donors "acceptors" (A) and the E atoms as "donors" by analogy with the jargon of the hydrogen bonding literature in which the Lewis acid centre (H) is the donor of the hydrogen bond and the Lewis base (Cl⁻ or whatever) is called the "acceptor".

It is noteworthy that in a class of organic crystal with exactly one hydrogen bond donor and one hydrogen bond acceptor per molecule (the monoalcohols) Brock and Duncan have shown[36] that unusual space groups are often encountered and that in many cases C(2) and Rn,n(2n) (n = 4, 6) networks are formed. It is striking that the same network types occur in the category of Bi(III) species with exactly one Lewis acid site and one available secondary bond acceptor (the ER$_2$X set).

Conclusions

From the CSD study outlined above it is clear that the analogy between hydrogen bonds and secondary bonding is a deep one. The prospects for its exploitation in crystal engineering are good and indeed in some respects the secondary bond may have significant advantages over the hydrogen bond. Thus there is every reason to expect that the availability of multiple Lewis acid sites at a single E atom will allow for bi- or tri-furcation of the secondary bond network in a way that is hard to engineer in hydrogen bond networks. In addition the availability of readily chemically modified sites *cis* to the secondary bond open the prospect of incorporating novel functionality into a secondary bond network in a way not possible for hydrogen bonds. The metric of the secondary bond is rather different form that of most hydrogen bonds - since the Bi-Cl bond for example is often about twice as long as an O-H hydrogen bond. Set against these observations is the fact that the enormous diversity and maturity of organic chemistry offers great practical advantages in exploiting the hydrogen bond in crystal engineering. In contrast heavy p-block chemistry of the sort required to engage in crystal synthesis is relatively under-developed.

Prospects

This paper reports a study in which an area of chemistry of potential application to crystal engineering is evaluated using CSD methods. This type of study has been very influential in the development of crystal engineering as a discipline but this is only the start of the process. Having established that secondary bonds are of potential utility is still to be proven that secondary bond networks and crystal structures of utility can actually be synthesised using this approach. Furthermore the analysis reported here ignores critical components of the crystal synthesis process - for example no explicit mention is made of the thermodynamics of the formation of networks or crystals or of the kinetics of their formation. Some evaluation of these aspects might be possible using computational approaches - for example to evaluate the energetics of secondary bonds for comparison with those of hydrogen bonds.

Finally, having identified some potential supramolecular synthons for exploitation in crystal synthesis it remains to develop useful reagents and preparative methods (or better still ranges of them) for synthesis and growth of crystals in this area. These challenges are being addressed in Bristol and elsewhere in the hope and expectation that systems based on secondary bonding will provide a useful new option in the preparation of molecular crystalline materials with useful properties.

Acknowledgements

The financial support of the UK EPSRC is gratefully acknowledged. My thanks also go to Drs N.C. Norman and F.H. Allen and Mr J. Starbuck for their vital contributions to this work. Finally, none of this work would be possible without the contributions of the many chemists and crystallographers who have synthesised and determined the structures of the crystal structures discussed here or of the staff of the CCDC who have made the CSD.

126

References
1. Desiraju, G. R. (1997) *Chem. Commun.*, 1475-1482.
2 MacDonald, J. C. and Whitesides, G. M. (1995) *Acc. Chem. Res*, **28**, 37-44.
3 Popper, K.R. (1963) *Conjectures and Refutations*, Routledge and Kegan Paul Ltd, London.
4 Desiraju, G. R. (1995) *Angew Chem. Int. Ed. Engl.*, **34**, 2311-2327.
5 Jeffery, G. A. and Saenger, W. (1991) *Hydrogen Bonding in Biology and Chemistry*. Springer Verlag, Berlin.
6 Aakeröy, C. B. and Seddon, K. R. (1993) *Chem. Soc. Rev.*, **22**, 397-407; Aakeröy, C.B. (1997) *Acta Cryst.* **B53**, 569-586.
7 Braga, D., Grepioni, F. and Desiraju, G. R. (1998) *Chem. Rev.*, **98**, 1375-1405.
8 Aullón, G., Bellamy, D., Brammer, L., Bruton, E. A., and Orpen, A. G. (1998) *Chem. Commun.*, 653-654.
9 Lewis, G.R. and Orpen, A. G. (1998) *Chem. Commun.*, 1873-1874.
10 Robson, R., Abrahams, B.F., Batten, S.R., Gable, R.W., Hoskins, B.F., and Liu, J. (1992) *Supramolecular Architectures*, 256-273, Am. Chem. Soc. Publication, Washington DC; Batten, S.R., Robson, R. (1998) *Angew. Chem., Int. Ed. Engl.*, **37**, 1461-1494; Zaworotko, M.J. (1994) *Chem. Soc. Rev.*, **23**, 283-288.
11 Dance, I. G. (1996) in *The Crystal as a Supramolecular Entity*, ed. Desiraju, G. R. *Perspectives in Supramolecular Chemistry*, **2**, Wiley, Chichester.
12 Allen F.H., Baalham, C.A., Lommerse, J-P.M. and Raithby, P.R. (1998) *Acta Cryst* **B54**, 320-329.
13 Lommerse, J-P.M., Stone, A.J., Taylor, R., and Allen F.H. (1996) *J. Amer. Chem. Soc.*, **118**, 3108-3116.
14 Allen, F.H., Goud, B.S., Hoy, V.J., Howard, J.A.K., Desiraju, G.R. (1994) *Chem. Commun.*, 2729-2730; Thalladi, V.R., Hoy, V.J., Allen, F.H., Howard, J.A.K., Desiraju, G.R. (1996) *Chem. Commun.*, 401-402.
15 Alcock, N.W. (1972) *Adv Inorg. Chem. Radiochem.*, **15**, 1-58.
16 (a) Landrum, G.A., Goldberg, N, and Hoffmann, R. (1997) *J. Chem. Soc., Dalton Trans*, 3605-3613. (b) Landrum, G.A. and Hoffmann, R. (1998) *Angew. Chem., Int. Ed. Engl.*, **37**, 1887-1890. (c) Landrum, G.A., Goldberg, N, Hoffmann, R. and Minyaev, R.M. (1998) *New J. Chem.*, **22**, 883-890
17 (a) Mitzi, D.B. (1999) *Prog. Inorg. Chem.*, **48**, 1-123; (b) Sobczyk, L., Jakubas, R., Zaleski, J. (1997) *Pol. J. Chem.*, **71**, 265-300.
18 Blake, A.J., Gould, R.O., Li, W.S., Lippolis, V., Parsons, S., Radek, C., and Schroder, M. (1998) *Inorg. Chem.*, **37**, 5070-5077.
19 Boucher, M., Macikenas, D., Ren, T. and Protasiewicz, J.D. (1997) *J. Amer. Chem. Soc.*, **119**, 9366-9376
20 Carmalt, C.J., Crossley, J.G., Knight, J.G., Lightfoot, P., Martin, A., Muldowney, M.P., Norman, N.C. and Orpen, A.G. (1994) *J. Chem. Soc., Chem. Commun.*, 2367-2368.
21 Pyykkö, P. (1997) *Chem. Rev.* **97**, 597-636
22 Alcock, N.W. (1990) *Bonding and Structure*. Ellis Horwood, Chichester.
23 (a) Allen, F. H., Davies, J. E., Galloy, J. J., Johnson, O., Kennard, O., Macrae, C. F. and Watson, D. G. (1991) *J. Chem. Inf. Comput. Sci.* 31, 187-204. (b) Allen, F. H. and Kennard, O. (1993) *Chem. Des. Automation News*, **8**, 1 & 31-37.
24 Jeffery, G. A. (1997) *Introduction to Hydrogen Bonding*, Wiley, Chichester.

25 Etter, M. C. (1990) *Acc. Chem. Res.*, **23**, 120; Etter, M.C., MacDonald, J.C., and Bernstein, J. (1990) *Acta Cryst* **B46**, 256-262, Etter, M. C. (1991) *J. Phys. Chem.*, **95**, 4601-4610; Bernstein, J., Davis, R. E., Shimoni, L. and Chang, N.-L. (1995) *Angew. Chem., Int. Ed. Engl.*, **34**, 1555-1573.

26 Allen, F. H., Raithby, P. R., Shields, G. P. and Taylor, R. (1998) *Chem. Commun.*, 1043-1044; Allen, F. H., Motherwell, W.D.S., Raithby, P. R., Shields, G. P. and Taylor, R. *New J. Chem.*, 1999, **23**, 25.

27 Clegg, W., Compton, N.A., Errington, R.J., Fisher, G.A., Hockless, D.C.R., Norman, N.C., Williams, N.A.L., Stratford, S.E., Nicholls, S.J., Jarrett, P.S. and Orpen, A.G. (1992) *J. Chem. Soc., Dalton Trans.*, 193-201.

28 (a) Compton, N.A., Errington, R.J., Fisher, G.A., Norman, N.C., Webster, P.M., Jarrett, P.S., Nicholls, S.J., Orpen, A.G., Stratford, S.E. and Williams, N.A.L. (1991) *J. Chem. Soc., Dalton Trans.*, 669-676. (b) Clegg, W., Errington, R.J., Fisher, G.A., Hockless, D.C.R., Norman, N.C., Orpen, A.G and Stratford, S.E. (1992) *J. Chem. Soc., Dalton Trans.*, 1967-1974. (c) Errington, R.J., Fisher, G.A., Norman, N.C., Orpen, A.G and Stratford, S.E. (1994) *Z. Anorg. All. Chem.*, **620**, 457-466.

29 Clegg, W., Compton, N.A., Errington, R.J., Fisher, G.A., Hockless, D.C.R., Norman, N.C., Orpen, A.G and Stratford, S.E. (1992) *J. Chem. Soc., Dalton Trans.*, 3515-3523.

30 Clegg, W., Compton, N.A., Errington, R.J., Norman, N.C., Tucker, A.J. and Winter, M.J. (1988) *J. Chem. Soc., Dalton Trans.*, 2941-2951.

31 Clegg, W., Compton, N.A., Errington, R.J., Norman, N.C. (1988) *J. Chem. Soc., Dalton Trans.*, 1671-1678.

32 Orpen, A.G., Starbuck, J. and Norman, N.C. (1999) unpublished results.

33 Data for relevant structures were retrieved from the CSD[23] and bond lengths (Å) and angles (°) calculated. The data sets were permuted[34] to reflect the symmetry of the potential energy hypersurface for the reaction path in question.

34 *Structure Correlation*. Eds Bürgi, H.-B., Dunitz, J. (1994) VCH, Weinheim.

35 Olofsson, I, Jönsson, P.G. in Schuster, P., Zundel, G., Sandorfy, C., Eds., (1976) *The Hydrogen Bond*, Vol. II, Ch. 8, North-Holland Amsterdam.

36 Brock, C. P. and Duncan, L. L. (1994) *Chem. Mater.*, **6**, 1307-1312.

MOLECULAR PACKING IN LIQUIDS, SOLUTIONS, AND CRYSTALS: ACETIC ACID AS A TEST CASE

A.Gavezzotti
Dipartimento di Chimica Strutturale e Stereochimica Inorganica, University of Milano, Milano, Italy

1. Crystal engineering and crystal structure prediction

Several papers in recent times [1-4] have discussed the general difficulties encountered in the completely *ab initio* prediction of full crystal structures by molecular mechanics calculations. There is almost general consensus that this is possible only for rigid molecules, and only for crystals with one single molecular entity in the asymmetric unit. Even for this severely restricted subset of organic molecular crystalline systems, success is measured by different degrees of ambition and is sometimes more serendipitous than robust and reproducible. Current methods cannot in fact go beyond the generation of a number of plausible crystal structures, always very similar in packing energy and density, among which usually, but not always, the observed crystal structure is found. The essential difficulty is just that close-packing is necessary but not sufficient, while the discriminating power of current force fields is still low. Prediction of the structure when cell parameters and space group are known – i.e., solution of the phase problem in X-ray crystallography – is more within reach of computer simulations, but, alas, the task can be much more comfortably and quickly carried out by direct methods.

In another line of thought [5], crystal structures are considered as determined by some key, structure-defining atom-atom or site-site interactions, while the cooperative nature of the structure-defining chemical potential of the crystalline phase is neglected. Success in prediction is here judged by much less ambitious criteria, like the

D. Braga et al. (eds.), Crystal Engineering: From Molecules and Crystals to Materials, 129–142.
© 1999 *Kluwer Academic Publishers. Printed in the Netherlands.*

identification of some recurring motifs or persistent intermolecular linkages. The main problems are connected with the consistent and objective identification of the postulated synthons.

In our opinion, these two approaches i) aim at different goals, crystal engineering (CE) being other than crystal structure prediction (CSP); ii) use different methods, CE relying on geometry, CSP on energy; iii) view success or failure in a different context and by different criteria. When the same concept is used, it may mean different things; for example, C-H...O interactions may be relevant or even crucial to CE, while they are worthless to CSP, given the inherent theoretical difficulties in using as attractors intermolecular interactions whose energetic contribution is within the kT range [6], or short intermolecular distances, which are by definition repulsive [7]. One should not be tempted, therefore, to mix these two languages or to compare performances.

This paper is an essay for the exploration of the methods, merits and pitfalls of the molecular mechanics approach to the study and prediction of intermolecular packing. Methods for the construction of crystal structures and for the comparison of their energies will be reviewed in tutorial form. At the present stage, their exploitation and improvement is to some extent challenging at least in a technical sense, but the conceptual challenge is elsewhere. The liquid state is introduced here as a frontier topic, since the roots of molecular recognition and crystal building lay there, but even present-day outstanding computer resources are dwarfed by the sheer size of the problem. The hope is that some skillfully designed computer experiments may point the way and provide conceptual clues, while waiting for either a computational breakthrough in present methods, or for a further, dramatic advance in computing resources which could allow a satisfactory sweep of phase space. Acetic acid is an easy choice as a test system, given its relatively high melting point for the small size of the molecule, and the fact that its intermolecular force field is reasonably well parameterized.

2. Human-aided computer search for structures

If one were to guess at the crystal structure of acetic acid from just a knowledge of the molecular shape and of what elementary chemistry may pertain to a beginner, the obvious choice would be the formation of a hydrogen-bonded cyclic dimer, followed by the search for a way to condense and close-pack together such dimers. The objects in question are the acetic acid molecule, or a central carbon atom surrounded by three substituents, the methyl group, the =O oxygen, and the OH group: and the dimer, formed over a center of symmetry, with its double O-H...O hydrogen bond and an overall planar, disklike shape with two minor spikes at the methyl groups.

In a first approximation, then, what is left to do is to pile up these disks, perhaps allowing for a minor offset to prevent the spikes from bumping into each other. Pure translation does the job nicely, so one would say P1 space group, except that pure translation of a centrosymmetric dimer leads to space group P1-, Z=2, instead. The computer program PROMET [8] can be very easily used to perform an actual crystal structure construction along the above lines. The structure of the dimer is first obtained by moderate optimization of its cohesive energy (11 kcal/mole), and then, using molecular size as a guideline, cell translation vectors are guessed at, raw crystal structures are generated, with a moderately cohesive lattice energy, and final optimization of cell parameters and molecular position parameters [9] leads to the P1- crystal structure described in Table 1, within a few hours, human time, or a few minutes, computing time.

This result is by no means an absolute prediction of crystal structure. One might guess, for example, that the dimer would use a screw axis to repeat itself into space, rather than just pure translation; this would lead to space group $P2_1/c$, Z=4. Indeed, PROMET can be instructed to produce cohesive clusters formed by applying a screw axis to the centrosymmetric dimer, yielding in fact double ribbons, or layers. Repetition of one of these layers, which are precursors of the $P2_1/c$ space group, by pure translation along an arbitrary direction (hence the monoclinic angle and an offset between layers) yields, after a brief search, the $P2_1/c$ crystal structure also shown in Table 1. The next attempt might be to apply to the above layers, instead of translation, a second screw axis perpendicular to the first, moving then to orthorhombic and yielding a three-dimensional crystal structure in space group Pbca. PROMET duly produces such a crystal structure, seen again in Table 1.

132

Table 1. Crystal structures for acetic acid as generated by amateur guessing followed by moderate computer help (Å, degrees and kJ/mole units).

Space group	a	b	c	cell angles	V_{cell}/Z	-E
P1-	4.17	4.62	10.25	113 123 96	140	62.5
P2$_1$/c	7.70	4.64	8.04	78	140	62.1
Pbca	16.50	4.68	7.29		141	60.8

The human-computer collaboration might continue and more and more elaborate crystal structures could be generated and tested, if desired. A large part of the task could be and has been automatized in the new user-friendly version of the computer program, called ZIP-Promet [10]. Other methods and computer programs could be used (see ref. [11] for a list and a review). The chemical essence of the answer would not change:
a) the lattice energies of crystal structures generated by different procedures and in different space groups differ by marginally significant amounts, and also their densities are quite similar;
b) once the main requests are satisfied, i.e. formation of the hydrogen bond and avoidance of confrontation between spikes, a great many packings are compatible with crystal compactness and high cohesive energy;
c) different force fields would give slightly different answers in what is the essence of the CSP exercise, that is, the ranking of generated crystal structures in descending order of lattice energy, assuming that the most cohesive energy corresponds to the real structure;
d) packing diagrams (Figure 1), which are so complicated even for a simple molecule such as acetic acid, do not lend themselves to an easy classification of packing motifs (some of which are even mutable according to the projection direction);

133

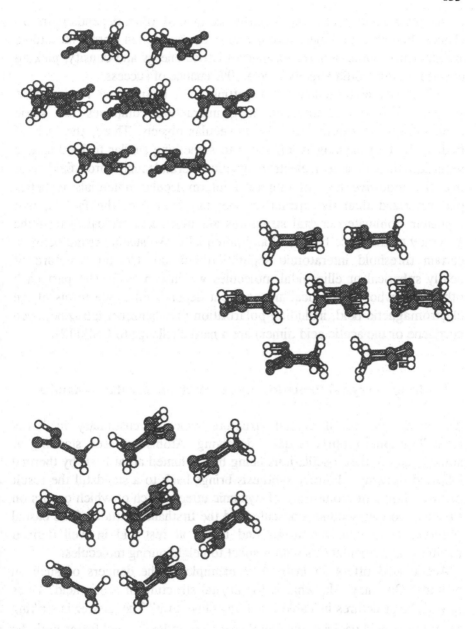

Figure 1. Arbitrary projections of the three crystal structures described in Table 1. Top to bottom: P1-, P2₁/c, Pbca.

e) the determination of a crystal structure by CSP often depends more on chance than on sound methodology, since, once say ten crystal structures are identified within a narrow range of lattice energy and density, picking one at random would already give a 10% chance of success.

CSP stands better chances of getting to a unique and reproducible answer when the structure-defining interaction comprises the steric compatibility of grossly irregular molecular objects. There, the task of finding the best packing is left more to something similar to hard-sphere repulsion than to a real electromagnetic intermolecular force field. Not that the repulsive part of empirical intermolecular potentials is better parameterized than the attractive one; far from that; the fact is, that repulsion-dominated crystal structures are much less critical against the accuracy of the force field, and any potential curve steeply rising below a certain threshold interatomic separation will do. Crystal structure of nearly spherical or ellipsoidal molecules which can easily slip past each other, and whose reciprocal arrangement depends on subtle terms of the electromagnetic field, including polarization (say benzene, tetrazine, even coronene or the acetic acid dimer) are a hard challenge to CSP [12].

3. Looking at crystal structures more closely: molecular dynamics

The static picture of crystal structure which is customary in X-ray crystallographic reports is quite deceiving. Atoms are seen standing at some location, their oscillations being barely hinted at, at best, by thermal ellipsoid drawings. Fourier synthesis brings back to a standstill the result of many hours of monitoring of dynamic effects, each of which occurs on the pico- to nanosecond timescale, and the firsthand attitude of the casual observer is to consider atoms and bonds at rest and in well defined positions and orientations with respect to neighbouring molecules.

Acetic acid offers an instructive example of the dangers of such an attitude. One may ask, what is the crystal structure of acetic acid? Does any of the structures in Table 1 get any close to it? The answer is striking [13]: acetic acid packs in the chiral space group $Pna2_1$, and forms periodic hydrogen-bonded chains (catemers) instead of cyclic dimers. The standard picture of the main structural determinant in that structure is seen in Figure 2. Also standard is the rationalization of this structure in terms of

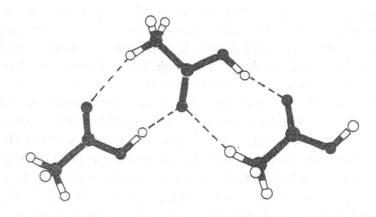

Figure 2. The main determinant in the real crystal structure of acetic acid.

a O-H...O strong hydrogen bond plus a weak C-H...O hydrogen bond.

No need to say, the lattice energy and density calculated for the real crystal structure of acetic acid are close to the ones shown in Table 1 within a few percent unit. The reader interested in detail is addressed to a very careful CSP study of acetic acid [3]. We address here a different question: is there a way of describing by a theoretical model and a computer the actual dynamics of the crystal? The answer is yes, and the method is of course molecular dynamics (MD).

A computational box is set up by repeating the molecules in space according to crystal symmetry, like in normal packing energy calculations, but then the system is given the thermal kinetic energy pertaining to the chosen temperature, and atoms are made free to move according to classical dynamics under the action of the intermolecular potential. The system is made periodic by imposing chain-like boundary conditions (any atom drifting out of the pristine box in one direction re-enters from the opposite direction), and, through the calculation of the virial, also pressure can be calculated. The box volume can then be equilibrated against pressure, and the whole machinery results in an evaluation of the cohesive energy and density of the system as a function of temperature, plus a representation of the actual trajectory of each atom in time.

Running a MD calculation on a crystal is an experience that any crystallographer or solid-state chemist should make. The movie-like display obtained by plotting atomic trajectories is most instructive: at room temperature (indeed, for molecular systems, a very hugh temperature) atoms and molecules are seen to undergo strong thermal agitation, and one better realizes how diffuse and dynamic the pattern of intermolecular interactions is. To be more specific, a MD simulation was run [6] on the acetic acid crystal, and the distribution of the H-C-C=O torsion angle was plotted against the intramolecular potential (Figure 3). To be sure, the intermolecular barrier to rotation is zero, so that the C-H...O "hydrogen bond" exists only as, presumably, a small excess residence in the colinear configuration. A reconsideration of the original crystallographic paper [13] reveals that during the structure refinement the high oscillation amplitude of the methyl group had become apparent. How the C(methyl)-H...O interaction should be reinterpreted in the light of these findings is left to the reader's judgement.

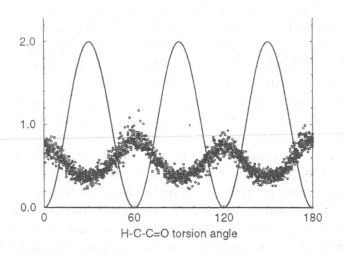

Figure 3. MD run on the acetic acid crystal: the distribution of intramolecular H-C-C=O angles (circles) against the intramolecular torsional potential (kJ/mole, dotted line). After ref. [6].

The MD trajectories contain full information on atomic libration, and therefore a normal mode decomposition of this motion (at low temperatures) would recover the usual description of lattice vibrations. Aside from technicalities, which may not be trivial, this could represent a valid alternative to harmonic lattice-dynamical treatments. But MD has the further advantage of not being constrained to a harmonic environment; in fact, appropriate calculations may simulate phase transitions.

For acetic acid, MD simulations of melting have been carried out [6]; the detailed atomic and molecular motions in the premelting stage have been monitored through the MD trajectories, and this reveals that the first large-amplitude motions to occur within the crystal upon increasing kinetic energy are molecular rotations around the central carbon atom, with breaking of the original hydrogen bond and possible cross-linking by formation of new hydrogen bonds among the catemer chains (Figure 4). Thus, the stronger intermolecular linkage is broken first, while dispersion-repulsion forces still keep the bulk structure intact, although with an increasing amount of orientational defects. Eventually, upon further increasing the kinetic energy, the ensuing lattice swelling reduces the rotational and translational barriers to almost zero, and the crystal collapses to the liquid. The MD simulation also reveals that vacancies in the lattice greatly enhance the rate at which this whole process occurs, so that, if the melting temperature is so reproducible as it is in fact, there must be an *equilibrium* distribution of vacancies in the premelting regime for each organic crystal.

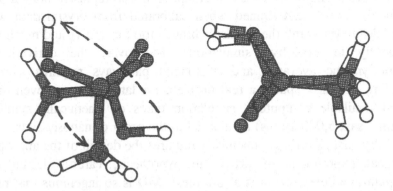

Figure 4. A premelting defect in the acetic acid crystal (after ref. [6]).

4. Force fields for dynamic simulations

Computer packages for the dynamic simulation of organic phases work on the basis of a generalized intra/intermolecular force field of the form:

$$U = \Sigma K_B (R-R_0)^2 + \Sigma K_\theta (\theta-\theta_0)^2 + \Sigma K_{UB}(S-S_0)^2 +$$
$$+ \Sigma K_\chi [1+\cos(n\chi-\delta)] + \Sigma K_{IMP}(\varphi-\varphi_0)^2 + \Sigma [AR^{-12} - BR^{-6} + q_iq_jR_{ij}^{-1}]$$

where the potential terms refer , respectively, to stretching, bending, Urey-Bradley 1-3, torsional, improper dihedral, and intra- and intermolecular non-bonded interactions. The problem of a consistent parametrization is at once evident, since Urey-Bradley, torsional, and intra- and intermolecular non-bonded terms are strongly coupled. For the results presented in this paper, the GROMOS computer package [14] has been used.

Dynamic simulations are an incredibly versatile tool in the study of molecular phenomena. If static simulations may sometimes be useful for crystals, for any partially or totally disordered state, like the liquid crystalline, glassy or liquid states, MD or Monte Carlo (MC) simulations must be carried out for an appropriate sampling of configurational space.

These evolutive simulation techniques pose however some special problems. The parameterization of a force field for dynamic simulations is even more problematic than for static lattice energy calculations, and transferability may be a problem not only over different compounds, but over different phases of the same compound. The repulsive branch of the potentials, poorly determined when calibrated from crystal data where molecules never climb the repulsive branch too far, are visited much more frequently in evolutive simulations. Moreover, the derivation of thermodynamic averages and of kinetic pathways requires a robust sampling of phase space, a real problem for large systems; even on the fastest available computers, simulation times on chemically significant systems (say, 5,000 atoms) are limited to the order of nanoseconds.

Finally, just as ordinary chemistry requires the design of the appropriate chemical experiment to prove an hypothesis, careful design of a computational experiment is also a must. MD is so ingenious that results of its simulations are often (mis)taken for truth.

5. Crystals: how are they made?

A crystal structure, even if seen in the rich detail afforded by standard X-ray single-crystal diffraction work, really reveals very little on how it was actually made up from the consituent molecules; we see the puzzle reconstructed, without a glimpse at the attempts, mistakes, wrong paths that the crystallization process may have taken before the right route was entered. No experiment is nowadays available to probe the elusive reality of the pre-crystallization world in solution or in the melt, while dynamic simulations are severely restricted by the timescale and size problems: what happens in seconds over millions of molecules is simulated over nanoseconds for a few thousand molecules at best.

Some computational experiments, again on acetic acid, were however conducted [15]. A droplet consisting of a few thousand carbon tetrachloride solvent molecules plus 50 solute acetic acid molecules was set up, and its evolution under forced evaporation-concentration cycles was monitored by MD. Early solute aggregation modes were seen to be highly fluxional cyclic, hydrogen-bonded oligomers, which grow from 2- to about 7-membered cycles by capturing new molecules and incorporating them into the ring structure (some examples in Figure 5). Quite frequent are events in which hydrogen bonds are broken and molecules evolve back from the aggregate into bulk solution. Solute aggregates grow further (within the nanosecond timescale, at least in this highly surface-dominated system) into micelles, which approach the density of the pure liquid without a trace of the ordering to be achieved in the crystal structure.

The first step of the nucleation process is therefore, at least in this highly biased view, the formation of a sort of emulsion in which the solute packs into microscopic liquid-like droplets. The stronger intermolecular interaction, the hydrogen bond, stays fluxional until a late stage of aggregation. Cohesion is rather fostered by diffusion than controlled by intermolecular attractions, as results from computational experiments in which the force field was artificially manipulated. The overall picture, then, is one of unselective cohesion, while the details of the eventual ordering process that leads to the crystal structure remain obscure, and it could not be otherwise since the model is so simplified. The absence of order in the early stages of coagulation is nevertheless

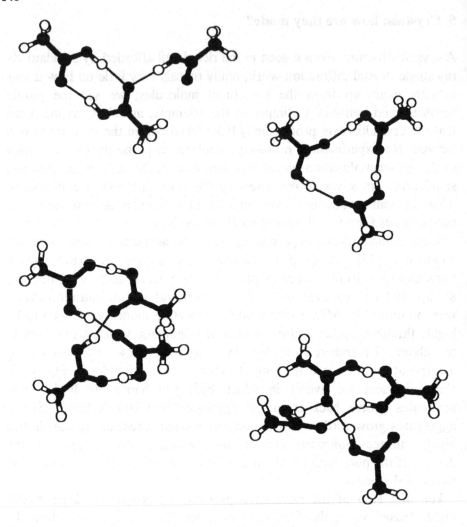

Figure 5. Some of the ring structures encountered in the early stages of aggregation of acetic acid molecules in solution (after ref.[14]).

a persistent and convincing feature of all calculations. Thus, one should not necessarily expect to see anything related to the crystal structure within the precursor solution, and the abundance of cyclic dimers in solution versus the catemer crystal structure of acetic acid testify to this. Conversely, test calculations showed that small clusters with the crystalline structure quickly collapsed in solution, being incapable of withstanding the extreme surface tension. The question of the critical size for nuclei to be stable remains open, but the formation of liquid-like particles seems a reasonable alternative. Not unexpectedly, the whole matter of elementary molecular recognition, nucleation and growth require more careful thinking and much more extensive computational effort.

6. References

1. Gavezzotti, A. (1991) Generation of possible crystal structures from the molecular structure of low-polarity organic compounds, *J.Am. Chem.Soc.* **113**, 4622-4629

2. Gavezzotti,A. (1994) Are Crystal Structures Predictable?, *Acc. Chem. Res.* **27**, 309-14.

3. Mooij, W.T.W.,van Eijck, B.P.,Price, S.L., Verwer, P., and Kroon, J. (1998) Crystal Structure Predictions for Acetic Acid, *J.Comput. Chem.* **19**, 459-74.

4. van Eijck,B.P., Mooij, W.T.M., and Kroon, J. (1995) Attempted prediction of the crystal structures of six monosaccharides, *Acta Cryst.* **B51**,99-103.

5. Desiraju, G.R. (1995) Supramolecular Synthons in Crystal Engineering – a New Organic Synthesis, *Angew.Chem.Int.Ed.Eng.* 34, 2311-27.

6. Gavezzotti, A. (1999) A Molecular Dynamics View of Some Kinetic and Structural Aspects of Melting in the Acetic Acid Crystal, *J.Mol. Struct.*, in the press.

7. Dunitz,J.D., and Gavezzotti, A. (1999) Attractions and Repulsions in Molecular Crystals: What can be Learned from the Crystal Structures of Condensed Ring Aromatic Hydrocarbons?, *Acc.Chem.Res.*, in the press.

8. Gavezzotti, A. (1997) PROMET 5.3, A Program for the Generation of Possible Crystal Structures from the Molecular Structure of Organic Compounds, University of Milano (Italy).

9. D.E.Williams (1983) PCK83, Quantum Chemistry Program Exchange Program no. 548, Indiana University, Bloomington, IN.

10. Gavezzotti, A. (1999) Zip-Promet, An Automatic Program for the Generation of Crystal Structures from Molecular Structure, University of Milano (Italy).

11. Gavezzotti, A. (Ed.) (1997) *Theoretical Aspects and Computer Modeling of the Molecular Solid State*, Wiley, Chichester.

12. Price, S.L. (1996) Applications of realistic electrostatic modelling to molecules in complexes, solids and proteins, *J. Chem. Soc., Faraday Trans.* **92**, 2997-3008.

13. Nahringbauer, I. (1970) Hydrogen Bond Studies. 39. Reinvestigation of the Crystal Structure of Acetic Acid (at +5°C and −190°C), *Acta Chem. Scand.* **24**,453-62.

14. van Gunsteren, W.F., Billeter, S.R., Eising, A.A., Hunenberger, P.H., Kruger, P., Mark, A.E., Scott, W.R.P. and Tironi, I.G. (1996) *Biomolecular Simulation: The GROMOS96 Manual and User Guide*, BIOMOS b.v., Zürich-Groningen.

15. Gavezzotti, A. (1999) Molecular Aggregation of Acetic Acid in a Carbon Tetrachloride Solution: A Molecular Dynamics Study with a View to Crystal Nucleation, *Chem. Eur. J.* **5**, 567-76.

CRYSTAL DESIGN

From Molecular to Application Properties

P. ERK
Colorants Laboratory of BASF AG
67056 Ludwigshafen/Rh.
Germany

1. Introduction

Modelling plays a central role in all design strategies mankind has developed. In the scientific world it is used as a bridging tool to connect macroscopic properties and microscopic structures. If this mapping is unique, modelling allows the prediction of properties. As a consequence of this attractive aspect, the major benefits of modelling-assisted materials design are:

(a) Inexpensive, quick screening by use of the computer.
(b) Exceptional insights into structure-property relations.
(c) Analytical and visual access to microscopic structures.

General limits to these goals (especially items (a) and (b)) are set by the size of the models and the time frame accessible by the appropriate modelling methods. These limits determine, which item of the listed benefits may be achievable. Additionally, the precision of the results and their experimental impact mark the value of the whole process, especially with respect to industrial needs.

TABLE 1. Microscopic properties of crystalline materials related to physical and technical performance.

Molecular,	crystal and	physical properties	technical performance
		hardness	agglomeration / flocculation
		refraction	wetting / dispersibility
	particle size	adsorption	bioavailability / dissolution rate
	crystal habit	ξ-potential	color / optical properties
crystal structure	surface structure	solubility	shock resistance / onset temperature
	band structure	light absorption	filtration / caking
	density	conductivity	slurry / dispersion properties
		magnetism	compaction properties
		brittleness	powder flow / density

In the chemical and pharmaceutical industries any process step that involves the handling of crystalline solids is crucial with respect to the crystal properties. For performance chemicals, crystal properties are usually defined (and hence requested) by the customer specification. In the case of drugs, legislation requires the investigation of

143

D. Braga et al. (eds.), Crystal Engineering: From Molecules and Crystals to Materials, 143–161.
© 1999 *Kluwer Academic Publishers. Printed in the Netherlands.*

possible performance changes related to crystal properties. Therefore, it is an old desire to trace the technical performance of crystalline solids via macroscopic physical properties back to the microscopic characteristics of the material, whenever possible.

A reasonable, but limited overview of some interrelated microscopic, macroscopic and technical features of crystalline products is given in table 1. Many examples for establishing relations across table 1 can be found in the literature, and intuitively all of the technical aspects – even in multi-component systems – may be at least explained through the structural and particulate properties of the crystalline materials involved. Additionally, computational material science has manifested itself throughout the last decades by creating a multitude of methods to derive crystal properties on the basis of structural knowledge only. Therefore the crystal structure is the key information and, subsequently crystal structure determination plays a key role in any design or analytical process.

In this chapter I will try to draw a pragmatic picture of computational approaches to calculate crystal structures. General pathways, computational algorithms, and methods of extracting validated information about the nature of crystalline organic solids will be discussed and exemplified.

2. General Remarks on Crystal Structure Calculation

Powered by the interest in crystal structure information and challenged by the perfect order of the crystalline state, solid state scientists have struggled for more than 50 years to find a way of calculating crystal structures *ab initio*, only on the basis of molecular information - without input from diffraction experiments.

The big practical drawback in this concept is simply called the *packing problem*. This term expresses the non-linearity of the packing energy as a function of the geometrical constraints of crystal structures (e.g. cell parameters, position and orientation of the asymmetric unit). As a consequence, attempts to simply optimize structures end usually in one of the many local minima of the chosen energy function. This crucial peculiarity has triggered the development of strategies and algorithms to tackle it, and as computer power increased, successful crystal structure calculations have left the level of occasional occurrences behind, thus justifying to be called predictions in some sense [1].

As with any prediction, reliability (or probability) is a major issue. In a crystal structure calculation several thousand possibilities (i.e. local minima) may be found. Which ones and how many is strongly depending on the strategy and the weighting function employed in the process, which is commonly the force field used for packing energy calculations. As packing energy is treated as a simplified equivalent to the free energy of the crystal [2], the calculated information may be highly beneficial in judging of potential polymorphs. Nevertheless, the physical meaning of any structure can not be ascertained even by the most accurately calculated packing energies, because the enthalpy of a developing crystal lattice refers only to one aspect of a real crystallization process, neglecting the influence of surfaces, physical conditions and crystallization kinetics.

Keeping these general limitations in mind, carefully validated crystal structure calculations are a valuable source of information for a variety of problem holders. Most frequently the computational approach is used to search for structural information that can not be obtained by experimental methods only. Typical candidates are metastable polymorphs (susceptible towards crystal growth conditions), twinned crystals, materials that are beam sensitive or extremely insoluble.

TABLE 2. Process steps and standard techniques for crystal structure calculation.

process step	applied techniques, algorithms, and knowledge
molecular geometry	quantum mechanics, force field, conformational analysis
space group determination	cluster analysis, nucleus generation, space group statistics
structure generation	cluster generation; random, discrete, systematic, Monte Carlo
structure optimization	energy minimization, scoring function, simulated annealing
redundancy check	comparison of density, energy, geometry
structure weighting	by energy (components), density, lattice geometry, scoring function
final minimization	full atomic force field; cell parameters from diffraction experiments
structure selection	comparison with x-ray or electron diffraction
Rietveld refinement	lattice refinement, rigid body refinement, atomic coordinates refinement

For many of these compounds, powder patterns are easily accessible. If they can be indexed, classical crystallographic [3] and elegant computational procedures [4,5] exist to solve the structures even of molecules with multiple conformational degrees of freedom within minutes on a PC. If indexing of the powder pattern fails, a solution via crystal structure calculation may be considered appropriate.

A general scheme for a crystal structure calculation is outlined in table 2. Entry point for all methods is the definition of the molecular geometry, which is kept fixed throughout the primary optimization of generated structures. In the case of flexible molecules or tautomers, every possible molecular structure requires in principle a separate treatment, thus multiplying the computational demands of the problem. Therefore, attempts of crystal structure calculation are usually restricted to molecules of defined geometry or limited flexibility and tautomerism (depending on the interest in the specific problem).

Choosing the space groups to be looked at is the next task. Space group assumptions may be guided by crystallographic knowledge and intuition, e.g. by the classification rules for aromatic hydrocarbons [6], or by searches for structures of similarly shaped molecules in the Cambridge structural database (CSD). Space group related information may also be derived from simulations of small clusters of molecules or in an elegant approach by calculating symmetry related pairs of molecules [7]. Mostly, one will come back to space group statistics [8] and simply search the most likely ones.

In the next step structure generation and optimization is performed. For this process several methods, approaches, strategies and programs have been developed during the last decade [7,9,10,11,12,13,14,15]. The currently published methods are excellently reviewed in great detail by Verwer and Leusen [16]. From the perspective of an industrial user, adaptability, reliability, flexibility, validity, transparentness, and speed are strongly requested attributes for such a method in order to be effective. From this perspective, commercial software or a well validated, user-friendly program that can be smoothly integrated in a material science modelling environment is the method of choice.

3. Systematic Crystal Structure Generation

3.1. DESCRIPTION OF THE METHOD

In 1992 I started to develop a method for structure generation of phthalocyanine polymorphs using the crystal builder and crystal packer module in the program CERIUS [17]. The crystal packer calculates packing energies and performs energy minimizations of crystal packings. Molecules are treated as rigid units and between these rigid units nonbond energies (van der Waals, Coulomb, H-bond) are calculated. Variables in the minimization are the cell parameters, the position of the molecule in the crystal lattice and the orientation expressed by the Eulerian angles (figure 1). Intuitively, these degrees of freedom can be split into two classes:

(1) The relative orientation(s) of the molecule(s) determines the quality of the intermolecular interactions. Its nature is purely vectorial, and it is merely trapped in a close packed crystal environment, resulting in a (local) minimum.

Cell angles also determine the relative orientation of molecules towards each other by changing the orientation of the coordinate system of the lattice relative to the internal coordinate system of molecule. Tentatively, the influence of cell angles is less pronounced compared to the molecular orientation

(2) The length of the cell axes scale only the quantity of the intermolecular interactions, and may therefore be optimized in a straightforward way.

The position of the molecule simply needs to follow any changes of the axes.

The apparent relation between molecular orientation and local minima suggests, that the phase space of a given space group with one molecule in the asymmetric unit might be scanned extensively by changing the molecular orientation in a systematic way [13,18].

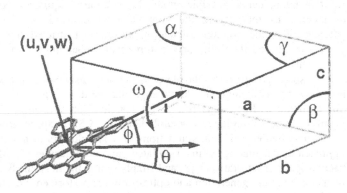

Figure 1. Degrees of freedom in a crystal structure: cell parameters a, b, c; cell angles α, β, γ; relative molecular position (u v w); relative molecular orientation (φ, θ, ω).

Practically this can be done in loose packings that allow free rotation of the molecule. The standard optimization tools are then used to construct a close packing while keeping the molecular orientation fixed.

These thoughts have led to the development of the following algorithm, which includes structure generation and optimization:

(1) *Generation of a start-up cell*: A rectangular cell with fully applied space group symmetry is constructed. The molecule is either placed at an appropriate special position or at non-special positions that guarantee an equal distribution of the molecular units, e.g. (¼ ¼ ¼) in P 2_1/c. The molecule must not necessarily be aligned to the cell axis by its moments of inertia, but it helps troubleshooting (in case). The lengths of the cell axes are adjusted to allow complete free rotation of the molecule within its van der Waals contact sphere, e.g. the minimum distance between adjacent molecules should be approximately 7 Å.

(2) *Variation of molecular orientation*: The Eulerian angles (ϕ,θ,ω) are varied stepwise within the symmetry-related limits to generate the initial structures. In practice, steps of 30° seem to be sufficient in most cases.

(3) *Generation of close packed structures*: First, the packing energy of the initial structures is minimized with respect to the cell axes only. Mostly it is advantageous to optimize the axes sequentially, permuting at least the first axis to be optimized. If an axis is subject to variation, the corresponding translational degree of freedom of the molecular position must be adjustable too. In any case the van der Waals energy is a sufficient criterion for this step.

Subsequently, the cell angle(s) are released for optimization. This involves the first change of the relative orientation of the molecules towards each other and therefore, in the case of polar molecules, Coulomb energy should be included in the packing energy minimization.

(4) *Optimization of the packing*: In the final setup, the molecular orientation or the complete structure including atomic coordinates, is set free for optimization. All possible contributions to the packing energy are included.

Depending on the space group, the increment in the variation of the Eulerian angles, and the depth of the axis-permutation up 10,000 packings are generated. To reduce calculation time, redundant packings may be sorted out, whenever it is appropriate. Usually, this is done before the final optimization, and repeated afterwards. Different criterions like energy, lattice parameters, density, or interatomic distances may be used for the redundancy check, which is always a tangle of reducing the vast amount of data and loosing information by poorly discriminating criteria. If the molecular structure is kept fixed during the whole process, redundant structures may be eliminated at low risk after the final optimization by comparing both, density and packing energy.

3.2. TEST RESULTS

In a first test series the method described has been employed to reproduce the known crystal structures of the molecules displayed in figure 2. These test cases have been chosen to cover various aspects of shape and polarity, including compact, flat aromatic systems, extreme lengthy hydrocarbons, paddle-like steroids, and some polar compounds with and without intra- or intermolecular H-bonds.

148

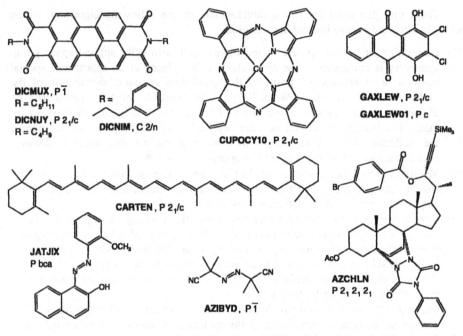

Figure 2. Selected test cases for systematic crystal structure calculation; for comparison the reference codes of the Cambridge structural database are given.

TABLE 3. Structure generation parameters and optimization pathways of test cases.

ref.-code of CSD[a]	space group	Z eff.[b]	DF[c]	start cell (Å)	molecul. position[d]	angular increment	sequence of released DFs for energy minimization[e]
AZIBYD	P$\bar{1}$	1	9	12	0 0 0	20	abc | αβγ | φθω
AZCHLN	P $2_1 2_1 2_1$	4	9	50	¼ ¼ ¼	30	(a+ | b+ | c+) | φθω
CARTEN	P 2_1/c	2	7	35	0 0 0	20	(a | b | c) | β | φθω
CUPOCY10	P 2_1/c	2	7	30	0 0 0	30	(a | b | c) | β | φθω
DICMUX	P$\bar{1}$	1	9	50	0 0 0	30	abc | αβγ | φθω
DICNIM	C 2/n	4	7	60	0 0 0	30	(a | b | c) | β | φθω
DICNUY	P 2_1/c	2	7	40	0 0 0	30	(a | b | c) | β | φθω
GAXLEW	P 2_1/c	4	10	30	¼ ¼ ¼	30	(a+ | b+ | c+) | β | φθω
GAXLEW01	P c	2	10	15,15,25	0 0 0	20	a | c | b | β | φθω
JATJIX	P bca	8	9	30	0 0.3 0.4	30	(a+ | b+ | c+) | φθω

a CSD = Cambridge Structural Database. *b* Z = number of molecules in the unit cell. *c* DF = degrees of freedom. *d* initial position of the molecular center of gravity. *e* a+ = axis plus corresponding translation; ( ) = permutation of parameters

All molecular structures were taken from the experimental data and kept fixed during packing optimization. Errors in the reference data have been corrected and hydrogen positions were recalculated using the Dreiding2.21 force field [19]. Charges were generated by the charge equilibrium method (QEQ) [20]. Unless otherwise noted the standard settings of CERIUS have been used.

For all structures listed in table 3, the specified procedures yielded the packing arrangements which correspond to the force field minimized experimental structure. These promising results indicate that the proposed concept seemingly works. The transparentness of this systematic structure generation method and its completely defined stages allow for a detailed analysis of every single minimization step and every single structure used during the procedure. This possibility of backtracking helps to identify critical steps and interdependencies of intermolecular forces, molecular shape and crystal packing. These complex relations must be crucial for all crystal structure prediction methods, that can not proof mathematically, how extensive phase space has been searched.

The procedures applied to the test structures of table 3 have all been carefully analyzed. In some cases parameters have been varied to check out dependencies.

AZIBYD: This is a true no problem case. Regardless of applied charges or the order of the axes released for optimization, the experimental packing is always found as the most dense, lowest energy structure.

AZCHNL: A real challenge because of the unusual shape of this molecule. Different molecular positions in the initial setup have been tested. The initial orientation has been chosen randomly. In all runs axes were released sequentially and the order of the release has been fully permutated. The correct structure has in all runs been found as the most dense, lowest energy packing. The cluster of hits spreads across a span of 4 kcal/mol showing the strong tendency of this ruffled molecule to be trapped in pseudo-minima (sometimes this problem may be solved by a few steps of molecular dynamics or applying and releasing external pressure). The other local minima are clearly set off by more than 10 kcal/mol and a density lowered by 10 %.

CARTEN: The lengthy shape of this hydrocarbon required an angular increment of 20° in the structure generation procedure to achieve the experimental structure (a test run using a 30° increment failed to produce the correct packing). A detailed evaluation of the pathway from structure generation to the target packing revealed that the tilt of the polyene chain with respect to the b-axis is crucial for success. The limits from which the global minimum may be reached are as narrow as 25°. For all other centrosymmetrical molecules a 30° increment was highly sufficient.

GAXLEW: In the centrosymmetrical polymorph the molecules form chains linked by intermolecular H-bonds. To prevent early formation of H-bonds, which may lock the packing in a false H-bond pattern, H-bond energy terms should not be included before Coulomb energy, i.e. after minimization with respect to all lattice parameters. Alternatively, a centrosymmetrical H-bonded dimer can be assumed and used as the molecular unit in space group P 2_1. Both approaches worked.

If the hydrogen bonded dimer is used as the principal unit, GAXLEW is found as the lowest energy and most dense packing. By the single molecule approach, the expected packing is the most dense found and ranked 4th in packing energy with a difference of 0.5 kcal/mol to the calculated global minimum. In terms of hydrogen bond energy the target packing is 5th, but only some outliers at total energy ranks 114 or higher are found to have a lower H-bond energy.

GAXLEW01: In this polymorph only intramolecular H-bonds are formed. Only the first axis to be optimized has been permutated. The experimental structure is located without problems, starting from various molecular positions. Among 99 unique packings, the target packing is the most dense one, but on rank 12 by energy (difference to 1st = 0.3 kcal/mol).

JATJIX: Also a no problem case. Several randomly chosen initial orientations or variations of the initial center of gravity did not affect the result and have led to the target structure as the most dense, lowest energy packing.

In all test cases redundant structures have been identified by comparing density and the components of packing energy. The limits in which packings are assumed to be redundant, have been set to 0.001 g/cm^3 in density and 0.1 kcal/unit cell for all energy components. In some of the test cases these limits turned out to be too discriminating, but no relevant structure has ever been eliminated.

4. Practical Aspects of Modeling-based Structure Determination

4.1. GENERAL EFFORT

Questions about the effort that has to be put into a crystal structure calculation are difficult to answer, except for the estimates on computing time based on the number of atoms, degrees of freedom etc. As increasing computer power and effective software tuning brought typical computing times for crystal structure calculations (incl. optimization) from weeks or days down to some hours, calculations in unusual space groups, checks of different force fields, and the testing of multiple conformations are no big deal any more. Thus a flood of crude structure information is easily accessible, but crystallographic expertise is particularly required to extract beneficial knowledge - if there is any. The likelihood of obtaining significant results increases with extra information from diffraction experiments, spectroscopy, or microscopy. Additionally, the number of pitfalls may be reduced by adjusting the computational procedure according to facts known from closely related structures.

4.2. ACCURACY OF CALCULATED CRYSTAL STRUCTURES

The accuracy of the force field methods must be considered as a fundamental problem. From a practical view this problem is less critical in the sense of thermodynamic accuracy, but geometrical accuracy is an essential requirement in order to solve a crystal structure.

Thermodynamically, packing energy is considered as a sufficient approximation of the free energy of a crystal, otherwise it could not be used as a criterion to determine the general stability of a packing. For molecules of low polarity its magnitude can be estimated from the molecular composition within the limits of the experimental error. The average deviation of force field derived packing energies from experimental sublimation enthalpies is approximately 2-3 kcal/mol or better, if the compilation is based on the experimental structure [21].

For packing energy calculations of force field minimized structures, as a rule of thumb, calculated sublimation enthalpies can be assumed to be accurate within 5-10 % of their absolute value. These error limits may expand significantly, if intramolecular contributions are included to describe packing related conformational changes [22]. As a consequence of this low thermodynamic precision, the information content of packing energy with respect to the stability of possible polymorphs needs to be considered with care [23].

The limited thermodynamic accuracy does not only originate from neglecting any entropic and vibronic contributions to the free energy. It is simply a tribute to generalized force fields using averaged, isotropic potential functions. In the case of polar compounds, electrostatics appeared to be a crucial issue, that may even cause the failure of finding any physical relevant structures at all. If polar compounds are subject to investigation it is usually worth the effort to test various charge distributions.

The critical influence of point charges is demonstrated by the packing energies of polymorphs GAXLEW and GAXLEW01 [24] (see figure 2, table 4). According to the literature data the system seems to act enantiotropic with GAXLEW being the stable polymorph at ambient conditions and transforming into GAXLEW01 at higher temperatures.

Starting from the experimental structures, packing energies have been minimized using Filippini and Gavezzottis interatomic parameters [21], and the Dreiding2.21 parameters [19] in combination with various charge models (table 4). In both cases molecules were treated as rigid units and for comparison molecular geometries were replaced by the PM3 optimized molecular structure [25].

The packing energies of the optimized structures vary within a range of 9 kcal for GAXLEW01, and 14.6 kcal for GAXLEW, respectively. The energy differences between both polymorphs amount from 0.5 to 7.4 kcal, revealing always GAXLEW as the more stable polymorph.

TABLE 4. Packing energies (PE [kcal/mol]) of polymorphs GAXLEW and GAXLEW01. Structures were minimized using different force fields and charge models. For geometrical comparison the normalized root mean square deviation (RMS [pm]) of calculated versus experimental lattice positions is given.

	vdW[ab]		F&G[bc]		QEQ[ade]		PM3-ESP[adf]		MNDO-ESP[adg]	
	PE	RMS	PE	RMS	PE	RMS	PE	RMS	PE	RMS
GAXLEW	-37.0	23.6	-40.0	20.4	-49.7	18.4	-35.1	22.7	-43.9	27.8
Δ PE	2.5		0.5		6.8		1.2		7.4	
GAXLEW01	-34.5	30.2	-39.5	36.1	-42.9	123	-33.9	9.4	-36.5	32.6
charges Cl/O[h]	-		-		-0.33 / -0.51		+0.19 / -0.3		+0.21 / -0.61	

a   van der Waals contribution using Dreiding2.21 parameters [19]
b   H-bond contribution by 12-10 potential ($r_0 = 275$ pm, $D_0 = 9$ kcal), approx. 5-6 kcal/mol
c   force field of Filippini and Gavezzotti [21]
d   H-bond contribution by 12-10 potential ($r_0 = 275$ pm, $D_0 = 4$ kcal), approx. 3 kcal/mol
e   point charges by charge equilibration method [20]
f   point charges by ESP calculation using the PM3 Hamiltonian [26,25]
g   point charges by ESP calculation using the MNDO Hamiltonian [26,25]
h   point charges of Cl and carbonyl O

Despite the huge differences in packing energy, geometrical accuracy of GAXLEW is acceptable for all force fields. A proposed measure for average geometrical differences between calculated and experimental lattices (of the same type of packing) is the root mean square deviation of the compared unit cells corner positions. For the purpose of comparing different space groups the RMS deviation has been normalized by the number of molecules per unit cell (Z). The quality of the fit may be judged by comparing with the length of the average space diagonal (ASD, also normalized by Z) of the unit cell, which is 641 pm for GAXLEW and 993 pm for GAXLEW01. As a rule of thumb geometrical accuracy is acceptable, if the RMS deviation is within 5 % of the ASD. For deviations exceeding 5 % of the ASD, XRD powder patterns of the compared lattices are completely out of match and their similarity (see 4.3.) may not be recognized - even by visual comparison.

The calculated packings of GAXLEW01 are mostly within the 5% range, except for the calculation based on equilibrated charges [20] (QEQ) which results in a shortening of the c-axis by 25 %. In contrast, atomic charges fitted to the electrostatic potential [26] (ESP), which is calculated by semi-empirical quantummechanics [25], lead especially in the case of the PM3 derived ESP to very satisfying conformity of experimental and calculated packing. These results indicate, that QEQ puts the wrong polarity on chlorine, which is reversed by the quantummechanical methods. It also seems logically, that the polar space group shows a stronger response to this change in polarity.

Other factors responsible for the mediocre performance of force fields in the case of GAXLEW may be attributed to the insufficient description of Cl by isotropic potentials [27]. A second drawback is the coarse estimate of H-bond contributions to the packing energy. On top of the electrostatic energy a relatively weak H-bond is assumed in the calculations, whereas for pure van der Waals based calculations stronger H-bond contributions are considered, that may amount up to 6 kcal/mol.

But even without obstacles like second row atoms or H-bonds, packing geometry appears to be extremely subtle to intermolecular forces in some cases. Filippini and Gavezzotti [21] already mentioned the failure of their force field to predict the γ-type structure of coronene, which transforms almost to a brick stone arrangement of parallel layered molecules, as can be seen from the shortening of the b-axis (table 5). They presume, that the lack of an electrostatic term in their force field may be responsible for this outlier. The similar behavior of the γ-phase of copperphthalocyanine (a highly symmetrical, planar molecule of similar size as coronene) required me to investigate these cases in more detail.

TABLE5. Packing energy components [kcal/mol] and geometrical parameters (b-axis, RMS, density) of coronene structures calculated with different force fields and charge models.

	PE	vdW	Coulomb	b [pm]	RMS [pm]	ρ [g/cm^3]
exptl. structure	-80.7	-79.1	-1.7	470.2	-	1.395
QEQ^a	-83.8	-86.8	+3.0	366	120	1.412
QEQ × 2^b	-82.5	-79.5	-3.0	469	8	1.370
F&G^c	-83.2	-	-	358	135	1.426

a   Dreiding 2.21 van der Waals parameters, point charges by charge equilibration. C(-H) -0.13, H +0.11
b   point charges by charge equilibration method scaled by 2.0, C(-H) -0.27, H +0.22
c   force field of Filippini and Gavezzotti [21]

The use of Dreiding 2.21 [19] parameters in connection with QEQ [20] charges does not improve the geometrical accuracy of the calculated packing of coronene. The RMS of calculated and experimental lattice exceeds 13 % of the average space diagonal. Surprisingly, doubling the QEQ charges (to quantities close to MNDO-ESP charges [26]) yields an almost perfectly fitting structure.

At this stage it is not clear, why molecules like coronene or copper phthalocyanine respond so strongly to (the scale of) charge distribution. A possible reason might be the polarization of these molecules in the crystalline environment, which is neglected through the use of static atomic charges derived from calculations of isolated molecules. This thought is supported by the observation, that the effect is at a different levels of susceptibility for other copper phthalocyanine polymorphs crystallizing in other space groups (e.g. the test structure CUPOCY10 seems to be unaffected [28]).

As one of many examples, this pitfall is pointing to the request for checking the sensibility of calculated structures with respect to force field settings. For attempted predictions of crystal structures this should be considered as a must.

4.3. STRUCTURE SELECTION

Because of the uncertain physical relevance of calculated packing energies and densities, these properties may only be considered as relative indicators for the possible existence of a polymorph. Only if the other packings are off by a significant amount of packing energy, the latter may be used as a suitable selection criterion, but without warranty. In any case estimates of the packing energy derived from similar structures or general formulas should be considered as expected values, unless experimental data exist.

Density is easy to obtain experimentally and opens usually the first possibility of comparing the model with reality. A combination of both, density and packing energy is already a strong indicator for the relevance of a calculated structure, usually leaving a handful of structures within the range of solutions to consider.

Finally, comparison of calculated structures with crystallographic data appears as the ultimate selection criterion and is also the final proof, if the selected packing converges in a subsequent Rietveld refinement. In the lucky cases, where careful optimization of the packing has left few structures, calculated and experimental powder patterns may be compared visually. If large numbers of trial packings are subject to this process, comparison has to be done computationally, based on quantitative criteria.

For this purpose the R-factor is the usual crystallographic similarity criterion, which requires a distinct overlap of peaks to become meaningful. A more versatile approach has been proposed by Karfunkel [29], who published an algorithm for comparing the folding of the diagrams within a given range. This method allows the identification of similar XRD curves, even if the peaks are nonoverlapping because of (the usual) lattice mismatch between calculated and experimental structure.

This similarity measure (CMACS) [17] allows the comparison of crude packings, before extensive optimization work (e.g. complete minimization of the atomic positions) is carried out. Thus it may help to reduce computational effort, and it may also allow the extraction of promising solutions from a broad base of structures.

4.4. STRUCTURE REFINEMENT

The final proof for any crystal structure calculation is the refinement with respect to experimental diffraction data. This process has been much facilitated through the possibility to perform Rietveld refinements in a semi-interactive way [17,30], that allows to follow the structural and energetic changes after each refinement cycle and to manipulate the structural features of the model accordingly.

Rietveld refinements include the correction of lattice parameters and atomic positions as well as the adjustment of instrumental and phase related peak shapes. However, the limited resolution shown by powder diagrams of low crystalline routine samples does usually not allow for the refinement of atomic coordinates or rigid bodies.

To bypass this problem, atomic positions have to be adjusted either manually or by energy minimization within the fixed unit cell. Thus energy minimization and the refinement of lattice parameters, background and profile functions are performed alternating until the R factors are minimized. Depending on the crystallinity and texture of the sample R factors of 0.1-0.2 may be expected for an acceptable structure solution by this procedure.

5. Application Examples

With few exceptions, structure calculations of unknown phases are described for comparatively small or rigid molecular units only, although modern times computer power allows for the treatment of molecules with higher flexibility. But as crystal structure calculations are subject to industrial interest it might be assumed that a lot of protagonistic work in the area is not open to the public.

Organic pigments like perylenes, phthalocyanines, azo pigments, isoindolines, diketopyrrolopyrroles and quinacridones match perfectly into the area. All of their properties are related to their crystalline nature and the rigidity of the molecules is in favor of the modellers CPU time. In the following section the prediction of the previously unknown structure of a perylene pigment will be discussed in detail and the crystal structure of an aromatic hydrocarbon is predicted as a challenge to future researchers.

5.1. N,N'-DIPHENYL-PERYLENE-3,4:9,10-TETRACARBOXYLIC ACID DIIMIDE

The diimides of perylene-3,4:9,10-tetracarboxylic acid are widely used as high performance pigments in plastics and coating applications. Depending on the substituent of the imido nitrogen they form solids whose colors range from orange to black in the crystalline state. However, regardless of any substituent they are red or orange (even the black ones) in the amorphous form or in solution, respectively. These striking color changes are attributed as crystallochromy and the class of perylene pigments probably shows this phenomenon in its most remarkable way.

It has been recognized early, that crystallochromy is related to the arrangement of the perylene chromophors in the crystal lattice [31]. The effect has been explained quantitatively by empirical correlations [32] and by quantummechanical calculations [33]. This unique possibility of mapping color properties to crystal structures has challenged several groups to find ways of calculating or predicting crystal structures of perylenes [34,35].

During the development of the structure calculation method presented in this article, diphenylsubstituted perylene-bis(dicarboximide) (DPPI) has been chosen as the first example of an unknown crystal structure to be calculated without the use of any information from diffraction experiments [36]. DPPI may be looked at as the parent compound for N-aryl substituted perylenes, which form brilliant red pigments that are of particular interest from a colorists point of view. Similar to most other perylenes DPPI is accessible in a one-step synthesis from perylene tetracarboxylic acid dianhydride and aniline. The resulting red solid has been used for this investigation without further treatment.

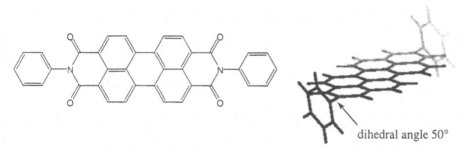

dihedral angle 50°

Figure 3: Molecular structure of DPPI, minimum AM1 geometry (right).

The powder pattern of DPPI was taken routinely. In the region of $2\Theta < 20°$ seven distinct peaks can be identified. Above 20° only overlapping and broadened peaks are found, making the powder pattern unsuitable for indexing.

The molecular structure of DPPI has been optimized with MOPAC using the AM1 Hamiltonian [25]. The phenyl ring is found to bend freely above dihedral angles of 50° with the perylene plane (figure 3). MNDO-ESP charges have been calculated with MOPAC and were scaled by 1.422 [26]. Atomic charges have been averaged for atoms related by inversion symmetry. As a response to the flexibility of the phenyl groups, dihedral angles have been set to 50°, 60° and 90° in models referred to as DPPI-50, -60 and -90.

Among the 18 structures studied by Klebe [32], 16 build up uniform stacks in the solid state. In all cases crystallographic and molecular inversion centers coincide. The stacking of perylenes has been studied intensively by Perlstein [34], who successfully used molecular dynamics to minimize the intermolecular energy of clusters of stacked perylenes. With respect to this background, it could be expected, that DPPI would be likely to follow the building principles of the other perylenes.

In the very first setup, DPPI-50 (the AM1 minimum) has been placed at the origin of a rectangular cell of 20×20×5 Å, with the perylene plane parallel to the a,b-plane and the N-N-axis aligned to the a-axis. The molecular orientation has been varied in steps of 15° by rotation around the b-axis and subsequently the c-axis within 0°-45°, thus generating 16 starting structures. For these starting structures the packing energy has been

156

optimized stepwise for the cell axes, cell axes plus angles, and finally all lattice parameters plus molecular orientation. The structure obtained from optimization of the 15°-30° starting model had the lowest packing energy, the highest density (both in reasonable quantities) and its powder pattern showed significant similarity to the experimental XRD curve.

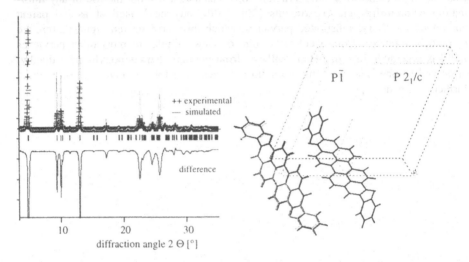

++ experimental
— simulated

difference

P Ī   P 2₁/c

diffraction angle 2 Θ [°]

Figure 4: Trial refinement of DPPI in space group P $\bar{1}$, R_p = 1.78 (left). Transformation of the refined triclinic lattice into space group P 2₁/c (right).

By alternating the refinement of the lattice parameters and the energy minimization of the packing (including torsions of the phenyl groups), the R-factor converged at 1.79 and packing energy rose by 1.6 kcal/mol during the refinement procedure. At the same time two of the cell angles refined to values close to 90° indicating a monoclinic space group (figure 4).

The setup of the unit cell allowed for a transformation of the model to space group P 2₁/c (figure 4). Optimization of the monoclinic structure yielded another 2 kcal/mol compared to the minimum obtained in P $\bar{1}$. The agreement of the calculated and experimental XRD powder pattern was exceptional for the P 2₁/c structure.

Some years later the structure solution of DPPI has been reexamined by the standard procedure, starting with a rectangular cell of 30x30x30 Å. The molecule has been placed on the origin and was roughly aligned with the N-N-direction parallel to the a-axis. 216 starting structures have been generated by rotating the molecule sequentially in steps of 30° around the a-, b- and c-axis within 0°-150°. For each starting structure the cell axes have been optimized sequentially (van der Waals energy only) with full permutation of the sequence, yielding 1296 pre-optimized packings. For each packing, first the cell angle(s), and in the following minimization run the molecular orientation has been released with van der Waals and Coulomb energy being optimized in both runs.

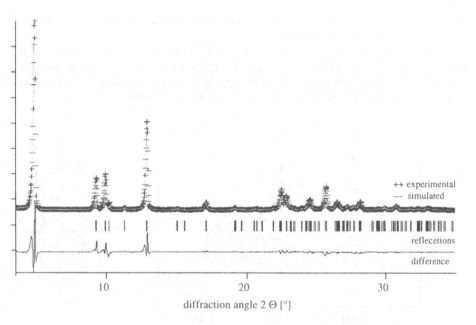

Figure 5: Rietveld plot of the refinement of DPPI, space group P 2_1/c, a=16.754(5), b=4.0108(4), c=18.326(5), β=81.55(2), R_p =0.179, R_{wp} =0.229, R_{Bragg} =0.106

The described procedure has been tested using DPPI-60 and DPPI-90, each in space groups P $\bar{1}$ and P 2_1/c. After eliminating redundant structures, by the combined energy/density criterion (see 3.2.), a final minimization including the torsions of the phenylgroups has been carried out on all packings. Regardless of the starting geometry, both calculations in space group P $\bar{1}$ yielded the minimum already known. For P 2_1/c the identical global minimum has been found with both starting geometries. This minimum differs significantly from the above constructed P 2_1/c packing, which is only obtained from DPPI-60 as a starting structure and ranked 4th in energy (difference 0.8 kcal/mol). Powder pattern similarity is clearly in favor of the already known packing, but the calculated global minimum should be considered as a possible polymorph.

In the combined Rietveld refinement / energy minimization procedure, a final R-factor of 0.24 has been reached, which was improved to 0.21 by full minimization of all atomic positions within the refined lattice parameters. This operation resulted in a slight distortion of the perylene, which is also observed in experimental structures. Finally, the (planar) molecule has been adjusted in the same way as the fully relaxed structure and was rotated manually while monitoring the changes in the R-factors, which converged at R_p = 0.179, R_{wp} = 0.229 and R_{Bragg} = 0.106 (figure 5).

5.2. PREDICTED CRYSTAL STRUCTURE OF $C_{54}H_{22}$

The rhombic shaped aromatic hydrocarbon $C_{54}H_{22}$ has recently been synthesized by Müllen et al. [37] (figure 6). Because of the material's insolubility, no single crystals could be obtained. Electron diffraction and STM investigations of monomolecular films

158

revealed a layered structure of close packed molecules, with a density of ≈1.5 g/cm³ if a layer distance of 3.5 Å is considered.

According to the classification of Desiraju and Gavezzotti, $C_{54}H_{22}$ should form a γ-type structure (S_g/S_{St} = 1.035, see lit. [6] for explanation). Therefore, space groups P 2_1/c and C 2/c have been chosen, to search for low energy structures of $C_{54}H_{22}$. The packing energy of $C_{54}H_{22}$ has been estimated to -79.6 ± 4.1 kcal/mol [38].

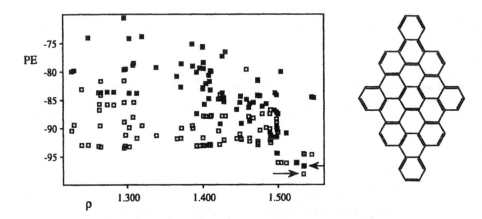

Figure 6: Scatter plot of packing energy PE [kcal/mol] versus density ρ [g/cm³] for calculated minimum energy structures of $C_{54}H_{22}$ (□ P2₁/c; ■ C 2/c), global energy minima are marked by arrows.

The search in space group P 2_1/c (increment 20°) yielded 206 local minima with a density ρ > 1.0 g/cm³, whereas in space group C 2/n (increment 30°) 136 packings were found in this density range (figure 6). The most dense packing (1.548 g/cm³) is identical for both space groups, a face-to-face structure with a packing energy of -84.7 kcal/mol. In general two types of structures are found: (a) layered structures, formed of splipped-stacked parallel molecules; (b) γ-type structures with edge- or corner-connected stacks. The lowest energy structures, which are most likely to have physical relevance, are found to be edge-connected γ-type (figure 4) and are waiting for confirmation (figure 7).

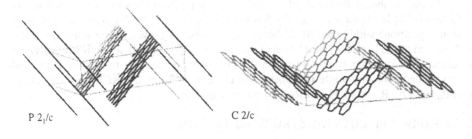

Figure 7: Predicted low energy packings of $C_{54}H_{22}$.
P 2₁/c: a=13.23 Å, b=5.26 Å, c=21.87 Å, β=72.04°, PE= -98:3 kcal/mol, ρ=1.538 g/cm³ (left).
C 2/c: a=29.55 Å, b=5.11 Å, c=19.84 Å, β=104.3°, PE= -96.8 kcal/mol, ρ=1.534 g/cm³ (right).

Comparison of both packings seems to point to a slight favor for P 2_1/c. This is also supported by the scatter plot of packing energy versus density of all calculated minima in figure 6. The distributions in the scatter plot suggest that packing energy in C 2/c weakens more pronounced as the density of the structures is reduced, whereas in P 2_1/c many low energy structures may be found even for $\rho < 1.3$ g/cm^3.

6. Conclusion and Outlook

Crystal structure calculations are a valuable tool in each crystallization laboratory. Their predictive potential must be considered with care, at least as long as force fields are not robust enough to yield unequivocal results. The progress made in solving crystal structures from powder patterns without indexing has been highly beneficial.

From the viewpoint of crystal engineering, the diversity of hypothetical packings is a pool of information about the supramolecular abilities of a molecule and it might be worth the effort, to analyze and use this data for the generation of new materials.

Increasing computer power has promoted many methods for crystal structure calculations close to routine operations. This foreseeable trend will certainly allow for the use of better force fields to get a more accurate view of the energetics of structures and compositions. It will also enable to calculate the kinetics of crystallization processes, which are usually even more important from an industrial point of view, than pure structural knowledge. A first step in that direction has been made by Gavezzotti [39], who used molecular dynamics to investigate solvent crystallizations.

Crystallization science is rich facetted with open questions and in some ways a mystic part of nature as crystallizations may be influenced by slightest changes of parameters [40, 41]. This subtle behavior appears almost as a contradiction compared to the well established static order, crystal structures stand for. On the other hand it is the challenge and the chance of crystal design.

7. References

1. Gavezzotti, A. (1994) Are crystal structures predictable ?, *Acc. Chem. Res.* **27**, 309-314.
2. Kitajgorodskij, A. I. (1965) The principle of close packing and the condition of thermodynamic stability of organic crystals, *Acta Crystallogr.* **18**, 585-590.
3. Cascarano, G., Favia, L., and Giacovazzo, C. (1992) SIRPOW91 - a direct-methods package optimized for powder data, *J. Appl. Cryst.* **25**, 310-317.
4. Shankland, K., David, W.I.F., and Csoka, T. (1997) Crystal structure determination from powder diffraction data by the application of a genetic algorithm, *Z. Kristallogr.* **212**, 550-552.
5. Harris, K.M.D. and Tremayne, M. (1996) Crystal structure determination from powder diffraction data, *Chem. Mater.* **8**, 2554-2570.
6. Desiraju, G.R. and Gavezzotti, A. (1989) Crystal structures of polynuclear aromatic hydrocarbons. Classification, rationalization and prediction from molecular structure, *Acta Crystallogr. Sect. B* **45**, 473-482.

7. Gavezzotti, A. (1991) Generation of possible crystal structures from the molecular structures for low-polarity organic compounds, *J. Am. Chem. Soc.* **113**, 4622-4629.
8. Wilson, A.J.C. (1990) Space Groups Rare for Organic Structures. II. Analysis by Arithmetic Crystal Class, *Acta Crystallogr. Sect. A* **46**, 742-754.
9. Gdanitz, R.J. (1992) Prediction of molecular crystal structures by Monte Carlo simulated annealing without reference to diffraction data, *Chem. Phys. Lett.* **190**, 391-396.
10. Perlstein, J. (1992) Molecular Self-Assemblies:Monte Carlo Prediction for the structure of one-dimensional translation aggregate, *J. Am. Chem. Soc.* **114**, 1955-1963.
11. Holden, J.R., Du, Z., and Ammon, H.L. (1993) Prediction of possible crystal structures for C-, H-, N-, O-, and F-containing organic compounds, *J. Comput. Chem.* **14**, 422-437.
12. van Eijck, B.P., Mooij, W.T.M., and Kroon, J. (1995) Attempted prediction of the crystal structures of six monosaccharides, *Acta Crystallogr. Sect. B* **51**, 99-103.
13. Chaka, A.M., Zaniewski, R., Youngs, W., Tessier, C., and Klopman, G. (1996) Predicting the crystal structures of organic molecular materials, *Acta Crystallogr. Sect. B* **52**, 165-183.
14. Schmidt, M.U. and Englert, U. (1996) Prediction of crystal structures, *J. Chem. Soc., Dalton Trans.*, 2077-2082.
15. Hofmann, D.W.M. and Lengauer, T. (1997) A discrete algorithm for crystal structure prediction of organic molecules, *Acta Crystallogr. Sect. A* **53**, 225-235.
16. Verwer, P. and Leusen, F.J.J. (1998) Computer simulation to predict possible crystal polymorphs, in K.B. Lipkowitz and D.B. Boyd (eds.), *Reviews in Computational Chemistry, Volume 12*, Wiley-VCH, New York, pp.327-365.
17. CERIUS2, Version 1.6-3.9, © Molecular Simulations Inc., 1994-1999.
18. Scaringe, R.P. and Perez, S. (1987) A novel method for calculating the structure of small-molecule chains on polymeric templates, *J. Phys. Chem.* **91**, 2394-2403.
19. Mayo, S.L., Olafson, B.D., and Goddard III, W.A. (1990) Dreiding: A generic force field for molecular simulations, *J. Phys. Chem.* **94**, 8897-8909.
20. Rappe, A.K. and Goddard III, W.A. (1991) Charge equilibration for molecular dynamics simulation, *J. Phys. Chem.* **95**, 3358-3363.
21. Filippini, G. and Gavezzotti, A. (1993) Empirical intermolecular potentials for organic crystals: the '6-exp' approximation revisited, *Acta Crystallogr. Sect. B* **49**, 868-880.
22. Karfunkel, H.R., Wu, Z.J., Burkhard, A., Rihs, G., Sinnreich, D., Buerger, H.M., and Stanek, J. (1996) Crystal packing calculations and Rietveld refinement in elucidating the crystal structures of two modifications of 4-amidinoamidanone guanylhydrazone, *Acta Crystallogr. Sect. B* **52**, 555-561.
23. Gavezzotti, A. and Filippini, G. (1995) Polymorphic forms of organic crystals at room conditions: Thermodynamic and structural implications, *J. Am. Chem. Soc.* **117**, 12299-12305.
24. Hall, R.C., Paul, I.C., and Curtin, D.Y., (1988) Structures and interconversion of polymorphs of 2,3-dichloroquinizarin. Use of second harmonic generation to follow the change of a centrosymmetric to a polar structure, *J. Am. Chem. Soc.* **110**, 2848-2854.

161

25. Stewart, J.J.P., MOPAC6.0, QCPE # 455.
26. Besler, B.H., Merz, K.M., and Kollman, P.A. (1990) Atomic charges derived from semiempirical methods, *J. Comput. Chem.* 11, 431-439.
27. Rodger, P.M., Stone, A.J., and Tildesley, D.J. (1988) The intermolecular potential of chlorine. A three phase study, *Mol. Physics* 63, 173-188.
28. Erk, P. (1999) Force, shape and structure: Modeling and making copper phthalocyanine polymorphs, *J. Porphyrins Phthalocyanines*, in preparation.
29. Karfunkel, H.R., Rohde, B., Leusen, F.J.J., Gdanitz, R.J., and Rihs, G. (1993) Continuous similarity measure between nonoverlapping X-ray powder diagrams of different crystal modifications, *J. Comput. Chem.* 14, 1125-1135.
30. Young, R.A. (1995) The Rietveld method, Oxford University Press, Oxford.
31. Graser, F. and Hädicke, E. (1980) Kristallstruktur und Farbe bei Perylen-3,4:9,10-bis(dicarboximid)-Pigmenten, *Liebigs Ann. Chem.*, 1994-2011.
32. Klebe, G., Graser, F., Hädicke, E., and Berndt, J. (1989) Crystallochromy as a solid-state effect: Correlation of molecular conformation, crystal packing and colour in perylene-3,4:9,10-bis(dicarboximide) pigments, *Acta Crystallogr. Sect. B* 45, 69-77.
33. Kazmaier, P.M. and Hoffmann, R. (1994) A theoretical study fo crystallochromy. Quantum interference effects in the spectra of perylene pigments, *J. Am. Chem. Soc.* 116, 9684-9691.
34. Perlstein, J. (1994) Molecular self-assemblies. 3. Quantitative predictions for the packing geometry of perylendicarboximide translation aggregates and the effects of flexible end groups. Implications for monolayers and three-dimensional crystal structure prediction, *Chem. Mater.* 6, 319-326.
35. Buncel, E., McKerrow, A.J., and Kazmaier, P.M. (1992) Molecular modelling of photoactive pigments. Investigation of polymorphism using the Dreiding force field, *Mol. Cryst. Liq. Cryst.* 211, 415-422.
36. Erk, P. and Henning, G. (1993) Von der Strukturformel zur Kristallstruktur - ab initio Modelling organischer Molekülkristalle, GVC-Fachausschuß Kristallisation, Freiberg.
37. Müller, M., Petersen, J., Strohmaier, R., Günther, C., Karl, N., and Müllen, K. (1996) Polybenzoide C$_{54}$-Kohlenwasserstoffe - Synthese und Strukturcharakterisierung in geordneten monomolekularen Aufdampfschichten, *Angew. Chem.* 108, 947-950.
38. Charlton, M.H., Docherty, R., and Hutchings, M.G. (1995) Quantitative structure-sublimation enthalpy relationship studied by neural networks, theoretical crystal packing calculations and multilinear regression analysis, *J. Chem. Soc. Perkin Trans.* 2, 2023-2030.
39. Gavezzotti, A. (1999) Molecular aggregation of acetic acid in a carbon tetrachloride solution: A molecular dynamics study with a view to crystal nucleation, *Chem.-Eur. J.* 5, 567-576.
40. McBride, J.M. and Carter, R.L. (1991) Spontane assymetrische Kristallisation durch Rühren, *Angew. Chem.* 103, 298-300.
41. Libbrecht, K.G. and Tanusheva, V.M. (1998) Electrically induced morphological instabilities in free dendritic growth, *Phys. Rev. Lett.* 81, 176-180.

THERMODYNAMICS AND KINETICS OF
CRYSTALLINE INCLUSION COMPOUNDS

L.R. NASSIMBENI
Chemistry Department, University of Cape Town,
Rondebosch, 7701, South Africa

This brief review will focus on the Physical Chemistry of crystalline inclusion compounds and will discuss their thermodynamic properties, the dynamics of enclathration and desolvation and attempt to relate them to their crystal structure. We will list a number of reviews and papers which outline the major concepts and some important techniques in the field of Inclusion Chemistry, with a view to introducing new researchers to this area and to facilitate their introduction to the extensive literature in this field. The bibliography does not pretend to be exhaustive, but concentrates on publications which give details of useful experimental techniques and illustrative examples.

Excellent historical introductions are given by Powell in the first chapter of Vol 1 of the series entitled "Inclusion Compounds" [1] and by Davies, Kemula, Powell and Smith in the first article that appeared in the Journal of Inclusion Phenomena [2]. Several books and monographs are available, culminating in the authoritative eleven volume publication of Comprehensive Supramolecular Chemistry which appeared in 1996 [3]. An historical overview is given which spans the years 1881 to 1998. This is taken from Anita Coetzee's PhD thesis (University of Cape Town, 1996) with permission of the author.

Historical Overview

1811	H. Davey prepares a chlorine hydrate [4]
1823	M. Faraday confirms Davy's observation and determines the composition of the chlorine hydrate [5]
1840	A. Damour observes the reversible dehydration of zeolite crystals [6]
1841	C. Schafhäutl prepares the graphite intercalates [7]
1849	F. Wohler prepares a β-quinol·H₂S molecular complex [8]
1891	A. Villiers prepares the first cyclodextrin inclusion compounds [9]
1893	S. U. Pickering publishes a study of alkylamine hydrates [10]

163

D. Braga et al. (eds.), Crystal Engineering: From Molecules and Crystals to Materials, 163–179.
© 1999 *Kluwer Academic Publishers. Printed in the Netherlands.*

164

	1894	E. Fischer introduces the "lock and key" principle [11]
Discovery of X-rays by Röntgen [12]	1895	
	1897	preparation of Hofmann's benzene inclusion compound: $Ni(CN)_2 \cdot NH_3 \cdot C_6H_6$ [13]
	1903	preparation of Hofmann's compound with aniline and phenol [14]
	1906	Preparation of inclusion compounds of triphenylmethane [15]
	1909	Preparation of tri-o-thymotide (TOT) benzene inclusion compound [16]
Nobel prize for Physics M.T.F. von Laue	1914	Synthesis of Dianin's compound: 4-(4-hydroxyphenyl)-2,2,4-trimethylchroman [17]
Nobel prize for Physics: W.H. Bragg and W.L. Bragg	1915	
World War 1	1916	H. Wieland and H. Sorge prepare the first choleic acid inclusion compounds [18]
	1918	
	1930	gas hydrates block natural gas pipelines in USA and USSR at temperatures higher than for normal ice formation – this results in a renewed interest in gas hydrates [19]
	1932	The term "molecular sieve" is coined to describe the behaviour of some microporous charcoals and zeolites [20]
	1935	E. Terres and W. Vollmer publish the preparation of phenol molecular complexes [21]
	1936	O. Kratky and G. Giacomello elucidate the crystal structure of choleic acid [22]

World War 2	1939	
	1940	M.F. Bengen patents the preparation of urea inclusion compounds [23]
	1945	
	1946	F.F. Mikus, R.M. Hixon and R.E. Rundle publish the preparation of amylose inclusion compounds [24]
	1947	D.E. Palin and H.M. Powell publish the crystal structures of the β-quinol·H$_2$S inclusion compound [25]
	1948	H.M. Powell coins the word "clathrate" [26]
	1951	D.J. Cram and H. Steinberg publish the synthesis of [2.2]paracyclophane [27]
	1951	M. von Stackelberg and H.R. Müller [28]; W.F. Claussen [29]; L. Pauling and R.E. Marsh [30] all propose crystal structures for the two types of gas hydrate and recognise them as clathrates [10]
Launch of Sputnik 1	1957	W.D. Schaeffer et al. publish an article entitled: "Separation of xylenes, cymenes, methylnaphthenes and other isomers by clathration of inorganic complexes", using Werner clathrates [31]
	1964	Publication of "Non Stoichiometric Compounds", L. Mandelcorn [32]
	1967	C.J. Pedersen publishes the preparation of crown ethers [33]
Man lands on the moon	1969	J.L. Atwood "accidentally" prepares first liquid clathrates [34]
	1969	J.-M. Lehn and co-workers publish the preparation of cryptands [35]
	1972	R.J. Argauer and G.R. Landolt describe zeolite ZSM5 [36]
	1974	D.J. Cram and J.M. Cram introduce the terms "host", "guest" and "host-guest complexation" [37]
	1976	D.D. MacNicol synthesises first hexa-hosts [38]
	1978	J.-M. Lehn introduces the concept and term "supramolecular chemistry" [39]

	1979	G.D. Andreetti, R. Ungaro and A. Pochini report the crystal structure of p-tert-butyl-calix[4]arene with toluene included in the cavity [40]
	1980	J. Lipkowski describes the physical chemistry of Werner clathrates [41]
	1980	Institute of Physical Chemistry of the Polish Academy of Science hosts the First International Symposium on "Clathrate Compounds and Molecular inclusion phenomena" in Jachranka, near Warsaw.
	1983	First issue of J. of Inclusion Phenomena is published, editors J.L. Atwood and J.E.D. Davies [2]
	1983	E. Weber and H.-P. Josel propose nomenclature and classification of host-guest type compounds [42]
	1984	First volume of Inclusion Compounds is published by Academic Press, London [1]
Nobel prize for Chemistry: J.Karle and H.A.Hauptman for developing direct phasing methods to determine X-ray crystal structures	1985	
	1987	J.F. Stoddart and co-workers publish the synthesis of the first catenanes and rotaxanes [43]
	1987	Nobel prize for Chemistry: D.J. Cram [44], J.-M. Lehn [45], C.J. Pedersen [46], for their respective work in the field of supramolecular chemistry
	1992	First issue of Supramolecular Chemistry is published, editors J.L. Atwood and G.W. Gokel [47]
	1993	F.H. Herbstein proposes principles for classification of crystalline binary adducts [48]

1994	First issue of Supramolecular Science is published, editor-in-chief W. Knoll [49]
1996	Publication of Comprehensive Supramolecular Chemistry, Vol. 1-11, Executive editors: J.L. Atwood, J.E.D. Davies, D.D. MacNicol and F. Vögtle, Pergamon Press. [3]
1998	First issue of Crystal Engineering is published, editors R.D. Rogers and M. Zaworotko [50]

Molecular recognition lies at the heart of Inclusion chemistry and is based on the intermolecular interactions which occur between the molecules which make up the host-guest framework. Herbstein [48] has given a classification scheme for binary adducts A-B, which distinguishes between "enclosure, segregated stack and packing complexes", and "molecular compounds", depending on the dominance of A...A or A...B interactions.

A more general approach has been proposed by Weber [42], who suggested a classification based on the host-guest type, their interaction, as well as topological and stoichiometric considerations. Hydrogen bonding remains the most important inter-molecular interaction, and a recent book by Jeffrey [51] reviews this important subject and has an extensive bibliography. It is these intermolecular forces which are ultimately responsible for the structure and thermal stability of the inclusion compound, the formation of which is presented by the equation

$$H\,(s,\alpha) + nG\,(g) \rightleftharpoons H \cdot G_n\,(s,\,\beta)$$

H is the non-porous solid α-phase of the Host compound, G is the guest molecule in vapour phase, and $H \cdot G_n$ is the solid inclusion compound or β-phase. Clearly different guest molecules give rise to different β-phases, but interestingly, when these desorb, they may give rise to different α-phases of the host. This is shown in Figure 1, which shows the process of induced polymorphism schematically. A variation of this is the formation of distinct polymorphs of 4,4'-Bis (diphenylhydroxymethyl) biphenyl which yields different structures when crystallised from diethylether or o-xylene. The polymorphs differ in their melting points by 6.4 °C and have different enthalpies of melting [52].

The question of polymorphism is of particular interest to the pharmaceutical industry, because the solubilities, and hence the bio-availabilities of different polymorphs vary. Giron discusses the thermal analysis and calorimeric methods used for the

168

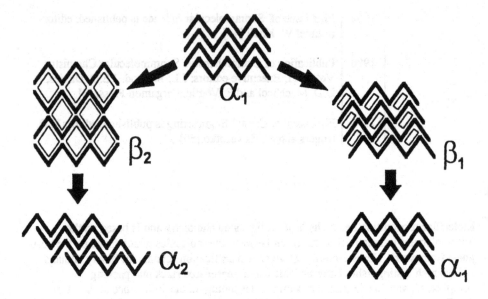

Figure 1: Induced polymorphism

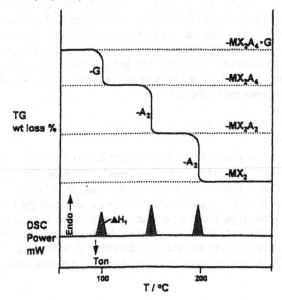

Figure 2

TG and DSC of a Werner clathrate $MX_2A_4 \cdot G$. The TG gives
the stoichiometry of each decomposition step, while the DSC
yields values of the onset temperatures and the ΔH associated
with each step.

characterisation of polymorphs and solvates [53]. This subject has also been recently reviewed by Caira [54].

Thermal analysis has become an important tool for the characterisation of inclusion compounds. There are several books and reviews on the subject [55-60]. With particular reference to inclusion compounds, McAdie analysed the thermal decomposition of urea-n-paraffin inclusion compounds [61] and of β-quinol clathrates [62,63]. The decomposition of Werner clathrates of the type $MX_2A_4 \cdot nG$ has been studied by various authors [64,65]. M is a transition metal such as Co, Ni, Cu, X is a monovalent anion such as halide or NCS⁻, and A is a substituted pyridine base, of which 4-picoline is the most efficient. The guest G is usually a small aromatic molecule. Thermal gravimetry (TG) has proved useful for elucidating the mechanism of the step-wise decomposition and Differential Scanning Calorimetry (DSC) yields the enthalpy changes associated with each step. A schematic result is shown in Fig. 2. An example of such a decomposition is given by Moore et al [66]. Values of ΔH devised from DSC measurements are not particularly reliable, because they are dependent on such experimental parameters as crystallite size, the heating rate, the flow rate of the purging gas and the geometry of the calorimeter [67]. We have therefore devised an apparatus which allows us to measure the vapour pressure of a volatile guest at various temperatures. The apparatus is shown diagrammatically in Figure 3 and consists of a transducer which converts pressure to a voltage signal which is amplified and fed to a computer. The temperatures of the oven and the specimen are monitored by thermocouples and the apparatus is fully automated. The programme is such that the temperature is ramped through a small temperature interval ΔT (typically 2°C) and this is followed by a long delay (typically 20 min) which allows the system to come to equilibrium. The pressure is then recorded and the process repeated. In this manner we obtained P vs T curves in the temperature range 20°C to 90°C which yielded straight semilogarithmic plots of ℓnp vs 1/T. These curves are shown schematically in Figures 4a and 4b. The values of ΔH derived from such experiments are highly reproducible, and we routinely obtained a precision of <2% in repetitive experiments. We note that the enthalpy change of the guest –release reaction is a complex quantity which involves the phase transformation of the β-phase to the "empty" clathrate structure βo, and its change to the non-porous α-form as well as the enthalpy of desorption [68], the various components of which can be measured by calorimetry. Recently Aoyama et al. [69] have discussed the importance of binding isotherms of guest vapours on solid hosts and the dynamics of lattice inclusion. The study extended to using porous metalated hosts which have remarkable catalytic properties [70].

KINETICS

We shall discuss the kinetics of enclathration or of desolvation as heterogeneous processes, and most processes analysed to date involve the decomposition of the solid inclusion compound to yield a solid host and a gaseous guest. The fraction decomposed (α) may be calculated from the mass loss $\alpha = (m_o - m_t) / (m_o - m_f)$ where m_o is the

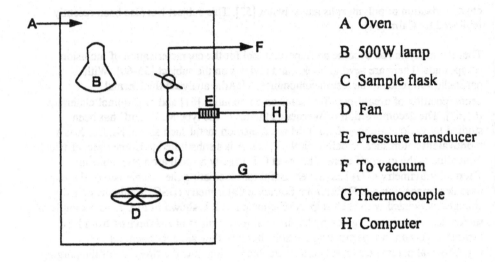

A	Oven
B	500W lamp
C	Sample flask
D	Fan
E	Pressure transducer
F	To vacuum
G	Thermocouple
H	Computer

Figure 3: Vapour Pressure Apparatus

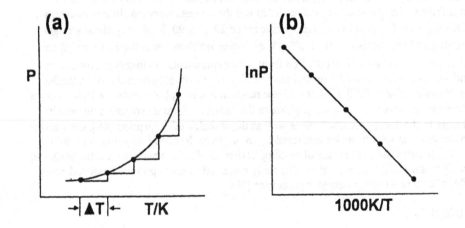

Figure 4

(a) Vapour pressure vs temperature for an inclusion compund
   with a volatile guest

(b) Semilogarithmic plot, whose slope = -ΔH/R

initial mass of the sample, m_t is the mass at time t and m_f is the final mass, after decomposition.

A generalised α - time curve is shown in Figure 5. Various kinds of apparatus are available for carrying out such experiments, such as commercial TG balances, or for very small samples, quartz micro balances (QMB) can be easily constructed [71]. The preferred procedure is that of isothermal runs at various fixed temperatures. Various models can be fitted to the ensuing α - time curves which correspond to particular mechanisms of the desolvation reaction [56]. We have found that the most common of these are:

$-\ln(1-\alpha) = kt$	First order	F1
$1-(1-\alpha)^{1/2} = kt$	Contracting area	R2
$1-(1-\alpha)^{1/3} = kt$	Contracting volume	R3

With increasing temperature, induction periods become shorter and rate constants, k, increase. Activation energies for both the induction period and the main part of the reaction may then be derived by application of the Arrhenius equation. An example of this form of kinetic decay is exhibited by the desolvation of trans- 9,10-Dihydroxy-9,10-diphenyl-9,10-dihydroanthracene (DDDA) with cyclohexanone [72].

Isothermal methods of analysis yield reliable kinetic parameters but are time consuming. Differential methods, which depend on mass losses measured at fixed heating rates, are quicker but less accurate. Nevertheless if precautions are taken with the particle size distribution, non-isothermal methods can be made to yield reproducible results of the activation energy. An example is the decomposition of DDDA with acetonitrile, which gave an activation energy ranging between 83 and 115 kJ mol^{-1} [73]

Measuring the kinetics of enclathration is experimentally more difficult, and not many results have been reported. We employ an automated suspension balance which measures the mass change of a host compound under controlled conditions of guest pressure and temperature [74]. Using this we analysed the kinetics of clathrate formation of DDDA with acetone vapour and noted that at a given temperature, a minimum threshold pressure of Po of the acetone vapour was required in order for reaction to occur. Values of Po were temperature dependent [75]. This phenomenon has also been observed by Aoyama [69]. The kinetics of solid-liquid reactions have also received some attention, particularly with respect to layered structures. The construction details of a thermostatted dilatometer have been published [76], which promises to prove useful for such measurements.

The application of the Arrhenius equation:

$$k = A e^{-Ea/RT}$$

to heterogeneous reactions is controversial, but Galway and Brown [77, 78] discuss this topic and give a theoretical justification for its use. An excellent paper by Brown [79], entitled "Steps in a Minefield. Some Kinetic aspects of thermal analysis" reviews

Figure 5

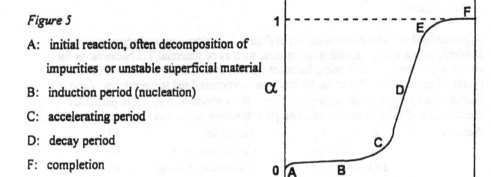

A: initial reaction, often decomposition of impurities or unstable superficial material

B: induction period (nucleation)

C: accelerating period

D: decay period

F: completion

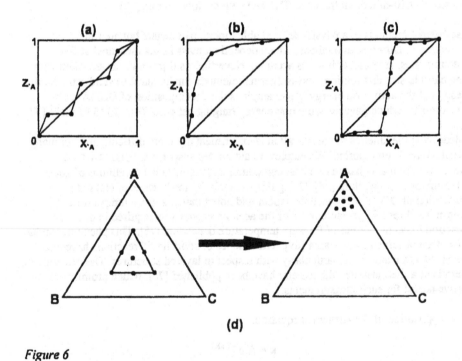

Figure 6

Competition experiments for two- or three-guest system

X_A = mole fraction of guest A in starting solution
Z_A = mole fraction of guest A in the crystals
6(a): poor selectivity 6(b): good selectivity 6(c): concentration dependent selectivity
6(d): three-guest system

controversial aspects of this field with particular reference to distinguishability and extent of fit of kinetic models, complementary experimental evidence and the compensation effect in Arrhenius parameters.

SELECTIVE ENCLATHRATION

One of the important applications of Inclusion Chemistry is the separation of close isomers by enclathration. This involves the choice of a suitable host compound which, when exposed to a mixture of guests, combines selectively with a particular guest to form a crystalline inclusion compound. The latter is filtered and the guest released by gentle warming, enabling the host to be recycled. The selectivity of the process depends on the efficiency of the molecular recognition between host and guest and selectivity > 99% is usually achieved in three or four cycles. In a typical experiment eleven vials are set up and the mole fraction of A is varied systematically from 0 to 1, taking care to keep the total guest content in ten fold excess of the host. Typical results are shown schematically in Figure 6, in which three situations occur:
Figure 6a: There is virtually no discrimination between A and B and the points lie close to the diagonal line, which corresponds to zero selectivity.
Figure 6b: A is strongly preferred to B.
Figure 6c: Selectivity of A and B is concentration dependent.

We have extended this technique to analyse the simultaneous competition of three guests, A, B and C, and the example shown gives the result in which A is preferentially enclathrated over B and C, as shown in Figure 6d. We have studied a number of systems using this technique [80, 81].

LATTICE ENERGY CALCULATIONS

We have attempted to rationalise the results of competition experiments by molecular mechanics calculations. We reasoned that the lattice energy of a given inclusion compound is an important parameter, which defines the potential energy environment of the guest in the lattice defined by the host molecules. We have had some success in that lower lattice energies generally correlate with preferred selectivity, but we have found some exceptions. One of the difficulties is the selection of appropriate coefficients of the atom-atom potentials and of suitable hydrogen-bonding potentials. Progress in this field has been made by Gavezzotti [82]. Regarding the prediction of the structure of host-guest complexes, the author remains sceptical that this can be achieved except in the most trivial cases. For example when an inclusion compound H·G1 does not yield suitable single crystals for structure elucidation, but its powder x-ray diffraction shows it to be isomorphous with a known structure H·G2 where G1 and G2 are similar, then potential energy calculations can yield good approximations of the unknown structure. We have used this methodology in a study of the structures formed between 1,1- bis (4-hydroxyphenyl) cyclohexane and the isomers of benzene diol [83].

174

However, predictions of crystal structure, particularly in the case of host-guest systems, remains an on-going challenge, and the reader is referred to an interesting paper by Gavezzotti [84], entitled "Are crystal structures predictable ?", to which he gives a succinct answer with the first word in the paper: "No".

References

1. Powell, H.M. (1984) Introduction, in J.L. Atwood, J.E.D. Davies, D.D. MacNicol (eds.) *Inclusion Compounds*, 1, Academic Press, London, pp 1-28.
2. Davies, J.E.D., Kemula, K., Powell, H.M. and Smith, N.O. (1983) Inclusion Compounds – Past, present and future, *J. Incl. Phenom.*, 1, 3-44.
3. Comprehensive Supramolecular Chemistry (1996) Executive Editors: J.L. Atwood, J.E. Davies, D.D. MacNicol and F. Vögtle, Vols 1-11. Pergamon, Elsevier Science Ltd, Oxford.
4. Davy, H. (1811) On some of the Combinations of Oxymuriatic gas and oxygene, and on the chemical relations of these principles, to inflammable bodies, *Philos. Trans. R. Soc. London.*, 101, 1-35.
5. Faraday, M. (1823) On hydrate of chlorine, *Quart. J. Sci.*, 15, 71-74.
6. Damour, A. (1840) *Ann. Mines.*, 17, 191.
7. Schafhäutl, C. (1841) Ueber die verbindungen des kohlenstoffes mit silicium, eisen und anderen metallen, welche die verschiedenen gattungen ron roheisen, stahl und schmiedeeisen bilden, *J. Prakt. Chem.*, 21, 129-157.
8. Wöhler, F. (1849) Ueber einige verbindungen aus der chinonreihe, *Ann. Chem. Liebigs.*, 69, 294-300.
9. Villiers, A. (1891) Sur la fermentation de la fécule par l'action du ferment butyrique, *C. R. Hebd. Seances Acad. Sci.*, 112, 536-538.
10. Pickering, S.U. (1893) The hydrate theory of solutions. Some compounds of the alkylamines and ammonia with water, *J. Chem. Soc. Trans.*, 63, 141-195.
11. Fischer, E. (1894) Einfluss der configuration auf die wirkung der Enzyme, *Ber. Deutsch. Chem. Ges.*, 27, 2985-2993.
12. Lima de Faria, J. (ed.) (1990) *Historical Atlas of Crystallography*, Kluwer Academic Publishers, Dordrecht.
13. Hofmann, K.A. and Küspert, F. (1897) Verbindungen von kohlenwasserstoffen mit metallsalzen, *Z. Anorg. Allg. Chem.*, 15, 204-207.
14. Hofmann, K.A. and Höchtlen, F. (1903) Abnorme verbindungen des nickels, *Chem. Ber.*, 36, 1149-1151.
15. Hartley, H. and Thomas, N.G. (1906) The solubility of triphenylmethane in organic liquids with which it forms crystalline compounds, *J. Chem. Soc.*, 1013 1033.
16. Spallino, R. and Provenzal, G. (1909) Sulla preparazione dell'acido ortotimotico e di alcuni suoi derivati, *Gazz. Chim. Ital.*, 39, 325-336.

17. Dianin, A.P. (1914) On the condensation of phenols with unsaturated ketones. Condensation of phenols with mesitylene oxide. *J. Soc. Phys. Chem. Russe*, **46**, 1310-1319.

18. Wieland, H. and Sorge, H. (1916) Untersuchungen über die gallensäuren. II. Mitteilung. Zur kenntnis der choleinsäure, *Z. Physiol. Chem. Hoppe-Seyler's*, **97**, 1- 27.

19. Hammerschmidt, E.G. (1934) Formation of gas hydrates in natural gas transmission lines, *Ind. Eng. Chem.*, **26**, 851-855.

20. McBain, J.W. (1932) Sorption by chabasite, other zeolites and permeable crystals, *Absorption of Gases and Vapours by Solids*, Routledge and Sons, Chapter 5, 167-176.

21. Terres, E. and Vollmer, W. (1935) *Z. Petroleum*, **31**, 1.

22. Kratky, O. and Giacomello, G. (1936) Der kristallbau der paraffin carboncholeinsäuren, *Monatsh. Chem.*, **69**, 427-436.

23. Bengen, M.F. *German Patent Application* 02123438, March 18, 1940.

24. Mikus, F.F. , Hixon, R.M. and Rundle, R.E. (1946) The complexes of fatty acids with amylose, *J. Amer. Chem. Soc.*, **68**, 1115 - 1123.

25. Palin, D.E. and Powell, H.M. (1947) The structure of molecular compounds. Part III. Crystal structure of addition complexes of quinol with certain volatile compounds, *J. Chem. Soc.*, 208 - 221.

26. Powell, H.M. (1948) The structure of molecular compounds. Part IV. Clathrate compounds, *J. Chem. Soc.*, 61-73.

27. Cram, D.J. and Steinberg, H. (1951) Macro Rings. I. Preparation and spectra of the paracyclophanes, *J. Am. Chem. Soc.*, **73**, 5691 - 5704.

28. Von Stackelberg, M. and Müller, H.R. (1951) On the structure of gas hydrates *J. Chem. Phys.*, **19**, 1319-1320.

29. Claussen, W.F. (1950) Suggested structures of water in inert gas hydrates, *J. Chem. Phys.*, **19**, 259 - 260.

30. Pauling, L. and Marsh, R.E. (1952) The structure of chlorine hydrate, *Proc. Natl. Acad. Sci. USA*, **38**, 112-118.

31. Schaeffer, W.D., Dorsey, W.S., Skinner, D.A., and Christian, C.G. (1957) Separation of Xylenes, Cymenes, Methylnaphthalenes and other isomers by clathration with inorganic complexes, *J. Am. Chem. Soc.*, **79**, 5870-5876.

32. L. Mandelcorn (ed.) (1964) *Non Stoichiometric Compounds*. Academic Press, New York.

33. Pedersen, C.J. (1967) Cyclic polyethers and their complexes with metal salts. *J. Am. Chem. Soc.*, **89**, 7017-7036.

34. Atwood, J.L., Milton, P.A. and Seale, S.K. (1971) Thermal behavior of anionic organoaluminum thiocyanates, *J. Organomet. Chem.*, **28**, C29-C30.

35. Dietrich, B., Lehn, J.-M. and Sauvage, J.-P. (1969) Diaza-polyoxa-macrocycles et macrobicycles, *Tetrahedron Lett.*, 2885-2888.

36. Argauer, R.J. and Landolt, G.R. (1972) US Patent, 3, 702, 886.

37. Cram, D. J. And Cram, J. M. (1974) Host-guest chemistry. *Science*, **183**, 803-809.

38. MacNicol, D.D. and Wilson, D.R. (1976) New strategy for the design of inclusion compounds: Discovery of the 'Hexa-hosts', *J. Chem. Soc. Chem. Comm.*, 494-495.

39. Lehn, J.-M. (1978) Cryptates: Inclusion complexes of macropolycyclic receptor molecules, *Pure Appl. Chem.*, **50**, 871-892.
40. Andreetti, G.D., Ungaro, R. and Pochini, A. (1979) Crystal and molecular structure of Cyclo{quarter[(5-t-butyl-2-hydroxy-1,3-phenylene)methylene]} Toluene (1:1) Clathrate, *J. Chem. Soc. Chem. Comm.*, 1005-1007.
41. Lipkowski, J. (1980) Structure and physico-chemical behavior of clathrates formed by the Ni(NCS)$_2$(4- Methylpyridine)$_4$ complex, *Accademia Polacca deele Scienze, Biblioteca e centro di studi a Roma. Conferenze*, **81**, 1-27.
42. Weber, E. and Josel, H.-P. (1983) A proposal for the classification and nomenclature of host-guest-type compounds, *J. Incl. Phenom.*, **1**, 79-85.
43. Allwood, B.L., Spencer, N., Shairiari-Zavarech, H., Stoddart, J.F. and Williams, D.J. (1987) Complexation of Paraquat by a bisparaphenylene-34-crown-10 derivative, *J. Chem. Soc. Chem. Commun.*, 1064-1066.
44. Cram, D.J. (1988) The design of molecular hosts, guests and their complexes, *Angew. Chem. Int. Ed Engl.*, **27**, 1009 - 1020.
45. Lehn, J.-M. (1988)Supramolecular chemistry-scope and perspectives molecules, supermolecules and molecular devices (Nobel lecture), *Angew. Chem. Int. Ed Engl.*, **27**, 90-112.
46. Pedersen, C.J. (1988)The discovery of crown ethers (Nobel lecture) *Angew. Chem. Int. Ed Engl.*,**27**, 1021 - 1027.
47. Supramolecular Chemistry (1992) J.L. Atwood and G.W. Gokel (eds.) Gordon and Breach, New York.
48. Herbstein, F.H. (1993) Structural principle in the classification of crystalline binary adducts (molecular compounds and complexes), *Acta Chim. Hung.*, **130**, 377-386.
49. Supramolecular Science (1992) Elsevier Science Ltd., Oxford.
50. Crystal Engineering (1998) R.D. Rodgers and M. Zaworotko (eds.)Pergamon, Elsevier Science, Oxford.
51. Jeffrey, G.A. (1997)*An introduction to Hydrogen Bonding*, Oxford University Press, Oxford.
52. Weber, E., Skobridis, K., Wierig,A., Nassimbeni, L.R. and Johnson, L. (1992) Complexation with diol host compounds. Part 10. Synthesis and solid state inclusion compounds of bis (diarylhydroxymethyl) –substituted benzenes and biphenyls: X-ray crystal structures of two host polymorphs and of a non-functional host analogue. *J. Chem. Soc. Perkin Trans.2*, 2123 –2130.
53. Giron, D. (1995) Thermal analysis and calorimeric methods in the characterisation of polymorphs and solvates. *Thermochim. Acta*, **248**, 1 –59.
54. Caira, M.R. (1998) Crystalline polymorphism of organic compounds in E. Weber (ed.) *Topics in Current Chemistry*, Vol 198, Springer Verlag, Berlin, pp163 – 208.
55. Wendlandt, W.W. (1986) *Thermal Analysis* (3rd ed), Wiley, New York.
56. Brown, M.E. (1988) *Introduction to Thermal Analysis*, Chapman and Hall, London.
57. Wunderlich, B. (1990) *Thermal Analysis*, Academic Press, San Diego.
58. Haines, P.J. (1995) *Thermal Methods of Analysis. Principles, Applications and Problems*, Chapman and Hall, London.
59. Höhne, G., Hemminger, W. and Flammersheim, H.-J. (1996) *Differential Scanning Calorimetry. An introduction for Practitioners*, Springer-Verlag, Berlin.

178

60. Cammenga, H.K. and Eppel, M. (1995) Basic principles of thermoanalytical techniques and their applications in preparative chemistry, *Angew. Chem. Int. Ed. Engl.*, **34**, 1171 – 1187.

61. McAdie, H.G. (1962) Thermal decomposition of molecular complexes. I. Urea-n-paraffin inclusion compounds, *Can. J. Chem.*, **40**, 2195 – 2203.

62. McAdie, H.G. (1963) Thermal decomposition of molecular complexes. II. β-quinol clathrates, *Can. J. Chem.*, **41**, 2137 – 2143.

63. McAdie, H.G. (1996) Thermal decomposition of molecular complexes. IV. Further studies of β-quinol clathrates, *Can. J. Chem.*, **44**, 1373 – 1385.

64. Gavrilova, G.V., Kislykh, N.V. and Logvinenko, V.A. (1998) Study of the thermal decomposition processes of clathrate compounds, *J. Therm. Anal.*, **33**, 229 – 235.

65. Lipkowski, J. (1996) Clathration and solvation of molecules, in G. Tsoucaris, J.L. Atwood and J. Lipkowski (eds.), *Crystallography of Supramolecular Compounds*, Kluwer Academic Publishers, Dordrecht, pp. 265 – 283.

66. Moore, M.H. , Nassimbeni, L.R. and Niven, M.L. (1987) Studies in Werner Clathrates. Part 5. Thermal analysis of bis (isothiocyanato) tetra (4-vinylpyridine) nickel (II) Inclusion Compounds. Crystal structure of the $Ni(NCS)_2(4$-Vipy$)_4$·$2CHCl_3$ clathrate, *Inorg. Chim. Acta*, **131**, 45 – 52.

67. Caira, M.R. and Nassimbeni, L.R. (1996) Phase transformation in inclusion compounds, kinetics and thermodynamics of enclathration in D.D. MacNicol, F. Toda and R. Bishop (eds.), *Comprehensive Supramolecular Chemistry, Vol 6. Solid State Supramolecular Chemistry: Crystal Engineering*, Pergamon, Elsevier Science Ltd., Oxford, pp. 825 – 850.

68. Starzewski, P., Zielenkiewicz, W. and Lipkowski, J. (1984) A thermokinetic study of the clathration of isomeric xylenes by the $Ni(NCS)_2$ (4-Menthylpyridine)$_4$ Host, *J. Incl. Phenom.*, **1**, 223 – 232.

69. Dewa, T., Endo, K. and Aoyama, Y. (1998) Dynamic aspects of lattice inclusion complexation involving a phase change. Equilibrium, kinetics and energetics of guest-binding to a hydrogen-bonded flexible organic network., *J. Amer. Chem. Soc.*, **120**, 8933 – 8940.

70. Sawaki, T., Dewa, T. and Aoyama, Y. (1998) Immobilisation of soluble metal complexes with a hydrogen-bonded organic network as supporter. A simple route to microporous solid Lewis acid catalysts., *J. Amer. Chem. Soc.*, **120**, 8539 – 8540.

71. Coetzee, A., Nassimbeni, L.R. and Achleitner, K. (1997)A quartz microbalance for measuring the kinetics of guest uptake from the vapour, *Thermochim. Acta*, **298**, 81 – 85.

72. Coetzee, A., Nassimbeni, L.R. and Su, H. (1999) Desolvation of trans-9,10-Dihydroxy-9,10-diphenyl-9,10-dihydroanthracene. Cyclohexanone: kinetic compensation effect. *J. Chem. Res.*, in press.

73. Caira, M.R., Nassimbeni, L.R. and Schubert, W.-D. (1992) Complexation with diol host compounds. Part 9. Structures and thermal analysis of inclusion compounds of trans-9,10-dihydroxy-9,10-diphenyl-9,10-dihydroanthracene with acetonitrile and 3- hydroxyproprionitrile, *Thermochim. Acta*, **206**, 265 – 271.

74. Barbour, L.J., Achleitner, K. and Greene, J.R. (1992) A system of studying gas-solid reaction kinetics in controlled atmospheres, *Thermochim. Acta*, **205**, 171-177.

75. Barbour, L.J., Caira, M.R. and Nassimbeni, L.R. (1993) Kinetics of Inclusion, *J. Chem. Soc. Perkin Trans. 2*, 2321 – 2322.
76. Votnisky, J., Kalousova, J. Benes, L., Bandysova, I. and Zima, V. (1993) Volumetric method for following the rate of intercalation of liquid molecular guests into layered hosts, *J. Incl. Phenom. Mol. Recog. Chem.*, 15, 71 – 78.
77. Brown, M.E. and Galway, A.K. (1989) Arrhenius parameter for solid-state reactions from isothermal rate-time curves, *Anal. Chem.*, 61, 1136 – 1139.
78. Galway, A.K. and Brown, M.E. (1995) A theoretical justification for the application of the Arrhenius equation to kinetics of solid state reactions (mainly ionic crystals), *Proc. R. Soc. Lond. A*, 450, 501 – 512.
79. Brown, M.E. (1997) Steps in a minefield. Some kinetic aspects of thermal analysis. *J. Therm. Anal.*, 49, 17 – 32.
80. Caira, M.R., Horne, A., Nassimbeni, L.R. and Toda, F. (1997) Inclusion and separation of picoline isomers by a diol host compound, *J. Mater. Chem.*, 7, 2145 – 2149.
81. Caira, M.R., Horne, A., Nassimbeni, L.R. and Toda, F. (1998) Selective inclusion of aliphatic alcohols by a diol host compound, *J. Mater. Chem.*, 8, 1481 – 1484.
82. Gavezzotti, A. (1998) The crystal packing of organic molecules: Challenge and fascination below 1000 Da, *Crystallography Reviews*, 7, 5 – 121.
83. Caira, M.R., Horne, A., Nassimbeni, L.R. and Toda, F. (1997) Complexation with diol host compounds. Part 25. Selective inclusion of benzenediol isomers by 1,1-bis (4-hydroxyphenyl) cyclohexane, *J. Chem. Soc. Perkin Trans. 2*, 1717 – 1720.
84. Gavezzotti, A. (1994) Are crystal structures predictable?, *Acc. Chem. Res.*, 27, 309 – 314.

AN APPROACH TO THE CRYSTAL ENGINEERING OF COORDINATION NETWORKS

MIR WAIS HOSSEINI
Université Louis Pasteur
Laboratoire de Chimie de Coordination Organique
Institut Le Bel,
F-67000 Strasbourg, France

1. Introduction

Molecular networks, in principle infinite architectures, are large size molecular assemblies (10^{-6}-10^{-3} m scale) or hyper molecules composed of molecular components with defined structures and coonectectivity patterns between the components. The construction of molecular networks with predicted and programmed structure may hardly be envisaged through stepwise classical synthesis using covalent linkages. However, the preparation of such higher-order materials may be attained through iterative process based on self-assembly [1-11] of individual tectons [8] (from Greek TEKTON, builder) [3,4]. In principle, tectons may be organic modules and/or metallic centres bearing in their structure specific informations (molecular algorithms) dealing with the formation processes and the final structure of molecular networks. Thus, molecular tectonics [9], based on molecular programming, deals with the formation of molecular networks in solution or in the solid state. Operational concepts in molecular tectonics are based on molecular recognition (molecular recognition is employed in a rather broad sens including any type of molecular interaction, in particular metal coordination processes) operating at the level of the complementary tectons and on geometrical and topological features allowing the iteration or repetition of the recognition pattern. Based on the nature of interactions involved in the assembling core, one may classify networks into three classes. When the recognition pattern between tecons is mainly based on inclusion processes, inclusion networks are obtained [12]. When the assembling core is formed by hydrgen bonds, the networks obtained may be called hydrogen bonded networks [6-8. 13-15] engaging metal cations, the network formed may be nemed coordination networks [16-22]. However, this classification is rather simplistic since many different type of interactions may operate at the same time and in concert.

In marked contrast with discrete molecules, molecular networks are non covalent hypermolecules obtained by repetition with translational symmetry of assembling cores which are defined as specific interaction patterns taking place between building blocks. The dimensionality of a molecular network depends on the number of translations operating at the level of the assembling cores [14]. Thus, the α-networks are formed when a single translation is operational. Similarly, β- and γ-networks are defined when two or three translations respectively of the same or different assembling cores take place (Figure 1).

181

D. Braga et al. (eds.), Crystal Engineering: From Molecules and Crystals to Materials, 181–208.

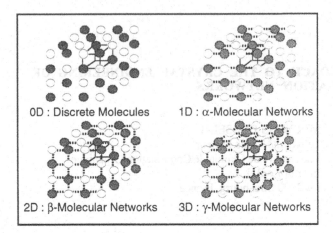

Figure 1. Schematic representation of discrete molecules and molecular networks in the crystalline phase. The dimensionality of the network depends upon the number of translation of assembling cores which may be defined as specific interaction motifs between modules defining the solid.

In the solid state, molecular crystals are defined by the chemical nature of their molecular components and by their interactions with respect to each other. Whereas molecular chemistry deals with the design and synthesis of the individual units, their assembly is governed by concepts of supramolecular chemistry [1].

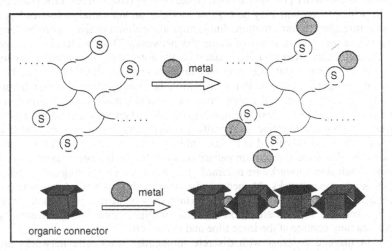

Figure 2. In marked contrast with coordinating polymers, *i.e.* polymers bearing ligating sites allowing the coordination of metals (top), for coordination polymers, metal centres are incorporated within the structure of the polymer behaving thus as structural points (bottom).

Although with the present level of our knowledge, we are not able to predict all types of interactions, in particular weak van der Waals interactions which are neccessarly present in the crystalline phase, by proper design and thus manipulation of interactions energy between some of molecular components (for example, often the presence of solvent molecules in the crystal is not predicted) some molecular networks with predicted structures and dimensionality (assembling, core, connectivity and translation) may be designed and obtained.

Because metals offer a vast variety of coordination geometries, oxidation states as well elctronic, photonic and magnetic properties, the design and preparation of coordination networks with controlled structures and, thus, properties, are currently under active investigation [16-22].

Since some years, we are engaged in investigations dealing with the design and formation of molecular networks in the crystalline phase. In particular, we have reported inclusion- [12] as well as hydrogen bonded- networks [14]. In the present contribution. we shall mainly focuss on our approach on coordinations networks [18-22].

Coordination polymers, equaly called coordination networks, are metallo-organic entities composed of metal centres bridged by organic connectors. As stated above, they may be described as molecular networks. In marked contrast with coordinating polymers, *i.e.* polymeric structures bearing coordination sites allowing the binding of metal cations, for coordination polymers, metal centres are incorporated within the structure of the polymer behaving thus as structural knots (Figure 2).

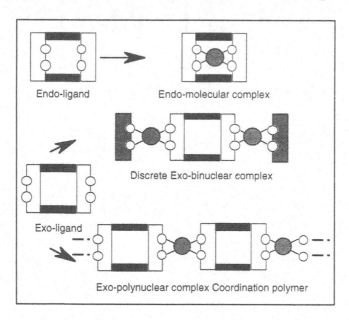

Figure 3. Whereas endo ligands, by conception, are designed to form only discrete molecular complexes with selected metals (top), exo-ligands may form either discrete exobinuclear complexes in the presence of metals and stopper ligands (middle) or molecular networks in the presence of connecting metals (bottom).

184

As stated above, in order to achieve an iterative assembling process, the building blocks or tectons must fulfil both structural and energy criteria. In particular, the complementary tectons (in the context of coordination polymers metals and organic ligands) must be able to interact with each other (molecular association) and furthermore should allow the repetition of the interaction pattern (iteration). For the connecting ligand, these two requirements lead to molecular modules possessing coordination sites oriented in a divergent fashion (Figure 3). Whereas endo ligands, by design, may form only discrete molecular complexes with selected metals, exo-ligands may form either discrete exobinuclear complexes in the presence of metals bearing stopper ligands blocking the translation of the assembling core, or molecular networks *i.e.* coordination polymers in the presence of bridging metals.

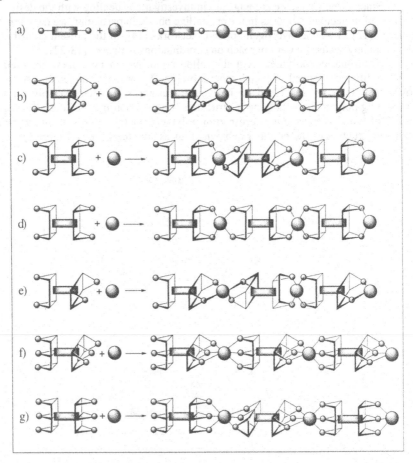

Figure 4. Schematic representation of linear coordination networks which may be formed when using bis-mon-, -di- and -tri-dentate ligands;

The formation as well as the control of dimensionality of coordination polymers are guided by the choice of the metal in conjunction with the design of the ligand. In the present chapter, we shall largely focuss on the formation of α-coordination networks. However, a recent example of a β-coordination well also be presented at the end of the contribution.

In order to obtain α-coordination networks, many different designs may be explored. Dealing with ligands bearing two coordination sites oriented in a divergent fashion (exobismondentate ligands), linear coordination networks may be obtained using metal cations accepting two coordination sites in their coordination sphere and permitting linear coordination (Figure 4a). On the other hand, with metal centres adopting either tetrahedral (Figure 4b, 4c) or square planar (Figure 4d, 4e) coordination geometries are used, ligands possessing four coordination sites forming two sets of two binding sites in which the two faces of the backbone bear each one set of bidentate site (exobisbidentate) may be used. Howeve, the two sets of coordination sites may be localised in the same plane (Figure 4c, 4d) or in two perpendicular planes (Figure 4a, 4e). Finally, exobistridentate ligands containing six coordination sites composing two sets or tridentate motifs may be used with metal cations adopting octahedral coordination geometry (Figure 4f, 4g). Again, the two sets of three coordination sites may be in the same plane (Figure 4g) or localised in two perpendicular planes (Figure 4f).

2. Linear coordination networks based on calixarenes

As stated above, for the formation of molecular networks, the design of exo-ligands in which the coordination sites are oriented in a divergent fashion is crucial. Molecular units possessing four coordination sites occupying the apices of a pseudo-tetrahedron may be of interest for construction of linear coordination polymers using metals requiring a tetrahedral coordination geometry. The design of such a ligand may be based on a preorganised backbone offering the possibility of anchoring four coordination sites in an alternating mode below and above of its main plane. This aspect was previously demonstrated in the case of mercaptocalix[4]arene derivatives for which the OH groups were replaced by SH moieties at the lower rim [23].

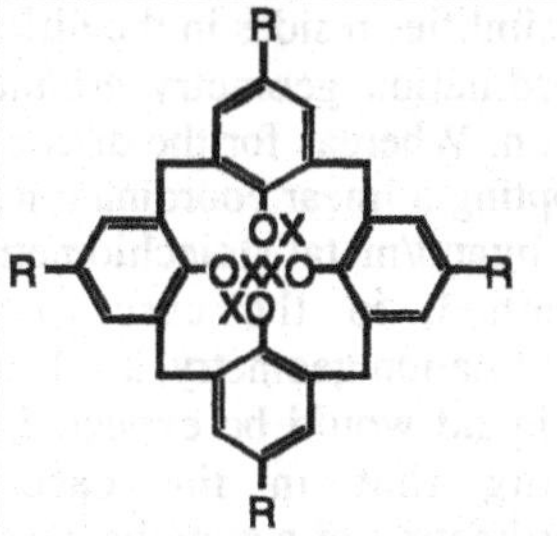

1 R = CN, X = CH₂CH₂OCH₃
2 R = C(CH₃)₃, X = H
3 R = H, X = H
4 R = H, X = CH₂CH₂OCH₃
5 R = Br, X = CH₂CH₂OCH₃

Scheme 1

An other design may be based on the use of both the upper and lower rims. Indeed, one may impose the needed 1,3-alternate conformation by proper transformation of all four hydroxy groups, and on the other hand, using *para* positions one may set-up, in a controlled manner, coordination sites. This has been demonstrated in the case of catechol units as the coordination sites [24]. Using this strategy, the exo-ligand 1 bearing four nitrile groups was designed (scheme 1) [19]. It is worth noting that for 1, due to the donor effect of oxygen atoms, the binding ability of nitriles is considerably enhanced.

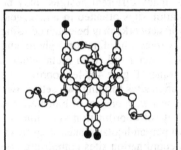

Figure 5. X-ray structural of the free ligand 1 adopting the 1,3-alternate conformation. H atoms are not represented for sake of clarity.

The synthesis of **1** was achieved as follows. The starting material was the *p-tertbutyl*calix[4]arene **2** [25] which after dealkylation in toluene in the presence of phenol and AlCl₃ afforded the compound **3** as a mixture of conformers [26]. The O-alkylation of the latter using 2-methoxyethyltosylate in DMF in the presence of Cs₂CO₃ afforded the compound **3** which, after recrystallisation, was shown to adopt the 1,3-alternate conformation [27]. The desired tetracyano compound **1** was obtained after bromination of **4** using NBS in butanone leading to the tetrabromo compound **5** and followed by treatment of the latter by CuCN in N-methylpyrolidone [28]. The structural assignment of **1** was achieved by classical NMR studies as well as by X-ray diffraction method which indeed confirmed the 1,3-alternate conformation (Figure 5).

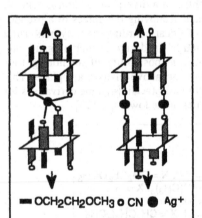

■OCH₂CH₂OCH₃ o CN ● Ag⁺

Figure 6. Schematic representation of two types of linear coordination networks which may be envisaged by the self-assembly of the ligand 1 and AgI cation adopting a tetrahedral- (left) or linear- (right) coordination geometries.

Dealing with the metal cation, Agⁱᴵ, alreday extensively used for the formation of coordination networks [29], was chosen because it forms kinetically labile complexes. In principle, for a combination of the ligand **1** and AgⁱᴵI, one may envisage two types of linear coordination polymers (Figure 6). The difference between the two possibilities resides in the difference in the coordination geometry around the silver cation. Whereas for the di-coordinated Ag adopting a linear coordination geometry, a 1/2 ligand/metal stoiechiometry would be obtained, in the case of tetrahedral coordination geometry, a 1/1 ratio of metal to ligand would be expected. It is worth noting that in the case of linear coordination of a metallic centre, one may obtain a tubular structure of the metallatubulane type.

Upon slow diffusion of a CH₂Cl₂ solution of the ligand **1** into a EtOH solution of AgAsF₆ in large excess, colourless crystalline material was obtained.

The analysis of single-crystals by X-ray crystallography revealed the presence of disordered H_2O and EtOH molecules in the lattice. In addition to the solvent molecules the crystal (tetragonal, space group *P 4/n n c*) was composed of linear coordination polymers and disordered AsF_6 anions. The ligand 1, as in the absence of Ag^I cation (Figure 5), adopted a 1,3-alternate conformation (Figure 7).

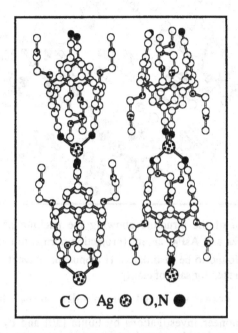

C ◯   Ag ⊕   O,N ●

Figure 7. A portion of the X-ray structure showing fragments of two parallel cationic linear coordination networks obtained by mutual bridging of Ag^I cations and ligands 1 (projection down the polymer axis). H atoms, solvent molecules and anions are not represented for sake of clarity.

The ether fragments adopted a gauche conformation with a OCCO dihedral angle of 73.8°. In order to bind two Ag^I cations in the tetrahedral mode of coordination, the calix unit was slightly pinched at the upper rim, *i.e.*, the N···N and O···O distances between nitrogen and oxygen atoms located within the same side of the calix unit were 3.533 and 5.670 Å respectively. The nitrile groups were almost linear with a CCN angle of 176.8° and CN distance of 1.104 Å. The cationic network was formed by mutual bridging between ligands 1 and Ag^I cations (Figure 7). The silver cations were tetrahedrally coordinated to four nitrile groups with the NAgN angle varying from 100.9 to 121.9° (average 106.2°) and CNAg angle of *ca* 144.0° and Ag-N distance of *ca* 2.292 Å. The packing of the cationic and anionic components (Figure 8) showed parallel strands of linear coordination polymers with columns of disordered AsF_6^- anions. Within the anionic columns, the AsF_6^- units were separated by water molecules.

188

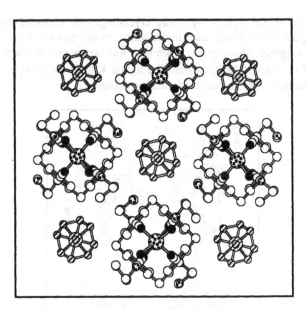

Figure 8. A portion of the structure showing the packing of the cationic linear coordination networks and AsF$_6^-$ anions (projection normal to the polymer axis). The AsF$_6^-$ anions were found to be disordered. H atoms and solvent molecules (H2O and EtOH) are not represented for sake of clarity.

3. α-coordination networks based on metallamacrocycles

Since the pioneer investigations by Fujita [30] and by Saalfranck [31] on metallamacrocycles considerable efforts have been focused on the design and synthesis of macrocyclic and macrobicyclic structures composed of metal centres connected by organic fragments [32]. The majority of metallmacrocycles reported so far are based on bis-monodentate ligands such as 4,4'-bipyridine, pyrimidine, bipyrazine, 4,7-phenanthroline or bis-pyrdines interconnected by rigid or flexible spacers.

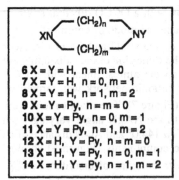

6 X = Y = H, n = m = 0
7 X = Y = H, n = 0, m = 1
8 X = Y = H, n = 1, m = 2
9 X = Y = Py, n = m = 0
10 X = Y = Py, n = 0, m = 1
11 X = Y = Py, n = 1, m = 2
12 X = H, Y = Py, n = m = 0
13 X = H, Y = Py, n = 0, m = 1
14 X = H, Y = Py, n = 1, m = 2

Scheme 2

p-dimethyaminopyridine (DMAP) shows strong alkaline character. We thought that one could take advantage of its peculiar electronic distribution for the design of ligands capable of forming metallamacrocycles. In order to interconnect two or more metals into a cyclic framework, the ligand needs to be of the exo type, i.e. it should possess at least two coordination sites oriented in divergent fashion. We have designed (scheme 2) and prepared exobis-monodentate ligands based on two DMAP derivatives as well as their binuclear silver and paladium metallmacrocycles [20].

The design of exo-ligands **9-11** (scheme 2) was based on the interconnection of two *p*-aminopyridine substructures at the 4 position by two tuneable alkyl bridges. Ligands **9-11** may also be regarded as diazacycloalkanes [33] **6-8** bearing two pyridines. We have previously used cyclic diamines for the design of cyclospermidines and cyclospermines [34] and as backbones connecting donor and acceptor groups [35]. The design of **9-11** is rather versatile since, by controlling the size and thus the conformation of the medium size cyclic core and by choosing metals with specific coordination demands, one may investigate structural aspects in the formation of metallamacrocycle under self-assembly conditions.

The synthesis of **9-11** was achieved by treating under reflux a mixture of 4-chloropyridine, PhLi and compounds **6-8**. In all cases studied, in addition to **9-11**, the monosubstituted compounds **12-14** were also isolated.

The formation of metallamacrocycles using Ag(I) was investigated by diffusion at room temperature of a EtOH solution of ligand **11** with an aqueous solution of $AgPF_6$. After several days colourless crystals of the complex were filtered and further washed with Et_2O. The complex was characterised by NMR and by FAB spectrometry. Furthermore, in the solid state the structure of $[11_2Ag_2, (PF_6)_2]$ complex was elucidated by X-ray crystallography.

In principle, due to the non symmetric nature of **11**, two different isomeric complexes, one with C_{2v} and the other with D_{2h} ideal symmetry may be envisaged (Figure 9). The difference between these two isomers originates from the mutual orientation of the cyclic cores.

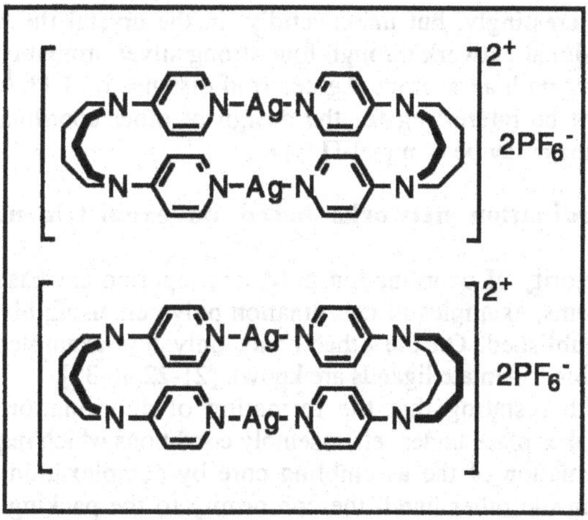

Figure 9. Two possible isomeric structures D_{2h} (top) and C_{2v} (bottom) for the $[11_2Ag_2]$ metallamacrocycle.

The crystal analysis showed the following features: i) the binuclear metallamacrocycle is formed through the bridging of two Ag(I) atoms by two pyridine units belonging to two ligands **11** which adopt a bent conformation; ii) both nitrogen atoms of the diaza core show a marked sp^2 character with an average CNC angle of *ca* 121.1°and N-C distance of 1.36 Å; iii) the coordination geometry around Ag atoms is almost linear with an average NAgN angle of ca 175.8°; iv) the average N-Ag distance is *ca* 2.10 Å; v) the Ag-Ag distance is 3.36 Å. in the solid state, only the complex with C_{2v} symmetry was observed (Figure 10).

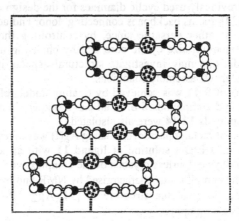

Figure 10. A portion of the crystal structure of the $[11_2Ag_2, (PF_6)_2]$ metallamacrocycle (see text for bond angles and distances) showing the formation of an infinite network formed by Ag-aromatic interactions between consecutive units. Anions and H atoms are not represented for sake of clarity.

Rather interestingly, but unexpectidly, in the crystal the metallamacrocyles form a one dimensional network through four strong silver-aromatic (pyridine) dihapto interactions per unit with an average Ag-centroid distance of 3.43 Å (Figure 10). This observation meight be interesting for the design of other coordination networks for which the interactions may be of metal-Π type.

4. α- and β-coordination networks based on exobistridentate ligands

Although, the majority of coordination polymers reported are based on bis- or tris-monodentate systems, examples of coordination polymers using bis-bidentate ligands have also been published. On the other hand, only few examples of characterised systems based on bis-tridentate ligands are known [21-22,36-37].

It is worth restating that the formation of coordination networks in the crystalline phase takes place under self-assembly conditions which require, on one hand, the reversible formation of the assembling core by complexation processes and its translation and, on the other hand, the conformity to the packing forces within the crystalline phase. Thus, for simple coordination networks, the choice of the two partners (metal, ligand) is the dominant feature. For porous coordination networks, in addition to the two partners mentioned before, one should also take into account the guest molecules initially occupying the pores. Due to the fact that tridentate centres form rather stable and non-labile complexes with transition metals, the choice of the metal and the ligand is not obvious and therefore, not many characterised examples are available so far.

Recently, we have succeeded in preparing the bis-tridentate ligand **15** and its self-assembly into 1- and 2-D coordination polymers using Cd(II) cation [21].

The design of the linear bis-tridentate ligand **15** (scheme 3) is based on a combination of two pyridine moieties and four thiaether groups. The two tridentate are interconnected by a phenyl ring. Examples of other bis-tridentate ligands have been reported [38].

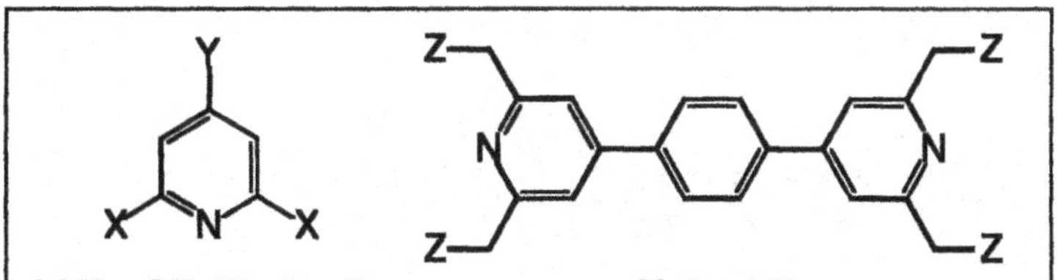

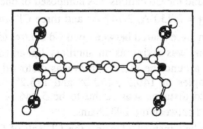

16 X = CH$_2$SEt, Y = H
17 X = CO$_2$H, Y = OH
18 X = CO$_2$Et, Y = Br
19 X = CH$_2$OH, Y = Br

20 Z = OH
21 Z = Br
15 Z = SMe

Scheme 3

Dealing with the metal cation, it has been shown previously that **16** forms a binuclear Cd(II) complex in which the two metal centres are bridged by two Cl⁻ anions [39].

The synthesis of **15** was achieved starting from chelidamic acid **17**, which was first transformed into **18** which was reduced to afford **19**. Pd(PPh$_3$)$_4$ catalysed coupling reaction of **19** with 1,4-phenyldiboronic acid in dry THF gave **20**. Bromination of the latter using afforded **21**. And finally, the treatment of **21** by NaSMe in dry THF afforded **15**. The latter, in addition to spectroscopic methods, was characterised by X-ray crystallography (Figure 11).

Figure 11. Solide structure of the ligand **15**. For sake of clarity, H atoms are not represented.

Upon slow diffusion at room temperature of a MeOH solution containing CdCl$_2$.xH$_2$O into a CHCl$_3$ solution of **15**, colourless crystals were deposited after 24 h. The X-ray study showed the following features (Figure 12) : the crystal (monoclinic) was composed of **15**, Cd(II), Cl⁻ anions, CHCl$_3$ and H$_2$O molecules. **15**, Cd(II) and Cl⁻ anions formed a neutral, infinite 1-D coordination network (Figure 12).

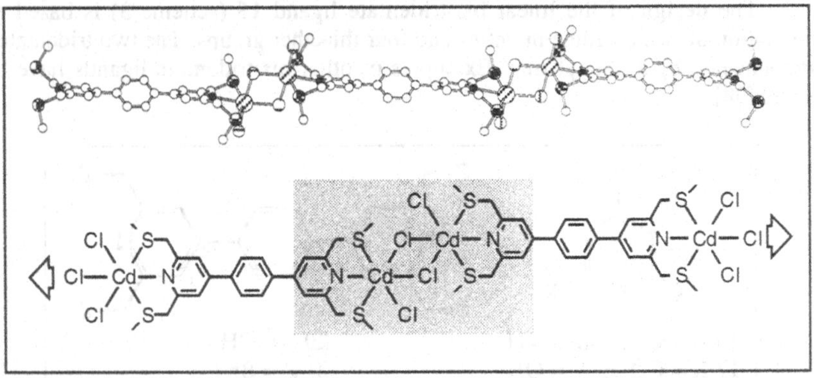

Figure 12. A section of the X-ray structure of the 1-D coordination network formed between 15 and $CdCl_2$. H atoms, anions and solvent molecules are not represented for clarity.

Depending on the definition of the assembling core, the 1-D network $(15Cd_2Cl_4)n$ thus obtained may be described in two different ways. Considering the metallic core as a binuclear Cd_2Cl_4 unit (Figure 12), the network may be regarded as resulting from the binding of two such units by the two PyS_2 fragments of 15 leading thus to the exo-tetranuclear Cd (II) complex and the repetition of the binding process leads to the 1-D coordination network. The same system may be also described as resulting from the bridging of consecutive exo-binuclear $15-Cd_2$ units by two Cl^- anions, and the remaining other two Cl^- anions behaving as ancillary ligands. For the ligand part, the pyridine rings were tilted by -37.6° and 38.3° with respect to the phenyl group. The CS and CN distances were roughly the same as those observed for the free 15. The coordination sphere around the Cd cations was composed of one N atom ($d_{NCd} = 2.426$ Å), two S atoms ($d_{SCd} = 2.730$ Å, 2.747 Å) and three Cl^- anions. Among the three Cl^- present, two of them were shared between two Cd centres ($d_{ClCd} = 2.591$ Å, 2.651 Å) whereas the third one was acting as an ancillary ligand ($d_{ClCd} = 2.511$ Å). The coordination geometry around the metal centre was distorted octahedral with SCdS, ClCdCl and NCdCl angles of 146.5°, 172.0° and 175.2° respectively. Within the Cd_2Cl_4 unit, the CdCd distance was found to be 3.946 Å. The $CHCl_3$ and H_2O molecules were localised between the 1-D chains.

Interestingly, when instead of using the Cl^- salt of Cd(II), a mixed Cl^- and BF_4^- salt was used, the same diffusion method afforded after two days another type of colourless crystals. The solid state analysis (Figure 13) showed that the crystal was composed of 15, Cd(II) cations, Cl^- and BF_4^- anions, $CHCl_3$ and MeOH molecules. 15, Cd(II) and Cl^- anions formed a cationic infinite 2-D coordination network (Figure 13). The latter may be described as the result of mutual interconnection of the 1-D coordination polymers described above through Cl^- anions acting now as bridging ligands. Thus, the 2-D network may be regarded as the translation into two different space directions of an assembling core which may be defined as $[15_4Cd_4Cl_7]^+$ (Figure 13). The charge neutrality being achieved by BF_4^- anions present in the lattice. Due to the aromatic nature of the ligand used, the 2-D network was composed of cavities delimited in a alternating fashion by organic and inorganic fragments. For **15**, the pyridine rings were tilted by -30.7° and 32.2° with respect to the phenyl group. The CS and CN distances were roughly the same as those observed for the free **15** and for the abovementioned 1-D network. The coordination sphere around the Cd cations was again composed of one N atom (average d_{NCd} = 2.385 Å), two S atoms (average d_{SCd} = 2.697 Å) and three Cl⁻ anions. Although all three Cl⁻, present in the coordination sphere of Cd cation, played a bridging role between the metallic centres, two of them were shared between two Cd centres (average d_{ClCd} = 2.608 Å) within the 1-D network, whereas the third one was bridged two consecutive linear networks (d_{ClCd} = 2.559 Å) leading thus to the two-dimensional polymer. The coordination geometry around the metal centre was again distorted octahedral with SCdS, ClCdCl and NCdCl angles of 147.6°, 169.0° and 163.6° respectively. With the Cd_4Cl_7 unit, the CdCd distances varied from 3.921 Å for the doubly bridged cations to 4.625 Å for the singly bridged centres.

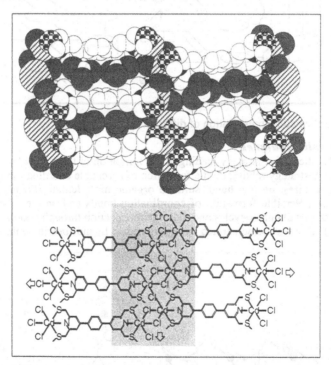

Figure 13. A section of the X-ray structure of the 2-D coordination network (see text). H atoms, anions and CH_3OH molecules are not represented for clarity.

Interestingly, whereas MeOH and BF_4^- anions were localised between the 2-D sheets. the $CHCl_3$ molecules were included within the cyclophane type cavities composing the 2-D network. The exchange of solvent molecules localised within the pores of the structure by other chlorinated solvents is under current investigation.

194

5. Tubular coordination networks

Tubular structures are interesting architectures, in particular with respect to their ability to transport ions or molecules (Figure 14). They may be formed by 1-D chains adopting a helical structure, as observed for polypeptides in natural systems [40] or for helical coordination polymers [41-42], by interconnection of ring modules as for example for cyclic peptides [40,43], by stacks of cyclic units in liquid crystalline phase [44], by poly-[45] or oligo-[46] meric structures bearing ring systems, or by rolling up 2-D sheets such as graphite leading to carbon nanotubes [47]. Dealing with tubular structures formed through multiconnection of cyclic units, the linking between the rings units may be achieved through either covalent bonds or other types of interactions such as H-bonding and stacking processes under self-assembly conditions [43].

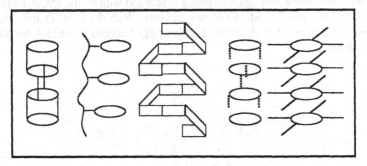

Figure 14. Schematic representation of different tubular systems.

5. 1. Metallatubulanes

The alternative strategy leading to the formation of tubes may be based on the self-assembly strategy relying on the formation of reversible coordination bonds. Such a strategy requires, on one hand, suitable organic multidentate ligands and metals allowing the reversible formation of coordination bonds and on the other hand, the formation of metallamacrocycles and their interconnection through coordination bonds (Figure 15). This was demonstrated using terpyridine ligands bearing a thioether group [22,48].

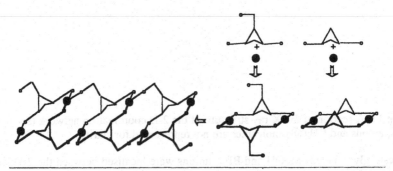

Figure 15. Schematic representation of a 2,2-metallamacrocycle (right) and metallatubes (left) obtained under self-assembly conditions using bi- and tri-dentate pyrazolyl based ligands 22, 23 and a silver cation.

Ligands **22** and **23** (scheme 4) are based on an aromatic core (pyrazine or phenyl) bearing two and three pyrazolyl units, respectively. Whereas **22** may act as bi-, tri- or even tetra-dentate ligand, **23** may be regarded as a tridentate unit. Ligand **22** was prepared upon treatment of pyrazole by NaH in DMF, followed by addition of 2,6-dichloropyrazine [22].

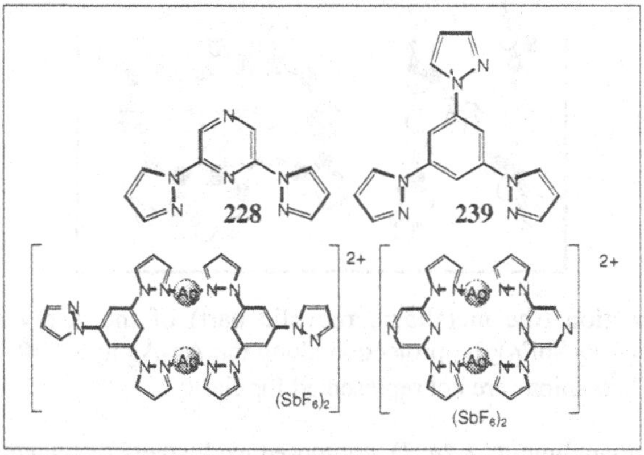

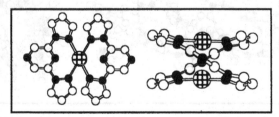

Scheme 4

The reaction under dark and argon of a equimolar mixture of **22** and $AgSbF_6$ in $CHCl_3$/EtOH leads to a white powder which was recrystallised and analysed in the crystalline phase by X-ray diffraction methods to be the binuclear silver metallamacrocycle $[22_2Ag_2(SbF_6)_2]$ (Scheme 4). The crystal (orthorhombic, $C\,c\,c\,a$,) was exclusively composed of discrete $[22_2Ag_2(SbF_6)_2]$ units (Figure 16); the cationic part was indeed a binuclear silver complex composed of two ligands and two Ag^+ cations forming a 2,2-metallamacrocycle implying that the **22** was acting as a bidentate unit; for the ligand **22**, the pyrazolyl and the pyrazine units were tilted by ca -21.6°, the two ligands were not in the same plane but twisted leading thus to a rather short Ag-Ag distance of 2.911 Å, both silver cations, bridging the two ligands and thus forming the cycle, were almost linearly coordinated (NAgN angle of 178.2°) to two pyrazolyl nitrogen atoms with a N-Ag distance of 2.160 Å, the SbF_6^- anions were neither disordered nor in specific interactions with the cationic units. The formation of such a metallamacrocycle is not unprecedented, analogous structures were obtained using Cu(I) and benzimidazol based ligands [49], and bis-pyridine and Ag(I) [50].

Figure 16. The X-ray structure of the metallamacrocycle $[22_2Ag_2(SbF_6)_2]$ (projection along the Ag-Ag axis (left) and lateral view (right)) obtained under self-assembly conditions between **22** and Ag^+. H atoms and anions are not represented for clarity.

Dealing with compound **23** bearing three pyrazolyl units, upon slow diffusion in at room temperature of a chloroforme solution of **23** into a EtOH solution of AgSbF$_6$, colourless crystals were obtained and studied by X-ray diffraction (Figures 17 and 18).

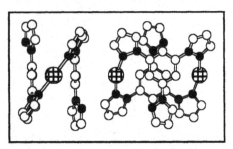

Figure 17. A section (the metallamacrocyclic part) of the X-ray structure of the metallatube [**23**$_2$Ag$_2$(SbF$_6$)$_2$]$_n$, projection along the Ag-Ag axis (left) and lateral view (right). H atoms and anions are not represented for clarity.

In the crystal (monoclinic, *C 1 2/c 1*), composed exclusively of ligand **23** and AgSbF$_6$ units, infinite cationic tubes formed by mutual bridging of binuclear silver metallamacrocycles [**23**$_2$Ag$_2$]$_n$ and SbF$_6^-$ anions were present. The tubular structure thus obtained may be described as follows: within the two tridentate ligands **23**, two out of three pyrazolyl units were bridged by two Ag$^+$ cations leading thus to a binuclear silver metallamacrocycle [**23**$_2$Ag$_2$] (Figure 17 and scheme 4); for the ligand **23**, all three pyrazolyl moieties were tilted by -41.2°, 32.4° and 23.0° with respect to the phenyl ring respectively; for the metallamacrocycle, the N-Ag and Ag-Ag distances were 2.150 Å, 2.154 Å and 7.785 Å, respectively; the metallamacrocycles were mutually and doubly interconnected (Figures 18) by the third pyrazolyl on each ligand **23** with a rather long N-Ag distance of 2.620 Å, thus the overall coordination geometry around the Ag$^+$ cation may be described as severely distorted trigonal (almost T-type) with NAgN angles of 162.4°, 99.0° and 98.6°; again, the SbF$_6^-$ anions were neither disordered nor in specific interactions with the cationic units.

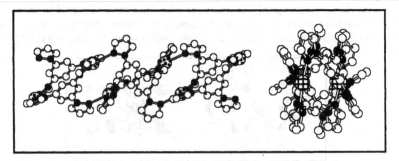

Figure 18. A portion of the X-ray structure showing the formation of the metallatube (Parallel (left) and perpendicular (right) views) between the tridentate ligand **23** and Ag$^+$ cations. H atoms and anions are not represented for clarity.

Although the topology the network formed may be described as tubular, due to the size of the metallamacrocycle formed the tubes internal space was rather small. Attempts to increase the diameter of the tube is under current investigation.

5.2. *Helical coordination networks*

The binding ability of the 2,2'-bipyridine unit towards transition metals has been widely used in coordination chemistry [51]. This ligand, in addition to its chelate structure, may be structurally modified at all position of its framework. Thus, for the preparation of coordination polymers one may use molecular modules based on 2,2'-bipyridine units.

The design of macrocyclic frameworks containing two of these chelates may be based on their interconnection by two bridges. The orientation of the coordination sites within the framework of the ligand results from the connection position on the 2,2'-bipyridine units (Figure 6).

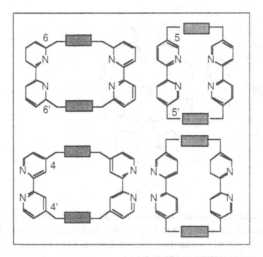

Figure 19. The orientation of the chelating units depends upon the connection position of the two bipyridines. In order to impose divergent localisation of the coordination sites, the 4 and 4' positions must be used as connection enters.

Whereas the connection at the 6 and 6' positions leads exclusively to endo-ligands in which the coordination sites are convergently oriented towards the interior of the macrocycle, the use of the 5 and 5' positions affords ligands which may act either as endo- or exo-ligands. Finally, the interconnection at the 4 and 4' positions leads exclusively to exo-ligands (Figure 19). Although, many bipyridine containing endo-ligands have been reported [51], only few examples of exo-ligands based on the interconnection at the 4 and 4' positions have been published (18,52-57).

We have reported [18-53-57] exoditopic ligands based on two 2,2'-bipyridine units (Scheme 5) interconnected at the 4 and 4' positions either by $(CH_2)_n$ (compounds 24-28) or by CH_2-Si-CH_2 (compound 29) or $(CH_2)_2$-Si-$(CH_2)_2$ (compound 30) fragments.

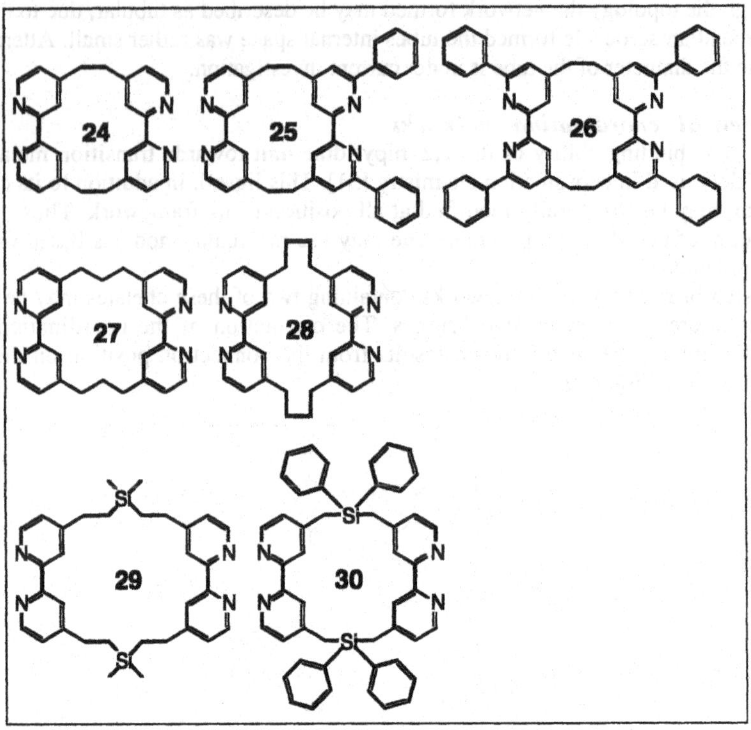

Scheme 5

The differences between ligands **24, 27** and **28** are due to the length of the spacer groups connecting the bipyridine units. Indeed, as demonstrated bellow, by varying the distance between the two bipyridine units, one may control the overall shape of the macrocyclic ligand in the complex

On the other hand, in principle, whereas ligands **24, 27** and **28-30** should lead to coordination polymers with metals adopting either square planar or tetrahedral coordination geometry, for ligands **25** and **26**, due to the presence of phenyl moieties, one may envisage coordination of tetrahedraly coordinated metals.

Some of the free ligands mentioned above were studied by X-ray diffraction methods (Figure 20). In all cases, the 2,2'-bipyridine unit adopts a trans conformation. As expected, the shape of the free ligands depends upon the length of the interconnecting spacers. All three compounds **24-26** for which the two bipyridine units are interconnected by two ethylene chains adopt an "oblong-shape" conformation.

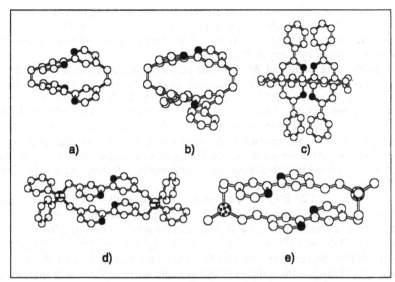

Figure 20. Crystal structures of : **24** (a), **25** (b), **26** (c) **29** (d) and **30** (e).

In the case of ligand **24**, homo complexes with Ru(II), Rh(I), and Os(II) were prepared (Scheme 6). On the other hand, using either two external 2,2'-bipyridine units as stoppers, (Ru,Os) heterobinuclear complexes were obtained, whereas in the case of rhodium the heterobinuclear complex was formed with CP* and Cl⁻ as stopper ligands. Other Ru(II) homobinuclear complexes were also prepared with ligands **28** and **30** (scheme 7).

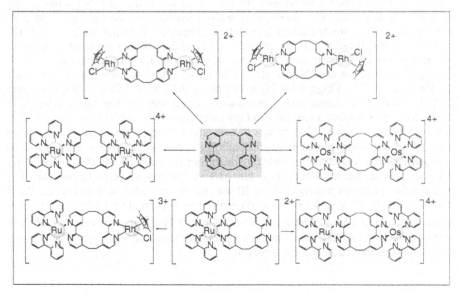

Scheme 6

Since octahedral metal complexes bearing three chelating ligands are chiral, for a binuclear octahedral complexes a mixture of Δ, Δ; Λ,Λ and Δ,Λ; Λ,Δ isomers were expected and obtained. All three Ru(II) binuclear complexes formed with ligands 24, 28 and 30 were analysed by X-ray diffraction methods.

In the case of ligand 24, the unit cell contains both the Δ,Δ and the Λ,Λ isomers. Both Ru(II) are hexacoordinated with an almost octahedral geometry. The Ru-N distances ranging from 2.052 to 2.064 Å, and the N-Ru-N angles for the three bipyridine ligands ranging from 78.7 to 79.8°, are similar to those observed for the well-known $Ru(bipy)_3(PF_6)_2$ complex. Within the complex, the two Ru cations are distant by 8.2 Å. The conformation of the macrocycle 24 in the complex is completely different from the one observed in the free ligand. The cyclophane structure is no longer present and is replaced by a roof-shape type conformation which allows the coordination of both metal cations (scheme 7).

In the case of 28-Ru_2-$4PF_6$ complex, the X-ray analysis revealed the presence of the meso (Δ,Λ) stereoisomer (scheme 7). Both Ru(II) centres are hexacoordinated with an almost octahedral geometry. Again, as in the previous case, the Ru-N distances ranging from 2.04 to 2.09 Å are similar to those observed for $Ru(bipy)_3(PF_6)_2$ complex. Within the complex, the two Ru cations are distant by 13.1 Å. In marked contrast with 24-Ru_2-$4PF_6$ complexes and as expected from the design principle, the ligand 28 adopts a zig-zag type conformation in which the chelating nitrogen centres of both bipyridine units are located in almost the same plane. Rather unexpectedly though, among the four PF_6^- present, two of them are located close to the centre of the complex above and bellow the main plane.

In the case of the silicon containing ligand 30, the solid state structure of the meso form of $(30$-$Ru_2)^{4+}$-$4PF_6^-$ complex was also studied by X-ray diffraction (scheme 7) which revealed the following features : i) in addition to $(30$-$Ru_2)^{4+}$-$4PF_6^-$ complex, $6CH_3CN$ and $2CH_3OH$ molecules are present in the solid state but without any interaction with the metallic centres; ii) in marked contrast with the free ligand 30, both bipy units in the complex adopt a cisoid conformation allowing, as expected, the chelation of Ru(II) cations; iii) the cationic moiety of the complex possesses a centre of symmetry; iv) the coordination geometry around Si atoms was tetrahedral with an average Si-C distance and CSiC angle of 1.86 Å and 109.4° respectively; v) due to the conformation of the -$(CH_2)_2$-Si-$(CH_2)_2$- fragments (CCCSi, and CSiCC dihedral angles of 171.0° and -87.2° respectively), in marked contrast with the free ligand adopting a cyclophane type conformation, the two bipy units oppositly oriented are almost parallel but not coplanar; vi) both CH_3 groups of both silane moieties are, as in the case of the free ligand, outwardly oriented; vii) for both Ru(II) centres, the coordination sphere is composed of 6 nitrogen atoms amongst which four are belonging to the two auxiliary bipy units and the remaining two are part of the macrocyclic ligand 30. The coordination geometry around each Ru(II) was almost octahedral with an average Ru-N distance of 2.05 A°; viii) the two Ru(II) centres were separated by 13.58 A°; ix) again, quiet interestingly and unexpectedly, among the four PF_6^- anion present, two of them were localised below and above the mean plane of the macrocyclic core and pointing almost towards the centre of the macrocycle.

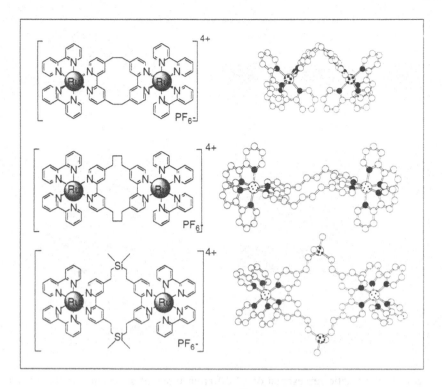

Scheme 7

Using exo-ligands such as **24-30** bearing two divergent sets of bipyridine and tetracoordinated metal centres, one may envisage the formation of linear coordination polymers which may be described as α-networks (Figure 21).

Furthermore, one may control the overall topology of coordination polymers by the coordination preferences of the linking metal as well as by the topology of the bridging ligand. Thus, using exobidentate ligands adopting a flatt conformation, and tetracoordinated metal centres adopting a square planar or tetrahedral coordination geometry, one may envisage the formation α-networks (Figure 21a). On the other hand, using metal cations adopting tetrahedral coordination geometry and ligands possing a bent conformation, one may envisage the formation of either discrete cyclic or infinite linear polynuclear species (Figure 21). Depending on the geometrical features of the ligand, a selection may be induced between the different possibilities. Indeed, with non planar bis-bidentate exo-ligands adopting a "roof-shape" conformation, in addition to discrete cyclic polynuclear species such as the tetranuclear complex presented in figure 21b, infinite 1-D networks of the "stair type" (Figure 21 c) or single stranded helical species (Figure 21d) may be formed. However, by imposing constraints such as steric hindrance, one may favour one of the alternative structures.

202

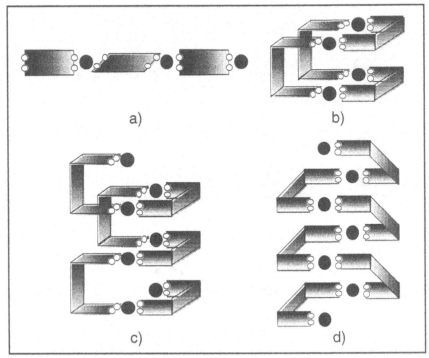

Figure 21. Schematic representation of different types of arrangements that may be obtained by self-assembly of exo-ditopic ligands and tetrahedrally coordinated metal cations: a) linear, c) "stair type", d) helical coordination polymer and b) discrete tetranuclear species.

The formation of a 1-D coordination polymer was achieved under self-assembly conditions using ligand **26** and Ag^I cations (Figure 22).

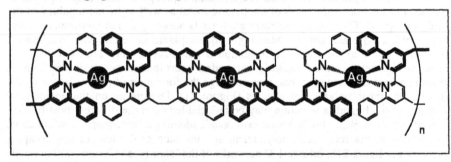

Figure 22. Representation of an infinite coordination polymer obtained by the self-assembly of the exo-ditopic ligand **26** and silver cations.

In the solid state, the structure of the coordination network was elucidated by an X-ray analysis [18] which revealed that the cationic component of the structure was a single stranded helical network formed by mutual bridging of silver atoms and ligand **26** (Figure 23). The pitch of the single helix was composed of four Ag atoms and four ligands. The interior of the helix was obviously not empty but occupied by phenyl groups (Figure 24). The interstices in the solid were occupied to different degrees of intrusiveness by CH₃CN molecules and PF₆⁻ anions.

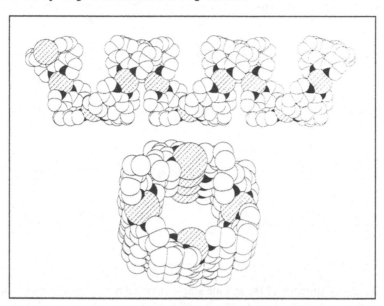

Figure 23. X-ray structure of the infinite single stranded helical coordination polymer obtained upon self-assembly of ligand **26** and silver cation : projection of the cationic array down (top) and normal (bottom) to the polymer axis. For clarity reasons, the Ph groups at the 6 and 6' positions, hydrogen atoms, anions and CH₃CN molecules are not represented.

The observed helical structure, a variety of tubular architecture, was derived from the primary structure of the ligand **26** adopting a "roof-shape" conformation in the complex. However, since by design, no intrinsic chirality was coded within the structure of the connecting ligand, a racemic mixture of both the right- and left-handed helical coordination polymers was expected and observed. The infinite network crystallised in the centrosymmetric space group $C_{2/c}$. The "roof-shape conformation adopted by **26** has been previously observed in the solid state for the Ru^{II} exobinuclear complex formed with the analogous ligand **24** (see above). This conformation was due to the short nature of the ethylene bridges connecting the two 2,2'-bipyridine units at the 4 and 4' positions. Indeed, it has been shown by X-ray studies on Ru^{II} exobinuclear complexes formed with ligands **28** and **30** that upon increasing the length of the spacer group, both ligands adopt a planar conformation (scheme 7).

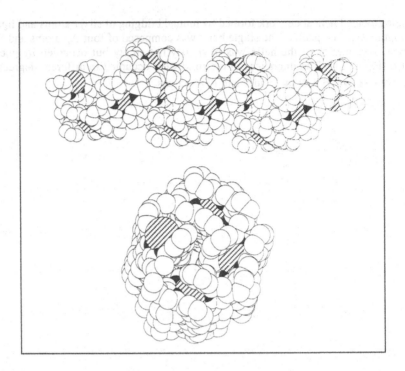

Figure 24. X-ray structure of the infinite single stranded helical coordination polymer obtained upon self-assembly of ligand **26** and silver cation : projection of the cationic array down (top) and normal (bottom) to the polymer axis. For the sake of clarity, hydrogen atoms, anions and CH3CN molecules are not represented.

All silver atoms were tetra-coordinated to two sets of bipyridine units affording a AgN4 coordination sphere with NAgN angles varying from 73.5° to 139.5° for Ag1: from 74.4° to 149.7° for Ag2 and from 73.6 to 153.9 for Ag3. For all three silver atoms the Ag-N distance varies from 2.30 (esd : 0.8)Å to 2.47 (esd : 0.2)Å. The bite angles for pair of ligands at each silver is between 73.5(9) and 74.4(6) and the interplanar (N2Ag/AgN2) dihedral angles are: 70(1), 65.2(7), 68(1).

Inspection of CPK models revealed that the formation of the helical structure with ligand **26** was probably due to the presence of phenyl groups at the 6 and 6' positions. Indeed, for both discrete cyclic polynuclear species such as the tetranuclear complex (Figure 11b), and the infinite "stair type" network (Figure 11c), the substitution of hydrogen atoms at the 6 and 6' positions in ligand **24** by phenyl groups in ligand **26** could generate severe steric hindrance between phenyl moieties belonging to consecutive ligands assembled around silver atoms.

6. Conclusions

In this chapter, the rational behind our approach of the design and preparation of coordination networks was presented. Using a variety of ligands with a wide range of architectures and coordination ability (denticity, the nature and number of coordination sites) and metal cations, the viability of the molecular tectonics approach based on self -assembly of complementary tectons was demonstrated. Although, we and others are not able to predict all the features of crystal formation or composition, it was nevertheless shown that by appropriate design of connecting ligands et the choice of metals. coordination polymers could be designed, prepared and structurally characterised in the crystalline phase. However, so far, using our approach, we have not been able to prepare functional coordination polymers. This aspect, *i.e.* shifting from structural to functional networks, remains a challenge.

Acknowledgements

The results presented were generated by my very talented co-workers and students their names appear in the literature cited. Thanks also to the Université Louis Pasteur, CNRS and the Institut Universitaire de France for financial support.

References

1. J.-M. Lehn, (**1995**) *Supramolecular Chemistry, Concepts and Perspectives,* VCH, Weinheim.
2. G. R. Desiraju, (**1995**)*Angew. Chem. Int. Ed. Engl., 34,* 2311.
3. X. Delaigue, E. Graf, F. Hajek, M. W. Hosseini, J.-M. PLaneix (**1996**), NATO ASI Series, Ed. G. Tsoucaris, J. L. Atwood, J. Lipkowski, Kluwer, Dordrecht. Netherlands, *480,* 159.
4. G. Brand, M. W. Hosseini, O. Félix, P. Schaeffer, R. Ruppert (**1996**), NATO ASI Series, Ed. O. Kahn, Kluwer, Dordrecht, Netherlands, *484,* 129; C. Kaes. M. W. Hosseini (1999), NATO ASI Series, Ed. J. Viciana, Kluwer, Dordrecht. Netherlands, *518,* 53.
5. J. S. Lindsey, (**1991**) *New. J. Chem., 15,* 153.
6. D. S. Lawrence, T. Jiang, M. Levett, (**1995**) *Chem. Rev., 95,* 2229.
7. J. F.Stoddart, D. Philip, (**1996**) *Angew. Chem. Int. Ed. Engl., 35,* 1155.
8. M. Simard, D. Su, J. D. Wuest, (**1991**) *J. Amer. Chem. Soc., 113,* 4696.
9. S. Mann, (**1993**) *Nature, 365,* 499.
10. J.-M. Lehn, (**1990**) *Angew. Chem. Int. Ed. Engl.,29,* 1304.
11. G. M. Whitesides, J. P. Mathias, C. T. Seto, (**1991**) *Science, 254,* 1312.
12. M. W. Hosseini, A. De Cian, (**1998**) *J. C. S. Chem. Comm,* 727 and references therein.
13. M. C. Etter (**1990**) *Acc. Chem. Res., 23,* 120; C. B. Aakeröy, K. R. Seddon (**1993**) *Chem. Soc. Rev.,* 22, 397; S. Subramanian, M. J. Zaworotko (**1994**) *Coord. Chem. Rev.,* 137, 357; D. Braga, F. Grepioni (**1994**) *Acc. Chem. Res.. 27,* 51; V. A. Russell, M. D. Ward (**1996**) *Chem. Mater.,* 8, 1654.
14. F. W. Fowler, J. W. Lauher (**1993**) *J. Amer. Chem. Soc., 115,* 5991.
15. G. Brand, M. W. Hosseini, R. Ruppert, A. De Cian, J. Fischer, N. Kyritsakas (**1995**) *New J. Chem.,* 19, 9; M. W. Hosseini, R. Ruppert, P. Schaeffer, A. De Cian, N. Kyritsakas, J. Fischer (**1994**) *J. C. S. Chem. Comm.,* 2135; M. W. Hosseini, G. Brand, P. Schaeffer, R. Ruppert, A. De Cian, J. Fischer (**1996**) *Tetrahedron Lett.* , 37, 1405; O. Félix, M. W. Hosseini, A. De Cian, J. Fischer (**1997**) *Angew. Chem. Int. Ed. Engl., 36,* 102; O. Félix, M. W. Hosseini, A. De Cian, J. Fischer (**1997**) *Tetrahedron Lett.,* 38, 1755; O. Félix, M. W. Hosseini, A. De Cian, J. Fischer (**1997**) *Tetrahedron Lett.,* 38, 1933; O. Félix.

M. W. Hosseini, A. De Cian, J. Fischer (**1997**) *New J. Chem.*, *21*, 285: O. Félix, M. W. Hosseini, A. De Cian, J. Fischer (**1998**) *New J. Chem.*, 22. 1389.

16. R. Robson (**1996**) in *Comprehensive Supramolecular Chemistry*, Eds; J. L. Atwood, J. E. D. Davies, D. D. Macnicol, F. Vögtle, Pergamon, Vol. 6 (Eds. D. D. Macnicol, F. Toda, R. Bishop), p. 733;O. Kahn, Y. Journaux, C. Mathoniere, in *Magnetism : A Supramolecular Function, Vol. C484* (Ed.: O. Kahn), Kluwer, Dordrecht, Netherlands, **1996**, pp. 531; S. Decurtins, R. Pellaux, A. Hauser, M. E. von Arx in *ibid*, pp. 487; O. M. Yaghi, H. Li, C. Davis, D. Richardson, T. L. Groy (**1998**), *Acc. Chem. Res.*, *31*, 474.

17. E. C. Constable, A. M. W. Cargill Thompson (**1992**), *J. C. S. Dalton Trans.*, 3467; M. Ferigo, P. Bonhôte, W. Marty, H. Stoeckli-Evans (**1994**), *ibid.*, 1549; T. L. Hennigar, D. C. MacQuarrie, P. Losier, R. D. Rogers, M. J. Zaworotko (**1997**), *Angew. Chem., Int. Ed. Engl.*, *36*, 972, and references therein; U. Velten, M. Rehahn (**1996**), *J. C. S. Chem. Comm*, 2639.

18. C. Kaes, M. W. Hosseini, C. E. F. Rickard, A. H. White, B. W. Skelton (**1998**), *Angew. Chem. Int. Ed. Engl.*,*37*, 920.

19. G. Mislin, E. Graf, M. W. Hosseini, A. De Cian, N. Kyritsakas, J. Fischer (**1998**), *J. C. S., Chem. Comm*, 2545.

20. R. Schneider, M. W. Hosseini, J.-M. Planeix, A. De Cian, J. Fischer (**1998**). *J. C. S., Chem. Commun.*, 1625.

21. M. Loï, E. Graf, M. W. Hosseini, A. De Cian, J. Fischer, (**1999**) *J. C. S.. Chem. Comm.*, 603.

22. M. Loï, M. W. Hosseini, A. Jouaiti, A. De Cian, J. Fischer, (**1999**) *E. J. Inorg. Chem.* in press.

23. X. Delaigue, M. W. Hosseini (**1993**), *Tetrahedron Lett.*, *34*, 8111; X. Delaigue. J. McB. Harrowfield, M. W. Hosseini, A. De Cian, J. Fischer, N. Kyritsakas (**1994**), *J. C. S., Chem. Comm.*, 1579; X. Delaigue, M. W. Hosseini, A. De Cian, N. Kyritsakas, J. Fischer (**1995**), *J. C. S., Chem. Comm.*, 609.

24. G. Mislin, E. Graf, M. W. Hosseini (**1996**), *Tetrahedron Lett.*, *37*, 4503.

25. C. D. Gutsche, J. A. Levine (**1982**) *J. Am. Chem. Soc.*, *104*, 2653.

26. C. D. Gutsche, J. A. Levine (**1982**) *J. Am. Chem. Soc.*, *104*, 2653.

27. W. Verboom, S. datta, Z. Asfari, S. Harkema, D. N. Reinhoudt (**1992**), *J. Org. Chem.*, *57*, 5394.

28. M. Conner, V. Janout, S. L. Regen (**1992**), *J. Org. Chem* , *57*, 3744.

29. J. Blake, N. R. Champness, S. S. M. Chung, W-S. Li, M. Schröder (**1997**) *J. C. S., Chem. Comm.*, 1675; M. A. Withersby, A. J. Blake, N. R. Champness. P. Hubberstey, W-S. Li, M. Schröder (**1997**) *J. C. S., Chem. Comm.*, 2327: M. J. Hannon, C. L. Painting, W. Errington (**1997**) *J. C. S., Chem. Comm..* 1805; D. Perreault, M. Drouin, A. Michel, P. D. Harvey (**1992**) *Inorg. Chem..* *31*, 3688; K. A. Hirsch, S. R. Wilson, J. S. Moore (**1997**) *ibid*, *36*, 2960; K. A. Hirsch, S. R. Wilson, J. S. Moore (**1997**) *Chem. Eur. J.*, *3*, 765; G. B. Gardner, D. Venkataraman, J. S. Moore, S. Lee (**1995**) *Nature.*, *374*, 792; G. B. Gardner, Y.-H. Kiang, S. Lee, A. Asgaonkar, D. Venkataraman (**1996**) *J. Am. Chem. Soc.*, *118*, 6946; B. F. Abrahams, S. J. Egan, B. F. Hoskins, R. Robson (**1996**), *J. C. S., Chem. Comm.*, 1099; F.-Q. Liu, T. Don Tilley (**1997**) *Inorg. Chem.*, *36*, 2090.

30. M. Fujita in *Comprehensive Supramolecular Chemistry*, Eds. : J. L. Atwood, J. E. D. Davies, D. D. MacNicol, F. Vögtle, *Vol. 9* (Eds. J. P. Sauvage, M. W. Hosseini), Elsevier, **1996**, 253.

31. R. W. Saalfrank, A. Stark, K. Peters, H.-G. von Schnering (**1988**), *Angew. Chem. Int. Ed. Engl.*, *27*, 851.

32. J. K. Sanders, *Comprehensive Supramolecular Chemistry*, Eds. : J. L. Atwood, J. E. D. Davies, D. D. MacNicol, F. Vögtle, *Vol. 9* (Eds. J. P. Sauvage, M. W. Hosseini), Elsevier, **1996** 131-164; P. J. Stang, B. Olenyuk (**1997**) *Acc. Chem. Res.*, *30*, 502; C. M. Drain, J.-M. Lehn (**1994**) *J. C. S., Chem. Comm.*, 2313; T. Beissel, R. E. Powers, K. N. Raymond (**1996**) *Angew. Chem. Int. Ed. Engl.*, *35*, 1084; C. V. Krishnamohan Sarma, S. T. Griffin, R. D. Rogers (**1998**) *J. C. S., Chem. Comm.*, 215; R.-D. Schnebeck, L. Randaccio, E. Zangrando, B. Lippert (**1998**) *Angew. Chem. Int. Ed. Engl.*, *37* 119; J. R. Hall, S. J. Loeb, G. K. H. Shimizu, G. P. A. Yap (**1998**) *Angew. Chem. Int. Ed. Engl.*, *37*, 121.

33. R. W. Alder (**1983**), *Acc. Chem. Res*, *16*, 321.

34. G. Brand, M. W. Hosseini, R. Ruppert (**1994**) *Tetrahedron Lett.*, *35*, 8609.

35. R. Schneider, M. W. Hosseini, J.-M. Planeix (**1996**) *Tetrahedron Lett.*, *35*, 4721.

36. S. J. Loeb, G. K. H. Shimizu (**1993**) *J. C S., Chem. Comm..*, 1395; A. Neels, B. Mathez Neels, H. Stoeckli-Evans (**1997**) *Inorg. Chem.*, *36*, 3402.

37. E. C. Constable, A. J. Edwards, D. Philips, P. R. Raithby (**1995**) *Supramolecular Chem.*, *5*, 93.

38. H. A. Goodwin, F. Lions (**1959**) *J. Am. Chem. Soc.*, *81*, 6415; J.-P. Sauvage, J.-P. Collin, J.-C. Chambron, S. Guillerez, C. Coudret, V. Balzani, F. Barigelletti, L. De Cola, L. Flamigni (**1994**) *Chem. Rev.*, *94*, 993; P. Steenwinkel, H. Kooijman, W. J. J. Smeets, A. L. Spek, D. M. Grove, G. van Koten (**1998**) *Organometallics*, *17*, 5411 and references therein.

39. F. Teixidor, L. Escriche, I. Rodrigez, J. Casabo (**1989**) *J. C. S., Dalton Trans.*, 1381.

40. D. H. Lee, M. R. Ghadiri in *Comprehensive Supramolecular Chemistry*, Eds: J. L. Atwood, J. E. D. Davies, D. D. Macnicol, F. Vögtle, Pergamon, Vol. 9 (Eds. J.-P. Sauvage, M. W. Hosseini), 1996, p. 451; J. D. Lear, Z. R. Wasserman, W. F. DeGrado (**1988**) *Science*, *240*, 1177.

41. P. De Santis, S. Morosetti, R. Rizzo (**1974**) *Macromolecules*, *7*, 52; M. R. Ghadiri, J. R. Granja, R. A. Milligan, D. E. McRee, N. Khazanovich (**1993**) *Nature*, *366*, 324.

42. O. J. Gelling, F. van Bolhuis, B. L. Feringa (**1991**) *J. C. S. Chem. Commun.*, 917, Y. Dai, T. J. Katz, D. A. Nichols (**1996**) *Angew. Chem., Int. Ed. Engl.*, *35*, 2109; B. Wu, W.-J. Zhang, S.-Y. Yu, X.-T. Wu (**1997**) *J. C. S. Chem. Commun.*, 1795.

43. C. Kaes, M. W. Hosseini, C. E. F. Rickard, B. W. Skelton, A. White (**1998**) *Angew. Chem. Int. Ed. Engl.*, *37*, 920.

44. J.-M. Lehn, J. Malthête, A.-M. Levelut (**1985**) *J. Chem. Soc., Chem. Commun.*, 1794; V. Percec, G. Johansson, J. A. Heck, G. Ungar, S. V. Betty (**1993**) *J. Chem. Soc., Perkin Trans 1*, 1411, T. Komori, S. Shinkai (**1993**) *Chem. Lett.*, 1455.

45. U. F. Kragten, M. F. M. Roks, R. J. M. Nolte (**1985**) *J. Cem. Soc., Chem. Commun.*, 1275; C. Mertesdorf, H. Ringsdorf, (**1989**) *Mol. Cryst. Liq. Cryst.*. *5*, 1757.

46. J.-P. Behr, C. J. Burrows, R. J. Heng, J.-M. Lehn (**1985**) *Tetrahedron Lett.*. *26*, 215; A. Nakano, Q. Xie, J. V. Mallen, L. Echegoyen, G. W. Gokel, (**1990**) *J. Amer. Chem. Soc., 112*, 1287.

47. S. Iijima (**1991**) *Nature, 354*, 56.

48. M. J. Hannon, C. L. Painting, W. Errington (**1997**) *J. Chem. Soc., Chem. Comm.*, 1805.

49. C. Piguet, G. Bernadinelli, A. F. Williams (**1989**) *Inorg. Chem., 28*, 2920; S. Rüttimann, C. Piguet, G. Bernardinelli, B. Bocquet, A. F. Williams (**1992**) *J. Am. Chem. Soc., 114*, 4230.

50. C. M. Hartshorn, P. J. Steel, (**1996**) *Inorg. Chem., 35*, 6902.

51. A. Juris, V. Balzani, F. Barigelletti, S. Campagna, P. Belser, A. Von Zelewsky, *Coord. Chem. Rev.* **1988**, *84*, 85; V. Balzani, A. Juris, M. Venturi, S. Campagna, S. Serroni (**1996**), *Chem. Rev., 96*, 759.

52. O.Kocian, R. J. Mortimer, P. D. Beer (**1990**), *Tetrahedron Lett.*, 31, 5069; J. Bolger, A. Gourdon, E. Ishow, J.-P. Launay (**1995**), *J. Chem. Soc., Chem. Commun.* 1799; A. Rudi, D. Gut, F. Lellouche, M. Kol (**1997**), *J. Chem. Soc., Chem. Commun.*, 17; M. Schmittel, A. ganz (**1997**), *J. C. S., Chem. Comm.*, 999.

53. C. Kaes, M. W. Hosseini, R. Ruppert, A. De Cian, J. Fischer (**1994**), *Tetrahedron Lett.*, *35*, 7233.

54. C. Kaes, M. W. Hosseini, R. Ruppert, A. De Cian, J. Fischer (**1995**), *J. C. S., Chem. Comm.*, 1445.

55. C. Kaes, M. W. Hosseini, A. De Cian, J. Fischer (**1997**), *Tetrahedron Lett.*. *38*, 3901.

56. C. Kaes, M. W. Hosseini, A. De Cian, J. Fischer (**1997**), *Tetrahedron Lett.*, *38*, 4389.

57. C. Kaes, M. W. Hosseini, A. De Cian, J. Fischer (**1997**), *J. C. S., Chem. Comm.*, 2229.

HOW TO LOOK AT BONDS IN EXTENDED SOLIDS

MARIA JOSÉ CALHORDA

ITQB, Quinta do Marquês, EAN, Apart. 127, 2781-901 Oeiras and
Departamento de Química e Bioquímica, Faculdade de Ciências,
Universidade de Lisboa, 1749-016 Lisboa, Portugal

1. Introduction

There are several ways of studying bonds in solids and they depend of the kind of
solid. A broad classification includes both molecular solids, where discrete entities
exist and bind to each other by weak interactions such as van der Waals or
hydrogen bonds, or extended solids, consisting of infinite arrays of structural
motifs. Many publications have addressed the study of the bonding in extended
solids, starting from one-dimensional chains, to two-dimensional layers, and three-
dimensional solids. Although the bonds may exhibit a character ranging from
completely covalent to almost completely ionic, the same model can be successfully
employed in many solids [1].

In order to give an introduction to the subject, we start from the molecular orbital
theory model of bonding for simple diatomic molecules, such as C_2 and O_2, and
move toward more complicated species, namely C_2 and O_2 complexes, and finally
metal carbides. Interpretations will be based on the results of extended Hückel [2]
tight-binding calculations [3].

209

D. Braga et al. (eds.), Crystal Engineering: From Molecules and Crystals to Materials, 209–228.
© 1999 Kluwer Academic Publishers. Printed in the Netherlands.

2. Bonding in Diatomic Molecules and Complexes

2.1. THE DIATOMIC X_2 MOLECULES

The bonding in homonuclear X_2 diatomic molecules can be well described by means of molecular orbital theory [4]. The linear combination of atomic orbital (LCAO) approach, along with symmetry arguments, is used to obtain molecular orbitals. For the O_2 molecule, the energy difference between s and p based orbitals is considerable, and no mixing between them occurs, resulting in a simple diagram (Scheme 1, left). For C_2 molecules, this difference is much smaller, so that orbitals with the same symmetry combine, giving rise to the diagram shown on the right side of Scheme 1, where an inversion in the order of of the σ_p and π_u orbitals has taken place. Another consequence of this mixing concerns the character of some orbitals, namely $2\sigma_g$, which becomes more strongly bonding, $2\sigma_u$ and $3\sigma_g$ which become almost non bonding, and $3\sigma_u$ which becomes much more antibonding.

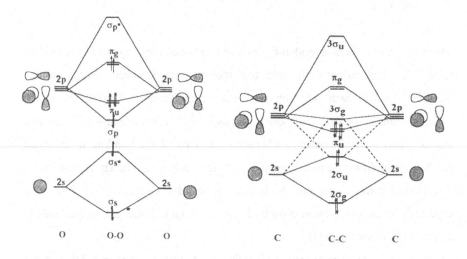

Scheme 1

Indeed, both σ_s and σ_p (left side of Scheme 1) have the same symmetry, the same happening to σ_s^* and σ_p^*. Therefore, they can mix, as sketched in Scheme 2.

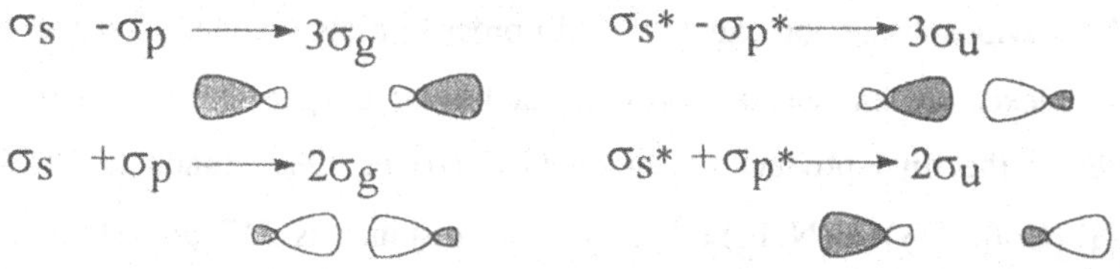

$$\sigma_s - \sigma_p \longrightarrow 3\sigma_g \qquad\qquad \sigma_s^* - \sigma_p^* \longrightarrow 3\sigma_u$$

$$\sigma_s + \sigma_p \longrightarrow 2\sigma_g \qquad\qquad \sigma_s^* + \sigma_p^* \longrightarrow 2\sigma_u$$

Scheme 2

The O_2 molecule is paramagnetic owing to the presence of two electrons with unpaired spins in π_g and the addition of electrons to this orbital to form O_2^- and O_2^{2-} leads to a longer internuclear distance. The O-O bond length increases from 121 pm in the neutral molecule to 126 pm in superoxide and 149 pm in peroxide. Removal of one π_g electron to form O_2^+ makes the bond stronger and the distance becomes only 112 pm [4]. The distribution of levels is different for the C_2 molecule, but the knowledge of internuclear distances is also extremely helpful in order to check the accuracy of the model and the nature of the levels. For instance, in the first excited state of C_2, one electron occupies the $3\sigma_g$ orbital: it leaves π_u (bonding) and occupies $3\sigma_g$ (almost non bonding, only slightly bonding), the distance becoming accordingly longer (124.22 pm for the ground state $^1\Sigma_g$ and 131.17 pm for the first excited state $^3\Pi_u$). On the other hand, it can be expected that the C_2^{2-} ion will exhibit a shorter C-C distance than the neutral species, as the two electrons occupy the slightly bonding $3\sigma_g$ orbital. The next four electrons occupy π_g, antibonding, so that C_2^{4-} and C_2^{6-} will have progressively longer C-C distances. Notice that O_2^{2-} and C_2^{6-} are isoelectronic and should have similar internuclear distances.

2.2. COMPLEXES CONTAINING X_2 MOLECULES

The X_2 molecules may be a part of more complicated entity (molecules or solids) and the previous discussion can be used in order to assign oxidation states. Cobalt (II) complexes are known to coordinate molecular oxygen, but it can discussed whether in the final product O_2 binds to Co(II) or O_2^{2-} binds to Co(III). In $[(NH_3)_5Co(\mu-O_2)Co(NH_3)_5]^{4+}$, the O-O distance is 147 pm, shortening to only 131 pm in $[(NH_3)_5Co(\mu-O_2)Co(NH_3)_5]^{5+}$. This indicates, by comparison, the presence of peroxide coordinated to Co(III) in the first species, and of superoxide coordinated to Co(III) in the second. The loss of one electron is felt by the oxygen ligand, rather than by the metal. The related $[(NH_3)_5Co(\mu-O_2)(\mu-NH_2)Co(NH_3)_5]^{4+}$ also contains a bridged superoxide, as shown by the O-O distance of 132 pm [5]. In these examples, we transfer directly our knowledge of the bonding in X_2 units to other species where they may be found.

Many similar examples are available for C_2. The simplest consists of the C_2 units in in C_2H_2, C_2H_4, and C_2H_6, which can be assigned as C_2^{2-}, C_2^{4-}, and C_2^{6-}. The discussion above would hint that as C_2^{6-} and O_2^{2-} are isoelectronic, the expected C-C distance in ethane should be close to 149 pm, a very good approach to 154 pm. In ethylene, the C-C distance is shorter, 134 pm, because there are two less π antibonding electrons (C_2^{4-}). The same trend leads to a C-C distance of 121 pm in acetylene.

Interestingly, these analogies can be further transferred to organometallic complexes. A nice series consists of the three binuclear complexes of ruthenium and zirconium, $[(\eta^5-C_5H_5)(PMe_3)_2Ru(\mu-X)ZrCl(\eta^5-C_5H_5)_2]$, where X=$C_2$, C_2H_2, and C_2H_4. They can be seen as containing respectively C_2^{2-}, C_2^{4-}, and C_2^{6-} units, as reflected by C-C distances increasingly longer (125.1, 130.4, 148.5 pm) [6]. The corresponding electron count indicates Ru(II) and Zr(IV) metal centers. The application of the same reasoning to the cluster $[Ru(\mu_5-C_2)(\mu-SMe)(\mu-PPh_2)(CO)_{12}]$ containing one C_2 unit (C-C distance 126 and 131 in two independent molecules) bridging more than two ruthenium atoms becomes in the least difficult [7].

3. Bonding in Solids Containing C_2 Units

3.1. THE SIMPLE APPROACH

The same model has also been used successfully in order to understand the bonding in carbides, at least partially. For instance, calcium carbide, CaC_2 (Scheme 3), can be simplistically described as $(Ca^{2+})(C_2^{2-})$, as calcium forms no other ion easily. The C-C distance is 119.1 pm, close to 121 pm as expected [8].

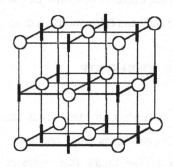

Scheme 3

Some more complicated solids containing well defined cations are good candidates for this exercise, as has been shown by Simon and coworkers for many ternary carbides [9]. In $Gd_2C_2Br_2$, the presence of Gd^{3+} and Br^- shows that C_2 is present as C_2^{4-}. The observed distance is 127 pm, only a little shorter than 134 pm. $Tb_2C_2Br_2$ can be described as $(Tb^{3+})_2(C_2^{4-})(Br^-)_2$ and the C-C distance is 129 pm, in the same range. In $Gd_{10}C_4Cl_{18}$, the C_2^{6-} ion is found, according to $(Gd^{3+})_{10}(C_2^{6-})(Cl^-)_{18}$, and the C-C distance of 147 pm is consistent with it. The decomposition of Gd_2CBr_2 in the ions $(Gd^{3+})_2(C^{4-})(Br^-)_2$ or, in a more revealing way, in $(Gd^{3+})_4(C_2^{8-})(Br^-)_4$, hints that all the levels of the C_2 unit have been filled and the C-C bond thus broken. Only isolated C atoms are found in this structure. UC_2 has two C-C distances of 134 and 135 pm, as expected from $(U^{4+})(C_2^{4-})$.

This reasoning can be applied in reverse, as for instance in $Gd_{10}C_4I_{16}$, where the C-C distance of 145 pm suggests the presence of C_2^{6-}, but there are two extra electrons, $(Gd^{3+})_{10}(C_2^{6-})(I^-)_{16}(e)_2$. These may occupy nonbonding levels of Gd, as in this compound, or Gd-Gd bonding levels, in others. Indeed, while $Gd_{10}C_4Cl_{18}$ mentioned above agrees with the presence of C_2^{6-}, in $Gd_{10}C_4Cl_{17}$ the C-C bond length is the same, but there is an extra electron, $(Gd^{3+})_{10}(C_2^{6-})(Cl^-)_{17}(e)$. The assignement of the extra electron to Gd-Gd bonding levels is based on the existence of shorter Gd-Gd bonds (374, 321, 355 pm decrease to 371, 312, 346) and band calculations, as discussed below.

3.2. THE LANGUAGE OF SOLID STATE

A good description of the bonding in these solids requires a different type of approach [1,3]. Normally, one starts from one atom, such as hydrogen, the simplest, and moves on to two atoms (the hydrogen molecule), three, four, five....and a very large number of hydrogen atoms equally spaced forming a chain. The numbers of molecular orbitals increase from one, to two, three, four, five... very large, and they span a successively larger energy window, which contains more and more levels. The lowest energy level has no nodes, and the number of nodes increases from each level to the next, the lowest level being totally bonding and the highest one completely antibonding. For the very large chain, the levels in the middle have a varying number of nodes and are partially bonding/antibonding. In the limit, there is a continuum of levels. This can be qualitatively represented as in Scheme 4.

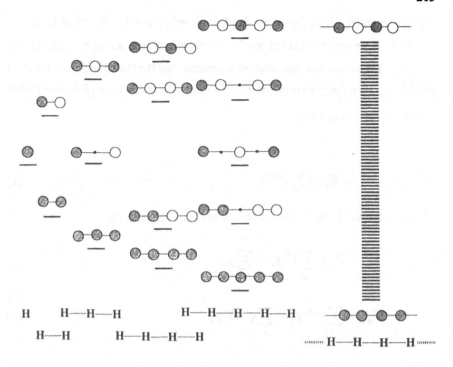

Scheme 4

The composition of the levels in the last situation can be described by Bloch functions. If we assume that the spacing between hydrogen atoms is a, an index n (0, 1, 2, ...) is assigned to each atom in a position, and the basis function is the H 1s orbital, χ_n, then the symmetry adapted linear combinations for a given irreducible representation k is given by

$$\Psi_k = \sum_n e^{ikna}\chi_n \tag{1}$$

We can calculate Ψ for k=0 and k=π/a, and obtain respectively

$$\Psi_0 = \sum_n e^0 \chi_n = \sum_n \chi_n \tag{2}$$

$$\Psi_{\pi/a} = \sum_n e^{i\pi n}\chi_n = \sum_n (-1)^n \chi_n \tag{3}$$

or in a more familiar representation $\Psi_0 = \chi_0 + \chi_1 + \chi_2 + ...$ and $\Psi_{\pi/a} = \chi_0 - \chi_1 + \chi_2 - ...$ as seen in Scheme 5.

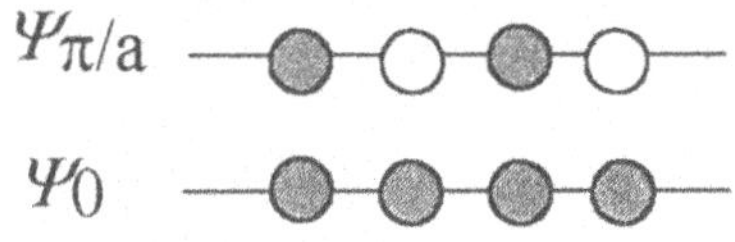

Scheme 5

For k=0, our well known lowest energy level is obtained, while k=π/a gives the highest energy, totally antibonding, level. It can be anticipated that intermediate k values will generate the other levels with different numbers of nodes.

As there are so many levels, almost a continuum of levels, it becomes more useful to depict the number of states which exist within a certain energy window, E and E + dE, the Density of States (DOS), defined as DOS(E)=DOS(E)dE. This plot of E as a function of k is called a band; for this specific chain of hydrogen atoms, the energy of each level increases with k, as the number of nodes also increases.

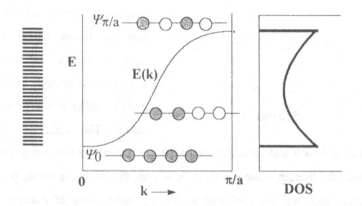

Scheme 6

It can be different. Imagine a chain of carbon atoms (z axis) and the interaction between z orbitals of adjacent atoms. The level for k=0 (all orbitals in phase) is antibonding, while the level for k=π/a, with all orbitals out of phase, is completely bonding (Scheme 7). The band runs in exactly the opposite way.

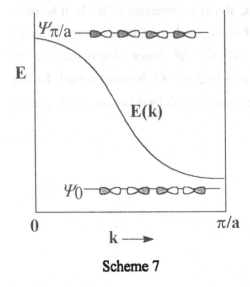

Scheme 7

If we go back to the infinite chain of hydrogen atoms, we can imagine a spacing between atoms which is close to the H-H distance in the H_2 molecule. The interactions between neighboring atoms are strong, so that the lowest energy level is strongly bonding and the highest energy one is strongly antibonding, and they differ by a large amount of energy. The continuum of levels, the band, covers a wide energy window. If the distance between adjacent atoms increases, the overlap between orbitals is weaker, the levels are not so bonding or so antibonding, and are contained within a smaller range of energies. In the limiting situation where the atoms are too far apart, there is no overlap between orbitals in adjacent atoms, and all the levels are found at the same energy as the energy of the isolated H atom. The bands under these different conditions are shown in Scheme 5, from left to right. Even when the band is flat (last case) a peak is normally seen for the DOS instead of a sharp line, but this is an artifact of the plotting (right side panel).

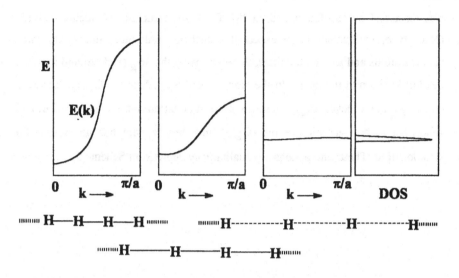

Scheme 8

The bands described above are qualitatively valid for chains of hydrogen atoms or a chain of an atom having a p orbital. In most situations a much more complicated representation will appear. Consider a chain of square planar PtX_4 complexes. Instead of forming Bloch functions from one 1s orbital, all the molecular orbitals of PtX_4 must now be considered. The remaining missing information concerns the indication of how many levels are filled or, in a different way, up to what energy is the band occupied. This highest occupied level is called the Fermi level and can be compared to the HOMO. It is normally marked with an arrow, as will be seen below.

3.3. THE BONDING IN SOME CARBIDES USING THE LANGUAGE OF SOLID STATE

The ideas developed above will be applied to some of the carbide compounds discussed earlier and we start with CaC_2 and UC_2. The bonding in these carbides

220

was discussed by Hoffmann and Li [8]. From the structure of calcium carbide (NaCl type) in Scheme 3, we expect that each C_2 unit binds end on to some calcium atoms and side on to others, therefore using the $3\sigma_g$ (end on) and π_u (side on) for donation to the metal. In the uranium carbide, where the C_2 unit carries a 4- charge, the π^* levels (π_g) is also involved in donation to the metal. This orbital might act as a π acceptor in the $(C_2)^{2-}$ unit, but calcium has no orbitals for backdonation. These interactions are qualitatively depicted in Scheme 9.

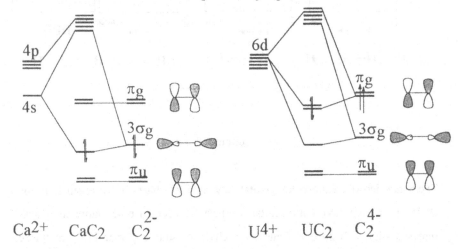

Scheme 9

There are two kinds of three-dimensional sublattices in the solids, one formed by the metal atoms (Ca or U) and the other by $(C_2)^{2-}$ or $(C_2)^{4-}$ units. When each of the C_2 occupies the respective position in the sublattice, the distance to adjacent units is large and the situation is as described for the chain of widely spaced hydrogen atoms. There is no interaction between the units, and each level of $(C_2)^{2-}$ or $(C_2)^{4-}$ gives rise to an almost isolated, well identifiable peak. In the left side of Figure 1, we show the projection of the three relevant orbitals of C_2, $3\sigma_g$, π_u, and π_g.

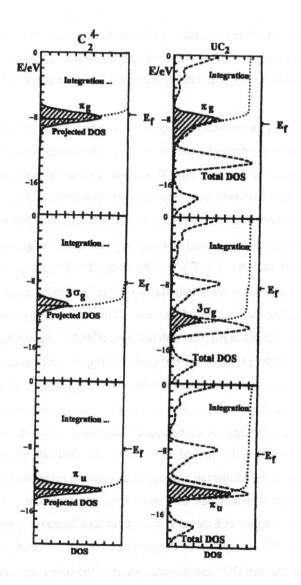

Figure 1. Projections and integrations (dotted line) of $3\sigma_g$, π_u, and π_g, for a three dimensional sublattice of $(C_2)^{4-}$ (left side) and total DOS (dashed line), projections and integrations (dotted line) of $3\sigma_g$, π_u, and π_g for UC_2 (right side) (Adapted from ref. 8).

By projection of π_u, for instance, we mean all the levels of the C_2 sublattice to which π_u contributes. There is only one: we see no spitting, no widening of the band. The dotted line indicates the integration, namely that 100% of the states projected occupy a narrow energy window between -12 and -14 eV. The stick represents the energy of the mo of isolated C_2 and the arrow represents the Fermi level.

From the qualitative diagram of Scheme 9 (right side) we know that the three groups of orbitals interact with metal orbitals and result in two groups of levels, bonding and antibonding. This can be seen on the panel of the right side of Figure 9. The shadowed peaks represent the projections of $3\sigma_g$, π_u, and π_g, from all the levels of UC_2, and the three behave similarly. Let us concentrate on π_u. The peak has been slightly pushed down by the interaction, compared to the projection of the same peak for free $(C_2)^{4-}$. The integration line shows that only up to ca. 90% of all the states fall within -12 to -15, the remaining states being part of the corresponding antibonding orbitals and with a high energy (higher than 0 eV). Something analogous happens with the other orbitals, $3\sigma_g$, and π_g.

All the panels in the right side of Figure 1 also contain a dashed line, which represents the total DOS for the whole UC_2 lattice. Let us go back to the bottom panel. The projection peak we analyzed above, containing ca. 90% of all π_u states, is smaller than the dashed peak. This indicates that the levels in that area are localized both in π_u and in U levels. To find out which uranium levels participate in the interaction, a search among several projections should take place.

The representation in Figure 1 is, in conclusion, very similar to the interaction diagram of Scheme 9, but a solid state language is used. The fragment mo's of $(C_2)^{4-}$ are replaced by the projections on the left panel, while the mo's are related to the total DOS and projected levels of the whole structure. Only the metal levels are missing here.

The total density of states can also give information concerning the conducting properties. The total DOS for UC_2 is given again, more clearly, in Figure 2 (left) and that of CaC_2 on the right. Here, the total DOS is given by the solid line and the dashed area represents all the levels up to the Fermi level. This level falls in the middle of a band for UC_2, revealing that it should behave as a metallic conductor, in accordance with our qualitative picture of a half filled set of degenerate orbitals (Scheme 9), while CaC_2 is clearly an insulator, owing to the position of the Fermi level, just below a large energy gap (ca. 2 eV).

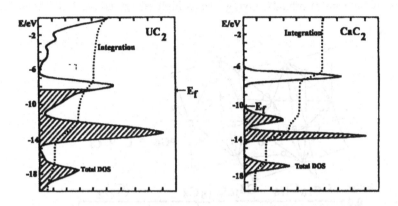

Figure 2. Total density of states (solid line) and integration (dotted line) for UC_2 (left side) and CaC_2 (right side). Occupied levels are indicated by dashed area (Adapted from ref 8).

We can go back and analyze our assumption that there was $(C_2)^{4-}$ in the uranium carbide and $(C_2)^{2-}$ in calcium carbide. Indeed, for CaC_2, we see that most of $3\sigma_g$, the peak around -10 to -12 eV, is filled, so that the description is a realistic one. For UC_2, both this peak and half of the next, π_g, are occupied, leading us to a formulation close to $(C_2)^{4-}$. The band structure shows in a different way what was expected from qualitative arguments.

Other authors have performed band calculations on carbides [10,11,12]. Just for comparison, with the ionic approach, we'll revisit $Gd_{10}Cl_{18}C_4$ [10,11] and $Gd_{10}I_{17}C_4$ [11], where the Gd atoms form octahedra sharing edges and containing the C_2 units in their centers. The band structure for the first compound

224

is shown in Figure 3, in the form of an interaction diagram between the C_2 fragment levels and those of the metal-chloride sublattice.

As can be seen, both $3\sigma_g$ and all of π_g fall below the Fermi level. The formulation is close to $(C_2)^{6-}$. It cannot be exactly this, because we know that parts of the $3\sigma_g$ and π_g have been pushed up by the antibonding interaction with metal orbitals, giving rise to high energy levels, well above the Fermi level.

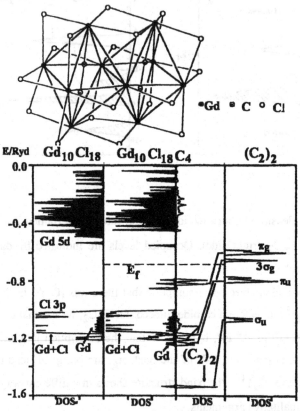

Figure 3. Total density of states with projections of some levels for $Gd_{10}Cl_{18}C_4$ (center), total DOS for the C_2 sublattice (right) and for $Gd_{10}Cl_{18}$ (left) (Adapted from ref. 10).

The same band structure has been calculated by other authors and the result is similar. They compare it with the band structure for $Gd_{10}I_{17}C_4$, which has one extra electron. This electron occupies half a level which is metal-metal bonding, explaining why some Gd-Gd bonds become shorter in this compound, instead of modifying the nature of the C_2 group.

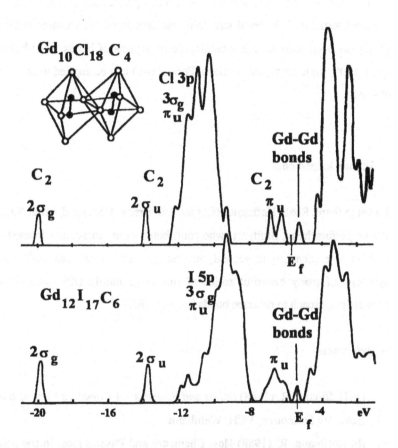

Figure 4. Total density of states for $Gd_{10}Cl_{18}C_4$ (top) and for $Gd_{10}I_{17}C_4$ (bottom), showing the Fermi level and the nature of several levels. Notice that the axes are reversed (energy in horizontal axis) (Adapted from ref 11).

226

4. Conclusions

We started from a bonding model for diatomic C_2 molecules and extended it to a very simplified solid state picture of bonding in solids, trying to show the relation between interaction diagrams in molecules and appropriate combinations of total DOS for a whole structure and for well chosen projections of levels from parts of the same structure. The band structure calculations reflect an ionic interpretation which has been used in understanding some aspects of bonds in carbides. These ideas (DOS, projections, integrations, Fermi level) can be applied to any other kind of solid.

5. Acknowledgements

I want to thank Roald Hoffmann (Cornell University, USA) and Arndt Simon (MPI Festkörperforschung, Stuttgart) who contributed to my increasing interest not only in solid state chemistry in general, but also specially in carbides, both through the synthesis of many beautiful solids of this large family (AS), and through the chemist's approach to describe bonds in solids (RH).

6. References

1. (a) Hoffmann, R. (1988) *Solids and Surfaces - A Chemist's View of Bonding in Extended Structures*, VCH, Weinheim.
    (b) Hoffmann, R. (1988) How Chemistry and Physics meet in the Solid State, *Angew. Chem. Int. Ed. Eng.*, **26**, 846-878.
2. (a) Hoffmann, R. (1963) An Extended Hückel Theory. I. Hydrocarbons, *J. Chem. Phys.*, **39**, 1397-1412.

(b) and Lipscomb, W. N. (1962) The Theory of Polyhedral Molecules. I. Physical Factorization of the Secular Equation, *J. Chem. Phys.*, **36**, 2179-2189.

(c) Ammeter, J. H., Bürgi, H.-B., Thibeault, J. C., Hoffmann, R. (1978) *J. Am. Chem. Soc.*, **100**, 3686-3696.

3. (a) Whangbo, M.-H., Hoffmann, R., Woodward, R. B. (1979) Conjugated one and two dimensional polymers, *Proc. R. Soc. Lond.*, **A366**, 23-46.

   (b) Whangbo, M.-H. (19##) Orbital interaction analysis for the electronic structures of low dimensional solids, in J. Rouxel (eds.), *Crystal Chemistry and Properties of Materials with Quasi One-Dimensional Structures*

   (c) Albright, T. A., Burdett, J. K., Whangbo, M.-H. (1985) *Orbital Interactions in Chemistry*, John Wiley & Sons, NY.

4. DeKock, R. L. and Gray, H. B. (1980) *Chemical Structure and Bonding*, The Benjamin/Cummings Publishing Company, Menlo Park.

5. Purcell, K. F. and Kotz, J. C. (1977) *Inorganic Chemistry*, W. B. Saunders Company, Philadelphia.

6. Lemke, F. R., Szalda, D. J., Morris Bullock, R. (1991) Ruthenium/Zirconium Complexes Containing C_2 Bridges with Bond Orders of 3, 2, and 1. Synthesis and Structures of $Cp(PMe_3)_2RuCH_nCH_nZrClCp_2$ (n=0, 1, 2), *J. Am. Chem. Soc.*, **113**, 8466-8477.

7. Adams, C. J., Bruce, M. I., Skelton, B. W., White, A. H. (1992) Synthesis and X-Ray Structure of Two Complexes Containing Dicarbon (C_2) Attached to Ru_5 Clusters with Unusual Core Geometries, *J. Chem. Soc., Chem. Comm.*, 26-29.

8. Li, J. and Hoffmann, R. (1989) How C-C bonds are formed and how they influence structural choices in some binary and ternary metal carbides, *Chem. Mater.*, **1**, 83-101.

9. (a) Simon, A. (1988) Clusters of Valence Electron Poor Metals - Structure, Bonding, and Properties, *Angew. Chem. Int. Ed. Eng.*, **26**, 159-183.

(b) Simon, A. (1987) Borderline of Clusters Chemistry with Lanthanides, *Inorg. Chim. Acta*, **140**, 57-86.

(c) Simon, A. (1995) From a molecular view on solids to molecules in solids, *J. Alloys Comp.*, **229**, 158-174

(d) Simon, A. (1997) Superconductivity and Chemistry, *Angew. Chem. Int. Ed. Eng.*, **36**, 1789-1806.

10. Satpathy, S. and Andersen, O. K. (1985) Intersticial Carbon Molecules, Metal-Metal Bonds, and Chemical Binding in $Gd_{10}C_4Cl_{18}$, *Inorg. Chem.*, **24**, 2604-2608

11. Bullett, D. W. (1985) Metal-Metal Bonding in Transition-Element and Lanthanoid Cluster Compounds, *Inorg. Chem.*, **24**, 3319-3323.

12. (a) Miller, G. J., Burdett, J. B., Schwarz, C., Simon, A. (1986) Molecular Intersticials: An Analysis of the $Gd_2C_2Cl_2$ Structure, *Inorg. Chem.*, **26**, 4437-4444.

(b) Long, J. R., Hoffmann, R., Meyer, H.-J. (1992) Distortions in the Structure of Calcium Carbide: A Theoretical Investigation, *Inorg. Chem.*, **31**, 1734-1740.

(c) Long, J. R., Halet, J.-F., Saillard, J.-Y., Hoffmann, R., Meyer, H.-J. (1992) Structural preferences among the rare earth dicarbides: the electronic structure of LaC_2 and ThC_2, *New J. Chem.*, **16**, 839-846.

(d) Meyer, G., Meyer, H.-J. (1992) Unusual Valences in Rare-Earth Halides, *Chem. Mater.*, **4**, 1157-1168.

DIVERSITY AND CERTAINTY — DATABASE RESEARCH IN CRYSTAL ENGINEERING

GAUTAM R. DESIRAJU

School of Chemistry, University of Hyderabad,
Hyderabad 500 046, India

ABSTRACT   Crystal structures are determined by a complex interplay of intermolecular interactions. Given the lack of easily identifiable correspondences between molecular and crystal structures, and the difficulties in theoretical methods of structure prediction, statistical analysis of existing structural data continues to be one of the most useful entries into crystal synthesis.

1. Introduction

Crystal engineering or the design of organic solids is one of the mainline areas in chemical research today. It is of interest to trace the historical development of this subject. The origins of crystal engineering lie in organic solid state chemistry, more specifically the study of topochemical reactions. Topochemistry was important in that it showed that properties of crystals are related to their internal structure. However, the topochemical emphasis was restrictive and the scope of crystal engineering rapidly outgrew what was originally envisaged by Schmidt. Another significant and distinct stream of thought was the contribution of Kitaigorodskii who developed the concept of close-packing for molecular crystals. This model provided the broad theoretical framework of the subject within which may be understood the grammar of crystal packing, even today. Kitaigorodskii's work, unlike that of Schmidt, was general enough so that the overall problems in understanding crystal packing could be identified.

The main problem with regard to anticipating and predicting crystal structures is that most hetero-atom interactions are directional and of a sufficiently long-range character that most molecular crystals are not *exactly* close-packed. These small deviations from close-packing are, however, very significant because they determine real crystal structures. It is in this context that database research became crucial in crystal engineering. Crystal structures are determined by a balance of a large number of weak intermolecular interactions. Databases offer a route to the understanding of weak interactions. The weaker the interaction, the more relevant is the database approach towards their understanding.

229

D. Braga et al. (eds.), Crystal Engineering: From Molecules and Crystals to Materials, 229–241.
© *1999 Kluwer Academic Publishers. Printed in the Netherlands.*

2. Extracting information regarding intermolecular interactions from databases — deconvolution and analysis.

The relevance of the Cambridge Structural Database (CSD) as a research tool in structural chemistry was highlighted in the important article of Kennard and co-workers in 1983. The article emphasised that the CSD could be used to obtain all relevant information on intra- and intermolecular geometrical parameters in crystal structures. For intramolecular bond lengths and angles, the CSD enables one to quickly obtain specific metrics. However, the range of covalent bond geometries formed by specific atoms in specific chemical situations is quite small. In principle therefore, one may also obtain such information on 'typical' or 'average' bond metrics manually and the use of databases is not absolutely critical. For intermolecular parameters, however, the situation is very different. Intermolecular interactions are, roughly speaking, an order of magnitude weaker than covalent bonds. This means that they are deformed far easier than covalent bonds. The range of geometries for intermolecular interactions formed by specific atoms in specific chemical situations is accordingly quite large. The weaker the interaction, the larger the range of geometries observed. Therefore a manual approach cannot be used to obtain estimates of 'typical' or 'average' interaction metrics. The use of databases is mandatory.

With the CSD, one scans the geometry of interest in several crystal structures and attempts to find trends and qualitative preferences in the interaction geometries. The idea is that if a sufficiently large number of interaction geometries is examined, the interfering effects of the 'other' interactions effectively cancel out and that what remains are the attributes of the interaction of interest. This approach has proved to be remarkably successful with weak interactions like C–H···X and X···X where X is a hetero-atom (O, N, S, P, halogen, metal) or even a π-system. The CSD has also been used to explore patterns of interactions between π-systems, for example the so-called phenyl embrace geometries. The use of the CSD in studying patterns of interactions in organometallic crystals is in its infancy but the early studies hint that much more can be done in this regard.

Any crystal structure is the result of a balance and compromise between many types of interactions. With the help of CSD-retrieved data on related structures, one tries to deconvolute the structure and to unravel the effects of the individual interactions. Ideally what one would like to have is a ranking of the interactions in terms of their efficacy in controlling crystal packing. This is not always possible, but approximate hierarchies of interaction effectiveness are already emerging from the mass of crystallographic data now available. The next step is to use the information on the individual interactions to identify and to then suitably extend a family of compounds, so that the effects of the various interactions can be reconfirmed even more clearly. In effect, what one does is to make a small database of compounds wherein the roles of the various intermolecular interactions is clearly identified. This is the classical approach to crystal engineering and justifies the description of the subject as a new form of physical organic chemistry.

3. Putting together intermolecular interactions to obtain pre-desired crystal structures — convolution and synthesis

In the process of analysing crystal structures, one recognises certain repetitive structural units. These are typically small in size and are composed of specific functionalities and their resulting interactions arranged in specific ways. The identification of these structural units is somewhat subjective but they are important in the sense that they may be associated with particular crystal packing characteristics. In this respect, these units are approximations of actual crystal structures. They are much closer to actual crystal structures than say, molecular functional groups or even single interactions. Considering the importance of these units in the process of building up a crystal structure from its components, these units have been termed *supramolecular synthons*.

A desirable attribute of a supramolecular synthon is the property of *robustness*, in other words with the likelihood that it can be formed from several different molecular structures. Identification of robust synthons greatly facilitates the synthetic aspects of the crystal engineering process. Designing a crystal structure effectively becomes then synthon design, and this in turn is only possible from a proper knowledge of the pertinent intermolecular interactions. One attempts, in other words, to put back together what has been taken apart in the process of analysis.

4. Diversity and certainty

Both the joy and sorrow of crystal synthesis lie in the fact that crystal structures cannot be easily or routinely derived from molecular structures. But within any crystal structure, lies the kernel, that is some combination of synthons that serves to economically yet completely define its essential structural content. Identification of this kernel is easy or difficult to varying degrees, depending on the degree of mutual interference between the important intermolecular interactions. It must be emphasised that all families of organic compounds are not equally amenable to crystal engineering strategies at the present time! Given, however, the complexity of the problem, the statistical approach to crystal structure analysis provides a vital breakthrough. A crystal structure may be considered to be generated from the molecular structure using some kind of code that the structural chemist understands only incompletely. This is why crystal structures are difficult to understand, rationalise and therefore to design. However, any code can be broken if a sufficient number of examples are available. This then is what is being attempted, with varying degrees of success in many laboratories worldwide. As in any other kind of database research, the ultimate goal of crystal engineering is to obtain structural predictions that are as certain as can be, for a group of compounds that are as diverse as possible.

5. Acknowledgements

Financial assistance from the Department of Science and Technology, Government of India and collaborations with Molecular Simulations Inc. (Cambridge and San Diego) are gratefully acknowledged. I would like to thank Dr. Ashwini Nangia for helpful suggestions and Mr. S. S. Kuduva for his assistance in the preparation of this chapter.

6. References

The following is a list of papers relevant to the development of crystal engineering, with special reference to database methods. Papers in the coordination polymer area have generally not been included. Owing to the very large number of papers in the crystal engineering area in recent years, it has not been possible to provide a comprehensive coverage here, especially with regard to host-guest complexes.

<u>General references</u>

Desiraju, G. R. (1989) *Crystal Engineering: The Design of Organic Solids*, Elsevier, Amsterdam.

Desiraju, G. R. (1995) Supramolecular synthons in crystal engineering – A new organic synthesis. *Angew. Chem. Int. Ed. Engl.*, **34**, 2311-2327.

Desiraju, G. R., (Ed.) (1996) *The Crystal as a Supramolecular Entity*, Perspectives in Supramolecular Chemistry, vol. 2, Wiley, Chichester.

Gavezzotti, A. (1996) Organic crystals: engineering and design. *Curr. Opin. Solid State Mater. Sci.*, **1**, 501-505.

Desiraju, G. R. (1997) Designer crystals: intermolecular interactions, network structures and supramolecular synthons. *Chem. Commun.* 1475-1482.

Aakeröy, C. B. (1997) Crystal engineering: strategies and architectures. *Acta Crystallogr.* B**53**, 569-583.

Desiraju, G. R. (1997) Crystal engineering: solid state supramolecular synthesis. *Curr. Opin. Solid State Mater. Sci.* **2**, 451-454.

Weber, E. (Ed.) (1998) *Design of Organic Solids*, Topics in Current Chemistry, vol. 198, Springer Verlag, Heidelberg.

Bürgi, H. -B., Hulliger, J., and Langley, P. J. (1998) Crystallisation of supramolecular materials *Curr. Opin. Solid State Mater. Sci.* **3**, 425-430.

Braga, D., Grepioni, F., and Desiraju, G. R. (1998) Crystal engineering and organometallic architecture. *Chem. Rev.*, **98**, 1375-1405.

Nangia, A., and Desiraju, G. R. (1998) Supramolecular structures – reason and imagination. *Acta Crystallogr.*, A**54**, 934-944.

Coordination polymers

Batten, S. R., and Robson, R. (1998) Interpenetrating nets: Ordered, periodic entanglement. *Angew. Chem. Int. Ed. Engl.* **37**, 1460-1494.

Champness, N. R., and Schröder, M. (1998) Extended networks formed by coordination polymers in the solid state. *Curr. Opin. Solid State Mater. Sci.* **3**, 419-424.

Yaghi, O. M., Li, H., Davis, C., Richardson, D., and Groy, T. L. (1998) Synthetic strategies, structure patterns and emerging properties in the chemistry of modular porous solids. *Acc. Chem. Res.* **31**, 474-484.

Zaworotko, M. J. (1998) From dissymetric molecules to chiral polymers: A twist for supramolecular synthesis? *Angew. Chem. Int. Ed. Engl.* **37**, 1211-1213.

Historical

Bernal, J. D., and Crowfoot, D. (1935) The structure of some hydrocarbons related to the sterols. *J. Chem. Soc.*, 93-100.

Robertson, J. M. (1951) The measurement of bond lengths in conjugated molecules of carbon centres. *Proc. R. Soc. London*, **A207**, 101-110.

Mustafa, A. (1952) Dimerisation reactions in sunlight, *Chem. Rev.*, **51**, 1-23.

Schmid, M. (1955) Neue organische Pigmentfarbstoffe, ihre Herstellung und Anwendung. *Deutsche Farben-Zeitschrift*, **9**, 252-255.

Topochemistry

Schmidt, G. M. J. (1971) Photodimerization in the solid state. *Pure Appl. Chem.*, **27**, 647-678.

Ginsburg, D., (Ed.) (1976) *G. M. J. Schmidt et al. Solid State Photochemistry*, Verlag Chemie, Weinheim.

Desiraju, G. R. (Ed.) (1987) *Organic Solid State Chemistry*, Elsevier, Amsterdam.

Kräutler, B., Müller, T., Maynollo, J., Gruber, K., Kratky, C., Ochsenbein, P., Schwarzenbach, D., and Bürgi, H.-B. (1996) A topochemically controlled, regiospecific fullerene bisfunctionalisation. *Angew. Chem. Int. Ed. Engl.*, **35**, 1204-1206.

Toda, F., Tanaka, K., Tamashima, T., and Kato, M. (1998) Stereoselective thermal conversion of s-trans-diallene into dimethylenecyclobutene via s-cis-diallene in the crystalline state. *Angew. Chem. Int. Ed. Engl.* **37**, 2724-2727.

Crystal packing theory

Kitaigorodskii, A. I (1973) *Molecular Crystals and Molecules*, Academic Press, New York.

Kitaigorodskii, A. I. (1984) *Mixed Crystals*, Springer Verlag, Berlin.

Pertsin, A. J., and Kitaigorodskii, A. I. (1987) *The Atom-Atom Potential Method*, Springer Verlag, Heidelberg.

Pauling L., and Delbrück M. (1940) Nature of the intermolecular forces operative in biological processes. *Science*, **92**, 77-79.

Nowacki, W. (1942) Symmetrie und physikalisch-chemische Eigenschaften krystallisierter Verbindungen. I. Die Verteilung der Krystallstruckturen über die 219 Raumgruppen. *Helv. Chim. Acta*, **25**, 863-878.

Bondi, A. (1964) van der Waals volumes and radii. *J. Phys. Chem.*, **68**, 441-451.

Desiraju, G. R. and Gavezzotti, A. (1989) Crystal structures of polynuclear aromatic hydrocarbons. classification, rationalisation and prediction from molecular structure *Acta Crystallogr.*, **B45**, 473-482.

Brock, C. P., and Dunitz, J. D. (1994) Towards a grammar of crystal packing. *Chem. Mater.* **6**, 1118-1127.

Gavezzotti A. (1994) Are crystal structures predictable? *Acc. Chem. Res.*, **27**, 309-314.

Perlstein, J., Steppe, K., Vaday, S., and Ndip, E. M. N. (1996) Molecular self assemblies. 5. Analysis of the vector properties of hydrogen bonding in crystal engineering. *J. Am. Chem. Soc.*, **118**, 8433-8443.

Lloyd, M. A. and Brock, C. P. (1997) Retention of $\overline{4}$ symmetry in compounds containing MAr$_4$ molecules and ions. *Acta Crystallogr.*, **B53**, 780-786.

Cambridge Structural Database

Allen, F. H., Kennard, O., and Taylor, R. (1983) Systematic analysis of structural data as a research technique in organic chemistry. *Acc. Chem. Res.*, **16**, 146.

Allen, F. H., Davies, J. E., Galloy, J. J., Johnson, O., Kennard, O., Macrae, C. F., Mitchell, G. F., Smith, J. M., and Watson, D. G. (1991) The development of versions 3 and 4 of the Cambridge Structural Database system. *J. Chem. Inf. Comput. Sci.*, **31**, 187-204.

Allen, F. H., and Kennard, O. (1993) Version 5 of the Cambridge Structural Database system. *Chem. Design. Autom. News*, **8**, 1 & 31-37.

Allen, F. H., Kennard, O., Watson, D. G., Brammer, L., Orpen, A. G., and Taylor, R. (1987). Tables and bond lengths determined by X-ray and neutron diffraction. Part I. Bond lengths in organic compounds. *J. Chem. Soc., Perkin Trans.* 2. S1-S19.

Orpen, A. G., Brammer, L., Allen, F. H., Kennard, O., Watson, D. G., and Taylor, R. (1989). Tables of bond lengths determined by X-ray and neutron diffraction. Part II. Organometallic compounds and coordination complexes of the d- and f-block metals. *J. Chem. Soc. Dalton Trans.*, S1-S83.

Rowland, R. S. and Taylor, R. (1996) Intermolecular nonbonded contact distances in organic crystal structures: comparision with distances expected from van der Waals radii. *J. Phys. Chem.*, **100**, 7384-7391.

Intermolecular interactions studied with the CSD

Rosenfield, R. E., Jr. Parthasarathy, R., and Dunitz, J. D. (1977) Directional preferences of non-bonded atomic contacts with divalent sulfur. 1. Electrophiles and nucleophiles. *J. Am. Chem. Soc.*, **99**, 4860-4862.

Row, T. N. G., and Parthasarathy, R. (1981) Directional preferences of nonbonded atomic contacts with divalent sulfur in terms of its orbital orientations. 2. S⋯S interactions and nonspherical shape of sulfur in crystals. *J. Am. Chem. Soc.*, **103**, 477-479.

Taylor, R., and Kennard, O. (1982) Crystallographic evidence for the existence of C–H⋯O, C–H⋯N and C–H⋯Cl hydrogen bonds. *J. Am. Chem. Soc.*, **104**, 5063-5070.

Murray-Rust, P., and Glusker, J. P. (1984) Directional hydrogen bonding to sp^2- and sp^3-hybridized oxygen atoms and its relevance to ligand-macromolecule interactions. *J. Am. Chem. Soc.*, **106**, 1018-1025.

Desiraju, G. R. (1989) Distance dependence of C–H⋯O interactions in some chloroalkyl compounds. *J. Chem. Soc., Chem. Commun.*, 179-180

Desiraju, G. R., and Parthasarathy, R. (1989) The nature of halogen⋯halogen interactions: are short halogen contacts due to specific attractive forces or due to close packing of non-spherical atoms? *J. Am. Chem. Soc.*, **111**, 8725-8726.

Desiraju, G. R. (1991) The C–H···O hydrogen bond in crystals. What is it? *Acc. Chem. Res.*, **24**, 290-296.

Pathaneni, S. S., and Desiraju, G. R. (1993) Database analysis of Au···Au interactions. *J. Chem. Soc., Dalton Trans.*, 319-322.

Price, S. L., Stone, A. J., Lucas, J., Rowland, R. S., and Thornley, A. E. (1994) The nature of –Cl···Cl– intermolecular interactions. *J. Am. Chem. Soc.*, **116**, 4910-4918.

Pedireddi, V. R., Reddy, D. S., Goud, B. S., Craig, D. C., Rae, A. D., and Desiraju, G. R. (1994) The nature of halogen···halogen interactions and the crystal structure of 1,3,5,7-tetraiodoadamantane. *J. Chem. Soc., Perkin Trans.* 2, 2353-2360.

Shimoni, L., and Glusker, J. P. (1994) The geometry of intermolecular interactions in some crystalline fluorine-containing organic compounds *Struct. Chem.*, **5**, 3843-397.

Braga, D., Grepioni, F., Biradha, K., Pedireddi, V. R., and Desiraju, G. R. (1995) Hydrogen bonding in organometallic crystals – 2. C–H···O hydrogen bonds in bridged and terminal first row metal carbonyls. *J. Am. Chem. Soc.*, **117**, 3156-3166.

Braga, D., Grepioni, F., Biradha, K., and Desiraju, G. R. (1996) Agostic interactions in organometallic compounds. A Cambridge Structural Database study. *J. Chem. Soc., Dalton Trans.*, 3925-3930.

Dance, I., and Scudder, M. (1998) Supramolecular motifs: Sextuple aryl embraces in crystalline [M(2,2'-bipy)₃] and related complexes. *J. Chem. Soc., Dalton Trans.*, 1341-1350.

Allen, F. H., Lommerse, J. P. M., Hoy, V. J., Howard, J. A. K., and Desiraju, G. R. (1996) The hydrogen bond C–H donor and π-acceptor characteristics of three-membered rings. *Acta Crystallogr.*, **B52**, 734-745.

Lommerse, J. P. M., Stone, A. J., Taylor, R., and Allen, F. H. (1996) The nature and geometry of intermolecular interactions between halogens and oxygen or nitrogen. *J. Am. Chem. Soc.*, **118**, 3108-3116.

Desiraju, G. R. (1996) The C–H···O hydrogen bond: structural implications and supramolecular design. *Acc. Chem. Res.* **29**, 441-449 (1996).

Steiner, T. (1997) Unrolling the hydrogen bond properties of C–H···O interactions. *Chem. Commun.* 727-734.

Dunitz, J. D., and Taylor, R. (1997) Organic fluorine hardly ever accepts hydrogen bonds. *Chem. Eur. J.*, **3**, 89-98.

Mascal, M. (1998) Statistical analysis of C-H···N hydrogen bonds in the solid state: There *are* real precedents. *Chem. Commun.*, 303-304.

Steiner, T., and Desiraju, G. R. (1998) Distinction between the weak hydrogen bond and the van der Waals interaction, *Chem. Commun.*, 891-892.

Allen, F. H., Raithby, P. R., Shields, G. P. and Taylor, R. (1998) Probabilities of formation of bimolecular cyclic hydrogen-bonded motifs in organic crystal structures: A systematic database study. *Chem. Commun.*, 1043-1044.

Aullón, G., Bellamy, D., Brammer, L., Vruton, E. A., and Orpen, G. A. (1998) Metal-bound chlorine often accepts hydrogen bonds. *Chem. Comun.*, 654-654.

Braga, D., Leonardis, P. D., Grepioni, F., and Tedesco, E. (1998) Crystalline dihydrogen complexes. Intramolecular and intermolecular interactions and dynamic behaviour. *Inorg. Chim. Acta*, **273**, 116-130.

Desiraju, G. R., and Steiner, T. (1999) *The Weak Hydrogen Bond in Structural Chemistry and Biology*, Oxford University Press, Oxford.

Hierarchies of interactions

Leiserowitz, L. (1976) Molecular packing modes. Carboxylic acids. *Acta Crystallogr.*, **B32**, 775-802.

Sarma, J. A. R. P., and Desiraju, G. R. (1986) The role of Cl···Cl and C-H···O interactions in the crystal engineering of 4Å - short axis structures. *Acc. Chem. Res.*, **19**, 222-228.

Etter, M. C. (1990) Uncoding and decoding hydrogen bond patterns of organic compounds. *Acc. Chem. Res.* **23**, 120-126.

Sharma, C. V. K., Panneerselvam, K., Pilati, T. and Desiraju, G. R. (1993) Molecular recognition involving an interplay of O-H···O, C-H···O and π···π interactions. The anomalous crystal structure of the 1:1 complex 3,5-dinitrobenzoic acid - 4-(N,N-dimethylamino)benzoic acid. *J. Chem. Soc., Perkin Trans. 2*, 2209-2216.

Aoyama, Y., Endo, K., Anzai, T., Yamaguchi, Y., Sawaki, T., Kobayashi, K., Kanehisa, N., Hashimoto, H., Kai, Y., and Masuda, H. (1996) Crystal engineering of stacked aromatic columns. Three-dimensional control of the alignment of orthogonal aromatic triads and guest quinones via self-assembly of hydrogen-bonded networks. *J. Am. Chem. Soc.*, **118**, 5562-5571.

Allen F. H., Hoy, V. J., Howard, J. A. K., Thalladi, V. R. Desiraju, G. R., Wilson, C.

238

C., and McIntyre, G. J. (1997) Crystal engineering and correspondence between molecular and crystal structures. Are 2- and 3-aminophenols anomalous? *J. Am. Chem. Soc.*, **119**, 3477-3480.

Desiraju, G. R. (1997) Crystal gazing: Structure prediction and polymorphism. *Science*, **278**, 404-405.

Deconvoluting interactions

Schwiebert, K. E., Chin, D. N., MacDonald, J. C., and Whitesides, G. M. (1995) Engineering the solid state with 2-benzimidazolones. *J. Am. Chem. Soc.*, **118**, 4018-4029.

Reddy, D. S., Ovchinnikov, Y. E., Shishkin, O. V., Struchkov, Y. T., and Desiraju, G. R. (1996) Supramolecular synthons in crystal engineering. 3. Solid state architecture and synthon robustness in some 2,3-dicyano-5,6-dichloro-1,4-dialkoxybenzenes. *J. Am. Chem. Soc.*, **118**, 4085-4089.

Lewis, F. D., Yang, J., and Stern, S. L. (1996) Crystal structures of secondary arene dicarboxamides. An investigation of arene hydrogen bonding relationships in the solid state. *J. Am. Chem. Soc.*, **118**, 12029-12037.

Coe, S., Kane, J. J., Nguyen, T. L., Toledo, L. M., Wininger, E., Fowler, F. W., and Lauher, J. W. (1996) Molecular symmetry and the design of molecular solids. The oxamide functionality as a persistent hydrogen bonding unit. *J. Am. Chem. Soc.*, **119**, 86-93.

Thalladi, V. R., Weiss, H.-C., Bläser, D., Boese, R., Nangia, A. and Desiraju, G. R. (1998) C–H···F interactions in the crystal structures of some flurobenzenes. *J. Am. Chem. Soc.*,**120**, 8702-8710.
Robinson, J. M. A., Kariuki, B. M., Harris, K. D. M. and Philp, D. (1998) Interchangeability of halogen and ethynyl substitutents in the solid state structures of di- and tri-substituted benzenes. *J. Chem. Soc. Perkin Trans 2.*, 2459-2469.

Kuduva, S. S., Craig, D. C., Nangia, A., and Desiraju, G. R. (1999) Cubanecarboxylic acids. Crystal engineering considerations and the role of C–H···O hydrogen bonds in determining O–H···O networks. *J. Am. Chem. Soc.*, **121**, (in press).

Crystal synthesis. Convolution

Leiserowitz, L., and Hagler, A. (1983) Generation of primary amide crystal structures. *Proc. R. Soc. London*, **A388**, 133-175.

Reddy, D. S., Craig, D. C., Rae, A. D. and Desiraju, G. R. (1993) N···Br mediated diamondoid network in the crystalline complex, carbon tetrabromide – hexamethylenetetramine *J. Chem. Soc., Chem. Commun.*, 1737-1738.

Reddy, D. S., Craig, D. C., and Desiraju, G. R. (1994) Organic alloys: Diamondoid networks in crystalline complexes of 1,3,5,7-tetrabromoadamantane, hexamethylene tetramine and carbon tetrabromide. *J. Chem. Soc., Chem. Commun.*, 1457-1458.

Reddy, D. S., Craig, D. C., and Desiraju, G. R. (1996) Supramolecular synthons in crystal engineering. 4. Structure simplification and synthon interchangeability in some organic diamondoid solids. *J. Am. Chem. Soc.*, **118**, 4090-4093.

Thalladi, V. R., Brasselet, S., Bläser, D., Boese, R., Zyss, J., Nangia, A., and Desiraju, G. R. (1997) Engieering of an octupolar non-linear optical crystal: Tribenzyl isocyanurate. *Chem. Commun.*, 1841-1842.

Swift, J. A., Pivovar, A. M., Reynolds, A. M. and Ward, M. D. (1998) Template-directed architectural isomerism of open molecular frameworks: Engineering of crystalline clathrates. *J. Am. Chem. Soc.*, **120**, 5887-5894.

Thalladi, V. R., Brasselet, S., Weiss, H.-C., Bläser, D., Katz, A. K., Carrell, H. L., Boese, R., Zyss, J., Nangia, A., and Desiraju, G. R. (1998) Crystal engineering of some 2,4,6-triaryloxy 1,3,5-triazines: octupolar non-linear materials. *J. Am. Chem. Soc.*, **120**, 2563-2577.

Langley, P. J., Hulliger, J., Thaimattam, R., and Desiraju, G. R. (1998) Supramolecular synthons mediated by weak hydrogen bonding: Forming linear molecular arrays via $C\equiv C-H\cdots O_2N$ recognition. *New J. Chem.*, 1307-1309.

Galoppini, E., and Gilardi, R. (1999) Weak hydrogen bonding between acelylenic groups: the formation of tetrakis(4-ethynylphenyl)methane. *Chem. Commun.*, 173-174. Jetti, R. K. R., Kuduva, S. S., Reddy, D. S., Xue, F., Mak, T. C. W., Nangia, A., and Desiraju, G. R. (1998) 4-(Triphenylmethyl)benzoic acid: a supramolecular wheel-and-axle host compund. *Tetrahedron Lett.*, **39**, 913-916.

MacGillivray, L. R., and Atwood, A. L. (1999) Unique guest inclusion within multi-component, extended-cavity resorcin [4] arenes. *Chem. Commun.*, 181-182.

Other recent articles

Coates, G. W., Dunn, A. R., Henling, L. M., Ziller, J. W., Lobkovsky, E. B., and Grubbs, R. H. (1998) Phenyl-perfluorophenyl stacking interactions: Topochemical [2+2] photodimerization of olefinic compounds. *J. Am. Chem. Soc.*, **120**, 3641-3649.

Ferguson, G., Glidewell, C., Lough, A. J., McManus, G. D., and Meehan, P. R. (1998) Crystal engineering using polyphenols. Host-guest behaviour of planar ribbons in C-methylcalix[4]resorcinarene-4,4'-trimethylene-dipyridine-methanol (1/2/0.5), and capture of 2,2'-bipyridyl molecules by paired calixarene bowls in C-methylcalix[4]-resorcinarene-2,2'-bipyridyl-methanol-water (1/1/1/1.16). *J. Mater. Chem.* **8**, 2339-2345.

Spaniel, T., Gorls, H., and Scholz, J. (1998) (1,4-Diaza-1,3-diene)titanium and -niobium halides: Unusual structures with intramolecular C-H···halogen hydrogen bonds. *Angew. Chem. Int. Ed. Engl.*, **37**, 1862-65.

Hulliger, J., and Langley, P. J. (1998) On intrinsic and extrinsic defect-forming mechanisms determining the disordered structure of 4-iodo-4'-nitrobiphenyl crystals. *Chem. Commun.*, 2557-2558.

Ranganathan, D., Haridas, V., Gilardi, R., and Karle, I. L. (1998) Self-assembling aromatic-bridged serine-based cyclodepsipeptides (Serinophanes): A demonstration of tubular structures formed through aromatic π-π interactions. *J. Am. Chem. Soc.*, **120**, 10793-10800.

Jones, P.G., and Ahrens, B. (1998) Bis(diphenylphosphino)methane and related ligands as hydrogen bond donors. *Chem. Commun.*, 2307-2308.

Krische, M. J., Lehn, J-M., Kyritsakas, N., and Fischer, J. (1998) Molecular-recognition-directed self-assembly of pleated sheets from 2-aminopyrimidine hydrogen-bonding motifs. *Helv. Chim. Acta*, **81**, 1909-1920.

Krische, M. J., Lehn, J-M., Kyritsakas, N., Fischer, J., Wegelius, E. K., Nissinen, M. J., and Rissanen, K. (1998) Exploring the 2,2'-diamino-5,5'-bipyrimidine hydrogen-bonding motif: A modular approach to alkoxy-functionalized hydrogen-bonded networks. *Helv. Chim. Acta*, **81**, 1921-1930.

Crabtree, R. H. (1998) A new type of hydrogen bond. *Science*, **282**, 2000-2001.

Aakeröy, C. B., Nieuwenhuyzen, M., and Price, S. L. (1998), Three polymorphs of 2-amino-5-nitropyrimidine: Experimental structures and theoretical predictions. *J. Am. Chem. Soc.*, **120**, 8986-8993.

Pedireddi, V. R., Ranganathan, A., and Chatterjee, S. (1998) Layered structures formed by dinitrobenzoic acids. *Tetrahedron Letters*, **39**, 9831-9834.

Holy, P., Zavada, J., Cisarova, I., and Podlaha, J. (1999) Self-assembly of 1,1'-biphenyl-2,2',6,6'-tetracarboxylic acid: Formation of an achiral grid with chiral compartments. *Angew. Chem. Int. Ed. Engl.*, **38**, 381-383.

Braga, D., and Grepioni, F. (1999) Complementary hydrogen bonds and ionic interactions give access to the engineering of organometallic crystals. *J. Chem. Soc., Dalton Trans*, 1-8.

Beketov, K., Weber, E., Siedel, J., Kohnke, K., Makhkamov, K., and Ibragimov, B. (1999) Temperature-controlled selectivity of isomeric guest inclusion: enclathration and release of xylenes by 1,1'-binaphthyl-2-2'-dicarboxylic acid. *Chem. Commun.*, 91-92.

Nangia, A., and Desiraju, G. R. (1999) Pseudopolymorphism. Occurrences of hydrogen bonding organic solvents in molecular crystals. *Chem. Commun.*, (in press).

THE DEVELOPMENT AND APPLICATION OF KNOWLEDGE-BASED APPROACHES TO MOLECULAR DESIGN

ROBIN TAYLOR, FRANK H ALLEN, IAN J BRUNO, JASON C
COLE, MAGNUS KESSLER, JOS P M LOMMERSE AND
MARCEL L VERDONK
CCDC, 12 Union Road, Cambridge CB2 1EZ, UK

1. Introduction

The development of high-performance computer graphics terminals in the late 1970's and early 1980's catalysed the introduction of computer-based molecular design methods into both academic and industrial research organisations. Twenty years later, virtually all pharmaceutical and agrochemical research departments have a computational chemistry group, and, increasingly, molecular design has found applications in other areas such as combinatorial chemistry, materials, oil and dyestuffs research.

However, despite the ever-growing interest in computational approaches to molecular and crystal design [1,2], there is a general feeling that they have under-performed. A common question posed to pharmaceutical molecular modellers is: "Can you name a commercial drug which was invented by molecular modelling?" While the question is simplistic (*no* commercial drug is ever the product of a single research technique) it does highlight the difficulty that people have found in getting rational design methods to work.

The main reason is the sheer complexity of the physical systems that must be modelled and the inadequacy of the theoretical methods that may be brought to bear on them. This is well illustrated by the theoretical methods available for predicting molecular conformations and intermolecular interactions. These basic properties are of fundamental importance in determining biological activity, crystal structures, crystallisation processes and many other phenomena, yet, even now, are difficult to predict reliably. Both of the two main theoretical approaches, molecular mechanics and ab initio molecular orbital techniques, have serious limitations. The former is fast but relies on empirical parameter sets whose extensibility to new types of molecules is always uncertain. The treatment of electrostatic interactions in molecular mechanics is particularly difficult, involving, for example, arbitrary choices of dielectric constant. While ab initio methods involve no empirical parameters, they are relatively slow – certainly too slow for exploring fully the conformational space of a large, flexible molecule. Molecules causing particular problems include hypervalent and anionic species, which typically require the use of large basis sets. Metal complexes are difficult, too, though the advent of density functional theory methods has improved the situation considerably.

D. Braga et al. (eds.), Crystal Engineering: From Molecules and Crystals to Materials, 243-260 .
© 1999 *Kluwer Academic Publishers. Printed in the Netherlands.*

244

Without doubt, the key difficulty in applying theoretical methods to real problems arises from solvation and entropic effects. For example, it might be possible to predict the preferred morphology of a crystal grown by sublimation by estimating the attachment energy of molecules to each of the possible growing faces. Usually, though, crystals are grown from solutions, so account must also be taken of the interactions between the solute molecules and the surrounding solvent. Because of the massive number of possible arrangements of solvent molecules around the growing crystal, and the dynamic nature of the medium, this increases the difficulty of the problem by orders of magnitude. Similarly, a large part of the free energy released when a small molecule binds to a protein is due to the release of "trapped" water molecules from hydrophobic cavities in the protein's binding site. Modelling this (largely entropic) effect again involves massive calculations which are beyond the reach of current technology.

These considerations do not negate the value of theoretical methods but do emphasise how important it is to use experimentally observed data as well. In particular, the reliability of predictions about molecular conformations and intermolecular interactions can be assessed by comparison with observed conformations and interactions in crystal structures. A crystal structure, of course, is no more realistic a model for a solution situation than is an in vacuo theoretical calculation, but it does give precise, experimental information relating to a condensed phase.

2. The Cambridge Structural Database

The most important source of crystal-structure data for small-molecule organic and organometallic compounds is the Cambridge Structural Database (CSD) [3]. This database was started in 1965 at the University of Cambridge and is now maintained and distributed by an independent, non-profit making company, the Cambridge Crystallographic Data Centre (CCDC). At the time of writing, the database contains results from almost 200,000 x-ray and neutron diffraction crystal-structure determinations, including, where available, full 3D coordinate data.

Using the CCDC program QUEST, it is possible to search the CSD for any desired substructure. Users may place constraints both on 2D properties, such as atom and bond types, atom coordination numbers, whether bonds or cyclic, etc., and on 3D properties – interatomic distances, valence and torsion angles, angles between least-squares planes, and so forth. Searches can also be performed for nonbonded contacts (e.g. hydrogen bonds) between molecules in the extended crystal structure. The results of searches can then be subjected to geometrical and statistical analysis with the program VISTA, which is provided as part of the CSD System. This allows users to prepare and display histograms and scattergrams of molecular geometry parameters.

The key value of the CSD is that it serves as a "reality check" on theoretical calculations. Thus, for example, an unexpected theoretical conformational prediction – say, a preference for a gauche rather than an anti orientation around a single bond – can be validated by looking at a histogram of observed torsion angles in related molecules from the CSD. This sort of technique has been used many times, e.g. in the modelling of the bioactive conformations of pyrethroid insecticides [4]. Similarly, the CSD can be searched for evidence that a particular nonbonded contact can exist. For example, a computed preference for short S...O contacts in certain systems was confirmed by database searching [5,6]. On a more mundane level, the CSD has been used to compile tables of "standard" bond lengths in organic and organometallic compounds [7,8]. The remarkable number of citations that these tables have attracted is testimony to their value both in checking and interpreting theoretical predictions and newly determined crystal structures, and in setting up molecules with correct geometries for molecular orbital calculations.

For all its usefulness, there is a key problem with the CSD: it takes a significant amount of time for users to, firstly, become familiar with the system and, secondly, to use it to solve a specific problem. For example, the time required to set up a QUEST search, run it, transfer the results to VISTA, prepare a histogram of, say, a torsion distribution, and then interpret the results, might typically be thirty minutes or more. Particularly in a pressured industrial context, this is often a significant barrier to the use of the CSD, with the lamentable result that crystallographic information is not exploited to its full potential.

Over the last few years, the CCDC has made moves to address this problem by creating "derived knowledge bases". The basic idea is to take a certain type of information from the CSD – say, information on intermolecular interactions – and package it in a form that is both easy and quick to use. Rather than having to spend an hour searching and analysing the CSD, the user is then able to get useful results in a few seconds. Moreover, if the knowledge base is made available with a well defined API (application programmer's interface), it can be coupled with applications programs. For example, one can imagine a derived knowledge base on intermolecular interactions being used to drive a program for predicting how small molecules might bind non-covalently to a protein binding site. In fact, such a program is currently under development and will be described below.

The remainder of this chapter will describe two knowledge bases that have been derived at CCDC. One is already released, the other is currently a prototype. Then, a knowledge-driven applications program will be described.

3. IsoStar

3.1 INTRODUCTION AND METHODOLOGY

The first derived knowledge base created at CCDC was called IsoStar [9], and was initially released in October 1997. IsoStar is a database of information about intermolecular interactions. Although derived mainly from the CSD, it also contains information from the Protein Data Bank [10] and from theoretical (intermolecular perturbation theory [11]) calculations.

The presentation of crystallographic information in IsoStar is best illustrated by an example. Imagine that we wish to investigate the directional preferences of hydrogen bonds to aliphatic ketones. Using the program QUEST, it is straightforward to search the CSD for all crystal structures containing such a hydrogen bond. Suppose that the first such structure contains a hydrogen bond to a cyclopentanone ring, with the donor hydrogen positioned roughly along the extension of the C=O bond (Figure 1, top left). Suppose, further, that the second "hit" structure contains an acetone solvate molecule forming a hydrogen bond approximately along the lone-pair direction of the carbonyl oxygen (Figure 1, bottom left). The two hydrogen bonds can be superimposed so that the ketone moieties are least-squares overlaid. This produces a composite picture (Figure 1, centre) in which both hydrogen bonds are shown simultaneously. Repetition of this process over all hit structures in the CSD produces a picture which shows all the ketone hydrogen bonds simultaneously (Figure 1, right). Essentially, this is the experimentally observed distribution of hydrogen-bonding protons around ketone carbonyls. Rather than showing multiple copies of the superimposed ketone group, a single average geometry is computed and plotted. The actual data shown in Figure 1, by the way, is imaginary – the real plot is shown later.

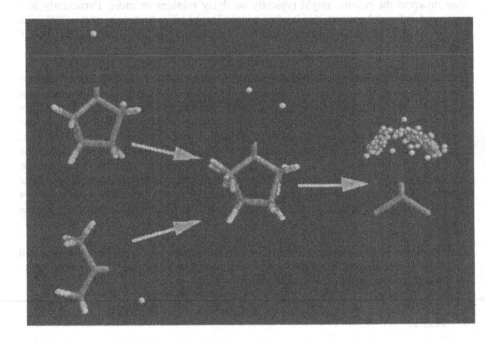

Figure 1 Method of constructing an IsoStar scatterplot

3.2 DATA CONTENT

IsoStar (Version 1.1, October 1998) contains over 10,000 of these "scatterplots", each showing a different type of contact. Four examples are shown in Figure 2. The first (Figure 2a) is the genuine distribution of hydrogen-bond donors (specifically, O-H groups) around aliphatic ketones. The plot reveals that this type of hydrogen bond acceptor shows lone-pair directionality, i.e. the donor groups have a small but clear preference to be positioned in approximately the directions of the oxygen lone pairs. This is not the case for all acceptor groups. For example, aliphatic ether oxygens show no tendency for lone-pair directionality (Figure 2b) although there is a distinct preference for hydrogen bonding in the lone-pair *plane*. Figure 2c shows the distribution of O-H groups around aliphatic esters. Apart from information about geometrical preferences, this plot also reveals something about the relative probabilities of different types of oxygens accepting H-bonds: the carbonyl oxygen is a frequent acceptor but the bridging oxygen almost never accepts. Finally, Figure 2d shows the geometrical distribution of an interaction to a hydrophobic group: chloride ions tend be positioned around the edges of phenyl groups, rather than above and below the plane. The hydrogen atoms of C-H groupings normally bear a small electropositive charge, so this phenomenon is presumably due to attractive interactions between the phenyl-group hydrogen atoms and the electronegative chlorides. By way of contrast, alkyl C-H groups are attracted to the pi-electron cloud of the aromatic system and tend to be positioned above and below the plane (scatterplot not shown).

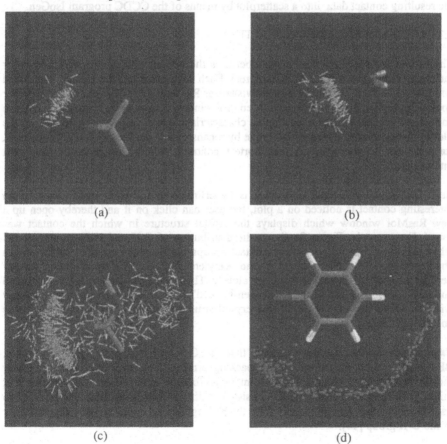

(a)

(b)

(c)

(d)

Figure 2 IsoStar scatterplots of *(a)* O-H groups around ketone carbonyl, *(b)* O-H around aliphatic ether oxygen, *(c)* O-H around aliphatic ester linkages, and *(d)* chloride ions around phenyl

In a plot such as Figure 2a, the superimposed group (ketone, in this case) is termed the *central group* and the surrounding atoms or groups (in this case hydroxyl) are termed the *contact groups*. All contact groups are reflected into the "asymmetric unit" of the plot; in the ketone...hydroxyl plot, this means that all the O-H groups are reflected into one quadrant since the ketone moiety has *mm* symmetry. The reader will notice that both the oxygen and the hydrogen atoms of the contact groups are displayed. This is usually the case in IsoStar scatterplots derived from the CSD. A CSD structure that contains a ketone...OH hydrogen bond, but in which the hydrogen atom of the OH group has not been located, will not contribute to IsoStar.

Currently in IsoStar there is information on over three hundred central groups interacting with up to forty five contact groups. Should a particular interaction be missing from the library, it can be added by running the appropriate QUEST search and then converting the resulting contact data into a scatterplot by means of the CCDC program IsoGen.

3.3 INTERFACE AND FUNCTIONALITY

The user of IsoStar is able to navigate between the various groups via a Web browser interface (Netscape or Microsoft Explorer). Each scatterplot can be viewed, translated and rotated in 3D in the public domain visualiser RasMol [12]. This is called as a helper application by the browser interface. A control window allows users to measure distance and angles on the plot, to alter display characteristics, and, importantly, to restrict the plot to distances within a specified range by means of a slider bar. This allows the user, for example, to focus in on the shortest contacts, which are generally the most interesting.

Each CSD-derived IsoStar scatterplot is hyperlinked to the CSD itself. Thus, if an interesting contact is noticed on a plot, the user can click on it and thereby open up a new RasMol window which displays the crystal structure in which the contact was found. For example, Figure 3a shows the distribution of O-H and N-H groups around ethynyl groups. One or two of the contact groups on the plot appear to be forming a "hydrogen bond" to the pi-cloud of the acetylene linkage. The hyperlinking facility permits the user to investigate this more closely. Thus, selecting the *Hyperlink* command and then clicking on one of the "hydrogen-bonded" O-H groups opens up a new display window which shows (Figure 3b) the crystal structure BETXAZ [13] from which the contact was taken.

Examination of this display confirms that the O-H...pi hydrogen bond is genuine. BETXAZ, in fact, has a noteworthy packing arrangement involving O-H...pi dimer motifs and C-H...O chains. These form in preference to the O-H...O-H chains that might have been expected. This is consistent with the hypothesis that it is difficult to pack monoalcohols so as to satisfy fully the hydrogen-bond donor and acceptor capacity of the O-H group [14].

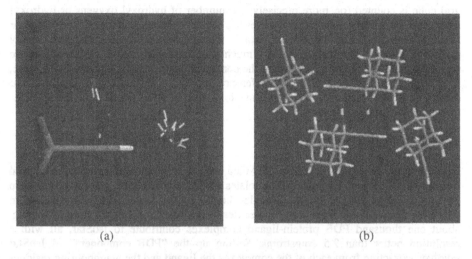

Figure 3 *(a)* IsoStar scatterplot of O-H and N-H groups around ethynyl; *(b)* packing arrangement observed in the crystal structure BETXAZ

Figure 4 IsoStar contour surface showing the density of O-H hydrogen atoms around aliphatic esters

Another important feature of IsoStar is an option to convert a scatterplot into a contoured surface. Consider, for example, the plot in Figure 2c showing the distribution of O-H groups around ester linkages. Because there are very many examples of this interaction in the CSD, the scatterplot shows many contacts. While this is obviously good, in that the user can draw conclusions about the interaction from a large, and therefore statistically significant data set, it does give the plot a crowded appearance and can make visual interpretation difficult. Thus, it is perhaps difficult to be convinced from the plot that there are appreciably more O-H groups along the lone-pair directions of the carbonyl oxygen than elsewhere. The problem is addressed within IsoStar by displaying the data, not as a raw scatterplot, but as a density surface. The space around the central group is divided into a regular cubic grid. The number of contact groups falling in each grid cube is counted (or, more precisely, the number of hydroxyl oxygens or hydroxyl hydrogens, depending on whether the user has chosen to base the calculation on the O or H atom positions). The atom-count data is then converted into a contoured density surface (Figure 4). The surface, being a much simpler representation, clearly shows the regions around the central group where the contact-group density is highest. In this case, it is now clear that there is a real preference for hydrogen bonding in the lone-pair directions of the ester oxygen – and one lone pair is more favoured than the other (presumably, for steric reasons).

3.4 PDB DERIVED DATA

In addition to scatterplots based on CSD data, IsoStar also contains information derived from the PDB. In fact, the information relates specifically to contacts observed between protein and ligands (i.e. small molecules bound to the proteins; ligand contacts to crystallographically observed water molecules are included too). At the time of writing, about one thousand PDB protein-ligand complexes contribute to IsoStar, all with a resolution better than 2.5 Angstroms. Setting up the "PDB component" of IsoStar involves extracting from each of the complexes the ligand and the neighbouring residues of the protein. These binding site – ligand complexes are transformed into CSD format. They can then be searched for any desired type of intermolecular protein-ligand contact using the QUEST program. The results are transformed into scatterplots in exactly the same way as described above for CSD data. The resulting scatterplots are accessible via the same Web browser interface and can be viewed in RasMol as usual. CSD-based and PDB-based scatterplots for the same type of contact can be viewed alongside each other and visually compared, e.g. Figure 5.

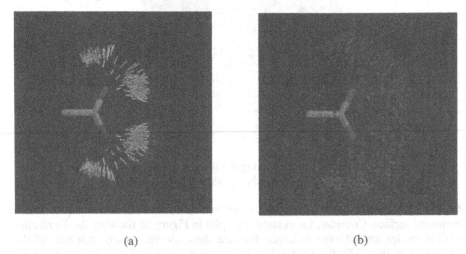

(a)   (b)

Figure 5 Distribution of O-H and N-H groups around *(a)* ionised carboxylates in the CSD, and *(b)* carboxylates and unionised carboxylic acids in PDB protein-ligand complexes

There are some obvious but trivial differences between the two types of plots. In the plots shown here, for example, there are three main differences. First, the PDB-based plot contains less data, a reflection of the fact that there are fewer structures in the PDB than the CSD. Secondly, there are no hydrogen atoms on the PDB-based plot; this is because hydrogen atoms are almost never located in protein crystal structure determinations. Finally, there is a more subtle difference. All contributors to the CSD-based plot in Figure 5a involve contacts between an N-H or O-H group and an ionised carboxylate group. We can be confident of this because CSD structures only contribute to IsoStar if they contain crystallographically located hydrogen atoms; therefore, ionisation states are known. However, the absence of hydrogens from PDB entries means that the ionisation states of acidic and basic groups cannot be inferred with confidence. In fact, it is likely that most of the contributors to the PDB-based plot in Figure 5b are ionised carboxylates (since the pKa of this group is relatively low), but some are probably unionised carboxylic acids.

One difference that is rarely observed between CSD- and PDB-based plots is a significant difference in the geometric distributions. In general, the geometries of intermolecular contacts in the small-molecule structures from the CSD appear to be similar to those observed between proteins and ligands [15].

As with the CSD-based plots, each scatterplot in IsoStar that is derived from the PDB is hyperlinked. Thus, clicking on a particular contact group will call up a display of the protein-ligand complex in which the contact was observed. Finally, it should be noted that the plot in Figure 5b shows the distribution of O-H and N-H groups from *protein residues* around carboxylates in *ligand molecules*. The reverse plot (O-H and N-H groups from *ligands* around carboxylates from *protein residues*) is also available in IsoStar.

3.5 THEORETICAL DATA

IsoStar also contains theoretically calculated data for several hundred model systems. Each model system consists of a pair of molecules, one representing an IsoStar central group, the other representing a contact group. For example, the crystallographic data in IsoStar for contacts between nitro (central group) and carbonyl oxygen (contact group) is supplemented by the results of theoretical calculations on the model system nitromethane ... acetone. Information is given about the calculated energies and geometries of the major minima in the potential energy surface of the bimolecular system. This information is computed in a two-stage process. First, the 6-31G** wave functions of the two molecules are calculated separately (i.e. for each molecule in isolation from the other). The calculated electron density distribution of each molecule is then expressed as a multipole expansion [16]. In combination with an empirical repulsion-dispersion term, this information can then be used to perform a rapid exploration of the potential energy hypersurface of the bimolecular system. One molecule is moved around the other, and, at each position, the energy of interaction between the molecules is calculated as the sum of the electrostatic interaction energy (estimated from the multipole representations of the molecular electron densities) and the repulsion-dispersion energy (estimated from the empirical function). The result is a computed surface that is accurate enough for identifying the major minima in the potential energy surface.

In the second stage of the procedure, energies of selected minima are calculated more accurately using intermolecular perturbation theory [11]. At this stage, the distance between the two molecules is allowed to optimise to correct for errors introduced in the first stage of the procedure by the rather poor anisotropic repulsion-dispersion term. However, the angular orientations of the molecules are fixed at the values obtained from the first stage of the procedure. This is because the electrostatic energy computed from multipole expansions gives a good estimate of the overall shape of the energy surface.

The programs used for these calculations are CADPAC [17] and ORIENT [18].

Both the energies and geometries of calculated minima can be viewed in IsoStar. The value of the information is, of course, limited because the calculations relate to an in vacuo situation. This produces some interesting differences between the crystallographic and the theoretical parts of IsoStar. For example, the calculated minimum energy geometry of the model system fluorobenzene ... N-methylacetamide contains an N-H...F interaction, with an energy of -12.5 kJ/mol. However, inspection of the relevant IsoStar scatterplots shows that such interactions almost never occur, presumably because O-H and N-H donors in condensed phases preferentially interact with stronger acceptors. More detailed inspection of the IsoStar scatterplots, incidentally, reveals that when hydrogen bonds to carbon-bound fluorines *do* occur, the fluorine is normally in a particular electron rich environment [19], e.g. in the calcium salt of 2-fluorobenzoic acid, CEVGUF [20]. The hyperlinking of IsoStar plots to the CSD is particularly useful for obtaining insights like this.

Figure 6 IsoStar plot showing a cluster of N...O contacts to the nitrogen atom of nitro groups.

For all their limitations, the theoretical data can be useful in confirming and adding insight to conclusions drawn from the crystallographic information. For example, the IsoStar scatterplot for the central group nitro and the contact group "terminal oxygen" (which means any sort of terminal oxygen, such as in carbonyl, sulphone and nitro itself) shows a very pronounced cluster of oxygens directly above and below the nitro-group nitrogen, and in a very narrow distance range (approximately 2.9-3.0 Angstroms) (Figure 6). The appropriate theoretical data confirm that this is a very attractive interaction; more attractive, in fact, than typical hydrogen bonds to the nitro group oxygens. Thus, the computed energy for the N...O "stacking interaction" between nitromethane and N-methylacetamide is –31.5 kJ/mol, compared with an energy of only –20.9 kJ/mol for the N-H...O hydrogen bond in the same system. On this basis, the tendency to form short N...O contacts is probably the dominant nonbonded interaction of nitro groups, rather than the more obvious tendency to hydrogen bond.

3.6 SUMMARY

Users of IsoStar can now get results in seconds that were previously available only after several hours work with the CSD itself. The results give insight into:

- the preferred geometries of intermolecular interactions;
- how common a particular interaction is;
- whether particular interactions only occur in specific situations (e.g. if the local intramolecular environment of one of the interacting groups is particularly electron deficient or electron rich);
- whether the geometries of contacts in small molecule crystal structures are similar to those in protein-ligand complexes;
- calculated energies of interactions.

IsoStar is therefore a valuable "quick reference" guide to nonbonded interactions for use by chemists, modellers and crystal engineers. However, its full power will probably only be realised when it is used through an API as a knowledge base for applications programs.

4. CSD Torsion Library

4.1 INTRODUCTION AND OBJECTIVE OF THE TORSION LIBRARY

Following the release of IsoStar, CCDC has turned its attention to the development of a new derived knowledge base aimed, this time, at intramolecular information. Probably the single most common use of the CSD is to investigate conformational preferences. Typically, a user's starting point will be a particular molecule whose conformation is of interest. The user will draw in QUEST an appropriate substructure, search the CSD for molecules containing that substructure, tabulate or plot as a histogram the torsion angles relating to the rotatable bond of interest, and draw inferences about the possible conformations which the starting molecule could adopt. While crystal packing forces occasionally have a systematic effect on torsional preferences in small-molecule crystal structures [21], this is not usually the case [22]. Thus, the experimental data is a valuable "reality check" on theoretical predictions. If the substructure as drawn found insufficient hits, the user might modify it so that it is less specific – for example, a substructure containing a phenyl group might be made more general so that it would allow *any* six-membered aromatic ring. This process may be referred to as substructure *generification*. The main aim of the "Torsion Library" project is to automate the above procedure so that it takes seconds, not minutes.

4.2 OPERATION OF THE TORSION LIBRARY PROTOTYPE

The current prototype is front-ended by the molecular modelling program SYBYL [23]. The assumed starting point is a molecule within SYBYL, containing a rotatable bond whose conformational preferences are of interest. The user selects with the mouse the four consecutive atoms in the molecule that define the relevant torsion angle. Novel software automatically converts this into a search substructure, e.g. Figure 7.

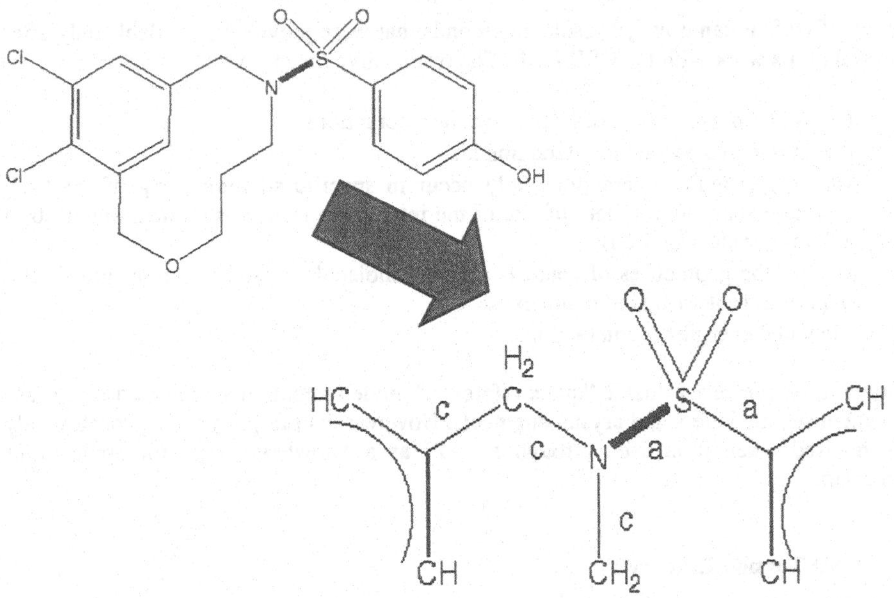

Figure 7 Automatic generation of a search substructure from a molecule

The software then accesses a file containing the torsion angles of all single, acyclic bonds in the CSD (there are several million of these). The data are indexed on substructure, so it is easy and rapid to go straight from the substructure definition to the corresponding torsion observations (Figure 8). The results are transferred back and displayed either as a polar histogram or as a set of overlaid positions in three dimensions. To construct the latter type of display, each observed torsion angle from the library is superimposed on the user's starting molecule, such that the first three atoms of the torsion angle (A,B,C of the angle A-B-C-D) are least-squares overlaid. The various observed positions of the fourth atom (D) are then displayed. This produces a plot in which clusters of "D atoms" indicate favourable positions for the fourth atom of the torsion. These can be directly compared by the user with the actual position in which this atom has been modelled in the starting molecule. As with IsoStar, data retrieved from the prototype torsion library is hyperlinked to the CSD, so that individual molecules containing the torsion angle of interest can be retrieved from the CSD and inspected.

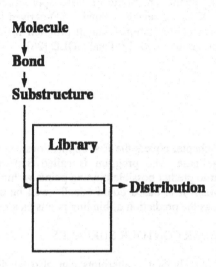

Figure 8 Overview of torsion library operation

4.3 AUTOMATIC SUBSTRUCTURE GENERIFICATION

In the event that there are insufficient molecules in the CSD containing the search substructure, the software will attempt to generify the search substructure according to a set of rules. For example, one of the rules might be: *change an O atom in the substructure to be either O or S.* In general, the terminal atoms A and D will be generified before the central atoms B and C. The software will retrieve (via a tree search) all of the torsion observations in the library corresponding to the generified substructure. If the resulting distribution contains more than a minimum number of observations (set by the user), it will be accepted. If not, further generification will be applied. The user interface then displays not only the torsion angle data but also the generified search substructure that was eventually chosen by the program, together with information about which torsion-angle observations came from which exact substructure.

For retrieval of an exact (i.e. non-generified) torsion distribution, the current prototype will take only a few seconds on a typical Unix workstation. Retrieval of generified substructures takes significantly longer, perhaps a minute or two. However, the current software is probably poorly optimised and it is envisaged that the final product will be substantially faster.

4.4 POSSIBLE APPLICATIONS

Apart from providing users rapid access to information for use in an interactive modelling session, the torsion library has obvious potential value as a knowledge base for a variety of applications. These include 2D to 3D structure conversion, conformational searching, structure validation, prediction of ligand conformations in protein-ligand docking, solution of crystal structures from powder diffraction data and de novo crystal structure prediction. Some of these applications are currently being investigated, both at CCDC and elsewhere. Small CSD-derived torsion libraries are, in fact, already in use in the conformational search program MIMUMBA [24] and the protein-ligand docking programs FlexX [25] and GOLD [26].

5. SuperStar

5.1 INTRODUCTION

The final section of this chapter covers the first applications program based solely on a CSD-derived knowledge base. The program is called SuperStar [27] and it uses information from IsoStar to predict possible binding points within an enzyme active site. The methodology on which it is based could equally well be applied to problems in crystal engineering such as the prediction of binding points on a growing crystal face.

5.2 SCALING OF ISOSTAR CONTOUR SURFACES

As was pointed out above, IsoStar scatterplots can also be displayed as contoured surfaces which portray the density of contact groups around the central group as a function of spatial position (Figure 4). An important aspect of this functionality is that it is possible to put two IsoStar surfaces – for different scatterplots – onto a common scale. This is done as follows. Consider the scatterplot of B (contact group) around A (central group). All crystal structures in the CSD that contain both groups are located. These are the structures where, theoretically, an A...B contact *could* occur. The density of B groups in each of these structures is computed from the number of B groups occurring in the unit cell and the volume of that cell. The average density of B groups in the structures can thus be determined. From this figure, it is then possible to compute how many B groups would be expected per unit volume of the IsoStar A...B scatterplot *purely by chance*, i.e. the random expectation given the stoichiometries of the structures from which the scatterplot was built. If the actual number of B groups in a unit volume element of the scatterplot is divided by this random expectation value, the resulting ratio X indicates that contacts in that region of space are X times more frequent than would be expected by chance. Clearly, a value of one indicates that contacts are neither more nor less common than would be expected from stoichiometric considerations. A value of two indicates that contacts are twice as frequent as would be expected; and so on.

The X ratios are referred to as *propensities*. Strictly speaking, they tell us about the *probabilities* of contacts forming rather than the *energies* of contacts. These two factors will usually be correlated – the more attractive an interaction is, the more likely it is to occur. However, there are circumstances in which the correlation might break down. For example, a short interaction between a phenyl ring and a carboxylate ion is attractive in some geometries, but such an interaction is unlikely to occur between, say, a protein and a bound ligand. This is because the carboxylate anion will preferentially form hydrogen bonds to O-H and N-H donors from water molecules, arginine side chains, and so on. This competition between functional groups is taken into account in the propensity values but not in calculated interaction energies. This is, in fact, a significant advantage in using propensities for molecular design.

5.3 COMBINING ISOSTAR SURFACES

SuperStar uses IsoStar propensities to predict where a given functional group will bind in a protein binding site. The first step is for the user to define the functional group of interest. This must be an IsoStar contact group – OH, for example. SuperStar first identifies all of the functional groups present in the protein binding site. Suppose that the first of these groups is an aspartate carboxylate. SuperStar retrieves from IsoStar the relevant scatterplot; in this case, this will be the plot of OH (contact group) around carboxylate (central group). The entire scatterplot is overlaid on the protein side chain, such that the carboxylate moieties are overlaid. The surface is than converted into a surface showing the propensity of OH oxygen or hydrogen (whichever the user chooses) around the aspartate carboxylate group. This procedure is then repeated for all the other functional groups present at the protein binding site. If the surfaces from two adjacent groups overlap, they are combined by multiplication, because propensities are related to probabilities. For example, suppose that one of the positions around the aspartate carboxylate has a high intrinsic propensity, but is too close to a functional group on a neighbouring residue. The high propensity from the aspartate IsoStar surface is multiplied by a very low propensity (almost zero) from the IsoStar surface of the neighbouring functional group, giving, of course, an overall propensity close to zero.

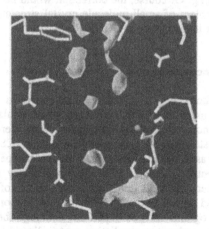

Figure 9 Example SuperStar surface showing where carbonyl oxygens bind in the active site of thermolysin

The result of applying this methodology to a complete protein binding site is a surface which highlights regions where the chosen contact group is likely to bind (Figure 9). The surface can be displayed and used as an aid to manual ligand design, rather like the energy-based surfaces provided by the well known program GRID [28]. Alternatively, dummy atoms can be positioned at the maxima of the SuperStar surface. These can then be used to generate pharmacophore hypotheses for, say, searching a 3D molecular database.

5.4 VALIDATION OF SUPERSTAR

Following its initial development, SuperStar was tested on over a hundred protein-ligand complexes from the PDB. Predictions based on SuperStar surfaces were compared with the experimentally observed positions of ligand atoms. SuperStar predictions appear to be very reliable. Of the ligand atoms studied, over eighty percent are in regions predicted by SuperStar to be favourable. One problem initially encountered was that hydrophobic contacts were predicted by SuperStar to be less likely than expected. Closer investigation revealed that this was due to a real difference between the contacts observed in small molecule crystal structures and those observed in protein-ligand complexes. Hydrophobic (i.e. CH...CH) contacts are much more common in the latter. This may well be due to the fact that most crystals of low molecular weight compounds are prepared from organic solvents rather than water. Therefore, when the solute molecules nucleate at the beginning of the crystal growing process, there is no strong driving force for hydrophobic groups to pack together in order to shield themselves from the solvent environment. This contrasts strongly with the binding of a small molecule to a biological macromolecule, where it is known that the "hydrophobic effect" (entropically favoured displacement of water molecules from hydrophobic cavities in the protein binding site) is a strong driving force. An empirical correction was introduced into SuperStar to compensate for this difference between small-molecule crystal packing and protein-ligand binding. Of course, the correction would not be needed if SuperStar were applied to the problem of small-molecule crystal structure prediction, an area of research currently under investigation at CCDC.

6. Summary and Conclusions

Knowledge-based methods are a valuable complement in molecular and crystal modelling to the more usual theoretical techniques of molecular mechanics, molecular dynamics, ab initio molecular orbital calculations and density functional theory. Databases of "raw" crystal-structure information such as the CSD can be used to derive secondary knowledge bases, each dedicated to a particular type of information, such as IsoStar (nonbonded contacts) and the CSD Torsion Library (molecular geometries). These have value in themselves, as quick and easy sources of reliable knowledge for molecular modellers and crystal engineers. Perhaps more importantly, they can be used to drive applications programs such as SuperStar. Although the emphasis so far in the SuperStar development has been on predicting protein-ligand binding, it is clear that a similar methodology could be applied to a variety of crystal engineering problems. It is also clear that the conformational information in the torsion library could be used to assist the modelling of crystal structures.

259

Like all methods, knowledge-based approaches have problems. The key one is that, even with a database as large as the CSD, there are still many systems of interest for which little experimental structural data exists. In this situation, knowledge-based methods are at a disadvantage although the problem will naturally be lessened as time goes by and more crystal structures are determined. Difficulties can also arise because of inherent biases in the PDB and CSD and the existence of common crystal-packing motifs [9]. Nevertheless, it is clear that knowledge-based methods are already useful in molecular and crystal design.

7. References

1. Leach, A.R. (1996) *Molecular Modelling, Principles and Applications*, Longman, Harlow, UK.
2. Desiraju, G.R. (1991) *Crystal Engineering: The Design of Organic Solids*, Academic Press, New York.
3. Allen, F.H and Kennard, O. (1993) 3D search and research using the Cambridge Structural Database, *Chem. Des. Automation News*, **8**, 31-33.
4. Mullaley, A. and Taylor, R. (1994) Conformational properties of pyrethroids, *J. Comput.-Aided Molec. Des.*, **8**, 135-152.
5. Burling, F.T. and Goldstein, B.M. (1992) Computational studies of nonbonded sulfur-oxygen and selenium-oxygen interactions in the thiazole and selenazole nucleosides, *J. Am. Chem. Soc.*, **114**, 2313-2320.
6. Baalham, C.A. (1996) M. Phil. Thesis, University of Cambridge, Cambridge, UK.
7. Allen, F.H., Kennard, O., Watson, D.G., Brammer, L., Orpen, A.G. and Taylor, R. (1987) Tables of bond lengths determined by X-ray and neutron diffraction. Part I. Bond lengths in organic compounds, *J. Chem. Soc., Perkin Trans. 2*, S1-S19.
8. Orpen, A.G., Brammer, L., Allen, F.H., Kennard, O., Watson, D.G. and Taylor, R. (1989) Tables of bond lengths determined by X-ray and neutron diffraction. Part II. Organometallic compounds and coordination complexes of the d- and f-block metals, *J. Chem. Soc., Dalton Trans.*, S1-S83.
9. Bruno, I.J., Cole, J.C., Lommerse, J.P.M., Rowland, R.S., Taylor, R. and Verdonk, M.L. (1997) IsoStar: A library of information about nonbonded interactions, *J. Comput.-Aided Molec. Des.*, **11**, 525-537.
10. Bernstein, F.C., Koetzle, T.F., Williams, G.J.B., Meyer, E.F., Brice, M.D., Rodgers, J.R., Kennard, O., Shimanouchi, T. and Tasumi, M. (1977) The Protein Data Bank: a computer based archival file for macromolecular structures, *J. Mol. Biol.*, **112**, 535-542.
11. Stone, A.J. (1996) *The Theory of Intermolecular Forces*, Clarendon Press, Oxford.
12. Sayle, R. (1995) *RASMOL – A Program for Structure Visualisation*, University of Edinburgh and Glaxo Group Research, Stevenage, UK.
13. Lin, S.Y., Okaya, Y., Chiou, D.M. and Le Noble, W.J. (1982) Structure of 2-ethynyl-2-adamantanol, *Acta Crystallogr., Sect. B*, **38**, 1669-1671.
14. Brock, C.P. and Duncan, L.L. (1994) Anomalous space-group frequences for monoalcohols C_nH_mOH, *Chem. Mater.*, **6**, 1307-1312.
15. Verdonk, M.L. (1998), unpublished results.
16. Stone, A.J. and Alderton, M. (1985) Distributed multipole analysis. Methods and applications, *Mol. Phys.*, **56**, 1047-1064.

17. Amos, R.D. (1996) *CADPAC 6.0: The Cambridge Analytical Derivatives Package*, Issue 6.0, University of Cambridge, Cambridge, UK.
18. Stone, A.J., Dullweber, A., Hodges, M.P., Popelier, P.L.A. and Wales, D.J. (1996) *ORIENT v.3.2*, University of Cambridge, Cambridge, UK.
19. Dunitz, J.D. and Taylor, R. (1997) Organic fluorine hardly ever accepts hydrogen bonds, *Chem. Eur. J.*, **3**, 89-98.
20. Karipides, A. and Miller, C. (1984) Crystal structure of calcium 2-fluorobenzoate dihydrate: Indirect calcium...fluorine binding through a water-bridged outer-sphere intermolecular hydrogen bond, *J. Am. Chem. Soc.*, **106**, 1494-1495.
21. Brock, C.P. and Minton. R. (1989) Systematic effects of crystal-packing forces: biphenyl fragments with H atoms in all four ortho positions, *J. Am. Chem. Soc.*, **111**, 4586-4593.
22. Allen, F.H., Harris, S.E. and Taylor, R. (1996) Comparison of conformer distributions in the crystalline state with conformational energies calculated by ab initio techniques, *J. Comput.-Aided Molec. Des.*, **10**, 247-254.
23. *SYBYL 6.5*, Tripos Inc., 1699 South Hanley Rd., St Louis, Missouri 63144, USA.
24. Klebe, G. and Mietzner, T. (1994) A fast and efficient method to generate biologically relevant conformations, *J. Comput.-Aided Molec. Des.*, **8**, 583-606.
25. Rarey, M., Kramer, B. and Lengauer, T. (1997) Multiple automatic base selection: Protein-ligand docking based on incremental construction without manual intervention, *J. Comput.-Aided Molec. Des.*, **11**, 369-384.
26. Jones, G., Willett, P., Glen, R.C., Leach, A.R. and Taylor, R. (1997) Development and validation of a genetic algorithm for flexible docking, *J. Mol. Biol.*, **267**, 727-748.
27. Verdonk, M.L., Cole, J.C. and Taylor, R. (1999) SuperStar: A knowledge-based approach for identifying interaction sites in proteins, submitted for publication.
28. Goodford, P.J. (1985) A computational procedure for determining energetically favourable binding sites on biologically important molecules, *J. Med. Chem.*, **28**, 849-857.

ORGANIC MATERIALS FOR SECOND-ORDER NONLINEAR OPTICS

CHRISTIAN BOSSHARD, ROLF SPREITER, URS MEIER, ILIAS LIAKATAS, MARTIN BÖSCH, MATTHIAS JÄGER, SABINE MANETTA, STEPHANE FOLLONIER, AND PETER GÜNTER
Nonlinear Optics Laboratory, Institute of Quantum Electronics, ETH Hönggerberg, CH-8093 Zürich, Switzerland

1. Introduction

Organic materials have become of great importance with regard to their nonlinear optical and electro-optic properties. They have attracted attention for applications in areas such as efficient frequency-conversion and high-speed light modulation. As a special feature their macroscopic properties arise from the properties of the constituent molecules. There are several possibilities to arrange these molecules macroscopically. The most common forms are single crystals [1-5], polymers [2, 6], and Langmuir-Blodgett films [7]. Besides these cases there exist other interesting concepts such as e.g. molecular beam deposition for the preparation of thin films. Since the macroscopic nonlinearity arises from the constituent molecules, the optimization of a material for a certain application consists of two steps. On the one hand, the nonlinear optical properties of the molecules have to be maximized and, on the other hand, the arrangement of these molecules in crystals or thin films has to be optimized.

In this work we focus on single crystals consisting of highly nonlinear molecules. Our interest in molecular crystals stems from the fact that the potential upper limits of macroscopic nonlinearities and long-term orientational stability of molecular crystals are significantly superior to those of polymers [1]. In order to exhibit non-vanishing macroscopic second-order nonlinear optical responses such as second harmonic generation, frequency mixing, linear electro-optic effect and photorefractive effects, the molecules in the crystalline solid are required to arrange noncentrosymmetrically. To achieve such an order within the crystal lattice is, however, a difficult task.

Below we discuss the current status and perspectives of nonlinear optical organic crystals. We start with a short introduction of the relevant nonlinear optical effects. We then discuss the requirements for organic molecules to exhibit large microscopic nonlinearities. After that we present crystal engineering strategies that were most successful in the last years to obtain crystals with large macroscopic nonlinearities. Specific examples will be discussed in terms of structure-property relationships and relevant figures of merit and we will point out the importance of intermolecular

261

D. Braga et al. (eds.), Crystal Engineering: From Molecules and Crystals to Materials, 261–278.

262

interactions. Finally, we will conclude with a summary and an outlook for further necessary work.

2. Nonlinear optical and electro-optic effects

Linear and nonlinear optical effects can be described in terms of the linear polarization P^L and the nonlinear polarization P^{NL} induced by the electric field

$$P_i = P_i^L + P_i^{NL} = \varepsilon_o\left(\chi_{ij}^{(1)}E_j + \chi_{ijk}^{(2)}E_jE_k+...\right).$$ (1)

Equation (1) defines the frequency-dependent susceptibility tensors $\chi^{(n)}$, which contain all the information about the optical properties of the respective material. ε_0 is the vacuum permeability. Several nonlinear and electro-optic effects can be distinguished (see Figure 1). Some of the more important ones are briefly introduced below.

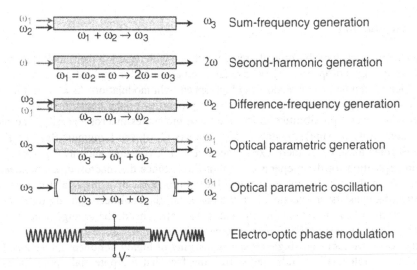

Figure 1. Schematic representation of important nonlinear optical and electro-optic effects.

Sum-frequency generation is the mixing of two incident light waves of frequencies ω_1 and ω_2 creating a wave of $\omega_1 + \omega_2 = \omega_3$. This situation is represented by the nonlinear polarization

$$P_i^{\omega_3} = \varepsilon_o\chi_{ijk}^{(2)}(-\omega_3,\omega_1,\omega_2)E_j^{\omega_1}E_k^{\omega_2}.$$ (2)

Optical frequency doubling or second-harmonic generation (SHG) is just a special case of sum-frequency generation. Only one light wave of frequency ω is incident which is "mixing with itself" thus generating a wave of twice the frequency. The nonlinear polarization for SHG can also be expressed using the nonlinear optical coefficient $d_{ijk}^{(2)}$, which is often used for the nonlinear optical characterization of macroscopic samples,

$$P_i^{2\omega} = \frac{1}{2}\varepsilon_o \chi_{ijk}^{(2)}(-2\omega, \omega, \omega)E_j^{\omega}E_k^{\omega} = \varepsilon_o d_{ijk}^{(2)}(-2\omega, \omega, \omega)E_j^{\omega}E_k^{\omega}. \qquad (3)$$

A corresponding equation can also be derived for the case of difference-frequency generation (DFG) which is characterized as the interaction of two input beams of frequencies ω_3 and ω_1 resulting in an optical field with the frequency $\omega_2 = \omega_3 - \omega_1$, i.e. the difference of the two. The beam with the frequency ω_3 is typically the strongest and therefore referred to as the pump beam. Optic parametric generation is a special case of difference frequency generation, where only the pump beam is incident on the nonlinear material generating two beams at the frequencies ω_1 and ω_2. In order to enhance the efficiency of either process, the nonlinear medium can be placed inside a cavity with highly reflecting mirrors for the frequencies ω_1 and/or ω_2. Optical parametric generation and oscillation are of particular importance because they allow to turn a single frequency laser into a broadly tunable laser system by adjusting the phase matching condition using e.g. angle or temperature tuning [8]. All of the above processes efficiently generate new wavelengths only if phase-matching is achieved. This means that the refractive index at frequency ω_3 should be equal to the average of the refractive indices that the waves at frequencies ω_1 and ω_2 experience [8]. Due to the normal wavelength dispersion of the refractive indices this condition cannot be fulfilled if the interacting light waves all have the same polarization (that is, a nonlinear optical coefficient d_{iii} cannot be phase-matched) and therefore birefringence is required [1]. The case of quasi-phase-matching is not treated here [9].

Electro-optic effects describe the deformation and rotation of the optical indicatrix if an electric field is applied to a noncentrosymmetric sample [10]. The linear electro-optic effect can also be expressed using the nonlinear $\chi^{(2)}$-tensor

$$P_i^{\omega} = 2\varepsilon_o \chi_{ijk}^{(2)}(-\omega, \omega, 0)E_j^{\omega}E_k^0. \qquad (4)$$

However, it is not considered a nonlinear optical effect, because one of the involved fields E_k^0 is not an optical but a 'quasi'-static electric field. Typically the linear electro-optic effect is described in terms of the change of the optical indicatrix

$$\Delta\left(\frac{1}{n^2}\right)_{ij} = \Delta\left(\frac{1}{\varepsilon}\right)_{ij} = r_{ijk}E_k^0. \qquad (5)$$

The above equation also defines the electro-optic tensor r_{ijk}. For small changes the linear refractive index change can be approximated by [8]

$$\Delta n_i \cong -\frac{n_i^3 r_{iik}E_k^0}{2}. \qquad (6)$$

The linear electro-optic effect is widely used in electro-optic modulators. These devices employ the induced phase change of an optical wave and convert it to a change in intensity. Therefore the optical intensity can be controlled by an electrical signal, an important application in telecommunications.

3. Molecular Hyperpolarizabilities

While equation (1) governs nonlinear effects on a macroscopic scale, one can also consider the molecular level. The dipole moment of the molecule p consists of its ground state dipole moment μ_g and the induced contribution. The corresponding expansion

$$p_i = \mu_{g,i} + \varepsilon_o(\alpha_{ij}E_j + \beta_{ijk}E_jE_k + \ldots) \tag{7}$$

defines the molecular coefficients: the linear polarizability α_{ij} and the first-order hyperpolarizability β_{ijk}. However, linking the macroscopic coefficients to the molecular ones is nontrivial because of interactions between neighboring molecules. However, often the macroscopic second-order nonlinearities of organic materials can be well explained by the nonlinearities of the constituent molecules (oriented gas-model [11]). Assuming one dominant tensor element β_{zzz} there is a simple relation between the nonlinear optical and electro-optic coefficients and β_{zzz}:

$$d(-2\omega, \omega, \omega) \propto N \times f \times g \times \beta_{zzz}(-2\omega, \omega, \omega)$$
$$r(-\omega, \omega, 0) \propto N \times f \times g \times \beta_{zzz}(-\omega, \omega, 0) \tag{8}$$

where N is the number of molecules per unit volume, f are local field corrections (mostly in the Lorentz approximation), g is a geometrical factor taking into account the angles between the dielectric and molecular axes, and β_{zzz} is the molecular first-order hyperpolarizability.

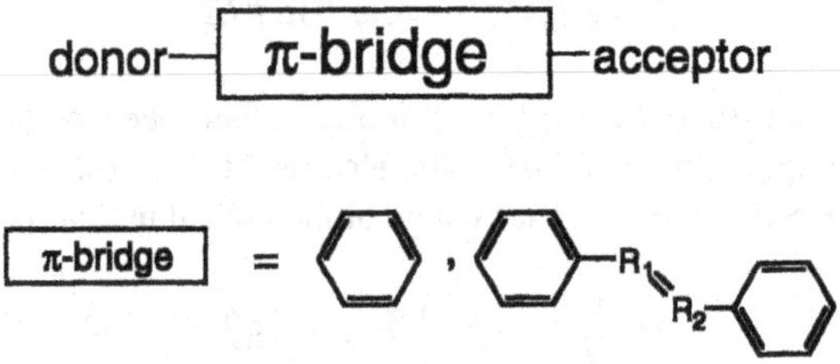

Figure 2. Typical organic molecules for second-order nonlinear optical effects. The electron donor group is connected to the electron acceptor group through a π electron system. A typical donor is e.g. $N(CH_3)_2$ and a typical acceptor is NO_2 (see also Table 1). The most common systems are those containing one benzene ring (benzene analogues) and those containing two benzene rings (stilbene analogues). R_1 and R_2 are usually carbon or nitrogen.

The design of nonlinear optical molecules is based on π bond systems. π bonds are regions of delocalized electronic charge distribution resulting from the overlap of π orbitals. This delocalization leads to a high mobility of the electron density. The electron distribution can be distorted by substituents at both sides of the π bond system. The extent of the redistribution is measured by the dipole moment, and the ease of redistribution in response to an externally applied field by the hyperpolarizability. The optical nonlinearity of organic molecules can be increased by either increasing the length of the π bond system or by using appropriate electron donor and electron acceptor groups (Figure 2). The addition of the appropriate functionality at the ends of the π system can enhance the asymmetric electronic distribution in either or both the ground state and excited state configurations (Figure 3).

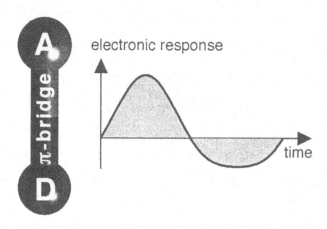

Figure 3. Simple picture of the physical mechanisms of the nonlinearity of donor-acceptor π-conjugated molecules. If we excite such molecules with an optical field we induce an asymmetric electronic response of the polarization: due to the nature of the substituents the electron cloud (i.e. the electronic response) favors the acceptor over the donor.

Although the relation between macroscopic nonlinearity and molecular structure is not yet completely understood and no control over the arrangement of the individual molecules is obtained, the mechanisms leading to large microscopic effects are well understood. During the last two decades, a large number of nonlinear optical molecules have been synthesized and investigated allowing scientists to gain insight into the chemistry and physics of optical hyperpolarizabilities. Improvement of the values of the hyperpolarizabilities, by using new electron donor and acceptor groups, led to many new nonlinear optical materials.

It has been shown that the molecular nonlinearity is strongly correlated to the transparency of the materials. The stronger the nonlinearity the more the wavelength of maximum absorption λ_{eg} is shifted to longer wavelengths (Figure 4) [12].

The first-order hyperpolarizability β_{ijk} (see equation (7)) depends on the strength of donor and acceptor substituents as well as on the extension of the π-electron system (conjugation length). This behavior is illustrated in Figure 4 which shows the connection of the dispersion free hyperpolarizability β_o (calculated by extrapolation to infinite optical wavelengths using the simple two-level model for the molecular hyperpolarizability [13]) with the wavelength of maximum absorption in solution. The shaded area schematically shows the experimentally determined range of available molecules.

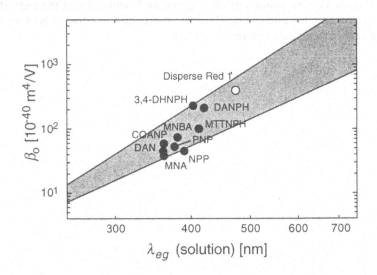

Figure 4. First-order hyperpolarizability β_o extrapolated to infinite wavelengths versus wavelength of maximum absorption of various molecules. The shaded area represents the range of values experimentally determined in various laboratories.

Figure 4 shows that there are already many molecules with very large nonlinearities. Values of β_o up to 3700 x 10^{-40}m^4/V have been measured. This has to be compared to values for e.g. 2-methyl-4-nitroaniline (MNA, $\beta_o = 38$ x 10^{-40}m^4/V)[14], 2-(N-prolinol)-5-nitropyridine (PNP, $\beta_o = 53$ x 10^{-40}m^4/V)[15] and Disperse Red 1 (DR1, $\beta_o = 400$ x 10^{-40}m^4/V)[12]. Measured macroscopic nonlinear optical coefficients are e.g. d_{111}= 150 pm/V for MNA and d_{211}= 51 pm/V for PNP, both at λ=1064nm (a comparison with DR1 is not possible since the molecules crystallize in a centrosymmetric point group). Note, that all of the above values were adjusted to the same convention and the same reference value (d_{111}=0.3pm/V of α-quartz at λ=1064nm) in order to allow a meaningful comparison. We see that much improvement is possible if solid state materials can be formed from molecules with highly increased nonlinearities. Therefore a great challenge is the crystal engineering, that is, the modification of the molecular structure with regard to the crystal structure to fully profit of the large nonlinearities of the new molecules.

4. Molecular Crystals

We have seen above that the guidelines to optimize molecules for second-order nonlinear optical effects are well established and that excellent materials are already available. The difficult part is the incorporation of these molecules in a crystal lattice in a noncentrosymmetric arrangement. There are only few chromophores with very large molecular hyperpolarizabilities such as donor-acceptor stilbenes and tolanes that form potentially useful crystalline materials. We have calculated upper limits with regard to electro-optic and nonlinear optical coefficients [1, 16]. This calculation showed that the macroscopic susceptibilities of crystalline materials based on highly extended π-conjugated donor-acceptor chromophores e.g. donor-acceptor disubstituted stilbenes have by far not reached the upper limit yet.

Of additional importance is the dependence of the precise orientation of the molecules in the crystal on the application. For the case of electro-optic modulation all molecules should be completely parallel to each other. On the other hand, for frequency conversion the angle between the molecular charge transfer axis (typically along the donor-acceptor axis) and the polar crystal axis should be roughly 55° to allow phase-matched interactions [11].

When a new material is developed, a Kurtz and Perry powder test is first performed to determine if the compound crystallizes noncentrosymmetrically or not [17]. With this technique the powder under investigation is illuminated with pulsed laser light and the generated signal at frequency 2ω (only if the sample is noncentrosymmetric) is detected. If the generated signal is sufficiently large in comparison with a reference material (often urea), the structure is determined by X-ray analysis and crystals of optical quality are grown to investigate linear optical, nonlinear optical, and electro-optic properties [1].

4.1 CRYSTAL ENGINEERING

Crystal growth is the prototype of self-assembly in nature [18]. However, crystallization of large organic molecules with desired optical properties is still a challenging topic. There are several routes to achieve optimized nonlinear optical organic crystals that are summarized here (see e.g. [19]).

4.1.1 Molecular asymmetry

Molecules tend to undergo shape simplification during crystal growth which gives rise to dimers and then higher order aggregates in order to adapt a close-packing in the solid state [20]. Since the high tendency of achiral molecules to crystallize centro-symmetrically could be due to such a close-packing driving force a reduction of the symmetry of the chromophores may lead to reduced probability for dimerization. Subsequently aggregation is no longer of advantage to the close packing and the probability of acentric crystallization increases. This symmetry reduction can be accomplished by either the introduction of molecular (structural) asymmetry or the incorporation of steric (bulky) substituents into the chromophore. These two approaches were widely and successfully applied to benzenoid chromophores. As an example, Tsunekawa and co-workers found that a substitution at the 3-position of 4'-

nitrobenzylidene 4-donor-substituted-aniline can promote a non-centrosymmetric packing for large optical nonlinearities [21]. This led to the discovery of 4'-nitrobenzylidene-3-acetamino-4-methoxyaniline (MNBA, Table 1).

4.1.2 Strong Coulomb interactions

Such forces can help to override the weak dipole-dipole interactions to induce a non-centrosymmetric packing. Meredith proved the validity of this concept for the case of 4-dimethylamino-N-methylstilbazolium salts which led to the development of 4-dimethylamino-N-methylstilbazolium methylsulfate (DMSM) [22], 1-methyl-4-(2-(4-hydroxyphenyl)vinyl)pyridium (or named as 4-hydroxy-N-methylstilbazolium) 4-toluene-sulfonate (MC-PTS) [23-25], and 4-dimethylamino-N-methylstilbazolium 4-toluene-sulfonate (DAST, Table 1) [26]. DAST is up to now the organic crystal with the largest nonlinear optical and electro-optic coefficients [27, 28].

4.1.3 Non-rod-shaped π-conjugated cores

In contrast to donor-acceptor disubstituted stilbene derivatives, hydrazone derivatives generally adopt a bent, non-rod-shaped conformation in the solid state because of the non-rigid nitrogen-nitrogen single bond ($-CH=N\backslash_{NH}-$). Donor-substituted (hetero)-aromatic aldehyde-4-nitrophenylhydrazones show an overwhelmingly high probability for a non-centrosymmetric packing [29-31]. Furthermore, most of the hydrazone crystals developed show large macroscopic nonlinearities, very good crystallinity, and high thermal stability. The flexibility of the hydrazone backbone poses, however, the problem of polymorphism (different crystalline structures for one molecule). This feature makes the development of useful crystals more difficult. Nevertheless, a proper control of the growth conditions such as selective choices of solvent and crystal growth method often leads to the desired acentric bulk crystal phase [32].

One of the best examples in this class is 4-dimethylaminobenzaldehyde-4-nitrophenyl-hydrazone (DANPH, Table 1). Another potential candidate is 5-(methylthio)thiophene-carboxaldehyde-4-nitrophenylhydrazone (MTTNPH) with large second-order nonlineari-ties in the crystalline state [33].

4.1.4 Co-crystal approach

In this approach molecular or ionic aggregates are designed by supramolecular synthesis to favor the desired crystal packing. It facilitates the conceptual design since one or both molecules may be tailored with respect to each other. Furthermore, physical properties such as melting point and solubility as well as crystal properties such as crystallinity and ease of crystal growth of the co-crystals can usually be improved compared to those of its starting components. Etter and co-workers first demonstrated the induction of a net dipole moment with a host-guest pair of 4-aminobenzoic acid and 3,5-dinitrobenzoic acid, unfortunately with small nonlinearities [34]. We found that co-crystals formed from the merocyanine dyes (M1 and M2) and phenolic derivatives in which the electron acceptor is *para*-related to the phenolic functionality together with a substituent either in the *ortho*- or *meta*-position (Figure 5) show the highest tendency of forming acentric co-crystals. In addition, 25% of acentric co-crystals based on M2 exhibit strong second-harmonic signals that are at least two orders of magnitudes larger than that of urea. One interesting example using this approach is M2-MDB (Table 1) which is optimized for electro-optic applications due to the parallel alignment of the nonlinear optical chromophores [35].

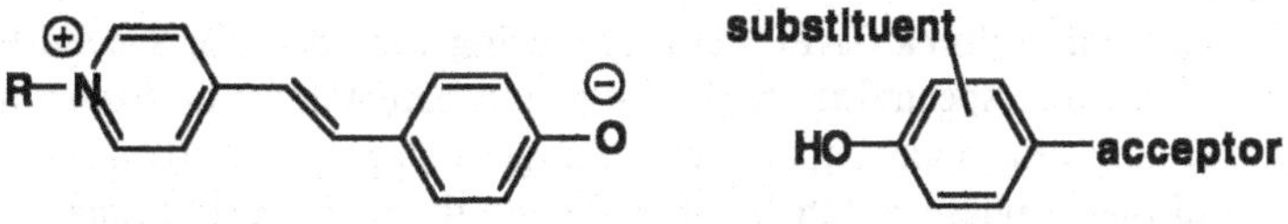

M1 : R = CH₃

M2 : R = CH₂CH₂OH

Figure 5. Chemical structures of M1, M2 and prototype phenolic derivatives.

4.1.5 Other strategies

Other crystal engineering approaches include e.g. the design of molecules with a deliberately low dipole moment. Due to negligible dipole-dipole interactions an almost parallel alignment of the molecules (ideal for electro-optics) can be achieved [36] with the drawback, however, that the conjugation is interrupted. No data on nonlinear optical experiments is available yet. Furthermore octupolar molecules having no dipole moment were synthesized in order to induce an asymmetric crystal packing due to the absence of dipole-dipole interactions [37]. Initial expectations of a higher occurence of noncentrosymmetric crystalline materials could not yet be confirmed likely due to the strong influence of steric forces (see above). Another route is to incorporate highly nonlinear optical molecules (with associated large dipole moments) into a lattice host. An example is e.g. perhydrotriphenylene which serves as a channel-like framework for highly active molecules [38, 39]. In this case a dilution of the chromophore density and a reduction of the nonlinear optical activity has to be taken into account. However, an optimum polar alignment of the molecules can be achieved in this way. Nonlinear optical measurements on crystals have to be performed in order to know whether this approach leads to competitive materials.

4.2 SPECIFIC EXAMPLES

Table 1 lists selected examples of electro-optic and/or nonlinear optical organic crystals. Only materials that are noncentrosymmetric and where the crystal structure is known are included. The table includes materials designed for different applications ranging from frequency conversion to electro-optics and therefore includes crystals of varying transmission from being completely colorless to being very dark. We will discuss the materials in the order of increasing cut-off wavelength. Only few materials with mostly recent results are listed.

We start with urea which was the first organic crystal in which interesting applications such as optic parametric oscillation (conversion efficiency up to 20%) and UV light generation by frequency-doubling (down to 213 nm) were demonstrated [40-43].

Furthermore it often serves as reference material in the powder test. HFB is a semiorganic crystal with the advantage of good mechanical and thermal properties. It is easy to grow and phase-matched parametric processes between 300 nm and 1300 nm have been predicted [44].

Isopropyl-4-acetylphenylurea (IAPU) is a promising candidate for second-harmonic generation into the blue spectral range [45]. It is transparent down to 380 nm and has a large diagonal nonlinear optical coefficient of $d_{222}=30.5$ pm/V ($\lambda=1064$nm). DIVA is another crystal that is phase-matchable (type I) into the blue spectral range [46]. At a wavelength of $\lambda=888$ nm an effective nonlinear coefficient 5.9 x d_{322} of $KNbO_3$, one of the best inorganic crystals for frequency-conversion into the blue, was measured making DIVA a very promising material for such applications.

TABLE 1. Examples of molecular crystals that have been investigated for their non-linear optical and/or electro-optic response. λ_c is the cut-off wavelength in the bulk, d (d_{eff}) is the nonlinear optical coefficient, and r is the electro-optic coefficient. References corresponding to the listed data can be found in the text.

material	point group	λ_c (nm)	d, r (pm/V)
urea	$\bar{4}2m$	200	$d_{123}(1064nm)=1.1$
HFB (L-histidine tetrafluoroborate)	2	250	$d_{eff}(1064nm)\approx2$
IAPU (isopropyl-4-acetylphenylurea)	2	380	$d_{222}(1064nm)=30.5$
DIVA (ortho-dicyanovinyl-anisole)	2	440	$d_{eff}(888nm)=81$

NPP (N-(4-nitrophenyl)-(L)-prolinol)	2	500	$d_{211}(1064nm)=51$

COANP (2-cyclooctylamino-5-nitropyridine)	mm2	490	$d_{322}(1064nm)=32$
			$r_{333}(514.5nm) = 28$

MNBA (4'-nitrobenzylidene-3-acetamino-4-methoxyaniline)	m	520	d_{111} (1064nm)=131
			$r_{111}(532nm)= 50$

DAST (4'-dimethylamino-N-methyl-4-stilbazolium tosylate)	m	700	d_{111} (1318nm)=1010
			d_{111} (1542nm)=290
			r_{111} (720nm) = 92

M2-MDB (4-{2-[1-(2-hydroxyethyl)-4-pyridylidene]-ethylidene}-cyclo-hexa-2,5-dien-1-one · methyl 2,4-dihydroxybenzoate)	m	phase I: 615	phase I: $d_{111}(1318nm)= 108$ r_{111} (1313nm)= 24
		phase II: 680	phase II: $d_{111}(1318nm)= 267$ r_{111} (1313nm)= 34

DANPH (4-dimethylaminobenzaldehyde-4-nitrophenyl-hydrazone)	m	670	d_{122} (1542nm)=190

NPP is one of the materials that is the farthest developed with respect to applications. Low-threshold parametric oscillation with large conversion efficiencies was demonstrated [47-49]. Recently, a detailed study of the crystal growth behavior of NPP in different solvents led to high-quality large crystals up to 2 x 0.8 x 0.8 cm^3 in size [50].

COANP is nitropyridine derivative with slightly lower nonlinearities than NPP. Doped with the sensitizer TCNQ (in the melt), COANP became the first photorefractive organic material [51, 52].

The materials discussed above were mainly investigated for frequency-conversion applications such as e.g. frequency-doubling since the cut-off wavelength is reasonably low. For most of the crystals described below the linear electro-optic effect is more important since an increased cut-off wavelength does not affect electro-optic effects in the near infrared (e.g. at the telecommunication wavelengths λ=1300nm and λ=1500nm).

MNBA illustrates the successful implementation of molecular asymmetry [53]. The small angle between the molecular charge transfer axis and the polar crystal axis (θ =18.7°) leads to both large diagonal elements d_{111} and r_{111}.

DAST is a successful example of the use of strong coulomb interactions [26]. It combines large nonlinear optical and electro-optic coefficients with rather easy crystal growth (crystals with a size of 1.8 x 1.8 x 0.4 cm^3 were obtained) [27, 54, 55]. Phase-matched frequency-doubling at telecommunication wavelengths was experimentally investigated and interesting conditions for parametric interactions were predicted [27]. The electro-optic coefficients were determined up to a modulation frequency of 1 GHz demonstrating the almost pure electronic origin of the large electro-optic effects [56]. In addition, photorefractive effects in the near infrared were observed and intensively investigated [57].

M2-MDB is an example of a co-crystal (see above) [35, 58, 59]. Interestingly, dependent on the growth conditions, two distinctively colored bulk crystals -phase I and phase II- were obtained for M2-MDB [35]. The single crystal X-ray structural analysis proved that the two differently colored crystals possess an identical molecular packing but with absorption cut-off wavelengths λ_c shifted by about 65 nm. Both phases were characterized with regard to their linear optical, nonlinear optical and electro-optic properties (Table 1). The observed discrepancies of the two nominally structurally identical materials might be due to the different location of the proton that constitutes the short hydrogen bond. However, this still needs confirmation. As the position of the hydrogen bonded proton varies, a different environment is induced for the merocyanine dye. Therefore the nonlinearities of this dye in a crystal lattice are anticipated to vary as its nonlinear optical response in different solvents. As an example the wavelength of maximum absorption of the dye M2 in solution shifts by more than 160nm when changing the solvent from water to pyridine! The case of M2-MDB nicely illustrates the increased importance of intermolecular interactions as the molecules become more nonlinear.

DANPH is a hydrazone derivative with very large nonlinear optical coefficients that is optimized for efficient parametric interactions [60]. The growth of large size DANPH crystals poses still some problems. If these problems could be solved, DANPH would be an extremely attractive candidate for nonlinear optical frequency conversion.

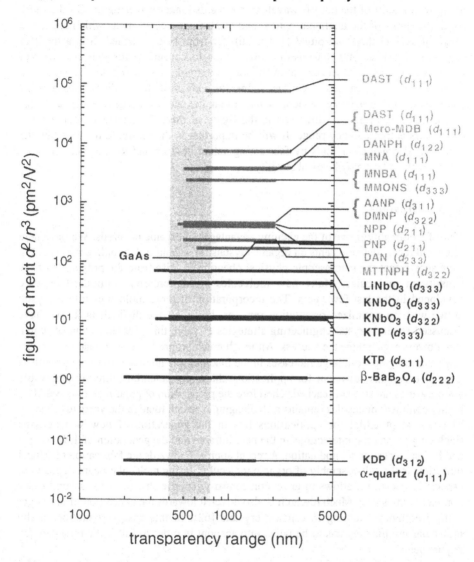

Figure 6. The figure of merit for frequency doubling, d^2/n^3, versus the transparency range of the crystals. The shaded area indicates the visible spectral range. Inorganic crystals are indicated in black, organic crystals in grey. It is clearly seen that organic crystals are by far superior in terms of the figure of merit. Also clearly visible is the shift of the transparency range towards longer wavelengths with increased nonlinearity.

What conclusions can be drawn from Table 1? On the one hand, as already mentioned above, an increase in the nonlinear optical and/or electro-optic coefficients usually goes along with a shift of the cut-off wavelength towards longer wavelengths. On the other hand, the limits of the nonlinear optical and electro-optic coefficients which are in the range of several thousand pm/V [16] is still far from being reached. It is a topic of further research to find out whether this is due to chromophores that are still not optimized or whether other reasons such as e.g. intermolecular interactions hinder these limits from being reached. As a last point, the values of the nonlinearities already achieved are nevertheless considerably larger than the ones of inorganic materials. This is illustrated in Figure 6 which shows the figure of merit for frequency-conversion, d^2/n^3, versus transparency range. It will be important in the future, to demonstrate the potential of organics crystals by fabricating actual devices and to test their thermal, mechanical, and (photo)chemical stability.

5. Conclusions

Multi-functionality is one of the exciting advantages of organic materials. We have here discussed recent developments on organic crystals for second-order nonlinear optics. We have shown that the macroscopic physical phenomena arise from the properties of the constituent molecules. Many new molecules with increased nonlinearities were developed in the last few years. The incorporation of these molecules into a crystal lattice with an optimized orientation remains, however, the difficult task. We have discussed several crystal engineering strategies to solve this problem, none of which can, however, guarantee the success. Although much progress has been made in the last few years ranging from large increases in the macroscopic nonlinearity to the growth of large size crystals sufficient for applications the upper potential limits of achievable nonlinearities are still not reached. Therefore the preparation of polar materials based on highly nonlinear molecules remains a challenging research topic in the years to come.

The largest potential for applications lies in the generation of new wavelengths (including parametric interactions in the near infrared and the generation of THz waves) and in fast electro-optic modulation. A recent approach towards the fabrication of optical thin films is the self-assembly of organic molecules during molecular beam deposition, especially because it allows to grow non-centrosymmetric structures for second-order nonlinear optics [61]. Much research is also directed towards the generation of blue light using frequency-doubling in organic crystals [62]. In this case a reduction of the molecular nonlinearity has to be taken into account in order to fulfill the transparency requirements.

The commercialization of the organic materials currently depends on the successful establishment of niche applications where they outperform traditional materials or at least offer a similar performance at a lower cost. As an example we can mention high-speed electro-optic light modulation (high frequency response in excess of 100 GHz). The acceptance of organic materials as reliable and high quality optical materials is also supported by the progress organics have made in other optical fields, such as linear optical polymers in photonics [63] and the recent advances in organic LEDs [64].

6. Acknowledgements

This work was supported by the Swiss National Science Foundation.

7. References

1. Bosshard, C., Sutter, K., Prêtre, P., Hulliger, J., Flörsheimer, M., Kaatz, P. and Günter, P. (1995) *Organic Nonlinear Optical Materials*, Garito, A. F. and Kajzar, F., Gordon and Breach Science Publishers, Amsterdam

2. Bosshard, C. and Günter, P. (1997) *Electro-optic effects in molecular crystals and polymers* in Nonlinear Optics of Organic Molecular and Polymeric Materials, Miyata, S. and Nalwa, H. S., CRC Press., Inc, Boca Raton, pp. 391-439

3. Bosshard, C., Wong, M. S., Pan, F., Spreiter, R., Follonier, S., Meier, U. and Günter, P. (1997) *Novel organic crystals for nonlinear and electro-optics* in Electrical and related properties of organic solids, Munn, R. W., Miniewicz, A. and Kuchta, B., Kluwer Academic Publishers, Dordrecht, pp. 279-296

4. Twieg, R. J. and Dirk, C. W. (1996) *Design, properties and applications of nonlinear optical chromophores* in Organic thin films for waveguiding nonlinear optics, Kajzar, F. and Swalen, J. D., Gordon and Breach Science Publishers, Amsterdam, pp.45-135

5. Zyss, J. (1994) *Molecular Nonlinear Optics: Materials, Physics, Devices* in Academic Press, Boston

6. Burland, D. (1994) *Optical Nonlinearities in Chemistry* in Chemical Reviews, American Chemical Society, Washington, D.C.

7. Bosshard, C. and Küpfer, M. (1996) *Oriented molecular systems* in Kajzar, F. and Swalen, J. D., Gordon and Breach Science Publishers, Amsterdam, pp. 163-191

8. Yariv, A. (1975) *Quantum Electronics*, John Wiley and Sons, New York

9. Bloembergen, N. and Sievers, A. J. (1970) Nonlinear optical properties of periodic laminar structures, *Appl. Phys. Lett.* 17, 483-485

10. Wemple, S. H. and DiDomenico, M., Jr. (1972) *Electrooptical and nonlinear optical properties of crystals* in Advances in Materials and Device Research, Wolfe, R., Academic, New York, pp. 263-381

11. Zyss, J. and Oudar, J. L. (1982) Relation between microscopic and macroscopic low-order optical nonlinearities of molecular crystals with one- or two-dimensional units, *Phys. Rev. A* 26, 2028-2048

12. Cheng, L. T., Tam, W., Stevenson, S. H., Meredith, G. R., Rikken, G. and Marder, S. (1991) Experimental Investigations of Organic Molecular Nonlinear Optical Polarizabilities. 1. Methods and Results on Benzene and Stilbene Derivatives, *J. Phys. Chem.* 95, 10631-10643

13. Oudar, J. L. and Chemla, D. S. (1977) Hyperpolarizabilities of the nitroanilines and their relations to the excited state dipole moment, *J. Chem. Phys.* 66, 2664-2668

14. Teng, C. C. and Garito, A. F. (1983) Dispersion of the nonlinear second-order optical susceptibility of organic systems, *Phys. Rev. B* 28, 6766-6773

15. Bosshard, C., Knöpfle, G., Prêtre, P. and Günter, P. (1992) Second-order polarizabilities of nitropyridine derivatives determined with electric-field-induced second-harmonic generation and a solvatochromic measurement: A comparative study, *J. Appl. Phys.* 71, 1594-1605

16. Bosshard, C., Sutter, K., Schlesser, R. and Günter, P. (1993) Electro-optic effects in molecular crystals, *J. Opt. Soc. Am. B* 10, 867-885

17. Kurtz, S. K. and Perry, T. T. (1968) A powder technique for the evaluation of nonlinear optical materials, *J. Appl. Phys.* **39**, 3798-3813

18. MacDonald, J. C. and Whitesides, G. M. (1994) Solid-state structures of hydrogen-bonded tapes based on cyclic secondary diamides, *Chem. Rev.* **94**, 2383-2420

19. Wong, M. S., Bosshard, C. and Günter, P. (1997) Crystal engineering of molecular NLO materials, *Adv. Mater.* **9**, 837-842

20. Kitaigorodskii, A. I. (1973) *Molecular Crystals and Molecules*, Academic Press, New York

21. Tsunekawa, T., Gotoh, T. and Iwamoto, M. (1990) New organic non-linear optical crystals of benzylidene-aniline derivative, *Chem. Phys. Lett.* **166**, 353-357

22. Meredith, G. R. (1983) *Design and Characterization of Molecular and Polymeric Nonlinear Optical Materials: Sucesses and Pitfalls* in Nonlinear Optical Properties of Organic and Polymeric Materials, Williams, D. J., American Chemical Society, Washington, D.C., pp. 27-56

23. Okada, S., Masaki, A., Matsuda, H., Nakanishi, H., Koto, M., Muramatsu, R. and Otsuka, M. (1990) Synthesis and Crystal Structure of a Novel Organic-Ion-Complex Crystal for Second-Order Nonlinear Optics, *Jpn. J. Appl. Phys.* **29**, 1112-1115

24. Okada, S., Masaki, A., Matsuda, H., Nakanishi, H., Koike, T., Ohmi, T. and Yoshikawa, N. (1990) Merocyanine-p-toluenesulfonic acid complex with large second-order nonlinearity, *Proc. SPIE* **1337**, 178-183

25. Duan, X. M., Okada, S., Nakanishi, H., Watanabe, A., Matsuda, M., Clays, K., Persoons, A. and Matsuda, H. (1994) *Evaluation of /β/ of stilbazolium p-toluenesulfonates by the hyper Rayleigh sattering method* in Organic, Metallo-Organic, and Polymeric Materials for Nonlinear Optical Applications, Marder, S. R. and Perry, J. W. (ed.), SPIE-The International Society for Optical Engineering, Los Angeles, pp. 41-51

26. Marder, S. R., Perry, J. W. and Schaeffer, W. P. (1989) Synthesis of organic salts with large second-order optical nonlinearities, *Science* **245**, 626-628

27. Meier, U., Bösch, M., Bosshard, C., Pan, F. and Günter, P. (1998) Parametric interactions in the organic salt 4-N,N-dimethylamino-4'-N'-methyl-stilbazolium tosylate at telecommunication wavelengths, *J. Appl. Phys.* **83**, 3486-3489

28. Pan, F., Knöpfle, G., Bosshard, C., Follonier, S., Spreiter, R., Wong, M. S. and Günter, P. (1996) Electro-Optic Properties of the Organic Salt 4-N,N-dimethylamino-4'-N'-methyl-stilbazolium tosylate, *Appl. Phys. Lett.* **69**, 13-15

29. Serbutoviez, C., Bosshard, C., Knöpfle, G., Wyss, P., Prêtre, P., Günter, P., Schenk, K., Solari, E. and Chapuis, G. (1995) Hydrazone Derivatives, an Efficient Class of Crystalline Materials for Nonlinear Optics, *Chem. Mater.* **7**, 1198-1206

30. Wong, M. S., Bosshard, C., Pan, F. and Günter, P. (1996) Non-classical donor-acceptor chromophores for second-order nonlinear optics, *Adv. Mater.* **8**, 677-680

31. Wong, M. S., Meier, U., Pan, F., Gramlich, V., Bosshard, C. and Günter, P. (1996) Five-membered hydrazone derivatives for second-order nonlinear optics, *Adv. Mater.* **7**, 416-420

32. Pan, F., Bosshard, C., Wong, M. S., Serbutoviez, C., Schenk, K., Gramlich, V. and Günter, P. (1997) Selective growth of polymorphs: an investigation of the organic nonlinear optical crystal 5-nitro-2-thiophenecarboxaldehyde-4-methylphenylhydrazone, *Chem. Mater.* **9**, 1328-1334

33. Pan, F., Wong, M. S., Bösch, M., Bosshard, C., Meier, U. and Bosshard, C. (1997) A highly efficient organic second-order nonlinear optical crystal based on a donor-acceptor substituted 4-nitrophenylhydrazone, *Appl. Phy. Lett.* **71**, 2064-2066

34. Etter, M. C. and Frankenbach, G. M. (1989) Hydrogen-bond directed cocrystallization as a tool for designing acentric solids, *Chem. Mater.* **1**, 10-12

35. Wong, M. S., Pan, F., Bösch, M., Spreiter, R., Bosshard, C. and Günter, P. (1998) Novel electro-optic molecular co-crystals with ideal chromophoric orientation and large second-order nonlinearities, *J. Opt. Soc. Am. B* **15**, 426-431

36. Chen, G. S., Wilbur, J. K., Barnes, C. L. and Glaser, R. (1995) Push-pull substitution versus intrinsic or packing related gauche preferences in azines. Synthesis, crystal structures and packing of asymmetrical acetophenone azines, *J. Chem. Soc. Perkin Trans.* **2**, 2311-2317

37. Zyss, J. and Ledoux, I. (1994) Nonlinear optics in multipolar media: theory and experiments, *Chem. Rev.* **94**, 77-105

38. Hoss, R., König, O., Kramer-Hoss, V., Berger, U., Rogin, P. and Hulliger, J. (1996) Crystallization of supramolecular materials: perhydrotriphenylene (PHTP) inclusion compounds with nonlinear optical properties, *Angew. Chemie* **35**, 1664-1666

39. Hulliger, J., König, O. and Hoss, R. (1995) Polar inclusion compounds of perhydrotriphenylene (PHTP) and efficient nonlinear optical molecules, *Adv. Mater.* **7**, 719-721

40. Bäuerle, D., Betzler, K., Hesse, H., Kapphan, S. and Loose, P. (1977) Phase-matched second-harmonic generation in urea, *phys. stat. sol. (a)* **42**, K119-K121

41. Donaldson, W. R. and Tang, C. L. (1984) Urea optical parametric oscillator, *Appl. Phys. Lett.* **44**, 25-27

42. Halbout, J. M., Blit, S., Donaldson, W. and Tang, C. L. (1979) Efficient phase-matched second-harmonic generation and sum-frequency mixing in urea, *IEEE J. Quantum Electron.* **QE-15**, 1176-1180

43. Kato, K. (1980) High-efficiency high-power UV generation at 2128 Å in urea, *IEEE J. Quantum Electron.* **QE-16**, 810-811

44. Marcy, H. O., Rosker, M. J., Warren, L. F., Cunningham, P. H., Thomas, C. A., DeLoach, L. A., Velsko, S. P., Ebbers, C. A., Liao, J.-H. and Kanatzidis, M. G. (1995) L-Histidine tertafluoroborate: a solution grown semi-organic crystal for nonlinear frequency conversion, *Opt. Lett.* **20**, 252-254

45. Yamamoto, H., Funato, S., Sugiyama, T., Jung, I., Kinoshita, T. and Sasaki, K. (1997) Organic crystal isopropyl-4-acetylphenylurea and waveguide second-harmonic generation, *J. Opt. Soc. Am. B* **14**, 1099-1108

46. Grossman, C. H., Schulhofer-Wohl, S. and Thoen, E. R. (1997) Blue light second-harmonic generation in the organic crystal ortho-dicyanovinyl-anisole, *Appl. Phys. Lett.* **70**, 283-285

47. Khodia, S., Josse, D., Samuel, I. D. W. and Zyss, J. (1995) Broadband pump wavelength tuning of a low threshold N-(4-nitrophenyl)-L-prolinol near infrared optical parametric oscillator, *Appl. Phys. Lett.* **67**, 3841-3843

48. Josse, D., Dou, S. X., Zyss, J., Andreazza, P. and P'rigaud, A. (1992) Near-infrared optical parametric oscillation in a N-(4-nitrophenyl)-L-prolinol molecular crystal, *Appl. Phys. Lett.* **61**, 121-123

49. Dou, S. X., Josse, D. and Zyss, J. (1993) Near-infrared pulsed optical parametric oscillation in N-(4-nitrophenyl)-L-prolinol at the 1ns time scale, *J. Opt. Soc. Am. B* **10**, 1708-1715

50. Shekunov, B. Y., Shepherd, E. E. A., Sherwood, J. N. and Simpson, G. S. (1995) Growth and perfection of organic nonlinear optical materials. Kinetics and Mechanism of the growth of N-(4-nitrophenyl)-L-prolinol (NPP) crystals from methanol and toluene, *J. Phys. Chem.* **99**, 7130-7136

278

51. Sutter, K., Hulliger, J. and Günter, P. (1990) Photorefractive effects observed in the organic crystal 2-cyclooctylamino-5-nitropyridine doped with 7,7,8,8-tetracyanochinodimethane, *Solid State Commun.* **74**, 867-870

52. Sutter, K. and Günter, P. (1990) Photorefractive gratings in the organic crystal 2-cyclooctylamino-5-nitropyridine doped with 7,7,8,8-tetracyanochinodimethane, *J. Opt. Soc. Am. B* **7**, 2274-2278

53. Knöpfle, G., Bosshard, C., Schlesser, R. and Günter, P. (1994) Optical, Nonlinear Optical, and Electrooptical Properties of 4'-nitrobenzylidene-3-acetamino-4-methoxyaniline (MNBA) Crystals, *IEEE J. Quantum. Electron.* **30**, 1303-1312

54. Pan, F., Wong, M. S., Bosshard, C. and Günter, P. (1996) Crystal growth and characterization of the organic salt 4-N,N-dimethylamino-4'-N'-methylstilbazolium tosylate (DAST), *Adv. Mater.* **8**, 592-594

55. Knöpfle, G., Schlesser, R., Ducret, R. and Günter, P. (1995) Optical and Nonlinear Optical Properties of 4'-Dimethylamino-N-methyl-4-stilbazolium tosylate (DAST) Crystals, *Nonlinear Optics* **9**, 143-149

56. Spreiter, R., Bosshard, R., Pan, F. and Günter, P. (1997) High-frequency response and acoustic phonon contribution of the linear electro-optic effect in DAST, *Opt. Lett.* **22**, 564-566

57. Follonier, S., Bosshard, C., Pan, F. and Günter, P. (1996) Photorefractive effects observed in 4-N,N-dimethylamino-4'-N'-methylstilbazolium toluene-p-sulfonate, *Opt. Lett.* **21**, 1655-1657

58. Wong, M. S., Pan, F., Gramlich, V., Bosshard, C. and Günter, P. (1997) Self-Assembly of Acentric Co-crystal of a Highly Hyperpolarizable Merocyanine Dye with Optimized Alignment for Nonlinear Optics, *Adv. Mater.* **9**, 554-557

59. Pan, F., Wong, M. S., Gramlich, V., Bosshard, C. and Günter, P. (1996) A Novel and Perfectly Aligned Highly Electro-Optic Organic Co-crystal of a Merocyanine Dye and 2,4-Dihydroxybenzaldehyde, *J. Am. Chem Soc.* **118**, 6315-6316

60. Follonier, S., Bosshard, C., Knöpfle, G., Meier, U., Serbutoviez, C., Pan, F. and Günter, P. (1997) A New Nonlinear Optical Organic Crystal: 4-Dimethylaminobenzaldehyde-4-Nitrophenyl-hydrazone (DANPH), *J. Opt. Soc. Am. B* **14**, 593-601

61. Cai, C., Bösch, M.M., Tao, Y., Müller, B., Gan, Z., Kündig, A., Bosshard, Ch., Liakatas, I., Jäger, M., and Günter, P. (1998) Self-Assembly in Ultrahigh Vacuum: Growth of Organic Thin Films with a Stable In-Plane Directional Order, *J. Am. Chem. Soc.* **120**, 8563-8564

62. Sagawa, M., Kagawa, H., Kakuta, A., Kaji, M., Saeki, M. and Namba, Y. (1995) Blue light emission from a laser diode pumped ring resonator with an organic second-harmonic generation crystal of 8-(4'acetylphenyl)-1,4-dioxa-8-azaspirol[4.5]decane, *Appl. Phys. Lett.* **66**, 547-549

63. Eldada, L. A., Shacklette, L.W., Norwood, R.A., Yardley, J.T. (1997) Next generation polymeric photonic devices, *SPIE Critical Reviews of Optical Science and Technology* **CR 68**, 207-227

64. Miyata, S. and Nalwa, H. S. (1997) *Organic Electroluminescent Materials and Devices*, Gordon and Breach, Amsterdam

TOWARD CRYSTAL DESIGN IN ORGANIC CONDUCTORS AND SUPERCONDUCTORS

U. GEISER

Argonne National Laboratory
Chemistry and Materials Science Divisions
9700 South Cass Avenue
Argonne, IL 60439, USA

1. Introduction

The vast majority of organic solids are electrical insulators with electrical conductivity values on the order of 10^{-20}–10^{-15} $\Omega^{-1}cm^{-1}$. This property is of course exploited in many everyday applications. Two principal reasons are responsible for this fact: (1) The highest occupied molecular orbital (HOMO) of most organic molecules is completely filled, and there is a significant energy difference to the lowest unoccupied molecular orbital (LUMO). (2) Organic solids are usually molecular, i.e., they do not possess a system of covalent bonds extending over macroscopic distances. Therefore the quantum mechanical interactions between the HOMOs of adjacent molecules are small. The valence band formed by these interactions remains therefore very narrow. Similarly, the conduction band arising from the interactions between the LUMOs is also small, and the band gap is essentially that of the free molecule. This holds even in the case of conventional polymers, e.g. polyethylene, that are σ-bonded.

Electrical conduction may be classified in a number of types. The two main ones are activated and metallic, see Table 1.

TABLE 1. Electrical conduction types.

	Activated	Metallic
Temperature Dependence (lowering T)	lower conductivity	higher conductivity
Band Situation	Filled valence band, empty conduction band	Partially filled bands
Fermi Level	In band gap	Intersects bands, Fermi surface

Activated behavior is found in semiconductors, of which insulators are an extreme case. Semiconductors with appreciable conductivities (10^{-10}–10^{-2} $\Omega^{-1}cm^{-1}$ at room

D. Braga et al. (eds.), Crystal Engineering: From Molecules and Crystals to Materials, 279–294.

temperature) are obtained by shrinking the HOMO–LUMO gap, e.g., with the use of extensive π-bonding, sometimes including heteroatoms with lone pair electrons. Examples of this kind are polyacetylene, polyaniline, polyaromatics, and in the extreme case, graphite. Conductivity is governed by an activated term, $e^{-Eg/kT}$, given by the band gap, E_g, and a pre-exponential factor which is more difficult to calculate. Conductivity may be increased by a reduction of the band gap, or by the introduction of impurity states within the gap.

In contrast, in a metal the Fermi level intersects a band, and the electrons possessing the proper momentum are able to delocalize freely. In a real system, conduction is not infinite, but reduced by scattering with other electrons, vibrations, and impurities. The temperature dependence of the resistivity (inverse of conductivity) is usually approximated by an power law with a small exponent (2–3). Partial band filling is achieved by the oxidation of electron donor molecules or the reduction of electron acceptor molecules to their respective radical ions and crystallization with charge-compensating counter-ions. It is also possible to combine electron donors and acceptors leading to spontaneous electron transfer. Provided these entities are spatially separated, e.g., in segregated stacks or layers, metallic conduction can occur within these substructures. Of necessity, the metallic properties of charge transfer salts will be low-dimensional, or at least highly anisotropic. In practice, this is also the case with most organic radical ion salts (see below).

Superconductivity is a special low-temperature regime of some metals and a sought-after property in our research program. It arises as a cooperative effect from the pairing of conduction electrons of opposite momentum (Cooper pairs) and is mediated in the conventional (Bardeen-Cooper-Shrieffer) theory by phonons. The detailed nature of the superconducting state in organic and many inorganic systems is the subject of current scientific debate.

Electrical conduction in organic solids was predicted in 1911 by McCoy and Moore [1]. It was achieved experimentally in 1954 when perylene was complexed with bromine [2]. Little in 1964 [3] predicted superconductivity in organic solids, which triggered more intense activity in the field. Milestones include the synthesis of the acceptor molecule tetracyano-p-quinodimethane (TCNQ) [4] and tetrathiafulvalene (TTF), see Figure 1. These were combined in 1972 to the charge transfer salt (TTF)(TCNQ), which was the first organic solid to show metallic conductivity over an extended temperature range [5, 6]. Superconductivity was found below 0.3 K in the polymer $(SN)_x$ in 1975. In the late 1970s, effort shifted to the exploration of the cation radical salts of TTF derivatives. Finally, superconductivity was observed in several salts of tetramethyltetraselenafulvalene (TMTSF) with simple anions such as ClO_4^- and PF_6^- [7]. Many more superconductors were found among the radical cation salts of a different TTF derivative, bis(ethylenedithio)tetrathiafulvalene (BEDT-TTF or ET), first synthesized by Mizuno et al. [8]. The salts of this donor molecule have been the subject of our research program for a number of years, and we have found the organic salts with the highest superconducting transition temperatures (T_c), i.e., κ-$(ET)_2Cu[N(CN)_2]Br$ $(T_c$ = 11.6 K at ambient pressure) [9] and κ-$(ET)_2Cu[N(CN)_2]Cl$ $(T_c$ = 12.5 K at 0.3 kbar applied pressure) [10]. Only the radical anion salts of C_{60} with alkali metal cations have recently achieved higher T_cs among molecular solids. The field has been reviewed in books [11, 12] and monographs [13].

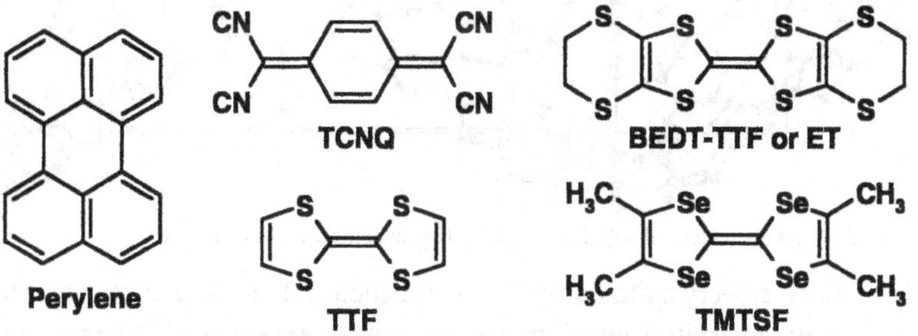

Figure 1. Typical electron donor and acceptor molecules.

2. Cation Radical Salts, a Type of Organic Conductors

Cation radical salts are composed of the radical cations of an electron donor molecule and charge-compensating counter anions. Most of the electron donor molecules (Figure 1) studied to date are related to TTF, and ET has achieved a prominent position with well over 200 salts characterized and the largest number of superconductors. The search for new superconductors encompasses on one hand the search for new donor molecules, and on the other the search for novel anions. In both cases, a more-or-less trial-and-error method is employed, and the field could use a more systematic approach.

The vast majority of ET salts are layered, i.e., the crystal structure is composed of layers of densely packed radical cations alternating with layers of anions, see Figure 2. Anisotropic properties thus result: the electrical conductivity is highest within the plane (often by several orders of magnitude), and in the superconductors, the critical field (field required to suppress superconductivity) is much higher when it is applied parallel to the plane than perpendicular to it. The stoichiometry of many ET salts with monovalent anions is 2:1, thus on average only one of every two ET molecules is oxidized. In 1:1 salts, the unpaired electrons tend to be localized (Mott insulator). Other stoichiometries, e.g., 3:2 and (insulating) 1:2 have also been observed.

282

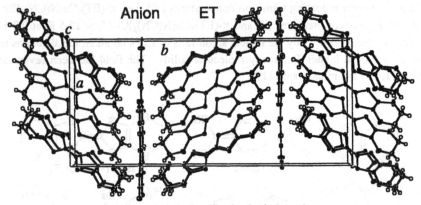

Figure 2. Crystal structure of κ-(ET)₂Cu[N(CN)₂]Br, showing its layered nature.

These salts are crystallized by an electrochemical method from an electrolyte solution containing neutral donor molecules and an excess of a suitable salt of the desired anion. Organic solvents such as 1,1,2-tricholorethane or tetrahydrofuran are frequently employed, thus anion salts with large organic counterions (e.g., tetra-*n*-butylammonium) are commonly used in order to achieve good solubility. Under constant-current conditions with applied potentials of a few volts, the donor molecules are oxidized and precipitate with the anions in solution. The occurrence of partial oxidation states is explained by the association of neutral donor molecules with cation radicals prior to the crystal growth process. After a few weeks, the salts are harvested as black, shiny, ca. millimeter size, and (usually) air-stable crystals. Standard characterization methods include single crystal X-ray diffraction analysis, ESR spectroscopy (temperature and angle dependent), electrical conductivity (temperature dependent), magnetic susceptibility (search for large diamagnetic signal in the superconducting state), and Raman spectroscopy. Physicists have subjected these crystals to a host of further experimental probes that are beyond the scope of this article.

3. ET-Based Superconductors

The first ambient-pressure organic superconductor based on the electron donor molecule ET was β-(ET)₂I₃ with T_c = 1.5 K [14]. In quick succession, two more salts of ET with linear, triatomic anions were found to be superconducting, β-(ET)₂IBr₂ (T_c = 2.8 K) [15] and β-(ET)₂AuI₂ (T_c = 4–5 K) [16]. At about the same time, it was realized that these two salts had to be compared with the high-pressure phase β*-(ET)₂I₃ (obtained by cooling at ca. 1.5 kbar pressure) with T_c = 7–8 K [17, 18]. Without applied pressure, β-(ET)₂I₃ undergoes a phase transition at 175 K to an incommensurately modulated structure which is not observed in the other compounds [19]. These three salts represented the first *isostructural* series of ET-based superconductors, and they varied primarily in the size of the anion, thus the volume of the unit cell which increases in the order $IBr_2^- < AuI_2^- < I_3^-$. ET salts with other linear triatomic anions were not superconducting, either because the anion was too small to form an isostructural salt, e.g., ICl_2^- and $AuCl_2^-$, or because it did not possess a center of inversion, e.g., I_2Br^-. The crystal structure of the β-salts contains the anion on a center of inversion. Therefore, non-centrosymmetric anions are disordered, and the random potentials generated by the disorder are usually (but not always!) sufficient to suppress superconductivity [20].

The unit cell volume has a direct connection to the electronic properties since it scales inversely with the width of the conduction band. Furthermore, in the context of standard Bardeen-Cooper-Schrieffer (BCS) theory [21, 22], the McMillan equation [23]

$$T_c \propto e^{-\lambda}, \text{ with } \lambda \propto \langle \omega^2 \rangle / n(e_F) \qquad (1)$$

relates the superconducting transition temperature via the electron-phonon coupling constant (λ) to the density of states at the Fermi level, $n(e_F)$, which is largest when the band with is narrow. $\langle \omega^2 \rangle$ is the mean-square coupling vibration frequency. Thus, keeping all other parameters constant, the isostructural salt with the largest cell volume should have the highest T_c. At that time, it was predicted that "puffing up the lattice" with the use of longer linear anions, e.g., $Cu(NCS)_2^-$, should lead to even higher transition temperatures.

Superconducting $(ET)_2Cu(NCS)_2$ indeed was synthesized [24] and held the T_c record (10.4 K) for a few years. However, its crystal structure contained a completely different way of packing the ET molecules (κ-type, see below), and the anion was not discrete-linear, but it polymerized into chains that contained three-coordinate copper(I) centers, with both a bridging and a terminal thiocyanate group. This discovery led to the search for ET salts with other cuprous complex anions.

In 1990, our group discovered the two isostructural superconducting salts κ-$(ET)_2Cu[N(CN)_2]Br$ [9] and κ-$(ET)_2Cu[N(CN)_2]Cl$ [10], which to this day hold the record for highest transition temperatures for organic cation radical salts (T_c = 11.6 K at ambient pressure for the Br salt, T_c = 12.5 K at 0.3 kbar applied pressure in the Cl salts). κ-$(ET)_2Cu[N(CN)_2]Cl$ was also found to possess a canted antiferromagnetic ground state [25] at ambient pressure with a rich phase diagram in applied magnetic fields [26]. In contrast, κ-$(ET)_2Cu[N(CN)_2]I$ was not found to become superconducting, even under pressure. The reason is probably the presence of a disordered ethylene end group which persists to low temperature. κ-$(ET)_2Cu(NCS)_2$ and κ-$(ET)_2Cu[N(CN)_2]Br$ are among the most studied organic salts in the literature, with at least 300 papers on their physical properties in print.

At the time of this writing, well over 50 organic superconductors based on electron donor cations are known, too many to mention all in this space. Noteworthy are a family of superconducting ET salts with organometallic $M(CF_3)_4^-$ (M = Cu, Ag, Au) anions that also incorporate 1,1,2-trihaloethane (with Cl and Br substituents) solvent

284

molecules in the crystal structure, reviewed by Schlueter [27]. In contrast to earlier experience, the presence of crystallographic disorder in these salts does not inhibit superconductivity. Furthermore, these salts are synthesized in two "flavors", a plate-like phase with well defined crystal structures and T_cs of 2–5 K and a needle-like phase that has not yet yielded crystals suitable for structural analysis, and T_cs up to 11 K [28]! It is tempting to speculate that the higher superconducting transition temperatures are due the absence of crystallographic disorder (compared to the low-T_c phases), but better samples are clearly needed for thorough characterization.

A common misconception for many years was that the presence of magnetic ions was incompatible with superconductivity. This was proven wrong by the discovery of a superconducting ground state in the salt β''-(ET)$_4$(H$_2$O)[Fe(C$_2$O$_4$)$_3$](C$_6$H$_5$CN) [29] and the corresponding chromium complex salt. Paramagnetism in the anion and superconductivity in the organic layers appear to coexist quite independently. It remains to be seen whether superconductivity can coexist with long-range magnetically ordered anions.

While most of the anions utilized in the synthesis of cation radical salts have been inorganic in nature, we have recently emphasized the search for new superconductors with organic anions such as carboxylates and sulfonates. We recently succeeded in observing superconductivity at 5 K in an all-organic salt, β''-(ET)$_2$(SF$_5$CH$_2$CF$_2$SO$_3$) [30]. We feel that ultimately, organic anions will be more versatile to obtain homologous series of salts because of the multitude of possible substitution patterns. However, as will be shown below, not all substitutions, not even seemingly minor ones, lead to isostructural salts, much less superconductivity.

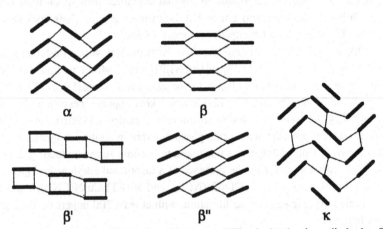

Figure 3. Schematic view of the main packing types of ET and related cation radical salts. The heavy bars represent the projection of the donor molecule along central C=C bond (which is usually inclined with respect to the layer normal, see Figure 2), and dashed lines short intermolecular contacts.

4. Crystal Structures of Cation Radical Salts

Over 200 crystal structures of ET salts are known to date (see the book by Williams *et al.* [11] for a list up to 1991), and a similar number of derivative salts have been determined. While there is great variety in packing motifs of the electron donor layers, a few principal prototypes have emerged. Mori, in a series of two articles, has recently reviewed the main packing types in great detail [31, 32]. A schematic representation is given in Figure 3.

A number of compositions exist in more than one crystallographic phase, sometimes even growing simultaneously. The "chameleon" with the most phases is probably the system ET:polyiodide. Not only are there at least five $(ET)_2I_3$ phases (α, β, θ, κ, and λ) under ambient conditions (not counting high pressure and low temperature phases!), but several other cation:anion ratios as well as solvated phases and compositions with larger polyiodide anions (I_5^-, I_8^{2-}) exist. The specific Greek lettering reflects traditional terminology rather than thermodynamic stability. The α- and various β-like phases all contain discernible stacks (running approximately vertically in Fig. 3). In the α-type packing, adjacent stacks contain donor molecules that are inclined in opposite directions with respect to the stack axis. The β-types all contain donor molecules that are in parallel orientation to each other. β' and β'' structures may be thought as two different distortions of the basic honeycomb-like standard β-type. The β'-packing is strongly dimerized (leading to nonmetallic properties), whereas the β''-type is a tilted variant of the β-type. Finally, the basic motif of the κ-type packing is a face-to-face dimer. Adjacent dimers are rotated by approximately 90° with respect to each other. The organic superconductors with the highest transition temperatures all contain the κ-packing type.

The vast majority of α- and β-like phases is triclinic, although not all are isomorphous. Depending on the size of the anion, structures with 2, 3, 4, or even more donor molecules per unit cell are found. On the other hand, most κ-phases possess higher crystallographic symmetry, with at least a 2_1 screw rotation axis within the donor plane connecting dimers of opposite tilt.

From the plethora of donor molecule packing types, even in salts of the same composition, it is clear that the specific crystal packing adopted is not intrinsic to the donor molecule, especially not for ET. Rather, it is the anion that determines the donor layer packing. The main interaction area between the anions and the donor radical cations is the ethylene end group (see Fig. 2). These interactions are discussed in the following section.

5. Design Principles: Donor–Anion Interactions

As is illustrated in numerous examples in this volume, the principal intermolecular interactions in organic solids are van der Waals dispersive forces (always present) and specific hydrogen bonding when suitable functional groups are present. $X–H \cdots Y$ ($X, Y =$ O, N, etc.) hydrogen bonding interactions, where present, are the strongest interactions possible in organic solids. Crystal engineering, i.e., the design and synthesis of crystal structures with predefined topology and functionality, makes heavy use of these hydrogen bonding interactions, as is evident from many contributions to this volume.

Organic conducting cation radical salts are by design composed of at least two different components: cations and counter-anions. One therefore has to consider at least three types of interactions: cation–cation, anion–anion, and cation–anion. Furthermore, at least two more bonding interactions must be considered in addition to van der Waals and hydrogen interactions, i.e., ionic (electrostatic) interaction, and metallic bonding due to the electron delocalization in the conducting cation layers. Let us consider each interaction in turn:

Van der Waals interactions are strongest between large, polarizable atoms, such as the sulfur or selenium atoms of the TTF derivative electron donor molecules. The packing of the ET cations into layers generally maximizes the number of intermolecular $S \cdots S$ contacts, which may be as short as 3.4 Å compared to the sum of the van der Waals radii of 3.6 Å [33]. The topology of the short $S \cdots S$ contacts is frequently indicative of significant electron overlap (leading to electron delocalization and ultimately electronic conduction), however, band electronic structure calculations are usually required in order to unambiguously determine the exact conduction pathways.

Hydrogen bonding interactions involving the electron donor molecules are usually restricted to the weak attraction of the aliphatic ethylene end group hydrogen atoms to electronegative atoms of the anion layer. While these hydrogen bonds are weak, they nevertheless provide the "glue" that binds the cation and anion layers together. The effect of deuterium substitution on the superconducting properties has been studied extensively, and the surprising result that at least in the κ-phase materials the deuterated salts have increased superconducting transition temperatures (first noted by Oshima *et al.* [34]) implicates the hydrogen–anion interactions in the electronic coupling between adjacent conducting layers. Stronger, more specific hydrogen bonding is sometimes found among anions or between anions and co-crystallized solvent molecules. Impurity water in the solvent used for crystallization has occasionally been found incorporated in the crystal structures of cation radical salts where it is invariably involved in a hydrogen bonding network with the anions (see the following section for an example).

The ionic interactions leading to lattice stabilization via the Madelung energy are difficult to quantify in these salts since the ionic charge is diffuse, and nowhere in the lattice does the latter attain large values. The layered packing found in most cation radical salts is not the minimum configuration based on the ionic forces alone. If ionic interactions were the only ones operative, the cations and anions would be much more intimately interspersed. The observation of self-organized cation and anion layers indicates that the ionic interactions are less than dominant in the stabilization of the crystal structures, but clearly they cannot be neglected completely.

The role of metallic bonding in lattice stabilization is likely to be small, based on the small carrier densities (less than one electron/hole per donor molecule containing 20–30 atoms), and it is only operative within the donor cation layer. Furthermore, topologically similar packing patterns are observed in metallic as well as semiconducting salts, thus further indicating that metallic bonding has a minor effect on the crystal structures.

In summary, it appears that van der Waals, C–H⋯anion hydrogen bonding, and ionic interactions are all weak but of similar magnitude in organic conductors and superconductors. Any attempt at *a priori* design or crystal structure prediction must take into account all of these interactions simultaneously. A detailed knowledge of the charge distribution over the atoms of the component molecules (made even more difficult due to the partial molecular charges and open electronic shells) is required for proper treatment, and it is no wonder that no calculations have been published yet. Any discussion of crystal engineering in these systems is currently limited to the crystallographic study of typical examples that illustrate the relative role of various intermolecular interactions.

6. Case Studies

6.1 ET SALTS WITH SF_5–R–SO_3^- ANIONS

As indicated above, β''-$(ET)_2SF_5CH_2CF_2SO_3$ was the first all-organic superconductor discovered. The anion offers a number of substitutions that might lead to isostructural salts and expected new superconductors: (1) The SF_5 head group may be replaced by CF_3, SiF_3, or similar. (2) The SO_3^- anionic end group may be replaced by carboxylate or similar. (3) The hydrogen and fluorine substitution pattern may be changed on the carbon backbone. (4) The number of carbon atoms on the backbone may be varied. We have obtained results on (3) and (4) so far, unfortunately with the result that none of the new anions led to an isostructural, superconducting salt.

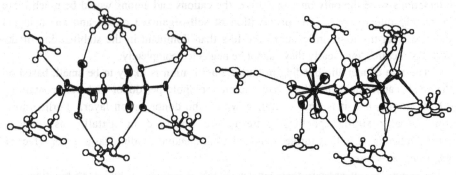

Figure 4. Anion and surrounding ET-ethylene groups in β"-(ET)$_2$SF$_5$CHRCF$_2$SO$_3$, R = H (left), F (right). Short H···F and H···O contacts are indicated by thin lines. In the right-hand structure, the anion is crystallographically disordered, with a superposition of the two possible enantiomers in an approximately 2:1 population ratio (the majority isomer is drawn darker).

The results of the simple replacement of one anion hydrogen atom by fluorine [35] are shown in Figure 4. While the unit cell metric of the two salts is quite similar (a=9.154 Å, b=11.440 Å, c=17.490 Å, α=94.32°, β=91.13°, γ=102.76° for the CH$_2$ derivative *vs.* a=9.246 Å, b=11.361 Å, c=17.714 Å, α=94.12°, β=94.91°, γ=103.14° for the CHF derivative), and the space group remains the same, $P\bar{1}$, the anion adjusts its conformation due to the involvement of the extra fluorine atom in weak hydrogen bonding with donor cation ethylene groups. This change in turn leads to a small distortion of the ET donor molecule network (note the change of the β-angle by 3.8° although the overall β"-type packing remains intact) which is sufficient to change the electronic properties. Not only is superconductivity suppressed (which by itself could be rationalized by the presence of anion crystallographic disorder), but the salt undergoes a gradual transition to a non-metallic ground state below 150 K which must have its origin in the subtle changes of the band electronic structure due to the small distortion. The anion site is a general position in the crystal structure, and the two enantiomers of the anions could occupy inversion-equivalent sites (a racemic mixture was employed in the synthesis). However, a superposition of the two enantiomers is found (in a ~2:1 ratio, opposite on the inversion-symmetric site) as the anion pocket is large enough to accommodate both with a similar number of C–H···F and C–H···O contacts, see Figure 4.

We have also studied the crystal structures of the ET salts with the corresponding anions containing only one carbon atom in the backbone (CH$_2$, CHF, and CF$_2$ derivatives) [36]. Of these, the former two also crystallize in β"-type structures, while the latter adopts the strongly dimerized β'-structure. Again, the β"-structures are distorted from that found in β"-(ET)$_2$SF$_5$CH$_2$CF$_2$SO$_3$, and no superconductors are found. In this case, the distortion derives from weak hydrogen bonding between pairs of anions, as shown in Figure 5 for (ET)$_2$SF$_5$CH$_2$SO$_3$. Note that this particular hydrogen bonding pattern is not possible for the two-carbon analogues described above, since they contain CF_2 moieties (lacking hydrogen) adjacent to the sulfonate groups.

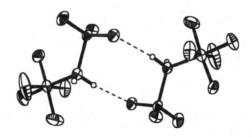

Figure 5. Hydrogen bonded anion dimer in $(ET)_2SF_5CH_2SO_3$.

6.2. TWO DIFFERENT PACKING PATTERNS IN $(ET)_2(C_6H_5CH_2SO_3)(H_2O)$

In our search for ET-based superconductors with all-organic anions, we also explored salts with the benzylsulfonate anion $(C_6H_5CH_2SO_3^-)$ [37]. We found that ET crystallized with this anion simultaneously in two distinct crystallographic phases (β and κ-4×4) of the same 2:1 stoichiometry, each incorporating a molecule of solvent water per anion. Unfortunately, neither salt is superconducting, and the one whose crystals are large enough to be examined by standard physical methods is not even metallic but a semiconductor. Nevertheless, their crystal structures are illustrative in the context of crystal engineering.

The packing of the ET donor molecules in the κ4×4 salt is shown in Figure 6. Compared to the standard κ-type packing, the basic unit is a tetramer (instead of a dimer). Neighboring tetramers in the c-direction are parallel, but in the β-direction, they are tilted in the opposite direction (by ca. 90°). On the other hand, in the β-salt, all ET molecules are parallel to each other, similar to the schematic diagram in Figure 4. However, compared to the standard β-phase salts, the unit cell is twice as large, and some tetramerization along the stack direction (**a+b**) is noted.

The differences between these packing types are related to the anion packing, see Figure 7. The strongest intermolecular interaction is the hydrogen bonding between the sulfonate oxygen atoms and the water protons. This interaction is so strong that even minute amounts of water present in the form as solvent and glass surface contamination is incorporated into the crystal structure. Thus, dimers of anion with two water bridges are formed, similar to those found in the monohydrates of carboxylate salts, involving O–H···O hydrogen bonds. Adjacent dimers line up in a tilted way into ribbons by the formation of much weaker hydrogen bonds involving C–H···O interactions with the benzylic and aromatic hydrogen atoms. This leaves the hydrophobic benzene rings

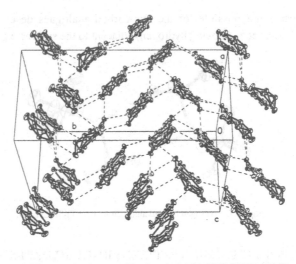

Figure 6. κ4×4-type ET molecular layer packing in κ4×4-(ET)₂(C₆H₅CH₂SO₃)(H₂O). The layer plane is parallel to *bc*, but this perspective view is oblique and approximately along the long molecular axis. Compare with Figure 4 for a schematic diagram of standard κ-packing. Most of the ethylene groups are conformationally disordered at room temperature.

sticking out of the side of the ribbons on alternating sides. Now there are two ways of interlocking the ribbons: one in parallel and the other with flipped neighbors. These two patterns are observed in the β- and κ-phases, respectively.

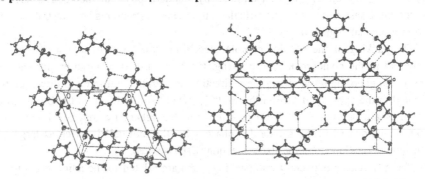

Figure 7. Anion network in β- and κ4×4-(ET)₂(C₆H₅CH₂SO₃)(H₂O). The water-bridged anion dimer is highlighted with shading in each case. Hydrogen bonds are indicated by dashed lines.

Coincidentally, the dimensions of the anion dimer, ca. 11.5 Å × 5.4 Å match the foot print of the ET tetramer. In each case, there is a center of inversion in the middle of the tetramer, thus the inner ET molecules are equivalent to each other, and the outer ones correspondingly. The ET-hydrogen-to-anion interactions of the inner ET molecules are primarily to the sulfonate and water oxygen atoms, whereas those of the outer ET molecules are with the benzene rings of the anion. This example shows that hydrogen bond building principles may be utilized to form the anion layer, which then in turn dictates the packing of the ET molecule conducting layer.

7. Conclusions

We have seen that many different types of intermolecular interactions in organic conducting cation radical salts. Hydrogen bonding between the donor molecules and the anions is weak but not negligible. The ionic Madelung energy is insufficient to completely intersperse anions and cations, thus the layers favored by the van der Waals interactions remain intact. The search for new conducting and superconducting salts has been mainly by trial-and-error methods, even though simple substitutions have been employed in order to obtain isostructural analogs of successful (e.g., superconducting) salts. However, even seemingly minor substitutions sometimes destroy the packing type, and different crystal structures result. Simulations with the aim at predicting crystal structures have not succeeded, mainly because the different interaction types are of comparable energy, and the delocalized and partial charges render the calculations of the ionic terms extremely unreliable. Clearly, the development of suitable crystal modeling techniques with predictive capabilities is one of the great needs of the field.

8. Acknowledgments

I especially wish to thank Jack M. Williams who has led the organic conductors effort at Argonne National Laboratory for many years, and my coworkers in the team: Aravinda M. Kini, John A. Schlueter, H. Hau Wang, James P. Parakka, and Brian H. Ward. Many aspects of this work are based on valuable discussion and collaboration with M.-H. (Mike) Whangbo of North Carolina State University. Work at Argonne National Laboratory is sponsored by the U. S. Department of Energy, Office of Basic Energy Sciences, Division of Materials Sciences, under Contract W-31-109-ENG-38. The Division of Educational Programs at Argonne National Laboratory has supported a large number of undergraduate student interns in our program.

9. References

1. McCoy, H. N. and Moore, W. C. (1911) Organic amalgams: Substances with metallic properties composed in part of non-metallic elements, *J. Am. Chem. Soc.*, 33, 273–292.
2. Akamatu, H., Inokuchi, H., and Matsunaga, Y. (1954) Electrical conductivity of the perylene–bromine complex, *Nature (London)*, 173, 168–169.
3. Little, W. A. (1964) Possibility of synthesizing an organic superconductor, *Phys. Rev.*, 134, A1416–1424.

292

4. Acker, D. S., Harder, R. J., Hertler, W. R., Mahler, W., Melby, L. R., Benson, R. E., and Mochel, W. E. (1960) 7,7,8,8-Tetracyanoquinodimethane and its electrically conducting anion-radical derivatives, *J. Am. Chem. Soc.*, **82**, 6408–6409.

5. Ferraris, J., Cowan, D. O., Walatka, V., Jr., and Perlstein, J. H. (1973) Electron transfer in a new highly conducting donor-acceptor complex, *J. Am. Chem. Soc.*, **95**, 948–949.

6. Coleman, L. B., Cohen, M. J., Sandman, D. J., Yamagishi, F. G., Garito, A. F., and Heeger, A. J. (1973) Superconducting fluctuations and the Peierls instability in an organic solid, *Solid State Commun.*, **12**, 1125–1132.

7. Jérome, D., Mazaud, A., Ribault, M., and Bechgaard, K. (1980) Superconductivity in a synthetic organic conductor $(TMTSF)_2PF_6$, *J. Phys., Lett. (Orsay, Fr.)*, **41**, L95–98.

8. Mizuno, M., Garito, A. F., and Cava, M. P. (1978) 'Organic metals': Alkylthio substitution effects in tetrathiafulvalene-tetracyanoquinodimethane charge-transfer complexes, *J. Chem. Soc., Chem. Commun.*, 18–19.

9. Kini, A. M., Geiser, U., Wang, H. H., Carlson, K. D., Williams, J. M., Kwok, W. K., Vandervoort, K. G., Thompson, J. E., Stupka, D. L., Jung, D., and Whangbo, M.-H. (1990) A new ambient pressure organic superconductor, κ-$(ET)_2Cu[N(CN)_2]Br$, with the highest transition temperature yet observed (inductive onset T_c = 11.6 K, resistive onset = 12.5 K), *Inorg. Chem.*, **29**, 2555–2557.

10. Williams, J. M., Kini, A. M., Wang, H. H., Carlson, K. D., Geiser, U., Montgomery, L. K., Pyrka, G. J., Watkins, D. M., Kommers, J. M., Boryschuk, S. J., Strieby Crouch, A. V., Kwok, W. K., Schirber, J. E., Overmyer, D. L., Jung, D., and Whangbo, M.-H. (1990) From semiconductor–semiconductor transition (42 K) to the highest-T_c organic superconductor, κ-$(ET)_2Cu[N(CN)_2]Cl$ (T_c = 12.5 K), *Inorg. Chem.*, **29**, 3272–3274.

11. Williams, J. M., Ferraro, J. R., Thorn, R. J., Carlson, K. D., Geiser, U., Wang, H. H., Kini, A. M., and Whangbo, M.-H. (1992) Organic Superconductors (Including Fullerenes): Synthesis, Structure, Properties and Theory, Prentice Hall, New Jersey.

12. Ishiguro, T., Yamaji, K., and Saito, G. (1998) Organic Superconductors, *Springer Series in Solid-State Sciences*, **88**, Springer-Verlag, Berlin, Heidelberg, New York.

13. Williams, J. M., Wang, H. H., Emge, T. J., Geiser, U., Beno, M. A., Leung, P. C. W., Carlson, K. D., Thorn, R. J., Schultz, A. J., and Whangbo, M.-H. (1987) Rational design of synthetic metal superconductors, in Lippard, S. J. (ed. *Prog. Inorg. Chem.*, **35**, John Wiley & Sons, Inc., New York, p. 51–218.

14. Yagubskii, É. B., Shchegolev, I. F., Laukhin, V. N., Kononovich, P. A., Kartsovnik, M. V., Zvarykina, A. V., and Buravov, L. I. (1984) Normal-pressure superconductivity in an organic metal (BEDT-TTF)$_2I_3$ [bis(ethylenedithiolo)tetrathiofulvalene triiodide], *Pis'ma Zh. Eksp. Teor. Fiz.*, **39**, 12–15 (Engl. Transl. *JETP Lett.*, **39**, 12).

15. Williams, J. M., Wang, H. H., Beno, M. A., Emge, T. J., Sowa, L. M., Copps, P. T., Behroozi, F., Hall, L. N., Carlson, K. D., and Crabtree, G. W. (1984) Ambient-pressure superconductivity at 2.7 K and higher temperatures in derivatives of (BEDT-TTF)$_2$IBr$_2$: Synthesis, structure, and detection of superconductivity, *Inorg. Chem.*, **23**, 3839–3841.

16. Wang, H. H., Beno, M. A., Geiser, U., Firestone, M. A., Webb, K. S., Nuñez, L., Crabtree, G. W., Carlson, K. D., Williams, J. M., Azevedo, L. J., Kwak, J. F., and Schirber, J. E. (1985) Ambient-pressure superconductivity at the highest temperature (5 K) observed in an organic system: β-(BEDT-TTF)$_2$AuI$_2$, *Inorg. Chem.*, **24**, 2465–2466.

17. Merzhanov, V. A., Kostyuchenko, E. É., Laukhin, V. N., Lobkovskaya, R. M., Makova, M. K., Shibaeva, R. P., Shchegolev, I. F., and Yagubskii, É. B. (1985) An increase in the superconducting transition temperature of β-(BEDT-TTF)$_2I_3$ to 6–7 K at a normal pressure, *Pis'ma Zh. Eksp. Teor. Fiz.*, **41**, 146–148 (Engl. Transl. *JETP Lett.*, **41**, 179).

18. Murata, K., Tokumoto, M., Bando, H., Tanino, H., Anzai, H., Kinoshita, N., Kajimura, K., Saito, G., and Ishiguro, T. (1985) High T_c superconducting state in (BEDT-TTF)$_2$ trihalides, *Physica B+C (Amsterdam)*, **135**, 515–519.

19. Leung, P. C. W., Emge, T. J., Beno, M. A., Wang, H. H., Williams, J. M., Petricek, V., and Coppens, P. (1984) Novel structural modulation in the first ambient-pressure sulfur-based organic superconductor (BEDT-TTF)$_2I_3$, *J. Am. Chem. Soc.*, **106**, 7644–7646.

20. Emge, T. J., Wang, H. H., Beno, M. A., Leung, P. C. W., Firestone, M. A., Jenkins, H. C., Carlson, K. D., Williams, J. M., Venturini, E. L., Azevedo, L. J., and Schirber, J. E. (1985) A test of superconductivity vs. molecular disorder in (BEDT-TTF)$_2$X synthetic metals: Synthesis, structure (298, 120 K), and microwave/ESR conductivity of (BEDT-TTF)$_2$I$_2$Br, *Inorg. Chem.*, **24**, 1736–1738.

21. Bardeen, J., Cooper, L. N., and Schrieffer, J. R. (1957) Microscopic theory of superconductivity, *Phys. Rev.*, **106**, 162–164.

22. Bardeen, J., Cooper, L. N., and Schrieffer, J. R. (1957) Theory of superconductivity, *Phys. Rev.*, **108**, 1175–1204.

23. McMillan, W. L. (1968) Transition temperature of strong-coupled superconductors, *Phys. Rev.*, **167**, 331–344.

24. Urayama, H., Yamochi, H., Saito, G., Nozawa, K., Sugano, T., Kinoshita, M., Sato, S., Oshima, K., Kawamoto, A., and Tanaka, J. (1988) A new ambient pressure organic superconductor based on BEDT-TTF with T_c higher than 10 K (T_c = 10.4 K), *Chem. Lett.*, 55–58.

25. Welp, U., Fleshler, S., Kwok, W. K., Crabtree, G. W., Carlson, K. D., Wang, H. H., Geiser, U., Williams, J. M., and Hitsman, V. M. (1992) Weak ferromagnetism in κ-(ET)$_2$Cu[N(CN)$_2$]Cl, where ET is bis(ethylenedithio)tetrathiafulvalene, *Phys. Rev. Lett.*, **69**, 840–843.

26. Sushko, Y. V., Murata, K., Ito, H., Ishiguro, T., and Saito, G. (1995) κ-(BEDT-TTF)$_2$Cu[N(CN)$_2$]Cl: Magnet and superconductor. High pressure and high magnetic field experiments, *Synth. Met.*, **70**, 907–910.

27. Schlueter, J. A., Geiser, U., Kini, A. M., Wang, H. H., Williams, J. M., Naumann, D., Roy, T., Hoge, B., and Eujen, R. (1999) Trifluoromethylmetallate anions as components of molecular charge transfer salts and superconductors, *Coord. Chem. Rev.*, in press.

28. Schlueter, J. A., Carlson, K. D., Geiser, U., Wang, H. H., Williams, J. M., Kwok, W.-K., Fendrich, J. A., Welp, U., Keane, P. M., Dudek, J. D., Komosa, A. S., Naumann, D., Roy, T., Schirber, J. E., Bayless, W. R., and Dodrill, B. (1994) Superconductivity up to 11.1 K in three solvated salts composed of [Ag(CF$_3$)$_4$]$^-$ and the organic electron-donor molecule bis(ethylenedithio)tetrathiafulvalene, *Physica (Amsterdam)*, **C233**, 379–386.

29. Kurmoo, M., Graham, A. W., Day, P., Coles, S. J., Hursthouse, M. B., Caulfield, J. L., Singleton, J., Pratt, F. L., Hayes, W., Ducasse, L., and Guionneau, P. (1995) Superconducting and semiconducting magnetic charge transfer salts: (BEDT-TTF)$_4$AFe(C$_2$O$_4$)$_3$·C$_6$H$_5$CN (A = H$_2$O, K, NH$_4$), *J. Am. Chem. Soc.*, **117**, 12209–12217.

30. Geiser, U., Schlueter, J. A., Wang, H. H., Kini, A. M., Williams, J. M., Sche, P. P., Zakowicz, H. I., Vanzile, M. L., Dudek, J. D., Nixon, P. G., Winter, R. W., Gard, G. L., Ren, J., and Whangbo, M.-H. (1996) Superconductivity at 5.2 K in an electron donor radical salt of bis(ethylenedithio)tetrathiafulvalene (BEDT-TTF) with the novel polyfluorinated organic anion SF$_5$CH$_2$CF$_2$SO$_3$$^-$, *J. Am. Chem. Soc.*, **118**, 9996–9997.

31. Mori, T. (1998) Structural genealogy of BEDT-TTF-based organic conductors I. Parallel molecules: β and β" phases, *Bull. Chem. Soc. Jpn.*, **71**, 2509–2526.

32. Mori, T., Mori, H., and Tanaka, S. (1999) Structural genealogy of BEDT-TTF-based organic conductors II. Inclined molecules: θ, α, and κ phases, *Bull. Chem. Soc. Jpn.*, **72**, 179–197.

33. Bondi, A. (1964) Van der Waals volumes and radii, *J. Phys. Chem.*, **68**, 441–451.

34. Oshima, K., Urayama, H., Yamochi, H., and Saito, G. (1988) Superconductivity and Deuteration Effect in (BEDT-TTF)$_2$Cu(NCS)$_2$, *Synth. Met.*, **27**, A473–A478.

35. Schlueter, J. A., Ward, B. H., Geiser, U., Wang, H. H., Kini, A. M., Parakka, J. P., Morales, E., Kelly, M. E., Koo, H.-J., Whangbo, M.-H., Nixon, P. G., Winter, R. G., and Gard, G. L. (1999) Crystal structure, physical properties and electronic structure of a new organic conductor: β"-(BEDT-TTF)$_2$SF$_5$CHFCF$_2$SO$_3$, submitted for publication in *Chem. Mater.*

36. Ward, B. H., Schlueter, J. A., Geiser, U., Wang, H. H., Morales, E., Parakka, J. P., Thomas, S. Y., Williams, J. M., Nixon, P. G., Winter, R. W., Gard, G. L., Koo, H.-J., and Whangbo, M.-H. (1999) Comparison of the crystal and electronic structures of three 2:1 salts of the organic donor molecule BEDT-TTF with pentafluorothiomethylsulfonate anions SF$_5$CH$_2$SO$_3$$^-$, SF$_5$CHFSO$_3$$^-$ and SF$_5$CF$_2$SO$_3$$^-$, submitted for publication in *Chem. Mater.*

294

37. Wang, H. H., Geiser, U., O'Malley, J. L., Ward, B. H., Morales, E., Kini, A. M., Parakka, J. P., Koo, H.-J., and Whangbo, M.-H. (1999) Unpublished results.

THEORETICAL PREDICTION OF CRYSTAL STRUCTURES OF RIGID ORGANIC MOLECULES

D. E. Williams
University of Louisville
Department of Chemistry
Louisville, Kentucky 40292

Introduction

In principle, when dealing with molecules of known molecular structure and known intermolecular force field, it should be possible to theoretically predict observed crystal structures by energy minimization. In practice, these predictions are difficult. The principal reason is that the portion of the energy hypersurface with downhill access to the global energy minimum is usually small; all other portions of the energy hypersurface lead downhill to a multitude of local minima. Obviously, these problems can be worsened if there are any inaccuracies in either the molecular structure or the intermolecular force field.

Molecular structure

The normal procedures used to find optimum structure(s) of an isolated molecule for input to crystal structure prediction are quantum mechanics, use of standard bond distances and angles, or by analogy to a similar molecule. In this work, we simplify our task by holding the molecular structure rigid during attempts to predict crystal structure. Obviously this procedure will not suffice if the molecule is flexible or is significantly affected by packing forces in the crystal, but addition of molecular flexibility complicates the problem of crystal structure prediction and further requires an intramolecular force field. A molecule may have several conformers of nearly equal energy, in which case each conformer can be tried separately. For our testing purposes molecular structures were taken from observed crystal structures, so any effects of packing on the molecular structure are automatically included. For crystal structures determined by x-ray diffraction, hydrogen atoms must be repositioned by lengthening the X-H bond to standard values.

D. Braga et al. (eds.), Crystal Engineering: From Molecules and Crystals to Materials, 295–310.

Intermolecular force fields

There are many intermolecular force fields available from the literature, of varying quality and applicability. When using force fields it is important to have compatible or self-consistent parameters. It is not a good idea to mix force fields derived from different sources.

In our tests we use the high quality W99 hydrocarbon intermolecular force field.[1] W99 was derived from crystal structures of 101 hydrocarbon molecules, with further tests on an additional 33 hydrocarbons. This force field is of the (exp-6-1) type which uses an exponential exchange repulsion, inverse sixth power dispersion attraction, and coulombic point charges. Coulombic point charges were obtained by fitting the molecular electric potential around the molecules found from the 6-31g** wavefunction. The W99 force field adds additional methylene bisector charge sites to the molecule.[2,3]

P1 crystal structure prediction method

In the P1 approach to ab initio crystal structure prediction no assumptions are made with regard to the space group of the crystal, or the position and orientation of the molecule in the unit cell. The most significant variable then is the number of molecules per cell, Z. Each molecule may translate and rotate in a triclinic cell in space group P1 (i.e., no symmetry other than a lattice). The energy hypersurface is then specified by $6Z+3$ variables: 6 cell constants, 3 rotations of a reference molecule which defines the origin of the cell, and 6 rotations and translations of the remaining $Z-1$ molecules.

This approach is very practical if $Z=1$; then there are only 9 variables specifying the energy hypersurface. No translational variables are needed, since space group P1 is polar in all three directions. One can begin by placing a reference molecule in random rotational orientation in a large cell, as if the substance were a dense ordered gas. Alternatively, one can carry out calculations on a rotational grid to be sure that all possible beginning rotational orientations are tried. Under energy minimization the cell will quickly decrease in volume, and molecules in surrounding cells will contact the reference molecule. The first plot below shows crystal energy (kJ mol^{-1}) of hexane versus refinement cycle. The slightly positive energy initially is an artifact of the accelerated convergence procedure. The energy decreases smoothly to the global minimum. The second plot shows the values of the three cell edge lengths (A) starting from a very large 30x30x30A cubic cell. The third plot shows values of the three cell angles; they were held to $90°$ until cycle 34. After cycle 34 frequent changes to a new reduced cell causes jumpy behavior of the cell constants. The calculation for a particular rotational grid point required 96 sec on a SGI Indigo2 computer. A scan of 26 uniformly spaced rotational grid points required 47 min; and the global minimum was achieved starting from 18 of these points. This example

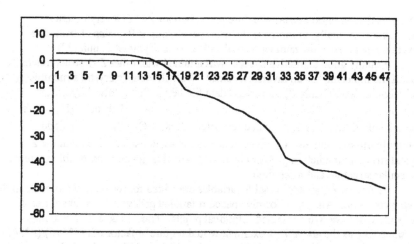

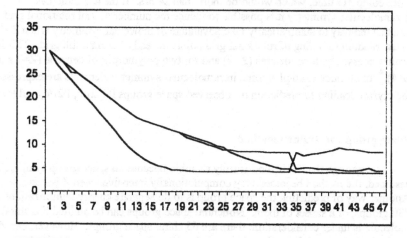

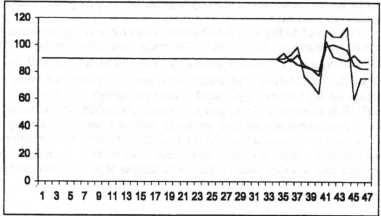

illustrates that ab initio crystal structure prediction for Z=1 is a very practical undertaking. Actually, in this case the true space group is P1_ since hexane has inversion symmetry; the number of molecules in the asymmetric unit (Z') is ½.

Equally good predictive success was attained with other Z=1 structures in P1, at least if the molecule has an inversion center. We will identify crystal structures by their six-character Cambridge structure data file label. Some solved hydrocarbon examples, using the W99 force field, are papvad (tetramethylethene, C_6H_{12}), bchxen (bicyclohexylidene, $C_{12}H_{20}$), cytdec (cyclotetradecane, $C_{14}H_{28}$)., and npdcbu (naptho[b,e]dicyclobutane, $C_{14}H_{12}$). It seems reasonable to conclude that the Z=1 crystal structure prediction problem is usually solvable, given a reasonably accurate molecular structure and force field.

If Z=2 in P1, an additional 6 variables are added for the second molecule in the asymmetric unit. Also, a uniformly spaced rotational grid with 26 points for each molecule will have 676 combined rotational points. Thus, to do a Z=2 calculation in the same way as the above Z=1 example would require a factor of (15*676)/(9*26) more computer time, which would be more than 24 hrs. If the molecule has intramolecular symmetry it is possible to reduce the number of trial rotational grid points. One way to automatically take advantage of molecular symmetry when Z>1 is to use random sampling of rotational grid points instead of using a full grid. This has been successfully done for urea (Z=2) and for two polymorphs of benzene (Z=2 and Z=4).[4] In all three examples some intramolecular symmetry elements are expressed in the crystal, leading to prediction of observed space groups P4_21m, P21/c, and Pbca.

Space group constrained method

Although the P1 method is truly ab initio because no space group symmetry is assumed, the method becomes very computationally intensive when Z is large. Another way of predicting the space group is to do separate energy minimizations in all 230 groups, or a subset of them. Nonchiral space groups can be omitted if a chiral molecule is under consideration; there are 65 chiral space groups. However, carrying out lattice energy minimizations in a very large number of space groups is also computationally intensive.

One way of dealing with this situation is to try lattice energy minimization only in popular space groups. Analysis of the Cambridge crystal structure data file shows that 93% of organic crystals are found in only 18 popular space groups; P21/c is the most popular.[5] An obvious disadvantage of this approach is that about 7% of crystal structures will not be correctly predicted. The above example of urea, which crystallizes in an unpopular space group, illustrates this problem. Nevertheless, crystal structure prediction in popular space groups has become a mainstream activity.

It must be recognized that, if Z is known, there are important restrictions on the number of space groups. If Z can be determined experimentally, for example from x-ray powder data, it is very valuable because the number of possible molecular packing groups is drastically reduced.[6] The following table shows the number of molecular packing groups possible for a given Z value. For instance, if Z=2, only the five space groups P1_, P2, P2₁, Pm, or Pc are possible. On the other hand, there are 63 space groups of order 8 (Z=8) so less simplification occurs. Nevertheless, it is very valuable to know the Z value before a crystal structure prediction is attempted.

Z	Number of groups	Z	Number of groups
1	1	16	41
2	5	18	4
3	3	24	15
4	26	32	6
6	18	36	2
8	63	48	11
9	1	96	8
12	22	192	4

In making up a trial model for the space group constrained method, the molecule comprising the asymmetric unit must be placed away from point symmetry elements such as inversion centers, rotation axes, and mirrors, because then it would rotate or reflect back onto itself. Space operations such as screw axes and glide planes do not require this restriction. Using standard settings for space groups, one often finds that ($\frac{1}{4}$,$\frac{1}{4}$,$\frac{1}{4}$) is a convenient starting location for the molecular center, which is distant from point symmetry operators. Space groups may be polar in one or more directions; molecular translation along a polar axis is not meaningful and must not be selected as a predictive variable.

Comparison of crystal structure prediction programs

Available computer programs for crystal structure prediction can be classified according to certain characteristics. Here is the listing of programs given by Verwer and Leusen,[7] modified to show the current status of mpa.. The mpa program will be discussed in detail below.

Program	a	b	c	d	e	f	g	h	i
mpa	S/R	Y	N	FF	Y	Y	Y	Y	U
ice9	S	Y	N	FF	MP	N	N	Y	1
flexcryst	S	N	N	S	N	N	Y	Y	1
mdcp	MD	Y	N	FF	Y	Y	N	N	4
molpak	S	Y	N	FF	N+Y	N	N	Y	1
msipp	MC	Y	Y	FF	Y	Y	Y	Y	U
promet	S	Y	N	FF	N+Y	Y	Y	Y	4
crysca	R	Y	N	FF	Y	N	N	Y	U
upack	S	Y	Y	FF	Y	N	Y	Y	1

a. Search type: systematic/random/MC/MD
b. Energy minimization: Y/N
c. MM included: Y/N
d. Energy: force field/scoring function
e. Electrostatics: Y/N; N+Y=in final
f. Accelerated convergence: Y/N
g. Clusters/docking: Y/N
h. Symmetry: Y/N
i. Maximum Z or Z' value/unlimited

Description of mpa/mpg[8]

Input files for mpa (molecular packing analysis) and mpg (molecular packing graphics) are the same, so that the structure being considered by mpa can easily be displayed by mpg. In contrast to many *molecular* display programs, mpg is specifically designed to display *crystal structures* and *molecular clusters* to best advantage. As indicated above, mpa is designed to find minimum energy structures for crystals or molecular clusters; the treatment of molecular clusters includes molecular docking as a special type of cluster. Mpa also includes the possibility of annealing the structures.

Input files for mpa are organized according to columns, and the sequence of input lines is organized by a control line. Description of the input lines will indicate how to use the program.

1. Title line(s). These allow the user to identify the structure and calculation protocol; they are only transferred to output with no action taken.

2. The control line specifies the number of atoms, number of space group symmetry operations, whether a cell-to-cartesian transformation is needed, initial rotation information, electrostatics off/on, maximum number of contact tables, bond

foreshortening information, agitation off/on, print full/partial, reduced cell off/on, initial cell edge refinement off/on, and number of Lattman grid sections.

There are several possibilities for the initial rotation control variable, NROT:

0   no initial rotation or translation
1   initial rotation about a specified center, followed by a specified translation
2   random rotational orientation
3   1 followed by 2
4   rotate to Lattman angle orientation
5   1 followed by 4
6   evaluate energies on a Lattman angle grid
7   1 followed by 6

The spacing of the Lattman rotation angle grid[9] is controlled by the number of sections:

3 sections=26 points, spacing 83 deg
4 sections=72 points, spacing 58 deg
5 sections=184 points, spacing 42 deg
6 sections=284 points, spacing 34 deg
7 sections=536 points, spacing 29 deg
9 sections=1280 points, spacing 21 deg

All first (gradient) and second (hessian) derivatives of the lattice energy are evaluated analytically by mpa. If the hessian is not positive definite OREM[10] energy minimization is selected. Otherwise, a modified Newton-Raphson (NR) minimization is used. If any difficulties are encountered with either OREM or NR, steepest descents (SD) is selected as a backup minimization procedure. If the starting model is an observed structure, the hessian will almost always be positive definite.

The contact table is a list of nonbonded atom pairs separated by distance r up to the summation limit. It is necessary to maintain a contact table because inclusion of new contacts or deletion of old contacts in refinement cycles causes convergence problems. The mpa protocol is to do three refinement cycles with a given contact table, checking for convergence, before a new table is made up. Starting from an observed structure, an energy minimum will usually be reached with one to three contact tables. For ab initio crystal structure prediction calculations starting from a random model 50 or more contact tables may be required.

During an ab initio crystal structure prediction a molecule may translate so that its center of mass lies outside the cell; mpa repositions the molecule with the appropriate lattice vectors so that it remains in the cell. Cell reduction (to the primitive conventional cell) may be selected during the calculation and it is generally very important for triclinic and monoclinic lattices. Normally, cell reduction is not necessary for lattices of higher symmetry. A primitive cell reduction is always done by mpa at the conclusion of a calculation; this makes it easier to compare apparently different sets of cell constants.

3. Atomic parameter lines. These specify the xyz of the atoms or sites (either cell or cartesian), net charge, molecule number (if there is more than one molecule in the asymmetric unit), group number (for specifying electrically neutral groups of atoms), force field reference, and subrotation reference.

4. Symmetry lines. P1 is the default space group; any other space group may be specified, either in the International Tables setting or in a custom setting.

5. Cell constants. These can be set at observed values, or they can be set at trial values for structure prediction runs.

6. Initial 3x3 rotation matrix (or matrices), if called for by NROT.

7. Initial rotation center, if called for by NROT.

8. Initial molecular translation, if called for by NROT.

9. Initial Lattman angles, if called for by NROT.

10. Subrotation information, if called for by atomic parameter lines; these specify the 4 atoms involved in a torsion, and the corresponding torsion potential.

11. Calculation conditions 1. Energy decrease exit criterion, summation limit, accelerated convergence data, reciprocal lattice summation limit, scale factor for net charges, and scale factor for hydrogen repulsion.

12. Calculation conditions 2. Option to use inertial coordinates, option to do neutral group summation, output reciprocal lattice terms off/on, list nonbonded contacts off/on, list full or blocked hessian off/on, cluster calculation off/on, number of negative eigenvalues to be considered, maximum number of nonbonded contacts, adjustment of van der Waals radii for nonbonded contact printout, maximum allowed parameter shift.

13. Calculation conditions 3. Read only for an annealing calculation. Initial effective temperature, decrement, maximum agitation amplitude. When energy is minimized with annealing the effective temperature is allowed to decrease to a set point where energy minimization is interrupted. The effective temperature is then increased by agitation of the structure and refinement is resumed. The process is iterated with lower set points and smaller agitations until the effective temperature is zero. Since annealing extends the length of a refinement, it is very demanding of computer power.

14. Parameter control lines. These lines contain integers which specify which parameters are to be varied. Parameters are ordered in the sequence: 6 cell constants, 3 rotations of molecule 1, 3 translations of molecule 1, 3 rotations of molecule 2, 3 translations of molecule 2, and so on.

15. Dependent parameter control lines. These line specify, for instance, that when refining in a tetragonal space group two cell edges must be kept equal.

Potential parameter file

The geometric mean combining law for heteroatomic interaction is assumed by mpa. Therefore this file needs to contain only homoatomic potentials. The potentials are listed first according to atomic number, then according to a particular potential type for this element. For instance, potential 6, type 28, could be a Dreiding (exp-6) potential for carbon; potential 7, type 14 could be a Biosym (12-6) potential for nitrogen. Atomic numbers are deduced from atomic labels, e.g., atom label Br23 is deduced to be element 35. The potential type is listed explicitly on the atomic parameter line; (exp-6) and (n-6) potentials cannot be mixed for the same structure. The potential parameter file may be edited by the user to include new potentials or to modify existing ones. All energies in mpa are in units of kJ/mol and all distances are in A.

Accelerated convergence

The idea of accelerated convergence[11] is to transform part of the direct lattice sum into reciprocal space via a Fourier series. Although mpa can include reciprocal lattice sums for the energy, it does not include contributions from reciprocal space terms to the first and second derivatives. In normal operation of mpa the reciprocal lattice sum is omitted by making it negligibly small. This is done by adjusting convergence parameters CONV1 and CONV6 so that the reciprocal sum RSUM is less than the desired error. Thus, if one wishes a lattice sum accuracy to 0.1%, the neglected reciprocal sum should contribute no more than that amount. Considerable benefit, it terms of improved accuracy at a given summation limit, is retained even if the reciprocal sum is negligibly small. For instance, consider the following lattice sum results for benzene with CONV1=0.125 and CONV6=0.15, and a 12A summation limit:

Converged values:	E1=-12.6675	E6=-77.8664
Without reciprocal sum	E1=-12.6698	E6=-77.8663
Without acceleration:	E1=-12.6371	E6=-77.2413

Thus the electrostatic sum (E1) is accurate to 0.0023 and the dispersion sum (E6) is accurate to 0.0001 without including any reciprocal lattice terms. If accelerated convergence is not used, E1 and E6 have errors of 0.0304 and 0.6251, respectively. Paradoxically, the error in E6 is larger than that of E1; this no doubt occurs because of the strong restriction for E1 that the sum of charges be zero and there is no analogous restriction for E6.

Of course, lattice sum errors will be larger if a smaller summation limit is used. With an 8A summation limit we commonly use in crystal structure prediction runs, E1=-12.8750 and E6=-77.8453. The energy decrease parameter, ZEND, should be assigned according to the accuracy desired for location of the energy minimum. Values of 0.001 to 0.00001 kJ/mol are usually appropriate, depending on the summation limit. Often it will be found that a summation limit of 8A is sufficient for practical calculations.

Relaxed crystal structures

A relaxed crystal structure is obtained by energy minimization starting from the observed structure, using a particular force field. For a hypothetically perfect force field, there should be no change in the structure. Real force fields will often give shifts of around 3% in the cell edge lengths, a degree or so in cell angles, several degrees in molecular rotation, and several tenths of A in molecular translation. The magnitudes of relaxation shifts are good indicators of the quality of the force field. Relaxed lattice constants are expected to be accurate only within the range of thermal expansion, since thermal effects are not explicitly included in the model. However, force fields do incorporate averaged thermal effects from their training data.

Here is an example comparing the observed and relaxed structures of hexane, using the W99 force field., where the maximum cell edge shift is -2.1%. Considering that the model does not explicitly consider thermal effects, the fit is excellent.

Observed cell	4.17	4.70	8.57	96.6	87.2	105.0
Relaxed cell	4.18	4.60	8.58	97.0	88.0	104.1
Molecular rotation	2.9					
Molecular translation	0.0					

Prediction of Z=2 crystal structures

As noted above, when Z=2 only 5 space groups are possible, so a reasonable procedure is to find minimum energy structures in these 5 groups: P1_, P2, P2₁, Pm, and Pc. It can be said in advance that P2 and Pm are unpopular space groups, usually only occurring if the molecule has intramolecular twofold or mirror symmetry which is expressed in the crystal. If the molecule is centrosymmetric, it may crystallize in molecular packing group $P2_1$ or Pc, in which case expression of molecular centrosymmetry yields the true space group $P2_1/c$.

Crystal structure fixwak (4,5-Dimethyl-9,10-dihydrophenanthrene, $C_{16}H_{16}$) provides an example of $Z=2$, $Z'=1$. The following table shows energies of predicted structures in the five possible space groups. A three section Lattman angle grid was used (26 points). The computer time required for scanning the five space groups was 6 hr.

Space group	Energy
P1_	-95.56
$P2_1$	-85.52
Pc	-81.71
P2	-65.92
Pm	-59.33

The correct space group and structure in P1_ is predicted by a wide margin. Note that space groups P2 and Pm are not energetically favored.

Crystal structure ticjeu (sym-hexahydropyrene, $C_{16}H_{16}$) provides an example of $Z=2$, $Z'=\frac{1}{2}$, molecular symmetry 1_. The lowest energy was found in $P2_1$, -98.2134; in this structure the molecule expresses its inversion symmetry to give true space group $P2_1/c$. Finding the correct crystal structure for this molecule required the use an expanded 4-section Lattman angle grid with 72 points. Computer time needed to search all 5 possible space groups was about 9 hr. In this case the energy of a wrong space group, P1_, is within 3 kJ/mol of the correct space group. We will see below that often the energy differences between alternative space groups are even smaller than this, which adds to the challenge of ab initio crystal structure prediction.

Space group	Energy
$P2_1$	-98.21
P1_	-95.83
Pc	-95.45
P2	-89.76
Pm	-77.50

When a molecule expresses inversion symmetry in molecular packing groups $P2_1$ and Pc the resulting structures are identical, in true space group $P2_1/c$. Therefore the lowest energy structure found here in Pc must be a local minimum. The narrow spacing of energies found in different space groups suggests a possible reason for the need to use closer spaced rotation grids.

Crystal structure prediction in popular space groups

Analysis of crystal structures in the Cambridge Structure Data file shows a great range of space group popularity. Baur and Kassner[5] found that 36.59% of all reported organic crystal structures were in space group $P2_1/c$. They found that 92.71% of organic crystal structures were found in only 18 space groups. These 18 popular space groups have $Z=1, 2, 4,$ or 8; note that $Z=3, 6,$ or greater than 8 is not represented in the popular list. The $Z=1$ and $Z=2$ cases can be thoroughly examined by procedures discussed above. At the present time it appears necessary to consider $Z'=1$ structures in space groups of order four or greater by the space group specific method. However, future increases in computer speed may extend the range of the $P1$ method.

The popular space groups and their frequencies are given in the following table from Baur and Kassner. If the crystal structure contains exclusively one enantiomer of a chiral molecule, space groups with improper symmetry operations can be eliminated; these are ranks 1, 2, 4, 6, 8, 10, 14, 16, 17, and 18, leaving only 8 popular space groups for pure enantiomers.

Space group	rank	frequency(%)	Space group	rank	frequency(%)
$P2_1/c$	1	36.59	Pbcn	10	1.01
P1_	2	16.92	Cc	11	0.97
$P2_12_12_1$	3	11.00	C2	12	0.90
C2/c	4	6.95	$Pca2_1$	13	0.75
$P2_1$	5	6.35	$P2_1/m$	14	0.64
Pbca	6	4.24	$P2_12_12$	15	0.53
$Pna2_1$	7	1.63	C2/m	16	0.49
Pnma	8	1.57	P2/c	17	0.49
P1	9	1.23	R3_	18	0.46

Therefore there is a good chance of successfully predicting the crystal structure of an organic molecule by energy minimization in only 18 space groups, or only 8 space groups if one is dealing with a single enantiomer. The following example shows popular space group method results for the molecule gohwab (1,2-dihydrocyclo-butabenzene, C_8H_8), which is not chiral. The table shows lowest energy in space group P1_, which is the observed space group. The structures predicted in P1_ with three or four Lattman sections are nearly identical, and both correspond to the observed structure. One can note that several competing space groups are close in energy: the perennial favorite $P2_1/c$, and also P1 and $Pca2_1$ give a low energy. Note that no minimum energy lower than that of the observed structure was found in any of the 18 space groups; this no doubt indicates that the molecular structure and intermolecular force field are sufficiently accurate for this application. The table shows that lower energy structures in unobserved space groups may be obtained when more Lattman sections were used. The question of how many rotational grid points are needed to successfully predict a crystal structure is further considered in the next section.

| Rank | Space group | Minimum energy | |
		3 sections	4 sections
1	$P2_1/c$	-58.45	-58.43
2	$P1\_$	-61.34	-61.45
3	$P2_12_12_1$	-56.34	-56.34
4	C2/c	-54.82	-56.80
5	$P2_1$	-56.95	-56.95
6	Pbca	-56.27	-56.36
7	$Pna2_1$	-56.57	-56.91
8	Pnma	-48.98	-48.73
9	P1	-58.40	-58.40
10	Pbcn	-56.13	-56.13
11	Cc	-54.75	-54.75
12	C2	-55.21	-56.17
13	$Pca2_1$	-57.40	-58.17
14	$P2_1/m$	-48.29	-48.66
15	$P2_12_12$	-53.10	-54.80
16	C2/m	-46.97	-47.74
17	P2/c	-56.09	-56.03
18	$R3\_$	-50.90	-50.90

Tests of space group specific energy minimization

If the popular space group crystal structure prediction method is to succeed, it must be possible to proceed from an initial random structure to the observed structure with space group constraints in place. This initial structure always has Z'=1, so that Z is equal to the order of the space group. If the method is successful, the correct structure will be found among the lowest energy minima; it is preferred that the lowest energy predicted structure be the correct structure, except in the case of polymorphs. The success (or lack of success) of predicting structures this way was tested with molecules selected from training data for the accurate W99 hydrocarbon force field. This test was unbiased with regard to selection of structures, their being taken systematically from Z'=1 structures in the training set in their order of listing; no structure was eliminated because of any difficulties encountered with it.

Perhaps the most critical consideration in the space group specific method is getting a correct molecular rotational orientation which will refine to the global energy minimum. The initial molecular rotation orientation must be fairly close to the observed value; if not, the calculation will probably refine to a local energy minimum instead of the global minimum. In the following tests initial trial models were generated with the molecular rotational orientation varied systematically on a Lattman grid. The unit cell (including any symmetry constraints) was collapsed around the molecule, with allowance for the molecule to translate but not rotate. When an intermediate energy minimum was reached, then the molecular rotation constraint was released. The follow table summarizes the results obtained, including cases where ab initio prediction was not successful.

Molecule	Space group	Lattman sections	Success?
vajgoc	R3_	5	yes
gesnib	$P2_12_12_1$	4	yes, but lower minimum found
heptan01	P1_	3	yes
nadvix	$P2_1/n$	7	no
sadhua	$P2_1/c$	5	yes
gohwab	P1_	3	yes
nadweu	$P2_1/c$	6	no
sadjem	$P2_1/c$	5	no
nadvuj	$P2_12_12_1$	3	yes
bulval02	$P2_1/c$	4	yes
hayxou	P1_	4	yes
mtannl	Fdd2	6	no
nadwaq	$P2_1/c$	4	yes
vajhap	$P2_1/c$	3	yes
birpoh	$P2_1/c$	4	yes

Correct observed structures were obtained for 11 out of 15 molecules. The case of gesnib was the only one where a deeper energy minimum was found, in addition to the minimum for the observed structure. This raises suspicion that gesnib may have a polymorph. For the successful trials, a significant difference was the number of Lattman grid sections required. For four of the structures, heptan01, gohwab, nadvuj, and vajhap, only three Lattman grid sections (26 points) were required. The structures of gesnib, bulval02, hayxou, nadwaq, and birpoh each required four Lattman grid sections (72 points) to locate the observed structure. Finally, the vajgoc and sadhua structure required five sections (184 points).

Observed structures were not obtained for nadvix (up to seven Lattman grid sections, 536 points, were tried), nadweu (up to six sections), sadjem (up to five sections), and mtannl (up to 6 sections). We could see no obvious reason why calculations for these structures always ended up in local minima. As a generality, it may be hypothesized that hydrocarbon crystal structures are especially difficult to predict because the overall intermolecular force field is weak. This hypothesis could be tested by carrying out analogous calculations, for instance, in hydrogen bonded crystals where there is a strong force field and a tendency to form the maximum number of hydrogen bonds which puts severe constraints on allowed molecular orientation.

Conclusions

The P1 ab initio crystal structure prediction method is generally successful when used for rigid organic molecules crystallizing with Z=1. The P1 method may also be successful, with reasonable computer effort, in predicting structures with Z=2 or Z=4, especially if the molecule expresses some of its symmetry in the crystal. For structures with Z'<1 the P1 method can successfully predict higher space group symmetry.

The space-group specific method is generally successful for crystal structures with Z=2, since only five space groups are possible and the calculations are straightforward. For larger Z values it is regretfully necessary, mostly because of computer limitations, to carry out calculations only in popular space groups. The popular space group method involves risk of failure for those molecules which do not crystallize in a popular space group.

Systematic unbiased tests of the space-group specific method showed a success rate of eleven out of fifteen hydrocarbon crystal structures using Lattman grids with up to five sections. Perhaps this score can be improved by further fine tuning of energy minimization conditions or by inclusion of annealing steps.

310

References

1. Williams, D. E. (1999). Improved Intermolecular Force Field for Crystalline Hydrocarbons Containing Four- or Three-Coordinated Carbon, J. Struct. Chem., in press.
2. Williams, D. E. (1994). Failure of Net Atomic Charge Models to Represent the van der Waals Envelope Electric Potential of n-Alkanes, J.Comp. Chem. 15, 719-732 .
3. Williams, D. E.; Abraha, A. (1999). Site Charge Models for the Molecular Electrostatic Potentials of Cycloalkanes and Tetrahedrane, J. Comp. Chem., in press.
4. Williams, D. E. (1996). Ab Initio Molecular Packing Analysis, Acta Cryst. A52, 326-328.
5. Baur, W. H.; Kassner, D. (1992). The Perils of Cc: Comparing the Frequencies of Falsely Assigned Space Groups with their General Population, Acta Cryst. B48, 356-369.
6. Gao, D.; Williams, D. E. (1999). Molecular Packing Groups and ab Initio Crystal Structure Prediction, Acta Cryst. B, in press.
7. Verwer, P.; Leusen, F. J. J. (1998). Computer Simulation to Predict Possible Crystal Polymorphs, Rev. Comp. Chem. 12, 327-365.
8. Williams, D. E. (1999). Mpa/mpg, Molecular Packing Analysis/Molecular Packing Graphics, Department of Chemistry, University of Louisville, Louisville, Kentucky 40292. email: dew01@xray5.chem.louisville.edu; internet: www.louisville.edu/~dewill01.
9. Williams, D. E. (1973). Optimally Spaced Rotational Grid Points, Acta Cryst. A29, 408-414.
10. Williams, D. E. (1992). OREMWA Prediction of the Structure of Benzene Clusters: Transition from Subsidiary to Global Energy Minima, Chem. Phys. Let. 192, 538-543.
11. Williams, D. E. (1971). Accelerated Convergence of Crystal-Lattice Nonbonded Potential Sums, Acta Cryst. A27, 452-455.

RATIONAL DESIGN OF POLAR SOLIDS

B. MOULTON and M. J. ZAWOROTKO
The University of Winnipeg
515 Portage Ave.
Winnipeg, Manitoba
CANADA
R3B 2E9

1.   Introduction

Solids that crystallize in acentric space groups are predisposed to exhibit useful bulk physical properties in the context of new materials for electrooptical applications[1], especially devices based upon second-order nonlinear optic (NLO), piezoelectric, pyroelectric or ferroelectric activity. It should therefore be unsurprising that the pursuit of new classes of acentric, or polar, solids has been ongoing for many years. Although a crystallographic center of inversion can be precluded by building materials from homochiral components, there are two significant limitations: (1) requirements for homochiral starting materials and/or products are significant hurdles for synthetic chemists; (2) the use of homochiral building blocks does not in any way ensure optimum alignment of dipoles. The potential importance of new design strategies that are independent of the need for chiral building blocks should therefore be apparent.

Although there are examples of achiral molecules that crystallize in acentric space groups[2] (roughly 24% of crystal structures adopt non-centrosymmetric space groups[3]), it is not yet well understood how to induce supramolecular chirality from an achiral molecular environment. However, crystal engineering[4] has delineated the concept of supramolecular synthesis: new strategies for designing materials with specific topological features that ultimately control the bulk physical properties of the material. For example, diamondoid[5], helical[6] and inclusion[7] architectures have recently been investigated, and all offer insight into developing new strategies for designing polar solids. In this contribution, we present results that suggest that supramolecular host-guest systems offer significant opportunities for generation of new materials that are predisposed to preclude a center of inversion. The primary reasons for optimism are that advances in crystal engineering have laid the foundation for predictable and persistent synthesis of new host architectures and it is becoming increasingly clear that both achiral hosts and achiral guests enhance the range of components available for rational design of new polar materials.

D. Braga et al. (eds.), Crystal Engineering: From Molecules and Crystals to Materials, 311–330.
© 1999 *Kluwer Academic Publishers. Printed in the Netherlands.*

312

2.   Design Strategies

2.1   STRUCTURES BASED ON HOMOCHIRAL COMPONENTS

Clearly, any material made from homochiral building blocks is guaranteed not to have a crystallographic center of inversion. However, few design strategies are based upon efforts to crystallize pure enantiomers, primarily because molecules or ions that are targeted for electrooptical applications must also satisfy criteria that are based on chemical and thermal stability, cost and physical properties (i.e. molecular hyperpolarizability). Furthermore, the need for optimum alignment of dipoles is not addressed. This imparts a significant challenge to synthetic chemists and crystal engineers: the creation of acentric crystals from achiral molecules. Structures based on homochiral components are not the focus of this contribution and will not be discussed further.

2.2   STRUCTURES BASED ON ACHIRAL COMPONENTS

Unfortunately, achiral molecules naturally adopt acentric space groups in the solid state only 24%[3] of the time. In order to rationally generate polar materials from achiral components, it is therefore necessary to develop strategies that facilitate the self-assembly of architectures that preclude, or at the very least mitigate against, a crystallographic center of inversion. In certain situations, this can be accomplished quite straightforwardly. For example, in 1D[6,8] structures, infinite linear chains of self-complimentary molecules or helical strands are inherently acentric, whereas in 3D[5] structures, tetrahedral moieties can be used to generate acentric diamondoid structures. Each of these strategies yields, at the very least, isolated acentric networks. Unfortunately, they are still not necessarily certain to pack in a manner that generates a polar solid. A study by Frankenbach and Etter[9] provides important insight in this context: roughly 50% occurrence of acentric space groups when non-centrosymmetric adducts are present; 0% occurrence if centrosymmetric adducts exist. The study clearly details the critical importance that supramolecular adducts have on whether or not bulk polarity is present in a crystal. It should be noted that most supramolecular adducts are non-centrosymmetric even if the components are themselves symmetrical. Polar 1D, 2D and 3D structures that are based upon acentric supramolecular architectures are discussed below.

2.2.1   1D Networks

Linear Networks. Perhaps the most obvious, and consequently most broadly studied, method of inducing achiral molecules to adopt an acentric space group is the formation of 1D linear aggregates based upon the "head-to-tail" motif illustrated by carboxylic acids[10], nitroanilines[8d-e] and hydrogensulfate salts[8f,11] (Figure 1). Table 1 lists other common "heads" and "tails" used to generate self-assembling linear chains[7a].

Inherent to the strategy is the need for individual polar stands to pack parallel rather than anti-parallel. Parallel alignment means that the dipole moments of the

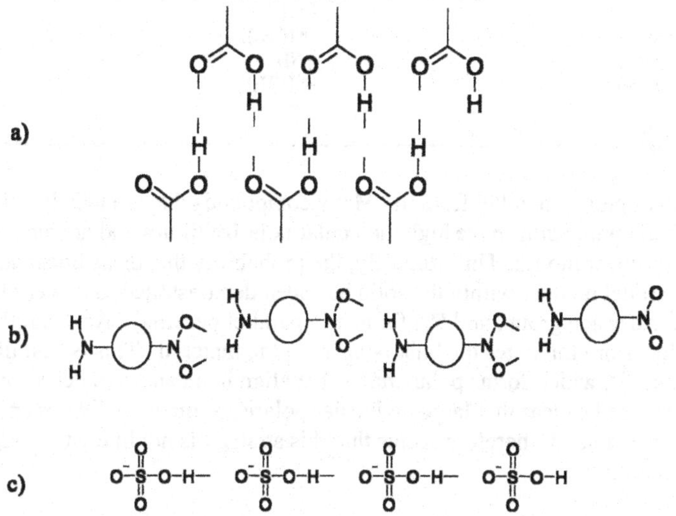

313

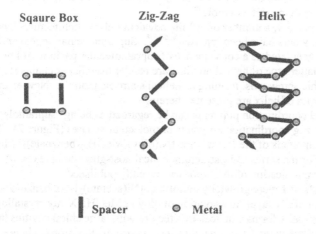

FIGURE 1. Schematic representation of the 1D polar aggregates that can result from the self-assembly of a) carboxylic acids, b) nitroaninilines and c) hydrogen sulfate anions.

Sqaure Box   Zig-Zag   Helix

▮ Spacer   ◉ Metal

FIGURE 2. Possible supramolecular isomers for compounds formed from a *cis*-octahedral or square planar metal and a linear "spacer" ligand.

TABLE 1.   Self-assembling motifs

Terminal Group 1	Terminal Group 2
$-NO_2$	$-NH_2$
$-NO_2$	$-N(CH_3)_2$
$-NO_2$	$=NH$
$-N(CN)_2$	$-N(CH_3)_2$
$-CN$	$-Br$
$=NH$	$-F$

strands also align in a parallel fashion. Many compounds that are ideally suited for electrooptical applications have high molecular polarizabilities and are amenable to formation of linear motifs. Unfortunately, the probability that these linear strands will pack in a parallel manner within the solid has been demonstrated to be very low[12] (25% for an acentric space group, and 5% for nearly parallel packing). Anti-parallel packing greatly reduces or eliminates the bulk polarity of the material. This is best illustrated by p-nitroaniline[8d-e], which forms polar chains that align in an anti-parallel manner. Indeed, it should be clear that large molecular polarity actually facilitates anti-parallel alignment of chains. It therefore seems that this strategy is not likely to work in a consistent manner.

Helical Networks. Helical architectures also provide opportunities for the development of non-centrosymmetric 1D networks since helices are inherently chiral. A number of helical architectures have been generated[6] in recent years, exemplified by discrete helical metal-ligand complexes[4b,6f,h-i] that are referred to as helicates. Helicates are viewed as prototypal examples of systems that are programmed to spontaneously assemble into helical architectures, and are considered important to the understanding of how biological systems self assemble[6b].

Most recently, a number of infinite covalent helical architectures based on coordination polymers have been reported[6a,d-e,k-l]. Supramolecular helical architectures have also been generated as a consequence of supramolecular packing of linear arrays[6c,j]. The latter type of helical architecture results from linear chains stacking along a crystallographic screw axis, forming a "helical staircase motif"[6j]. New examples of both of these types of helix are presented herein.

Helical coordination polymers can be regarded as being "supramolecular isomers" of zig-zag coordination polymers or molecular squares (Figure 2). Table 2 reveals that compounds of the formula cis-[Ni(dipy)(MeOH)$_2$(benzoate)$_2$] (dipy = 4,4'-dipyridine), 1, form a series of isostructural helical host-guest complexes. A preliminary communication of this work was recently published[6l].

Crystals of 1 were grown by dissolving [Ni(acetate)$_2$] and benzoic acid in methanol and carefully layering a solution of dipy in MeOH. X-ray crystallographic analysis reveals that 1 forms a helical architecture with large chiral cavities that can accommodate nitrobenzene, benzene, dioxane, veratrole, chloroform, phenol, methanol and p-nitroaniline. The helices are generated around crystallographic 4_3 or 4_1 screw axes and each coil of the helix therefore contains four residues. The distance between each coil corresponds to the unit cell length, which ranges from 27.02 to 27.91 Å. Examination of the crystal packing in 1 reveals edge-to-face stacking interactions between benzoate ligands from each helix and the dipy ligands on adjacent

TABLE 2. Crystallographic parameters for compounds of type 1

	1·nitrobenzene	1·benzene	1·dioxane	1·veratrole
Formula Weight	644.31	599.31	609.30	659.36
Space Group	P4₃2₁2	P4₃2₁2	P4₃2₁2	P4₃2₁2
a, Å	14.9890(8)	15.0406(9)	15.0643(8)	15.0789(6)
b, Å	14.9890(8)	15.0406(9)	15.0643(8)	15.0789(6)
c, Å	27.913(2)	27.442(2)	27.020(1)	27.167(2)
α, deg	90	90	90	90
β, deg	90	90	90	90
γ, deg	90	90	90	90
V, Å³	6271.1(6)	6207.9(6)	6131.8(5)	6177.0(5)
R, %	5.22	6.56	9.32	10.92

	1·CHCl₃	1·PhOH	1·MeOH	1·p-nitroaniline
Formula Weight	640.58	615.31	553.24	659.32
Space Group	P4₃2₁2	P4₃2₁2	P4₃2₁2	P4₃2₁2
a, Å	14.9608(9)	15.0211(8)	15.0449(5)	14.9536(6)
b, Å	14.9608(9)	15.0211(8)	15.0449(5)	14.9536(6)
c, Å	27.180(2)	27.092(1)	27.010(1)	27.164(2)
α, deg	90	90	90	90
β, deg	90	90	90	90
γ, deg	90	90	90	90
V, Å³	6083.5(6)	6112.9(6)	6113.7(4)	6074.2(5)
R, %	6.82	9.11	8.45	12.74

strands. The packing is inherently directional and thus induces spontaneous resolution of the chirality of the helices; meaning that the helices are all either left or right handed. In effect, the guest has little influence on the 3D helical superstructure. The structure of 1·nitrobenzene is illustrated in Figure 3. It reveals that large cavities are generated because adjacent helices are translated by half a unit cell. The cavities are inherently chiral and induce guest molecules to form chiral supramolecular adducts.

trans-[Ni(L)(MeOH)₂(benzoate)₂], (L = dipy, 4,4'-ethylenedipyridine (dipyeta), 2, and 4,4'-vinylenedipyridine (dipyete)), on the other hand, illustrates a prototypal example of supramolecular helical architectures resulting from the helical packing of linear aggregates. Although it should be clear that the helical architecture of 1 is a rational supramolecular isomer of the *cis*- conformation of octahedral and square planar metal moieties, that the 1D linear chains formed by the *trans*- conformation would pack in a helical array is much less obvious.

Crystals of 2 were grown according to the procedure described for 1. X-ray crystallographic analysis of 2 shows that each *trans*-Ni(benzoate)₂ is coordinated to two dipyeta ligands such that a linear chain moiety of Ni centers separated by alternating dipyeta ligands is generated. The chains align in a parallel fashion to form 2D sheets. Each Ni center is coordinated to two terminal benzoate ligands that are oriented orthogonally with respect to both the Ni-dipyeta-Ni axis and the 2D sheets. This imposes a pseudo square planar geometry at the Ni center, the plane of which is orthogonal to the plane of the 2D sheets. The helicity of 2 is not inherently present in the polymeric chains, rather it is the result of the 2D sheets stacking around a crystallographic 3₁ or 3₂ screw axis. Therefore, each chain is related to the next by a

316

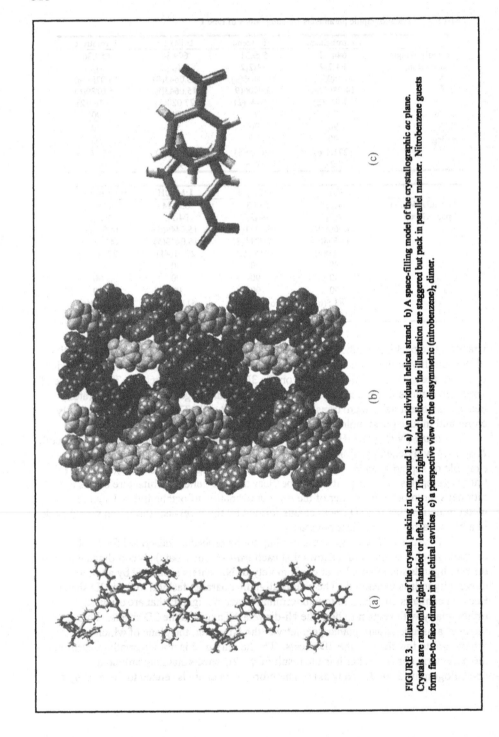

(a)   (b)   (c)

FIGURE 3. Illustrations of the crystal packing in compound 1: a) An individual helical strand. b) A space-filling model of the crystallographic *ac* plane. Crystals are randomly right-handed or left-handed. The right-handed helices in the illustration are staggered but pack in parallel manner. Nitrobenzene guests form face-to-face dimers in the chiral cavities. c) a perspective view of the dissymmetric (nitrobenzene)$_2$ dimer.

120° rotation and a separation of 4.53 Å. The helical 3D structure is directed by weak aromatic interactions between interdigitating benzoate ligands, which spontaneously resolve the stacking as either right handed or left handed and generate homochiral helices. The structure of 2 is illustrated in Figure 4.

TABLE 3. Crystallographic parameters for 2 and isostructural compounds

	[Ni(dipyeta)(MeOH)$_2$ (benzoate)$_2$]	[Ni(dipyete)(MeOH)$_2$ (benzoate)$_2$]	[Ni(dipyeta)(MeOH)$_2$ (salicylate)$_2$]
Formula Weight	549.25	547.24	581.25
Space Group	P3$_2$	P3$_1$	P3$_2$
a, Å	13.2983(6)	13.4737(7)	13.272(3)
b, Å	13.2983(6)	13.4737(7)	13.272(3)
c, Å	13.5925(9)	13.5878(10)	13.834(4)
α, deg	90	90	90
β, deg	90	90	90
γ, deg	120	120	120
V, Å^3	2081.7(2)	2136.3(2)	2110.5(8)
R, %	6.99	6.44	11.06

	[Ni(dipy)(MeOH)$_2$(m-nitrobenzoate)$_2$]	[Ni(dipy)(H$_2$O)$_4$]·(SO$_4$)(H$_2$O)(MeOH)	[Ni(dipy)(MeOH)$_2$(p-hydroxybenzoate)$_2$]
Formula Weight	611.19	433.06	553.20
Space Group	P3$_1$	P3$_1$	P3$_1$
a, Å	11.2989(5)	11.3269(7)	13.3062(5)
b, Å	11.2989(5)	11.3269(7)	13.3062(5)
c, Å	17.5432(10)	23.580(2)	14.0723(7)
α, deg	90	90	90
β, deg	90	90	90
γ, deg	120	120	120
V, Å^3	1936.6(2)	2620.0(3)	2157.8(2)
R, %	7.33	8.19	6.14

Table 3 lists the crystallographic parameters for 2 and a series of isostructural complexes. Closely related complexes have been made using dipy, dipyeta and dipyete as spacer ligands, and terminal groups including benzoate, salycilate, nitrobenzoate, and p-hydroxybenzoate. Of particular relevance is the observation that the same supramolecular helical architecture is generated regardless of the length of the spacer. It should also be noted that the helical structure is also generated if the benzoate ligands are functionalized. This is particularly surprising in the case of p-hydroxybenzoate, which is a strong H-bond donor. The latter structure reinforces the conclusion that the 3D architecture is directed by weak aromatic interactions between the interdigitating terminal ligands.

A common problem encountered in the synthesis of helical structures is that the metals and the ligands typically have no inherent chirality and left-handed and right-handed helices are often obtained in equal amounts, thereby affording non-polar, heterochiral crystals. This is not the case with the compounds described herein. However, even if homochiral helical architectures are obtained within a single crystal, the bulk material is often obtained as a racemate. It is not yet fully understood how to induce homochiral packing or spontaneously resolve homochiral crystals. Fortunately, it was recently demonstrated that seeding solutions with single homochiral crystals

318

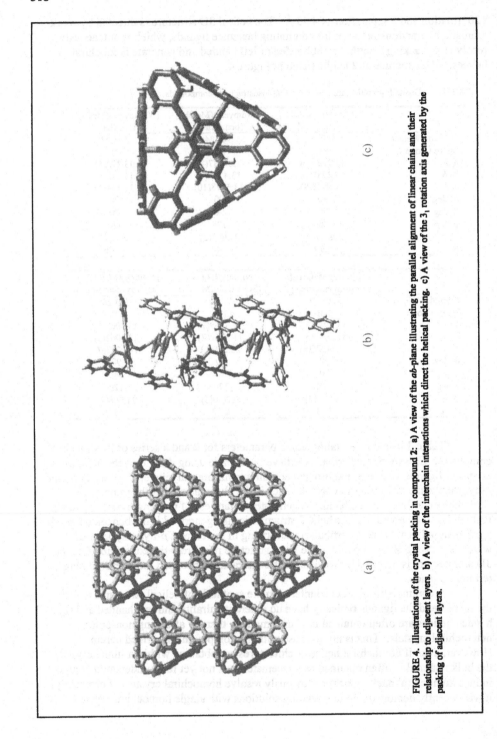

FIGURE 4. Illustrations of the crystal packing in compound 2: a) A view of the *ab*-plane illustrating the parallel alignment of linear chains and their relationship to adjacent layers. b) A view of the interchain interactions which direct the helical packing. c) A view of the 3₁ rotation axis generated by the packing of adjacent layers.

of a particular handedness[6r] could effect spontaneous resolution of enantiomorphically pure materials.

2D Networks.   Square grid networks (Figure 5) exemplify a particularly simple example of a predictable 2D network. The primary reason for interest in these compounds has been their ability to afford controllable cavities suitable for

FIGURE 5. A schematic illustration of the square grid motif generated from square planar metal ions and linear "spacer" ligands.

enclathration of organic guest molecules. In this context, a number of simple bifunctional spacer ligands have been examined[13], particularly dipy. However, a phenomenon that mitigates against, or even precludes, enclathration is the filling of voids by self-inclusion or interpenetration[5b]. Interpenetrated structures have limited potential in the context of host-guest chemistry but have significant potential in terms of other bulk properties[14]. [Ni(dipy)$_2$(NO$_3$)$_2$]·2Pyrene, 3·2Pyrene, represents to our knowledge the first example of a compound, which combines the above features by generating two very different types of 2D nets that, interpenetrate: a square grid coordination polymer and a noncovalent net of pyrene molecules.

    A realistic motif for self-assembly of aromatic molecules is a 2D net that is sustained by edge-to-face stacking interactions. A pyrene net of this type coexists with the square grid in 3·2pyrene. Blue crystals of 3·2pyrene were grown by carefully layering a MeOH solution of Ni(NO$_3$)$_2$·6H$_2$O over a MeOH solution of dipy and pyrene under ambient conditions. The coordination networks possess inner cavities of ca. 8 x 8 Å and stack parallel to one another with an interlayer separation of ca. 7.9 Å and bear a close resemblance to the coordination array in [Cd(4,4'-bipyridine)$_2$(NO$_3$)$_2$]·2C$_6$H$_4$Br$_2$[13a]. An additional factor that appears to influence the packing of the grids is weak CH---O hydrogen bonding between the dipy ligands and nitrate anions of the adjacent grids. C---O separations are in the range of 2.877-3.149 Å. The pyridyl rings of the dipy ligands are twisted by 32.5-48.9°.

    The pyrene nets are sustained by edge-to-face interactions. The shortest intermolecular C-C separations (3.518 Å) are similar to those reported for related compounds such as pyrene itself[15] and substituted pyrenes[16]. The planes of the neighboring molecules intersect at an angle of ca. 60° and there are no face-to-face stacking interactions between the molecules. The pyrene nets thread orthogonally and therefore generate the observed 3D crystal structure. The shortest carbon-carbon separation between the atoms of dipy ligands and pyrene molecules is ca. 3.396 Å and corresponds to face-to-face stacking (interplanar angle = 7.3°). The shortest C---O separations involving hydrocarbon moieties, 3.198 Å, may be attributed to CH---ONO$_2$ to hydrogen bonds. The structure of 3 is illustrated in Figure 6.

Interestingly, although both nets appear at first glance to be centrosymmetric, 3·2pyrene crystallizes in the non-centrosymmetric space group Pn. The polarity is a consequence of opposing pyrene molecules not being related by inversion, and thus they are crystallographically nonequivalent. In addition, they are not coplanar and adopt a "bowl" conformation, where the "bowls" are aligned in a parallel manner. The orientation of alternate pyrenes along the y direction is also slightly different, leading to doubling of the b-axis and the observed superstructure. Thus, the polarity of the crystal is determined by the disposition of the non-covalent pyrene network that intercalates

TABLE 4. Crystallographic parameters for 3·2Pyrene and related compounds

	3·2Pyrene	3·3Toluene	[Co(NO$_3$)$_2$(dipyete)$_2$]·2(1 -methylnaphthalene)	[Co(NO$_3$)$_2$(dipyete)$_2$]· nitrobenzene
Formula Weight	899.59	771.50	831.8	670.5
Space Group	Pn	Cc	Cc	Pna2$_1$
a, Å	11.3602(6)	47.0450(3)	24.3986(14)	16.9849(14)
b, Å	22.7707(13)	11.2791(1)	13.7567(8)	14.4083(12)
c, Å	15.8508(9)	22.6565(1)	12.1360(7)	15.5282(13)
α, deg	90	90	90	90
β, deg	93.956(1)	103.8757(2)	90.687(1)	90
γ, deg	90	90	90	90
V, Å^3	4092.39	11671.27	4073.1(4)	3800.1(5)
R, %	4.28	12.2	5.2	12.3

with the centric square grid network.

In summary, 3·2pyrene appears to be prototypal for a new class of hybrid compound that is based upon rationally interpenetrated 2D covalent and noncovalent nets. 3·2pyrene also illustrates how chirality in crystals can be generated from subtle packing of achiral components. Table 4 provides crystallographic parameters for additional compounds of this genre that also adopt acentric space groups because of an acentric network of guest molecules.

Although there are several other examples of 2D networks that crystallize in acentric space groups[17], the salient point of the structure of 3·2pyrene and related complexes is that the polarity arises from the generation of an acentric supramolecular network of achiral components. Other 2D networks rely on the lack of a center of inversion in either the metal or the bridging ligand to ensure a non-centrosymmetric network. In the case of bis{3-[2-(4-pyridyl)ethenyl]benzoato}cadmium[17a], the cis-octahedral metal center is inherently acentric, as are the m-pyridinecarboxylate spacer ligands. In another example, bis(4,4'-dihydroxyphenyl)sulfone[17b] exploits its tetrahedrally disposed complimentary hydrogen bonding sites to generate a unique doubly interwoven square grid architecture. This is an alternative motif to the more common diamondoid structure for the generation of structures based on tetrahedral moieties, which are inherently non-centrosymmetric. It is therefore unsurprising that in both cases the resultant 2D architecture is acentric.

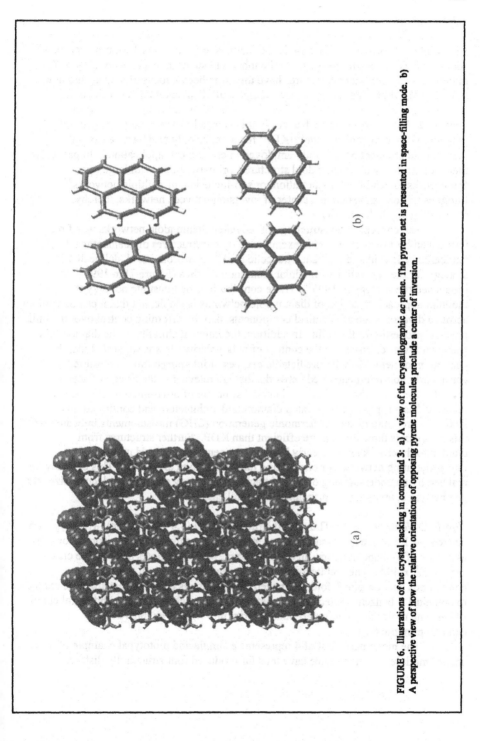

FIGURE 6. Illustrations of the crystal packing in compound 3: a) A view of the crystallographic *ac* plane. The pyrene net is presented in space-filling mode. b) A perspective view of how the relative orientations of opposing pyrene molecules preclude a center of inversion.

(a)

(b)

2.2.2   3D Networks

Diamondoid Networks   Diamondoid structures are at the very least predisposed to adopt acentric space groups owing to the inherent lack of inversion symmetry in T_d moieties. Diamondoid architectures have therefore been a focus of studies into new polar solids, and in this regard, many examples of 3D diamondoid coordination polymers based upon the tetrahedral or S_4 coordination environment have been generated. However, despite advances in this area, a low temperature polymorph of potassium dihydrogenphosphate[18] (KDP) remains the only NLO-active example of diamondoid architecture currently employed in commercial applications. In particular, there is a tendency for diamondoid structures, primarily those with large open frameworks to exhibit interpenetration, or self-inclusion, of multiple networks[5b]. Interpenetration can generate a center of inversion between networks, thereby eliminating any bulk polarity.

     The concept of constructing non-covalent diamondoid networks was first introduced by Ermer et al. and is exemplified by the structures of adamantane-1,3,5,7-tetracarboxylic acid and methanetetraacetic acid[19]. Zaworotko et al. delineated a strategy for the generation of modular diamondoid solids (Figure 7) in 1994[5c], and broadened the strategy in 1995[5l]. These contributions demonstrate not only that the modular approach to design of diamondoid networks is viable, but that it can be applied across a diverse range of chemical components, thereby affording control over the bulk physical properties of the solid. In addition, the inherent chirality of the diamondoid framework was discussed in the context of bulk polarity. It was suggested that the generation of networks with predictable degrees of interpenetration is feasible, particularly non-interpenetrated networks that are relevant to the design of acentric crystals. Recently, the synthesis and crystal structure of bis(isonicotinato)zinc was reported[5a] as a triply interpenetrated diamondoid architecture and confirmed this strategy. Preliminary second harmonic generation (SHG) measurements indicate that this complex is three times more efficient than KDP. Further structures from diamondoid networks must clearly be based on generating an odd number of interpenetrating networks, or eliminating interpenetration altogether. This can only be realized by an understanding of crystal packing efficiency and supramolecular size and geometry; i.e. applying the principles of crystal engineering.

Novel 3D Network.   The crystal structure of [Co(4,4'-dipyridine)$_{1.5}$(NO$_3$)$_2$]$_n$ · 1.5 benzene, 4·1.5 benzene, reveals the presence of a new supramolecular isomer for self-assembly of T-shape moieties, an open 3D framework with pores of effective cross-section 8Å x 40Å. The architecture[20] has no precedent in natural or synthetic compounds and, despite 3-fold interpenetration, contains channels that are large enough to contain 1.5 benzene molecules per metal ion. These channels surround chiral chains of benzene molecules that impart bulk polarity into what is an inherently centrosymmetric host.

     The monomeric unit of 4 represents a simple and prototypal example of a "T-shape" module. Such modules have thus far produced four structurally distinct

323

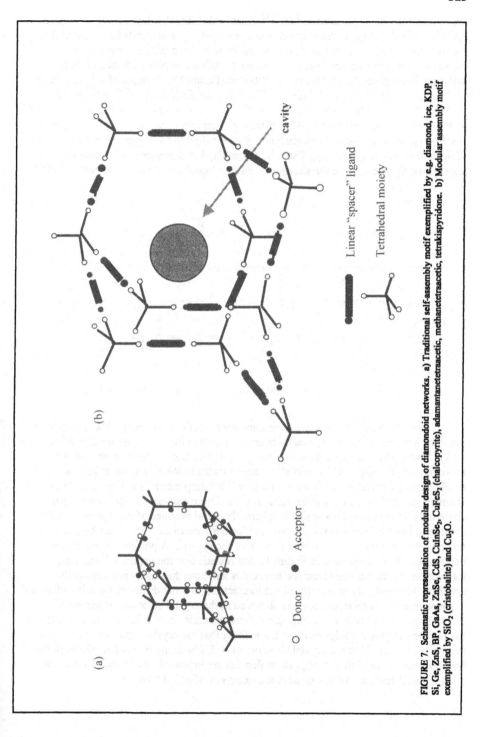

FIGURE 7. Schematic representation of modular design of diamondoid networks. a) Traditional self-assembly motif exemplified by e.g. diamond, ice, KDP, Si, Ge, ZnS, BP, GaAs, ZnSe, CdS, CuInSe₂, CuInSe₂, CuFeS₂ (chalcopyrite), adamantanetetraacetic, methanetetraacetic, tetrakispyridone. b) Modular assembly motif exemplified by SiO₂ (cristobalite) and Cu₂O.

Linear "spacer" ligand

Tetrahedral moiety

cavity

○ Donor ● Acceptor

(a) (b)

324

supramolecular isomers: ladder (1D)[21,22], brick wall (2D)[22a], bilayer (2D)[21,23], frame (3D)[24]. Indeed, using different crystallization conditions and guests has allowed 4 to exhibit two of these supramolecular isomers, ladder[22b] and bilayer[23] structures. However, if 4 is crystallized via intermediate pyridine complexes, a dramatically different 3D supramolecular isomer is obtained (Figure 8). Crystals of 4·1.5 benzene were prepared as follows: a solution of $Co(NO_3)_2 \cdot 6H_2O$ (0.873 g, 3.0 mmol) and pyridine (0.32 ml, 4.0 mmol) in 10ml MeOH was layered onto a solution of dipy (0.624 g, 4.0 mmol) in 10ml benzene. After standing overnight under ambient conditions, the reaction vessel was placed in an ice bath. This resulted in two types of crystals, colored violet and peach, respectively. The violet crystals, 4·1.5 benzene, are stable for approximately one hour in the absence of mother liquor and do not melt below 300°C.

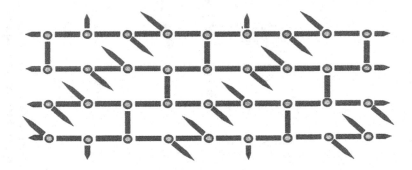

FIGURE 8. Schematic representation of novel 3D supramolecular motif for "T-shaped" modules.

Figure 9 reveals how 4·1.5 benzene exists as an open framework coordination polymer with remarkably large cavities and channels (effective cross-section of ca. 8Å x 40Å). This particular supramolecular isomer is to our knowledge unprecedented; however, it was suggested by Wells[20]. The networks in 4·1.5 benzene exhibit a 3-fold level of interpenetration[5b]. However, despite this interpenetration, there exist large channels parallel to the crystallographic z-axis. These channels contain an ordered chain of benzene molecules which is inherently chiral because of its supramolecular structure. Based upon a molecular volume for benzene of ca. 125Å^3, the benzene molecules represent ca. 30% of the volume of the crystal. A portion of one of the benzene chains is illustrated in Figure 10 and reveals that the expected[25] stacking interactions occur between benzene molecules and how the chains are inherently acentric. However, the geometry of the interactions is quite different from the idealized T-shape or edge-to-face interactions observed in the crystal structure of benzene[26]. Interestingly, the coordination polymer network contains a pseudo inversion center but the benzene chains are aligned in such a manner that the crystal structure is non-centrosymmetric. That the crystal structure of 4·1.5 benzene is acentric although the host framework and individual guest molecules are inherently centrosymmetric is a most unusual feature. Whereas there are examples of chiral host

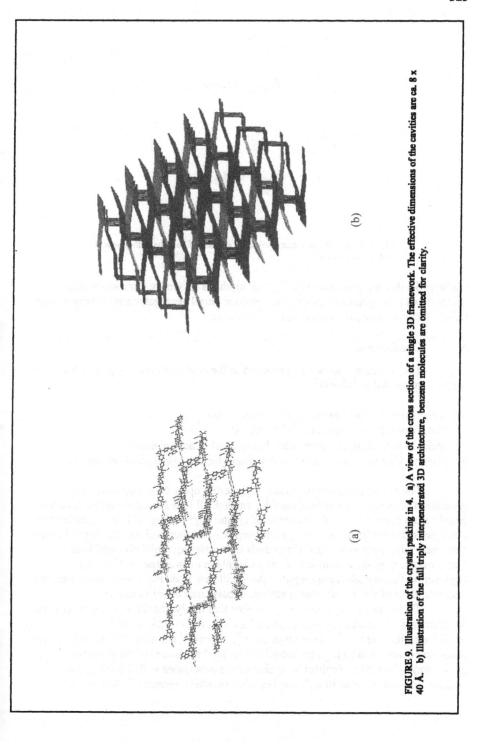

(a)

(b)

FIGURE 9. Illustration of the crystal packing in 4. a) A view of the cross section of a single 3D framework. The effective dimensions of the cavities are ca. 8 × 40 Å. b) Illustration of the full triply interpenetrated 3D architecture, benzene molecules are omitted for clarity.

FIGURE 10. The chiral supramolecular adducts of benzene as they appear in 4·1.5 benzene

frameworks that are generated from achiral modules[7c,e], we are unaware of other examples of high symmetry guests that generate chiral aggregates and thereby impart bulk polarity to centrosymmetric host frameworks.

3. Conclusions

The strategies that can be exploited in the context of the design of polar solids can be categorized as follows:

- Polar crystals from homochiral building blocks
- Polar crystals from achiral building blocks
- Polar host frameworks with either homochiral or achiral guests
- Achiral host framework with polar supramolecular aggregate of guests

The potential impact of crystal engineering upon the design of a new generation of polar solids is considerable. Indeed, all four strategies can be expected benefit from application of the concepts of crystal engineering. However, the latter two strategies offer particular promise since it appears that they will inevitably produce an extremely large number of new compounds from **existing molecules and ions**. Furthermore, they are well placed to take advantage of two aspects of crystal engineering that are advancing rapidly: design of new open framework structures; our understanding of supramolecular synthons, in particular chiral synthons.

There are already a number of studies that imply that there is particular reason for optimism. Host lattices of cylcophosphazene[27], deoxycholic acid[28], perhydrotriphenylene[7a,c-d,29], tri-o-thymotide[30], thiourea[31] and urea[7c-f,32] are inherently chiral and demonstrate a high probability of parallel alignment of the dipoles of 1D guest arrays (60-85%). Zeolites have also been investigated with the intention of isolating individual polar strands and inducing parallel alignment[33]. It therefore appears that isolation of polar strands in a host lattice reduces or eliminates the inter-strand attractions that induce anti-parallel alignment of dipoles. This would presumably also be the case for centrosymmetric host lattices as detailed herein. Indeed, it is conceivable that as the range of open framework structures increases rapidly that there will be no effective limit to the range of molecules or ions that can be incorporated into polar crystals. Furthermore, there are already several examples of host structures that are well documented and would appear to be ideal candidates for inclusion of 1D arrays. Extension of these concepts to inclusion of 2D arrays is a more complex design issue but offers even more possibilities. As detailed herein, 2D networks of complementary topology, but very different chemical nature, can coexist.

4. References

1. a) Agullo-Lopez, F., Cabrera, J. M. and Agullo-Rueda, F. (1994) *Electrooptics: Phenomena, Materials and Applications*, Academic Press, New York; b) Zyss, J. (1993) *Molecular Nonlinear Optics: Materials, Physics and Devices*, Academic Press, New York.

2. a) Koshima, H. and Matsuura, T. (1998) *J. Synth. Org. Chem. Jpn.*, **56**, 268-279; b) Koshima, H. and Matsuura, T. (1998) *J. Synth. Org. Chem. Jpn.*, **56**, 466-477; c) Whitesell, J. K., Davis, R. E., Wong, M.-S. and Chang, N.-L. (1994) *J. Am. Chem. Soc.*, **116**, 523-527; d) Green, B. S., Lahav, M. and Rabinovich, D. (1979) *Acc. Chem. Res.*, **12**, 191-197.

3. 46,030 of 190,306 compounds in the Cambridge Structural Database (v. 5.16, October 1998 release) crystallize in acentric space groups (24.2%).

4. a) Desiraju, G. R. (1989) *Crystal Engineering: The Design of Organic Solids*, Elsevier, New York; b) Lehn, J.-M. (1995) *Supramolecular Chemistry: Concepts and Perspectives*, VCH Publishers, New York; c) Desiraju, G. R. (1995) *Angew. Chem. Int. Ed. Engl.*, **34**, 2311-2327.

5. a) Evans, O. R., Xiong, R.-G., Wang, Z., Wong, G. K. and Lin, W. (1999) *Angew. Chem. Int. Ed. Engl.*, **38**, 536-538; b) Batten, S. R. and Robson, R. (1998) *Angew. Chem. Int. Ed. Engl.*, **37**, 1460-1494; c) Zaworotko, M. J. (1994) *Chem. Soc. Rev.*, **23**, 283-288; d) MacGillivray, L. R., Subramanian, S. and Zaworotko, M. J. (1994) *J. Chem. Soc., Chem. Commun.*, 1325-1326; e) Lopez, S., Kahraman, M., Harmata, M. and Keller, S. W. (1997) *Inorg. Chem.*, **36**, 6138-6140; f) Hirsch, K. A., Venkataraman, D., Wilson, S. R., Moore, J. S. and Lee, S. (1995) *J. Chem. Soc., Chem. Commun.*, 2199-2200; g) Kim, K.-W. and Kanatzidis (1992) *J. Am. Chem. Soc.*, **114**, 4878-4883; h) Munukata, M., Wu, L. P., Yamamoto, M., Kuroda-Sowa, M. and Maekawa, M. (1996) *J. Am. Chem. Soc.*, **118**, 3117-3124; i) Michaelides, A., Kiritsis, V., Skoulika S. and Aubry, A. (1993) *Angew. Chem. Int. Ed. Engl.*, **32**, 1495-1497; j) Sinzger, K., Hunig, S., Jopp, M., Bauer, D., Bietsch, W., von Shutz, J. U., Wolf, H. C., Kremer, R. K., Metzenthin, T., Bau, R., Khan, S. I., Lindbaum, A., Langauer, C. L. and Tillmanns, E. (1993) *J. Am. Chem. Soc.*, **115**, 7696-7705; k) Robson, R., Abrahams, B. F., Batten, S. R., Gable, R. W., Hoskins, B. and Liu, J. (1992) *ACS Symp. Ser.: Supramolecular Architectures*, **499**, 257-273; l) Copp, S. B., Holman, K. T., Sangster, J. O. S., Subramanian, S. and Zaworotko, M. J. (1995) *J. Chem. Soc., Dalton Trans.*, 2233-2243.

6. a) Bowyer, P. K., Porter, K. A., Rae, A. D., Willis, A. C. and Wild, S. B. (1998) *J. Chem. Soc., Chem. Commun.*, 1153-1154; b) Rowan, A. E., and Nolte, R. J. M. (1998) *Angew. Chem. Int. Ed. Engl.*, **37**, 63-68; c) Zaworotko, M. J. (1998) *Angew. Chem. Int. Ed. Engl.*, **37**, 1211-1213; d) Masciocchi, N., Ardizzoia, G. A., LaMonica, G., Maspero, A., and Sironi, A. (1998) *Angew. Chem. Int. Ed. Engl.*, **37**, 3366-3369; e) Wu, B., Zhang, W.-J., Yu, S.-Y. and Wu, X.-T. (1997) *J. Chem. Soc., Dalton Trans.*, 1795-1796; f) Piquet, C., Bernardinelli, G., Hopfgartner, G. (1997) *Chem. Rev.*, **97**, 2005-2062; g) Moore, J. S. (1996) *Curr. Opin. Solid State Mater. Sci.*, **1**, 777-787; h) Williams, A. F. (1997) *Chem. Eur. J.*, **3**, 15-19; i) Constable, E. C. (1992) *Tetrahedron*, **48**, 10013-10059; j) Withersby, M. A., Blake, A. J., Champness, N. R., Hubberstey, P., Li, W.-S. and Schroder, M. (1997) *Angew. Chem. Int. Ed. Engl.*, **36**, 2327-2329; k) Ranford, J. D., Vittal, J. J. and Wu, D. (1998) *Angew. Chem. Int. Ed. Engl.*, **37**, 1114-1116; l) Biradha, K., Seward, C. and Zaworotko, M. J. (1999) *Angew. Chem. Int. Ed. Engl.*, **38**, 492-495; m) Soghomonian, V., Chen, Q., Haushalter, R. C., Zubieta, J. and O'Connor, C. J. (1993) *Science*, **259**, 1596-1599; n) Gelling, O. J., van Bolhuis, F. and Feringa, B. L. (1991) *J. Chem. Soc., Chem. Commun.*, 917-919; o) Dai, Y., Katz, T. J. and Nichols, D. A. (1996) *Angew. Chem. Int. Ed. Engl.*, **35**, 2109-2111; p) Kaes, C., Hosseini, M. W., Rickard, C. E. F., Skelton, B. W. and White, A. H. (1998) *Angew. Chem. Int. Ed. Engl.*, **37**, 920-922; q) Fleming, J. S., Mann, K. L. V., Couchman, S. M., Jeffrey, J. C., McLeverty, J. A. and Ward, M. D. (1998) *J. Chem Soc., Dalton Trans.*, 2047; r) Endo, K. (1999) *The 62nd Okazaki Conference: Structural Hierarchy in Molecular Science From Nano- and Mesostructures to Macrostructures*, Okazaki, Japan (Abstract L6).

7. a) Konig, O., Burgi, H.-B., Armbruster, T., Hulliger, J. and Weber, T. (1997) *J. Am. Chem. Soc.*, **119**, 10632-10640; b) Ramamurthy, V. and Eaton, D. F. (1994) *Chem. Mater.*, **6**, 1128-1136; c) Hoss, R., Konig, O., Kramer-Hoss, V., Berger, U., Rogin, P. and Hulliger, J. (1996) *Angew. Chem. Int. Ed. Engl.*, **35**, 1664-1666; d) Hulliger, J., Langley, P. J., Konig, O., Quintel, A. and Rechsteiner, P. (1998) *Pure Appl. Opt.*, **7**, 221-227; e) Brown, M. E. and Hollingsworth, M. D. (1995) *Nature*, **376**, 323-327; f) Hollingsworth, M. D., Brown, M. E., Hillier, A. C., Santarsiero, B. D. and Chaney, J. D. (1996) *Science*, **273**, 1355-1359; g) Bishop, R. (1996) *Comprehensive Supramolecular Chemistry, Vol.6*, Pergamon, Oxford (pp. 85-115).

8.   a) Lehn, J.-M., Mascal, M., DeCian, A. and Fischer, J. (1990) *J. Chem. Soc., Chem. Commun.*,
      479-481; b) Zerkowski, J. A., Seto, C. T. and Whitesides, G. M (1992) *J. Am. Chem. Soc.*, **114**,
      5473-5475; c) Geib, J. S., Vicent, C., Fan, E. and Hamilton, A. D. (1993) *Angew. Chem. Int. Ed.
      Engl.*, **32**, 119-121; d) Etter, M. C. and Huang, K.-S. (1992) *Chem. Mater.*, **4**, 272-278; e) Panunto,
      T. W., Urbanczyk-Lipkowska, Z., Johnson, R. and Etter, M. C. (1987) *J. Am. Chem. Soc.*, **109**,
      7786-7797; f) Pecaut, J., LeFur, Y. and Masse, R. (1993) *Acta. Crystallogr.*, **B49**, 535-541.
9.   Frankenbach, G. M and Etter, M. C (1992) *Chem. Mater.*, **4**, 272.
10.   a) Leiserowitz, L. (1976) *Acta. Crystallogr.*, **B32**, 775-802; b) Gorbitz, C. M. and Etter, M. C.
      (1992) *J. Chem. Soc., Perkin Trans. 2*, 131-135.
11.   a) Melendez, R., Robinson, F. and Zaworotko, M. J. (1996) *Supermolecular Chemistry, Vol.7*,
      Gordon and Breach, Malaysia (pp. 275-293); b) Pepinsky, R., Vedam, K., Hoshino, S. and Okaya,
      Y. (1958) *Phys. Rev.*, **111**, 1508; c) Pepinsky, R. and Vedam, K. (1960) *Phys. Rev.*, **117**, 1502; d)
      Payan, F. and Haser, R. (1976) *Acta. Crystallogr.*, **B32**, 1875-1879.
12.   a) Bosshard, C., Sutter, K., Pretre, P., Hulliger, J., Florsheimer, M., Kaatz, P. and Gunter, P. (1995)
      Advances in Nonlinear Optics, Vol.1, Gordon and Breach, New York; b) Chemla, D. S. and Zyss,
      J. (1987) *Nonlinear Optical Properties of Organic Molecules and Crystals, Vol. 1, 2*, Academic
      Press, New York.
13.   a) Fujita, M., Kwon, Y. J., Washizu, S. and Ogura, K. (1994) *J. Am. Chem. Soc.*, **116**, 1151; b)
      MacGillivray, L. R., Groeneman, R. H and Atwood, J. L (1998) *J. Am. Chem. Soc.*, **120**, 2676; c)
      Gable, R. W., Hoskins, B. F. and Robson, R. (1990) *J. Chem. Soc., Chem. Commun.*, 1677; d)
      Subramanian, S. and Zaworotko, M. J. (1995) *Angew. Chem. Int. Ed. Engl.*, **35**, 2127; e) Lu, J.,
      Paliwala, T., Lim, S. C., Yu, C., Niu, T. and Jacobsen, A. J. (1997) *Inorg. Chem.*, **36**, 923.
14.   a) Thalladi, V. R, Brasselet, S., Weiss, H.-C., Blaser, D., Katz, A. K., Carrell, H. L., Boese, R.,
      Zyss, J., Nangia, A. and Desiraju, G. R (1998) *J. Am. Chem. Soc.*, **120**, 2563; b) Lin, W., Wang, Z.
      and Xiong, R.-G. (1998) Materials Research Society, Fall Meeting, Boston, Abstract U1.1.
15.   a) Allman, R. (1970) *Z. Kristallogr.*, **132**, 129; b) Kai, Y., Hama, F., Yasuoka, N. and Kasai, N.
      (1978) *Acta Crystallogr.*, Sect. B, **34**, 1263.
16.   a) Mague, J. T., Foroozesh, M., Hopkins, N. E., Gan, L. L.-S. and Alworth, W. L (1997) *J. Chem.
      Crystallogr.*, **27**, 183-189; b) Hazell, A. C. and Jagner, S. (1976) *Acta Crystallogr.*, Sect. B, **32**,
      682; c) Hazell, A. C. and Weigelt, A. (1975) *Acta Crystallogr.*, Sect. B, **31**, 2891.
17.   a) Lin, W., Evans, O. R., Xiong, R.-G. and Wang, Z. (1998) *J. Am. Chem. Soc.*, **120**, 13272-13273;
      b) Davies, C., Langler, R. F., Sharma, C. V. K. and Zaworotko, M. J. (1997) *J. Chem. Soc., Chem.
      Commun.*, 567-568.
18.   Endo, S., Chino, T., Tsuboi, S., and Koto, K. (1989) *Nature*, **340**, 452
19.   a) Ermer, O. and Eling, A. (1988) *Angew. Chem. Int. Ed. Engl.*, **27**, 829-833; b) Ermer, O. (1988)
      J. Am. Chem. Soc., **110**, 3747-3754.
20.   Wells, A. F. (1977) *Three-Dimensional Nets and Polyhedra*, Wiley, New York. (p. 27).
21.   Hennigar, T. L., MacQuarrie, D. C., Losier, P., Rogers, R. D. and Zaworotko, M. J. (1997) *Angew.
      Chem. Int. Ed. Engl.*, **36**, 972.
22.   a) Fujita, M., Kwon, Y. J., Sasaki, O., Yamaguchi, K. and Ogura, K. (1995) *J. Am. Chem. Soc.*,
      117, 7287; b) Losier, P. and Zaworotko, M. J. (1996) *Angew. Chem. Int. Ed. Engl.*, **35**, 2779.
23.   Power, K. N., Hennigar, T. L. and Zaworotko, M. J. (1998) *New J. Chem.*, 177.
24.   a) Robinson, F. and Zaworotko, M. J. (1995) *J. Chem. Soc., Chem. Commun.*, 2413; b) Yaghi, O.
      M. and Li, H. (1996) *J. Am. Chem. Soc.*, **118**, 295.
25.   Jorgensen, W. L. and Severance, D. L. (1990) *J. Am. Chem. Soc.*, **112**, 4768.
26.   Bacon, G. E., Curry, N. A. and Wilson, S. A. (1964) *Proc. R. Soc. London, Ser. A.*, **279**, 98.
27.   a) Roesky, H. W., Katti, K. V., Seseke, U., Schmidt, H.-G., Egert, E., Herbst, R. and Sheldrick, G.
      M. (1987) *J. Chem. Soc., Dalton Trans.*, 847; b) Lork, E., Watson, P. G. and Mews, R. (1995) *J.
      Chem. Soc., Chem. Commun.*, 1717; c) Vij, A., Staples, R. J., Kirchmeier, R. L. and Shreeve, J. M.
      (1996) *Acta Crystallogr.*, Sect. C (Cr. Str. Comm.), **52**, 2515.
28.   a) Popovitz-Biro, R., Tang, C. P., Chang, H. C., Lahav, M. and Leiserowitz, L. (1985) *J. Am.
      Chem. Soc.*, **107**, 4043; b) Giglio, E., Mazza, F. and Scaramuzza (1985) *J. Inclusion Phenom.*, **3**,
      437; c) Chang, H. C., Tang, C. P., Popovitz-Biro, R., Lahav, M. and Leiserowitz, L. (1981) *New J.
      Chem.*, **5**, 475; d) Chang, H. C., Popovitz-Biro, R., Lahav, M. and Leiserowitz, L. (1981) *J. Am.
      Chem. Soc.*, **109**, 3883; e) Weisinger-Lewin, Y., Vaida, M., Popovitz-Biro, R., Chang, H. C.,
      Mannig, F., Frolow, M., Lahav, M. and Leiserowitz, L. (1987) *Tetrahedron*, **43**, 1449; f) Sada, K.,
      Kitamura, T. and Miyata, M. (1994) *J. Chem. Soc., Chem. Commun.*, 905; g) Padmanabhan, K.,
      Ramamurthy, V. and Venkatesan, K. (1987) *J. Inclusion Phenom.*, **5**, 315.

29.   a) Harlow, R. L. and Desiraju, G. R. (1990) *Acta Crystallogr., Sect. C. (Cr. Str. Comm.)*, **46**, 1054; b) Luca, C., Popa, A., Bilba, N. and Mihaila, G. (1983) *Rev. Roum. Chim.*, **28**, 211.

30.   a) Arad-Yellin, R., Green, B. S., Knossow, M. and Tsoucaris, G. (1983) *J. Am. Chem. Soc.*, **105**, 4561; b) Gerdil, R. and Frew, A. (1985) *J. Inclusion Phenom.*, **3**, 335; c) Facey, G. A., Ratcliffe, C. I., Hynes, R. and Ripmeester, J. A. (1992) *J. Phys. Org. Chem.*, **5**, 670; d) Pang, L. and Brisse, F. (1996) *J. Chem. Cryst.*, **26**, 461.

31.   a) Qi Li, T. C. and Mak, W. (1995) *J. Inclusion. Phenom.*, **20**, 73; b) Harris, K. D. M (1990) *J. Solid State Chem.*, 280; c) Tam, W., Eaton, D. F., Calabrese, J. C., Williams, I. D., Wang, Y. and Anderson, A. G. (1989) *Chem. Mater.*, **1**, 128; d) Shindo, T., Shindo, M., Ohnuma, H. and Kabuto, C. (1993) *Bul. Chem. Soc. Jpn.*, **66**, 1914; e) Anderson A. G., Calabrese, J. C., Tam, W. and Williams, I. D. (1987) *Chem. Phys. Lett.*, **134**, 392.

32.   a) Uiterwijk, J. W. H. M., van Hummel, G. J., Harkema, S., Aarts, V. M. L. J., Daasvatn, K., Geevers, J., den Hertog Jr., H. J. and Reinhoudt, D. N. (1988) *J. Inclusion Chem.*, **6**, 79; b) Goldberg, I., Lin, W. and Hart, H. (1984) *J. Inclusion Chem.*, **2**, 377; c) Benetello, F., Bombieri, G. and Truter, M. R. (1987) *J. Inclusion Chem.*, **5**, 165; d) Mak, T. C. W. and McMullin, R. K. (1988) *J. Inclusion Chem.*, **6**, 473; e) Panunto, T. W. and Etter, M. C. (1988) *J. Am. Chem. Soc.*, **110**, 5896; f) Etter, M. C., Urbanczyk-Lipkowska, Z., Zia-Ebrahimi, M. and Panunto, T. W. (1990) *J. Am. Chem. Soc.*, **112**, 8415; g) Hollingsworth, M. D., Santarsiero, B. D. and Harris, K. D. M. (1994) *Angew. Chem. Intl. Ed. Eng.*, **33**, 649; h) Yeo, L., Harris, K. D. M. and Guillaume, F. (1997) *J. Solid State Chem.*, **128**, 273; i) Qi Li and Mak, T. C. W. (1996) *Supramolecular Chemistry*, **8**, 73; j) Brown, M. E., Chaney, J. D., Santasiero, B. D and Hollingsworth, M. D. (1996) *Chem. Mater.*, **8**, 1588.

33.   a) Girnus, I., Pohl, M.-M., Richter-Mendau, J., Schneider, M., Noack, M., Venzke, D. and Caro, J. (1995) *Adv. Mater.*, **7**, 711-714; b) Cox, S. D., Gier, T. E., Stucky, G. D. and Bierlein, J. D. (1988) *J. Am. Chem. Soc.*, **110**, 2986-2987.

ENERGY MINIMIZATION AS A TOOL FOR CRYSTAL STRUCTURE DETERMINATION OF INDUSTRIAL PIGMENTS

M.U. SCHMIDT

Clariant GmbH
Pigment Technology Research, G 834
D-65926 Frankfurt am Main, Germany
E-mail: martinulrich.schmidt@clariant.com

This chapter describes theory and applications of energy minimization for molecular crystals. The applications include prediction of crystal structures, search for polymorphic forms, and determination of crystal structures from powder diffraction data.

1. Organic Pigments

Industrial organic pigments are coloured crystalline powders. In their application media (coatings, plastics, printing inks) they are not dissolved, but finely dispersed. Properties like colour, heat stability, light and weather fastness, rheology, density etc. depend both on the molecular structure and on the crystal structure. A prominent example is quinacridone:

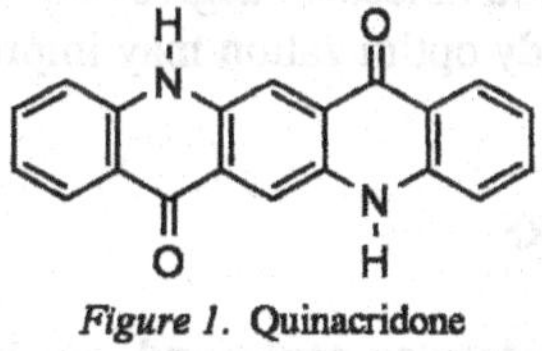

Figure 1. Quinacridone

Quinacridone exhibits a pale yellow-orange shade in organic solvents. In the solid state its shade is between deep red and violet, depending on the polymorphic form. The crystal structures of three modifications of quinacridone could be determined by single crystal structure analyses. For many other pigments the crystal structures are still unknown, because the poor solubility does not allow to grow single crystals from solution. Attempted recrystallization by melting or sublimation usually results in decomposition. On the other hand, the knowledge of the crystal structure is the basis for crystal modelling, structure-property relationships and crystal engineering, like planned syntheses of compounds having specific properties, or of mixed crystals or additives controlling the morphology of the crystallites.

D. Braga et al. (eds.), Crystal Engineering: From Molecules and Crystals to Materials, 331–348.
© 1999 *Kluwer Academic Publishers. Printed in the Netherlands.*

332

2. Lattice Energy Minimization: Method [1-3]

2.1. MOLECULAR GEOMETRY

For the calculation of the molecular packing it is necessary to know the approximate molecular geometry. The molecular geometry may be obtained from the connectivity of the atoms by a variety of computational or experimental methods (e.g. quantum mechanics, force fields; X-ray structures of similar compounds or polymorphs, electron diffraction etc.). The geometry can be idealized, e.g. by averaging over chemically equivalent distances. Hydrogen atoms may be placed on calculated positions. The C-H bond length should be set to the value used during the development of the force field. The geometry of a molecule in the solid state can be different from the geometry in the gas phase (so-called "packing effects"). Two examples are given: (1) In the gas phase biphenyl is twisted by about 45° [4], while in the crystalline state the biphenyl fragment is almost planar in about 50% of the investigated structures [5]. (2) The fragment Ph-O-CH_2-CH_3 can adopt various conformations in the gas phase, while in the crystalline state mostly a planar zigzag conformation is observed. If the conformation is not known, it is recommended to use preferably conformations, which are preferred in the solid state. This approach is particularly useful, when the molecule is large and exhibits degrees of freedom in its central part (like biphenyl), especially if the molecular geometry is kept fixed in the first step of the energy minimization. On the other hand test calculations show, that the precision of the molecular geometry is generally not crucial for finding the correct arrangement of the molecules, as long as the external shape of the molecule is approximately correct. E.g. the five-membered rings of pentamethylferrocene (C_5H_5)Fe(C_5Me_5) can be rotated by 36° without major changes in the molecular packing (see below).

During the minimization the molecular geometry can either be kept fixed, or be refined together with the arrangement of the molecules. For organic compounds with known geometry for most atomic groups, it is normally sufficient to keep the geometry fixed or to vary only a few intramolecular degrees of freedom. In the final step of the energy minimization a full-body optimization may improve the results - or make them worse.

2.2. PACKING PARAMETERS

The crystal structure of a molecular compound can be described by the molecular geometry, the crystal symmetry, and a set of packing parameters, i.e. the unit-cell dimensions a, b, c, α, β, γ, and the positions and orientations of each symmetry-independent molecule. For the position of a molecule the fractional coordinates x, y, z of the centre of gravity can be used. The spatial orientation of a molecule is described by three angles φ_x, φ_y, and φ_z. Unfortunately each program uses its own definition of angles (which even may change in program updates); thus it is a nightmare to compare the results of different programs, or to switch from one program to another, e.g. from an energy minimization to a Rietveld program.

2.3. SYMMETRY

In crystal structure calculations the space group symmetry is generally included from the beginning. For applications like structure prediction or search for possible poly-morphic forms, where no *a priori* space group information is available, possible space groups have to be tested separately. Fortunately only a very limited number of space groups are common. It is recommended to respect also space groups known to occur in similar compounds. If the molecule has internal symmetry, or may adopt internal symmetry during the minimization, supergroups of the tested space groups can be reached (Tab. 1).

Table 1. Reachable space groups for molecules having internal symmetry

Molecular symmetry	Reachable space groups	
	Calculations in $P\bar{1}$, $Z=2$ (molecules on general positions)	Calculations in $P2_1/c$, $Z=4$ (molecules on general positions)
$\bar{1}$	$P\bar{1}$, $Z=1$	$P2_1/m$, $P2/c$, $P2_1/c$, $Z=2$ $C2/m$, $C2/c$, $Z=4$
m	$P2/m$, $P2_1/m$, $Z=2$ $C2/m$, $Z=4$	$P2_1/m$, $Z=2$ $C2/m$, $Z=4$ $Pnma$, $Pmna$, $Pbam$, $Pbcm$, $Pnnm$, $Z=4$ $Cmce$, $Z=8$
2	$P2/m$, $P2/c$, $Z=2$ $C2/m$, $C2/c$, $Z=4$	$P2/c$, $Z=2$ $C2/m$, $C2/c$, $Z=4$ $Pnna$, $Pmna$, $Pcca$, $Pbam$, $Pccn$, $Pbcm$, $Pnnm$, $Pbcn$, $Z=4$ $Cmce$, $Z=8$
3	$P\bar{3}$, $Z=2$ $R\bar{3}$, $Z=2$	–
$2/m$	$P2/m$, $Z=1$ $C2/m$, $Z=2$	$P2/m$, $Z=1$ $C2/m$, $Z=2$ $Pbam$, $Pmna$, $Pnnm$, $Z=2$ $Cmcm$, $Cmce$, $Cmmm$, $Cccm$, $Cmme$, $Ibam$, $Imma$, $Z=4$ $Fmmm$, $Z=8$
$mm2$	$Pmmm$, $Pmma$, $Pmmn$, $Z=2$ $Cmcm$, $Cmmm$, $Cmme$, $Immm$, $Imma$, $Z=4$ $Fmmm$, $Z=8$	$Pmma$, $Pmmn$, $Z=2$ $Cmcm$, $Cmmm$, $Cmme$, $Immm$, $Imma$, $Z=4$ $Fmmm$, $Z=8$ $P4/mbm$, $P4_2/mnm$, $P4_2/ncm$, $Z=4$

2.4. ENERGY TERMS

The most sophisticated method for calculating the energy of a molecular crystal would be a high-level quantum mechanical calculation, taking into account the periodicity of the crystal lattice. Ab initio calculations on crystals of medium-sized molecules are not possible yet with a sufficiently high accuracy. Therefore force field methods are used.

Generally speaking, three assumptions are made for crystal structure calculations by energy minimization:

1. Entropic effects are neglected. The free energy of a crystal lattice, given by

$$F = U - T \cdot S, \tag{1}$$

is approximated by a temperature-independent energy E. The entropic term $T \cdot S$ is not small, but similar for different packings of a given molecule: Under the assumption, that the molecular geometry does not change drastically, the intramolecular contributions to the sum of states remain almost constant; furthermore the intermolecular contributions change only slightly, since in all packings the molecules are surrounded by other molecules. The entropic term differs mostly in the order of $T \cdot \Delta S = 0$ to 10 kJ/mol between polymorphic forms. The energy E includes an averaged entropic term; the force field parameters are adjusted in order to reproduce crystal structures at ambient temperature. If the entropic effects are neglected, the temperature of phase transitions etc. cannot be calculated.

2. It is assumed, that the experimental crystal structure corresponds to the absolute minimum of energy. In fact, experimental crystal structures can correspond to either the global or a local minimum of the free energy. Energy differences between different polymorphic forms are mostly in the order of $\Delta H = 0$ to 10 kJ/mol. Thus for a prediction of all possible polymorphic forms one should take into account not only the structure with the lowest energy, but also all packings having slightly higher energies. The energy range, which has to be considered, depends on the reliability of the force field, and on the other assumptions made.

3. The intermolecular energy is calculated by the atom-atom potential method [6]. In this approximation the interactions of the molecules are divided into a sum over individual atom-atom interactions:

$$E = \frac{1}{2} \sum_i \sum_j E_{ij}(r_{ij}) \tag{2}$$

The sums run over all atoms i of a reference molecule and all atoms j of all other molecules. The energy contributions of 3-body interactions are fitted empirically onto the energy of 2-body interactions. In the atom-atom potential approximation the atom-atom energy E_{ij} depends only on the distance r_{ij}. This works well for van der Waals and Coulomb interactions. In a first approximation, hydrogen bonds can be calculated with this approach, too [7].

The total energy of a molecular crystal can be described by four terms:

$$E_{total} = E_{vdW} + E_{el} + E_{intra} + E_{other} \tag{3}$$

with

E_{vdW}: van der Waals energy
E_{el}: electrostatic energy
E_{intra}: intramolecular energy (if the molecule is not treated as rigid)
E_{other}: energies of other interactions

All energy terms have to be carefully scaled in respect to each other.

For the van der Waals interactions different types of potentials are currently used. The calculations described in this chapter were performed with a Buckingham potential:

$$E_{vdW} = \frac{1}{2}\sum_i \sum_j (-Ar_{ij}^{-6} + Be^{-Cr_{ij}}).$$ (4)

A, B, and C are empirical parameters depending on the atom types (Tab. 2). The summation in principle extends to infinity and must in practice be limited to a user-defined cut-off. A cut-off of 20 Å accounts for about 99% of the total van der Waals energy. For A, B, and C different sets of empirical parameters have been published (e.g. [1,8]). Both their numerical values and the lattice energies obtained by their application differ considerably, whereas the minimizations converge to quite similar crystal structures. This behaviour is obviously due to high correlation between the parameters. Therefore care must be taken when combining different parameter sets from the literature.

Table 2. Parameters for the van der Waals potential $E_{vdW} = \frac{1}{2}\Sigma\Sigma (-Ar^{-6} + Be^{-Cr})$ [1,2]

Atoms	A ($Å^6 \cdot kJ/mol$)	B (kJ/mol)	C ($Å^{-1}$)	Atoms	A ($Å^6 \cdot kJ/mol$)	B (kJ/mol)	C ($Å^{-1}$)
C···C	2377.0	349908	3.60	H^a···O	0	0	0
C···H	523.0	36677	3.67	Si···Si	9702.0	1542969	3.46
H···H	144.2	11104	3.74	Si···C	4802.0	734777	3.53
N···N	1240.7	201191	3.78	Si···H	1053.0	130894	3.60
N···C	1483.6	247571	3.73	Fe···Fe	6463.0	1804000	4.00
N···H	407.4	64467	4.00	Fe···C	3920	794502	3.80
O···O	1242.6	372203	4.18	Fe···H	965.4	141533	3.87
O···C	1718.6	360883	3.89	Fe···O	2834.0	819423	4.09
O···H	423.3	64288	3.96	Cl···Cl	6000.0	1000000	3.56
O···N	1241.6	273649	3.98	Cl···C	3777.0	591530	3.58
H^a···H^a	0	0	0	Cl···H	930.2	105376	3.65
H^a···H	144.2	11104	3.74	Cl···N	2728.0	448543	3.67
H^a···C	523.0	36677	3.67	Cl···O	2730.0	610084	3.87
H^a···N	0	0	0				

^{a}) hydrogen of OH and NH groups

Generally the van der Waals interactions hold the major contribution to the lattice energy, and hence the molecular packings are always quite dense. Hydrogen bridges and Coulomb interactions are decisive for the preference of one molecular arrangement over others of comparable van der Waals energy.

The electrostatic interactions can be calculated using the Coulomb formula

$$E_{el} = \frac{1}{2}\sum_i \sum_j \frac{1}{4\pi\varepsilon\varepsilon_0} \frac{q_i q_j}{r_{ij}}. \tag{5}$$

The relative permittivity ε was set to 1.0. Point charges q_i and q_j were assigned to the atomic centres. This strategy is a rapid and simple one because both the atomic positions and the interatomic distances are also required for the calculation of van der Waals interactions. A better description of the electrostatic interactions would be achieved by using multipoles on the atomic positions, but at the expense of longer calculation times. In the present calculations the summation includes five unit cells in each direction. For polar molecules in polar space groups the summation range is extended to a larger number of unit cells in the directions of the polar axes. The total Coulomb energy of a polar crystal also depends on the surface charges and on the external shape of the crystal [9]. If the morphology is known (e.g. from electron microscopy), it can be taken into account for the energy calculations [10], but in general this effect is neglected.

It should be mentioned, that instead of energy terms also statistical potentials can be used [11,12].

2.5. MINIMIZATION

The expression for the lattice energy as a sum of several thousand individual interactions is too complicated to be minimized analytically. Therefore the minimization must be performed by numerical methods. In the last years a variety of different methods has been applied, like steepest descent [2], conjugate gradient, Newton-Raphson, truncated Newton, simulated annealing [13], molecular dynamics [14], diffusion-equation [15] and cluster methods [16]. Frequently combinations of these methods are used [16,17]. Since the energy hypersurface has a large number of local minima, the 'classical' minimization methods like steepest descent require several hundred runs starting from different points. These starting points can be randomly chosen [2], systematically varied [18,19], or calculated previously [17]. A review on different methods and their use to predict possible crystal polymorphs is given by Verwer and Leusen [20].

The accuracy of a calculated crystal structure depends on the force field. Even if the force field parameters are carefully adjusted, the calculated crystal structures are less accurate than crystal structures determined by single crystal X-ray diffraction. Typical deviations between calculated and experimental crystal structures are in the order of 0.3 Å / 1° for lattice parameters and 0.1 to 0.2 Å for intermolecular distances. This is about 100 times larger than the standard deviations coming out of a single crystal structure analysis. The calculation difficulties are mostly generated by the weak forces between the molecules, not by the strong, directed forces within them.

The structures described in this chapter have been calculated using the following method, which is implemented in the program CRYSCA [1,2,21]:

The minimization starts from random packings of the molecules; i.e. all packing parameters are assigned random values inside a user defined range. If the lattice parameters are known, they may be used as well. In contrast to several other methods the minimization procedure allows calculations in all space groups with molecules occupying every kind of special position, including 'complicated' cases like $Ni(CO)_4$ (space group $Pa\overline{3}$ with molecules on the <111> axes) [2]. Disorder and non-crystallographic symmetries can be handled as well. One of the most complex cases was the calculation of the disordered structure of $Si[Si(CH_3)_4]_4$ using a molecule consisting of one fully occupied and 624 partially occupied atomic positions per asymmetric unit [22].

The energy is minimized by a special steepest-descent procedure. After the minimization has located an energy minimum, new random values are generated for all free packing parameters. This procedure is repeated, until the best minima are found several times from different starting points. The reproducibility is <0.001 Å, which is by far better than the precision of the force fields. The minima are sorted according to energy and checked for higher symmetries, meaningful molecular conformations and reliable intermolecular interactions. The packing having the lowest energy corresponds to the 'predicted' crystal structure, other minima having slightly higher energies are possible polymorphic forms.

3. Prediction of Crystal Structures

The prediction of a crystal structure before the synthesis of the compound is a highly desired target, especially for the search of compounds having desired solid state properties. Examples include materials with non-linear optical properties (structures without inversion centres), explosives (crystals with high densities), pigments (coloured, stable, insoluble crystals) and pharmaceuticals (stable, bioavailable compounds). Furthermore new polymorphic forms can be patented. Therefore much work has been done with the aim to predict crystal structures from a given molecular structure; and it is not by fortune, that a considerable amount of this research has been done in the industry [10,17,21,23-25] or in collaboration with it. The prediction of crystal structures without reference to experimental data is sometimes called "*ab initio* prediction". This term might be misleading in the future, if quantum mechanical *ab initio* methods will be used for crystal structure predictions.

Are crystal structures predictable? Angelo Gavezzotti said "no" [26]. Other authors say "yes" or "sometimes" [13,27,28]. Apart from the approximations concerning the molecular geometry and the energy terms, there are two additional problems hindering a successful prediction of molecular crystal structures:
1. The calculations are incomplete: Most methods require either the space group symmetry and the number of independent molecules, or the number of molecules per unit cell as input. The number of space groups is limited, but there are unlimited packing possibilities, if molecules are disordered or if the crystal contains more than one independent molecule. Packings with 'exotic' crystal symmetries like in CD_4 (space

338

group $P\bar{4}\,m2$, $Z = 32$, 9 independent molecules [29]) would probably not be generated, unless the user of the program explicitly performs calculations with the corresponding symmetry operators.

2. Most organic compounds can crystallize in different polymorphic forms. In these cases it is not possible to predict "*the*" crystal structure, one can only try to calculate possible polymorphic forms.

3. Often several minima with comparable energies are found. (In the case of acetic acid about 100 minima were found within 5 *kJ/mol* above the minimum energy; the number of structures could be reduced by removing space group symmetry constraints, or by molecular dynamics, but many possible structures remained [30]). It is difficult to predict, which possible crystal structures may be realised experimentally, and how this could be achieved.

3.1. EXAMPLE: PENTAMETHYLFERROCENE

Pentamethylferrocene is one of the few examples, where crystal structure could be predicted successfully.

Figure 2. Pentamethylferrocene $(C_5H_5)Fe(C_5Me_5)$

The molecular geometry was constructed from the crystal data of decamethylferrocene $Fe(C_5Me_5)_2$ [31] and unsubstituted ferrocene $Fe(C_5H_5)_2$ [32]. Like other sandwich compounds pentamethylferrocene can adopt different conformations:

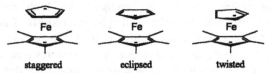

| staggered | eclipsed | twisted |

Figure 3. Possible conformations of pentamethylferrocene

From the literature it was known, that decamethylferrocene crystallizes with staggered molecules, whereas unsubstituted ferrocene consists of eclipsed, twisted or disordered molecules, depending on the polymorphic form. Gas phase electron diffraction showed, that decamethylferrocene is staggered while ferrocene is eclipsed. In both cases the rotational barriers are only about 4 *kJ/mol* [33,34]. This leads to the conclusion, that pentamethylferrocene may adopt both conformations, with an energy difference small enough, that in the crystalline state the molecular conformation is determined by the packing. The crystal structure calculations were performed separately for staggered and eclipsed molecules. Since the calculated packings were quite similar for both conformations, it was not necessary to perform separate runs with twisted molecules.

The energy minimizations were performed in the space groups $P1$, $P\bar{1}$, $P2_1/m$, $P2_1/c$ and $Pnma$. At that time (1992) computer performance was quite low, and about 9 months of CPU time on a VAXstation 3100 was needed, corresponding to a time of a few days on a modern workstation.

The lowest energy was found for a structure with eclipsed molecules in space group $P\bar{1}$. The difference to the corresponding packing with staggered molecules was 7 kJ/mol. The orientation of the small C_5H_5 ring is identical in both packings, whereas the C_5Me_5 fragment is rotated by 36°. This can be regarded as a hint for a disorder of the C_5Me_5 fragment. Separate calculations incorporating this disorder did not lead to a better energy.

We therefore predicted [35], that pentamethylferrocene should crystallize with eclipsed molecules in $P\bar{1}$ with lattice parameters given in the first line of Tab. 3. It cannot be excluded, that there might exist an additional packing with a similar or even lower energy in a space group, which has not been considered in the calculations. Other uncertainties include the entropic effects or the reliability of the force field. Therefore also this 'crystal structure prediction' could only be regarded as a prediction of the probable crystal structure.

Table 3. Calculated energy minima of pentamethylferrocene $(C_5H_5)Fe(C_5Me_5)$

Confor-mation	E (kJ/mol)	a (Å)	b (Å)	c (Å)	α (°)	β (°)	γ (°)	V/Mol. (Å³)
$P\bar{1}$, Z = 2:								
eclipsed[a]	-94.18	8.046	11.889	7.806	93.69	118.23	73.28	313.9
staggered	-89.85	8.057	12.962	7.506	97.12	120.81	72.85	321.5
disord.[b]	-84.41	8.032	12.171	8.093	102.52	120.18	75.26	328.8
$P2_1/m$, Z = 2:								
eclipsed	-86.60	8.110	12.153	6.807	90	99.16	90	331.2
staggered	-87.40	6.741	13.128	7.809	90	107.39	90	329.8
P1, Z = 1:								
eclipsed	-80.60	7.989	6.801	8.025	114.05	114.95	67.39	348.9
$P2_1/c$, Z = 4:								
eclipsed	-89.11	8.420	10.899	14.088	90	94.76	90	344.1
staggered	-86.15	7.672	13.000	14.340	90	112.69	90	329.9
Pnma, Z = 4:								
eclipsed	-87.72	8.345	11.889	12.970	90	90	90	323.9
staggered	-83.51	12.082	8.653	12.807	90	90	90	334.8

a) Predicted crystal structure
b) Disordered molecule (C_5Me_5 fragments with occupancies 0.5, eclipsed and staggered to the C_5H_5 ring)

An X-ray structure analysis by Struchkov et al. [36] confirmed the correctness of our prediction:

340

Table 4. Comparison between predicted and experimental crystal structures of pentamethylferrocene

Space group, Z	Conformation	a (Å)	b (Å)	c (Å)	α (°)	β (°)	γ (°)	V/Mol. (Å³)
Predicted:[a]								
P$\bar{1}$, Z = 2	eclipsed	7.806	8.046	11.889	73.28	86.31	61.77	313.9
Experimental (293 K):								
P$\bar{1}$, Z = 2	eclipsed	7.819 ±0.001	8.169 ±0.001	12.239 ±0.001	73.14 ±0.01	85.27 ±0.01	62.19 ±0.01	330.2 ±0.1
Deviation:		0.01	0.12	0.35	0.14	1.04	0.42	16.3
		Largest deviation in atomic positions: 0,3 Å						

a) Lattice parameters transformed

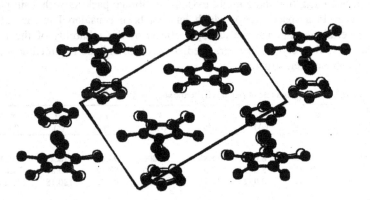

Figure 4. Calculated (full circles) and experimental (open circles) structures of pentamethylferrocene (SCHAKAL plot [37])

This example shows, that in some cases it is possible to predict the crystal structure successfully.

4. Calculation of Possible Polymorphic Forms

4.1. EXAMPLE: PIGMENT YELLOW 74

Pigment Yellow 74 (®Hansa-Brillantgelb 5GX) is an industrial yellow pigment used for coatings and printing inks [38]. Its crystal structure was solved from single crystal data several years ago [39]. We applied the energy minimization technique in order to search for further polymorphic forms having better properties or a different shade. Calculations were performed in generally popular space groups and in space groups known to occur in similar compounds.

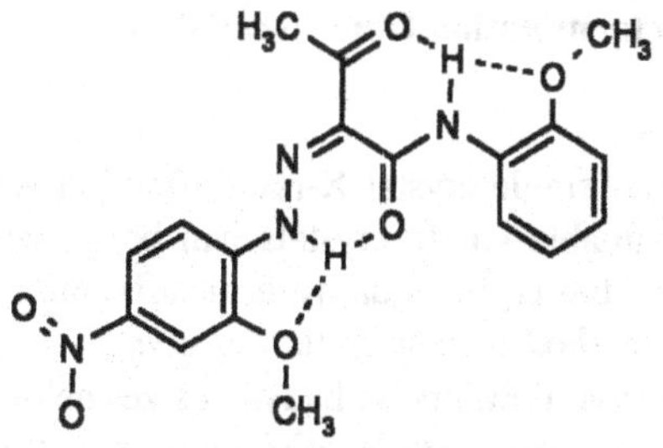

Figure 5. Pigment Yellow 74
(The numbering of pigments is standardized by the Colour Index [40])

Table 5. Experimental and calculated crystal structures of Pigment Yellow 74

Nr.	E (kJ/mol)	a (Å)	b (Å)	c (Å)	α (°)	β (°)	γ (°)	V (Å³)
Experimental:								
$P\bar{1}$, Z = 2		10.729	11.976	7.628	103.43	110.28	88.06	853.1
Calculated:								
$P\bar{1}$, Z = 2:								
1	-223.6	10.417	11.461	7.907	109.86	108.66	89.42	836.1
2	-210.8	10.439	11.697	8.013	103.69	105.13	100.97	883.8
$P2_1$, Z = 2:								
1	-199.6	4.745	10.325	19.412		102.57		928.3
2	-199.6	4.700	10.340	19.107	90	91.63	90	928.2
3	-199.2	12.607	6.320	11.327		89.26		902.4
4	-192.6	11.594	15.712	5.231		101.85		932.5
Pc, Z = 2:								
1	-184.9	8.461	17.156	6.391	90	90.61	90	927.6
$P2_1/c$, Z = 4:								
1	-208.0	16.112	17.075	6.426		100.21		1739.8
2	-207.9	11.465	6.778	22.425		95.03		1735.8
3	-211.2	8.341	24.289	9.099	90	106.88	90	1764.0
4	-198.5	14.746	8.451	15.380		111.12		1787.9
$Pna2_1$, Z = 4:								
1	-195.3	38.702	4.738	10.362	90	90	90	1899.9
$Pbca$, Z = 8:								
1	-205.5	7.256	24.709	19.858				3560.4
2	-197.1	11.315	8.734	38.378	90	90	90	3792.6
3	-194.5	6.865	33.263	16.738				3822.0

The energy minimization reproduced the experimental crystal structure quite well. In the calculations no other minimum with a comparable energy, i.e. no other possible polymorphic form, was found. This prediction was confirmed experimentally: Various methods for syntheses and recrystallizations were applied, but no other polymorphic form could be detected.

5. Crystal Structure Determination from X-ray Powder Data using Energy Minimization [10]

Structure determination by single crystal X-ray diffraction is nowadays a more or less routine procedure - if suitable single crystals can be grown. If the crystals are too small, too unstable, or if the crystal quality is insufficient, X-ray powder diffraction often remains the only method to determine the crystal structure. Good powder diagrams can be taken also from powders with crystal sizes below 1 μm.

The main differences between powder diffraction and single crystal analysis arise from the fact, that in single crystal analysis all diffraction spots can be observed separately, whereas in a powder diagram all peaks are superimposed in one dimension only. A single crystal of an organic compound with 20 to 50 atoms shows typically several thousand independent diffraction spots, while the corresponding powder diagram contains normally only between 50 and 100 distinguishable peaks. The high peak overlap, especially in the higher angle region of a powder pattern, causes not only problems in identifying and separating the individual reflections but also in determining the correct background and extracting reliable peak intensities. Additional problems arise from preferred orientation of crystals in the sample, from anisotropic lattice strain effects etc. Generally a powder diagram not only contains much less information than the data from a single crystal analysis, but it is also more difficult to extract these informations. This may be the reason, why only about 50 organic crystal structures have been solved from powder data so far (not counting isotypic structures) [41]. In contrast, 7000 to 8000 organic crystal structures are solved from single crystal data each year.

The 'classical' procedure to solve a crystal structure from powder data starts with the indexing of the powder diagram, followed by the determination of the possible space groups. Indexing is a crucial point. Although non-indexable powder diagrams are rarely published, they occur frequently in practice. After indexing, as many reflection intensities as possible are extracted from the diagram. For the structure solution itself a variety of different methods can be applied [42]. Among them are 'classical' approaches such as direct methods, the heavy-atom method, difference-Fourier analysis, Patterson methods and intuition (based upon similarity to known structures and upon chemical knowledge). Most methods often fail due to the effects of overlapping reflections, insufficient crystal quality, or unknown impurities in the sample. In such cases only a few methods for structure determination remain, such as guessing and energy minimization.

The procedure for crystal structure determination from powder data using energy minimization consists of six steps:

1. Indexing of the powder diagram and deduction of the possible space groups. If indexing is not possible, energy minimization can be performed as well, but larger calculation times are needed.
2. Set-up of the molecular geometry.
3. Calculation of the possible crystal structures by minimizing the lattice energy.
4. Calculation of the powder patterns for the possible crystal structures.

5. Selection of the correct solution by comparing the calculated with the experimental powder diagrams.
6. Fit onto the full powder diagram by Rietveld refinement [43].

Due to the limited information in a powder diagram, it is recommended to use all available additional information e.g. from IR, NMR, and electron diffraction analyses.

5.1. EXAMPLE: PERINONE PIGMENT $C_{18}H_{10}N_2O_3$ [10]

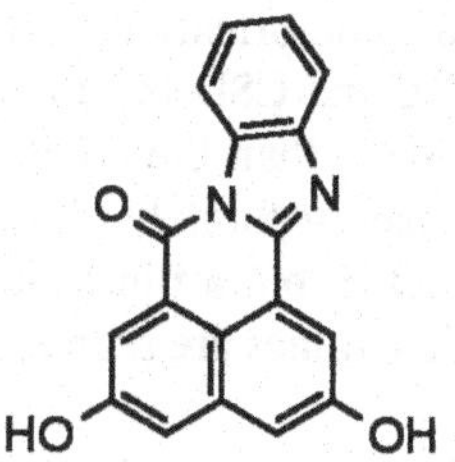

Figure 6. Perinone pigment $C_{18}H_{10}N_2O_3$,
2,5-Dihydroxy-benzo[de]benzo[4,5]imidazo[2,1-a]isoquinoline-7-one

$C_{18}H_{10}N_2O_3$ is a yellow pigment, belonging to the industrially important class of perinone pigments [39]. No crystal structures of this class of compounds have been published so far. For $C_{18}H_{10}N_2O_3$ all attempts to grow single crystals suitable for X-ray structure analysis failed. The largest crystals obtained were thin, bent needles with a diameter of about 1 μm. Therefore the crystal structure was determined from powder data.

The laboratory X-ray powder diagram contained 24 sharp peaks. It could be indexed on the basis of an orthorhombic lattice with the following lattice parameters: $a = 4.79$ Å, $b = 13.30$ Å, $c = 21.01$ Å, $\alpha = \beta = \gamma = 90°$, $V = 1337$ Å^3, $Z = 4$. The lattice parameters were confirmed by electron diffraction on small crystallites. The systematic absences from the X-ray powder pattern led to the extinction symbol $P - nb$, corresponding to the space groups $P2_1nb$ ($Pna2_1$) or $Pmnb$ ($Pnma$). Due to the low quality of the powder diagram it was somewhat uncertain if the observed extinctions were completely obeyed. Reduction of the number of systematic extinction rules increased the number of possible space groups ($P222$, $P222_1$, $P2_12_12$, $P2_12_12_1$, $Pmm2$, $Pmc2_1$, $Pma2$, $Pmn2_1$, $Pmmm$, $Pmma$, and $Pmmn$). Except for $P2_12_12_1$ these space groups are rarely or never observed for organic substances [44]. Therefore the initial calculations were performed in $Pna2_1$ and $P2_12_12_1$. Separate calculations in $Pnma$ were not necessary, since all packings with $Pnma$ ($Z = 4$) would automatically be found during the calculations in its *translationengleiche* subgroups $Pna2_1$ ($Z = 4$) and $P2_12_12_1$ ($Z = 4$) [2].

The number of molecules per asymmetric unit was confirmed by solid-state NMR experiments. In powders measured under magic-angle spinning conditions, crystallographically equivalent atoms are magnetically equivalent. A broadening or splitting of lines is an indication for more than one crystallographically independent molecule. In the hydrogen decoupled ^{13}C-CP-MAS NMR spectrum of the perinone no

344

broadening or splitting of lines was observed, thus giving no hint for more than one molecule per asymmetric unit.

The IR solid state spectrum indicated, that the carbonyl group is probably not involved in hydrogen bonds.

The molecular geometry of the perinone was constructed from crystal structures containing similar fragments (naphtholes, benzimidazoles, peryl imides) [5]. As a first approximation, for the calculations the perinone was assumed to be planar. The positions of the hydrogen atoms of the OH groups were not known (in-plane exo or endo). Therefore four different conformations were set up (Fig. 7). Quantum mechanical calculations (MNDO), and CSD [5] analyses gave no indication, that one of these conformations is more likely than the others. Thus the energy minimizations were performed for each conformer separately. An alternative would have been the release of the torsion angle C-C-O-H, but a flip from 0° to 180° is difficult to achieve during the minimization, and other values are scarce.

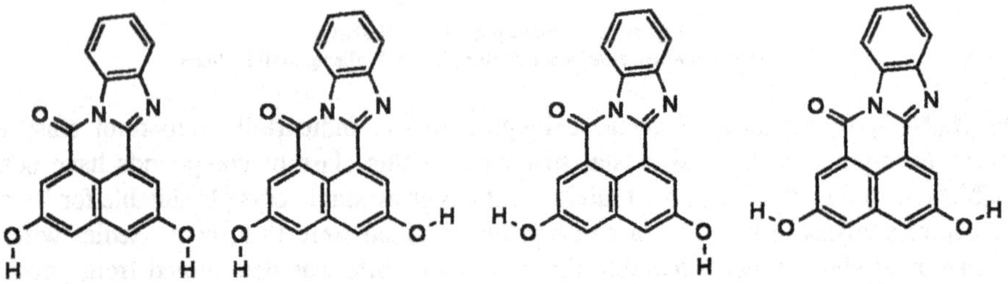

Figure 7. Conformations of the perinone $C_{18}H_{10}N_2O_3$ considered in the energy minimization

The minimization was performed as described above. The starting values for the lattice parameters were randomly chosen from the following ranges: a = 13 to 16 Å, b = 20 to 24 Å, c = 5 to 7 Å. These ranges were chosen in order to find the correct arrangement of the molecules more easily. Random values within reasonable ranges were also used for x, y, z, φ_x, φ_y and φ_z (φ_z = 0 for $Pna2_1$). In the first stage of each minimization all parameters were refined; in the subsequent stage the lattice parameters were set to the experimental values and kept fixed. The correct crystal structure was found within two days: prior to setting the lattice parameters to the experimental values, a packing in $Pna2_1$ with a = 13.56, b = 21.70, c = 4.48 Å was obtained with a calculated powder diagram close to the experimental one. This means that the crystal structure would also have been found if the lattice parameters and possible space groups were not previously known. After setting the lattice parameters to the experimental values the energy of this packing differed from the deepest minimum by a value of 6.6 kJ/mol (see Tab. 6). This difference probably arised from inadequacies in the description of the intermolecular interactions. From the good agreement between calculated and experimental powder diagrams it could be concluded, that the differences between calculated and correct crystal structures are in the order of 0.1Å, although there was no fitting to the experimental peak intensities in this stage (Fig. 8).

Table 6. Calculated energy minima, sorted by energy
(Lattice parameters fixed to $a = 13.30$, $b = 21.01$, $c = 4.79$ Å)

Nr.	E (kJ/mol)	Space group	Confor-mation[a]	Hydrogen Bonds	
1	-164.5	$P2_12_12_1$	a	O–H⋯O=C	
2	-163.0	$P2_12_12_1$	d	O–H⋯O–H,	O–H⋯O=C
3	-160.8	$P2_12_12_1$	d	O–H⋯O–H,	O–H⋯N
4	-158.4	$P2_12_12_1$	b	O–H⋯O=C,	O–H⋯N
5	-158.1	$P2_12_12_1$	c	O–H⋯O=C	
6	-157.9	$P2_12_12_1$	a	O–H⋯O–H	
7[b]	-157.9	$Pna2_1$	d	O–H⋯O–H,	O–H⋯N
11	-155.7	$Pna2_1$	a	O–H⋯O–H	
14	-154.3	$Pna2_1$	d	O–H⋯O–H,	O–H⋯O=C
17	-153.8	$Pna2_1$	b	O–H⋯O=C	
31	-149.9	$Pna2_1$	c	O–H⋯O–H,	O–H⋯O=C

a) See Fig. 7
b) Corresponding to the experimental crystal structure

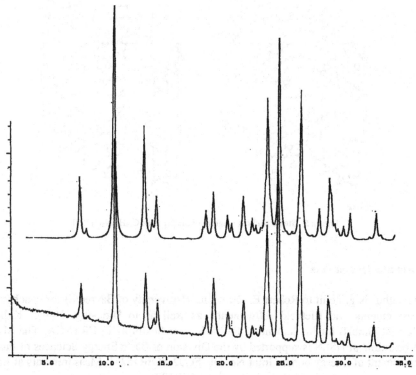

Figure 8. Powder diagrams of the perinone.
Top: powder diagram of calculated crystal structure (without fit to the experimental peak intensities)
Bottom: experimental X-ray powder diagram

346

Subsequently, the crystal structure was refined by Rietveld analysis. Since we were interested in structural details like the planarity of the molecule, the powder diagram was recorded again, this time using high resolution synchrotron radiation. The measurements were carried out at the SUNY X3B1 beamline at the National Synchrotron Light Source (NSLS) of the Brookhaven National Laboratory (USA), and at beamline BM16 of the European Synchrotron Radiation Facility (ESRF) in Grenoble.

During Rietveld refinement, the molecule moved and rotated only slightly from its calculated position. Four intramolecular degrees of freedom were refined, too. The Rietveld analysis confirmed, that the molecule is planar. The refinement converged at R_{Bragg} = 10.7 % (R_{wp} = 5.7%, χ^2 = 2.40) for 592 reflections and 13 parameters.

The crystal structure is shown in Fig. 9. The molecules are connected by a three-dimensional net of hydrogen bonds of the type O–H$\cdots$OH and O–H$\cdots$N. Every molecule is connected to four neighbour molecules. As indicated by IR spectroscopy, the carbonyl group is not involved in hydrogen bridges.

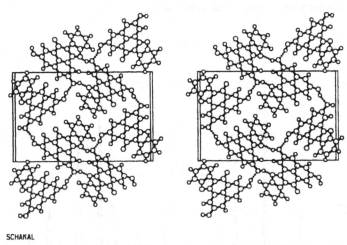

SCHAKAL

Figure 9. Stereo packing diagram of the perinone $C_{18}H_{10}N_2O_3$ [37]

6. Acknowledgements

The author is grateful to Robert E. Dinnebier (University of Bayreuth) for synchrotron measurements and Rietveld refinements, as well as to Dipl.-Ing. Holger Kalkhof (Clariant GmbH) for implementing the molecular flexibility in CRYSCA. The SUNY X3 beamline at NSLS is supported by the Division of Basic Energy Sciences of the US Department of Energy under Grant No. DE-FG02-86ER45231. Measurements at ESRF were carried out under general user proposal CH-189.

7. References

1. Schmidt, M.U. and Englert, U. (1996) Prediction of crystal structures, *J. Chem. Soc. Dalton Trans.* **1996**, 2077-82.
2. Schmidt, M.U. (1995) *Kristallstrukturberechnungen metallorganischer Molekülverbindungen*, Verlag Shaker, Aachen, Germany.
3. Timofeeva, T.V., Chernikova, N.Yu. and Zorkii, P.M. (1980) Theoretical calculation of the spatial distribution of molecules in crystals, *Успехи Химии* **49**, 966-997; *Russian Chemical Reviews* **49**, 509-525.
4. Tsuzuki, S. and Tanabe, K. (1991) Ab initio molecular orbital calculations of the internal rotational potential of biphenyl using polarized basis sets with electron correlation correction, *J. Phys. Chem.* **95**, 139-144.
5. *Cambridge Structural Database* (1999) Cambridge Crystallographic Data Centre, Cambridge, UK.
6. Pertsin, A.J. and Kitaigorodsky, A.I. (1987) *The Atom-Atom Potential Method*, Springer-Verlag, Berlin Heidelberg New York.
7. Gavezzotti, A. and Filippini, G. (1994) Geometry of the intermolecular X–H···Y (X, Y = N, O) hydrogen bond and the calibration of empirical hydrogen-bond potentials, *J. Phys. Chem.* **98**, 4831-37.
8. Gavezzotti, A. and Filippini, G. (1994) Non-covalent interactions in organic crystals, and the calibration of empirical force fields, *Computational Approaches in Supramolecular Chemistry*, NATO ASI Ser., **C426**, pp. 51-62.
9. van Eijck, B.P. and Kroon, J. (1997). Coulomb energy of polar crystals, *J. Phys. Chem.* **B101**, 1096-1100.
10. Schmidt, M.U. and Dinnebier, R.E. (1999) Combination of energy minimizations and rigid body Rietveld refinement: The structure of 2,5-dihydroxy-benzo[de]benzo[4,5]imidazo[2,1-a]isoquinolin-7-one, *J. Appl. Cryst.* **32**, in the press.
11. Hofmann, D.W.M. and Lengauer, T. (1997) A discrete algorithm for crystal structure prediction of organic molecules, *Acta Cryst.* **A53**, 225-235.
12. Motherwell, W.D.S. (1998) Crystal structure prediction and the Cambridge crystallographic database, International School of Crystallography: "Implications of Molecular and Materials Structure for New Technologies", Erice, 28 May - 7 June 1998, Poster Abstracts P. 40.
13. Gdanitz, R.J. (1992) Prediction of molecular crystal structures by Monte Carlo simulated annealing without reference to diffraction data, *Chem. Phys. Lett.* **190**, 391-96.
14. Tajima, N., Tanaka, T., Arikawa, T., Sakurai, T., Teramae, S. and Hirano, T. (1995) A heuristic molecular-dynamics approach for the prediction of a molecular crystal structure, *Bull. Chem. Soc. Jpn.* **68**, 519-527.
15. Wawak, R.J., Gibson, K.D., Liwo, A. and Scheraga, H.A. (1996) Theoretical prediction of a crystal structure, *Proc. Natl. Acad. Sci. USA* **93**, 1734-36.
16. Gavezzotti, A. (1991) Generation of possible crystal structures from the molecular structure for low-polarity organic compounds, *J. Am. Chem. Soc.* **113**, 4622-29.
17. Karfunkel, H.R. and Gdanitz, R.J. (1992) Ab initio prediction of possible crystal structures for general organic molecules, *J. Computat. Chem.* **13**, 1171-83.
18. van Eijck, B.P., Mooij, W.T.M. and Kroon, J. (1995) Attempted prediction of the crystal structures of six monosaccharides, *Acta Cryst.* **B51**, 99-103.
19. Williams, D.E. (1996) Ab initio molecular packing analysis. *Acta Cryst.* **A52**, 326-328.
20. Verwer, P. and Leusen, F.J.J. (1998) Computer simulation to predict possible crystal polymorphs, *Reviews in Computational Chemistry* **12**, 327-365.
21. Schmidt, M.U. and Kalkhof, H. (1997) CRYSCA, Program for crystal structure calculations of flexible molecules, Clariant GmbH, Frankfurt am Main, Germany.
22. Dinnebier, R.E., Dollase, W.A., Helluy, X., Kümmerlen, J., Sebald, A., Schmidt, M.U., Pagola, S., Stephens, P.W. and van Smaalen, S. (1999) Order - disorder phenomena determined by high-resolution powder diffraction: the structures of tetrakis(trimethylsilyl)methane C(Si(CH₃)₃)₄ and tetrakis(trimethylsilyl)silane Si(Si(CH₃)₃)₄, *Acta Cryst.* **B**, submitted.
23. Perlstein, J. (1994) Molecular self-assemblies. 2. A computational method for the prediction of the structure of one-dimensional screw, glide, and inversion molecular aggregates and implications for the packing of molecules in monolayers and crystals, *J. Am. Chem. Soc.* **116**, 455-70.
24. Frank, D. (1997) Hoechst AG, Corporate Research and Technology, Frankfurt am Main, Germany, unpublished results.
25. Erk, P. (1998) BASF AG, Colourants Laboratory, Ludwigshafen, Germany, unpublished results.
26. Gavezzotti, A. (1994) Are crystal structures predictable? *Acc. Chem. Res.* **27**, 309-314.

27. Chaka, A.M., Zaniewski, R., Youngs, W., Tessier, C., Klopman, G. (1996). Predicting the crystal structure of organic molecular materials, *Acta Cryst.* **B52**, 165-183.

28. Leusen, F.J.J. (1994) Ab initio prediction of possible crystal structures, *Z. Krist. Suppl.* **8**, 161.

29. Prokhvatilov, A.I., Isakina, A.P. (1980) An X-ray powder diffraction study of crystalline α-methan-d₄, *Acta Cryst.* **B36**, 1576-80.

30. Mooij, W.T.M., van Eijck, B.P., Price, S.L., Verwer, P. and Kroon, J. (1998) Crystal structure predictions for acetic acid, *J. Computat. Chem.* **19**, 459-474.

31. Freyberg, D.P., Robbins, J.L., Raymond, K.N. and Smart, J.C. (1992) Crystal and molecular structures of decamethylmanganocene and decamethylferrocene. Static Jahn-Teller distortion in a metallocene, *J. Am. Chem. Soc.* **101**, 892-7.

32. Seiler, P. and Dunitz, J.D. (1982) Low-temperatur crystallization of orthorhombic ferrocene: Structure analysis at 98K, *Acta Cryst.* **B38**, 1741-5.

33. Haaland, A. and Nilsson, J.E. (1968) The determination of barriers to internal rotation by means of electron diffraction. Ferrocene and ruthenocene, *Acta Chem. Scand.* **22**, 2653-70.

34. Almenningen, A., Haaland, A, Samdal, S., Brunvoll, J., Robbins, J.L. and Smart, J.C. (1979) The molecular structure of decamethylferrocene studied by gas phase electron diffraction. Determination of equilibrium conformation and barrier to internal rotation of the ligand rings, *J. Organomet. Chem.* **173**, 293-9.

35. Englert, U., Herberich, G.E. and Schmidt, M.U. (1993, submitted 1992) Vorhersage der Kristallstruktur von Pentamethylferrocen, *Z. Krist. Suppl.* **7**, 44.

36. Занин, И.Е., Антипин, М.Ю., Стручков, Ю.Т., Кудинов, А.Р. и Рыбинская, М.И. (1992) Молекулярная и кристаллическая структура пентаметилферроцена (η^5-C₅H₅)Fe(η^5-C₅Me₅) в интервале 153-293 К. Анализ теплого движения в кристалле по рентгеновским дифрактионным данным, *Металлорганическая Химия* **5**, 579-89. (Zanin, I.E , Antipin, M.Yu., Struchkov, Yu.T., Kudinov, A.R. and Rybinskaya, M.I. (1992) Molecular and crystal structure of pentamethylferrocene (η^5-C₅H₅)Fe(η^5-C₅Me₅) in the interval 153-293 K. Analysis of the thermal motion in the crystal using X-ray diffraction data, *Metallorganicheskaya Khimiya* **5**, 579-89).

37. Keller, E. (1997). SCHAKAL97, Kristallographisches Institut der Universität, Freiburg, Germany.

38. Herbst, W. and Hunger, K. (1997). *Industrial Organic Pigments.* 2nd ed., Verlag Chemie, Weinheim, Germany.

39. Whitaker, A. and Walker, N.P.C. (1987) CI Pigment Yellow 74, α-(2-Methoxy-4-nitrophenylhydrazono)-α-aceto-2'-methoxyacetanilide, *Acta Cryst.* **C43**, 2137-41.

40. Society of Dyers and Colourists (1982) *Colour Index,* 3rd ed..

41. LeBail, A. (1999). *Structure Determination from Powder Diffraction - Database.* Available from http://fluo.univ-lemans.fr:8001/iniref.html .

42. Harris, K.D.M. and Tremayne, M. (1996) Crystal structure determination from powder diffraction data, *Chem. Mater.* **8**, 2554-2570.

43. Rietveld, H.M. (1969) A profile refinement method for nuclear and magnetic structures, *J. Appl. Cryst.* **2**, 65-71.

44. Belsky, V.K., Zorkaya, O.N. and Zorky, P.M. (1995) Structural classes and space groups of organic homomolecular crystals: new statistical data, *Acta Cryst.* **A51**, 473-481.

THE PREDICTION AND PRODUCTION OF POLARITY IN CRYSTALLINE SUPRAMOLECULAR MATERIALS

A general principle of polarity formation

J. HULLIGER, S. W. ROTH, A. QUINTEL
Department of Chemistry and Biochemistry, University of Berne
Freiestrasse 3, CH-3012 Berne, Switzerland
Phone: +4131 6314241, Fax: +4131 6313993
Email: juerg.hulliger@iac.unibe.ch
Homepage: http://dcbwww.unibe.ch/groups/hulliger/

Abstract

Spontaneous polarity formation in molecular crystals is a key issue in the design of materials featuring tensorial properties of interest to technical applications. This review discusses theoretical models for the prediction of spontaneous polarity formation in channel-type zeolites, channel-type inclusion compounds and organic solid solutions. Theoretical models presented here are based on a general principle of polarity formation, typical for growth processes where dipolar compounds are attached to surfaces.

"Molecular Design of Materials. Science, guided by molecular understanding, takes up the challenge to create materials for the future" A. R. von Hippel, 1962

1. Introduction

There is a long tradition in the synthesis and crystal growth of dielectric materials showing *polar* properties called, e.g. the pyroelectric, the piezoelectric, the second order nonlinear optical (NLO) and the linear electro-optic (EO) effect [1, 2]. The development of property-directed theoretical and synthetic concepts is therefore one of the challenging issues facing modern crystallography, crystal physics and supramolecular chemistry. In this respect, the present review is a summary of recently developed theoretical concepts covering the *spontaneous formation of polarity* in crystals grown by assembling *dipolar* compounds (for more details, see [3-7]).

The linear EO effect (the electric field-induced change of the refractive index of a material; for a further discussion see C. Bosshard, this book) is one of the technologically promising properties which can result from particularly designed *organic* materials. At the molecular level, the presence of a high value of the molecular hyperpolarisability β is known to be a first basic (electronic) requirement for a 3rd rank NLO property. This is provided by a large difference in the acentric electron density distribution between the ground and first excited state [1]. Many organic molecules investigated for the purpose of polar solid state properties possess a framework of an

D. Braga et al. (eds.), Crystal Engineering: From Molecules and Crystals to Materials, 349–368.

acentric π-conjugation to which electronic donor (D) and acceptor (A) substituents are attached [2] (Tab. 1). For the EO effect in particular, *parallel alignment* of all molecular directions showing the largest β_{ijk} represents a second basic (geometrical) condition to achieve a maximum EO response [1, 2]. This may be accomplished by selecting *rod-shaped* molecules which (i) tend to pack closely, like pencils in a box, along their $\beta_{longitudinal}$ axis, and which (ii) thereby align in an acentric fashion.

At this point, readers may already notice one key issue of crystal engineering of polar materials: *how do you induce the acentric packing of acentric rod-shaped molecular bricks?*

To give an extensive account for the packing behaviour of organic molecules is beyond the scope of this contribution [8]. Instead, the focus will be on the tendency of rod-shaped molecules to pack closely in lateral dimension by aligning their long axis parallel. Hereby, terminal groups can promote chain formation. Topologically, such structures may be described by an architecture of *aligned chains*. In some respect, packing requirements are similar to the *nematic* state of liquid crystals.

Fig. 1: Schematic representation of a general principle of spontaneous polarity formation in condensed matter. →: dipoles, S: surface to which dipoles attach.

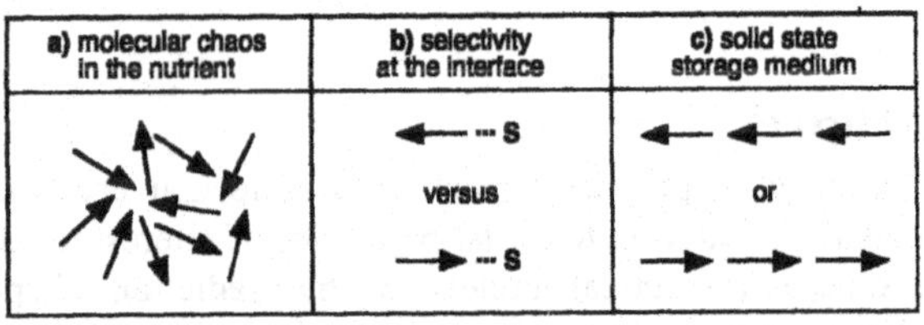

a) molecular chaos in the nutrient	b) selectivity at the interface	c) solid state storage medium
	← ··· S	← ← ←
	versus	or
	→ ··· S	→ → →

Once aligned this way, individual chains, as well as their assembly, may feature a *centric* or an *acentric* packing. Our theoretical discussion will focus on the possibility to induce polar ordering within such an architecture of aligned building blocks by establishing a model which takes into account (i) the *selective nature of interactions between terminal groups D and A*, and (ii) an energy parameter describing the lateral interaction. In more general terms, we address the phenomena related to the degree of orientational ordering or disorder ($0° \rightleftharpoons 180°$) within a topologically organised structure of aligned molecules. It is important to notice right at the beginning that all models discussed below, which describe spontaneous polarity formation, explicitly rely on a mechanism of *slowly growing crystals*. A summary of recent experimental approaches, which yield almost perfectly aligned molecules featuring polar properties, is provided by Tab. 1 and reviewed in [9].

2. A general principle of polarity formation

The fact that the 3rd rank electrical susceptibility of most molecular crystals is to a very good approximation proportional to the density of dipoles and to the hyperpolarisability $\beta_{longitudinal}$ of the molecular bricks, allows for a rather simple but far reaching conclusion: (i) the polarity of crystals made up by dipoles is due to the *acentric* building blocks; (ii) all we have to control is the formation of the largest possible *vector sum* of dipoles within a macroscopic volume at a given temperature. Here, an underlying question is emerging: *from the Braunian chaos in a nutrient to the final packing geometry in the bulk, is there a prevalent step preserving a vector property, independent of the 2D point symmetry of the growing crystal face?* The answer is yes: it is the *surface* of a growing crystal, where dipoles are attached to at an angle $\theta \neq 0$. Fig. 1 is a summary of the key steps to spontaneous polarity formation resulting from a *growth process*. First, let us consider a simple example where dipoles are attached to a surface (S) providing a homogeneous but different interaction to both ends of a dipolar molecule A-π-D. One orientational state will be favoured over the corresponding one, because of $\Delta E = E(\leftarrow \cdots S) - E(\rightarrow \cdots S) \neq 0$. At next we may distinguish between ad-layers attaching to a homogeneous surface where the *lateral* interaction energy between molecules (i) does ($\Delta E_\perp \neq 0$) or (ii) does not ($\Delta E_\perp = 0$) depend on the relative orientation of dipoles ($\Delta E_\perp \equiv E_{parallel} - E_{antiparallel}$). In the case of $\Delta E_\perp = 0$, the resultant polarity will simply be a function of the selective interactions of terminal groups (see Chap. 4). If, however, $\Delta E_\perp \neq 0$, competition can arise between effects of orientational ordering in 2D with respect to the interactions with S (1D). Before discussing these effects in terms of thermodynamics and statistical mechanics, let us introduce a simple *mechanical analogue* [10], demonstrating polarity formation on the macroscopic level.

Fig. 2: Schematic view of a mechanical analogue simulating the stochastic formation of polar ordering in a 1D storage medium. Plastic pearls were suspended in paraffin oil and kept in motion by a shaking machine. Each run induced the entrance of about 25 to 27 pearls (ca. 9% of total) into the storage cylinder within a few hours. A rather long running time ensured a statistical process

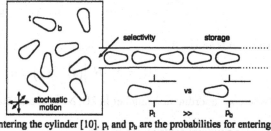

concerning an orientational selectivity for entering the cylinder [10]. p_t and p_b are the probabilities for entering the cylinder.

According to Fig. 2, a suspension of drop-shaped pearls is subject to random translational movement in all three directions. To one side of the vessel, a cylinder is attached. Pearls can enter through an appropriate hole, being then stored in the orientational state adopted at the entry. The diffusion of pearls from the main *vessel* (Fig. 1a) into an adjacent *storage medium* (Fig. 1c) is controlled by the difference in the aspects of the tip (t) and the back (b) section of the pearls (Fig. 1b). This is analogous to the orientational selectivity imposed on molecular dipoles within a nutrient source before they enter a solid state structure in which after the process of growth no 180° re-orientation is possible. In this example we obviously have a case where $\Delta E_\perp$ is strictly zero. In the course of our experiments (40 independent runs), an average fraction $x_{tip} = 0.98(3)$ was obtained, which clearly demonstrates the preference for the pearls to enter the cylinder tip-first [10].

Fig. 3: Schematic representation of a homogeneous substrate (where dipoles do not relax) with an attached ad-layer (where relaxation is allowed), containing dipoles oriented either ↓ or ↑ (molar fractions x_A, x_D, respectively). Dipoles in the nutrient have no preferred orientation.

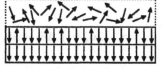

Now we come back to a *thermodynamic* description of a *single* ad-layer allocated to a homogeneous surface S. According to Fig. 3 the dipoles can adopt either an orientation pointing up- or downwards. Let us call x_A the molar fraction of dipoles pointing downwards (↓), whereas $x_D = 1 - x_A$ denotes the opposite orientation (↑).

Dipoles within the ad-layer are subject to accomodation of the lowest free energy F with respect to x_A, x_D. As for many basic descriptions, F is given by

$$F = U - T \cdot S , \tag{1}$$

where U accounts for the *intermolecular interactions* in the longitudinal (1D, ∥) and transversal (2D, ⊥) directions, whereas here, S represents the *configurational* entropy. Following a mean-field type approach, known from the discussion of solid solutions and surface roughening [11, 12], we introduce 2D interactions as follows:

$$\frac{z_\perp}{2} \cdot N \cdot x_A^2 \;\triangleq\; \downarrow\cdots\downarrow \;\triangleq\; E_{parallel} , \tag{2a}$$

$$\frac{z_\perp}{2} \cdot N \cdot (1-x_A)^2 \;\triangleq\; \uparrow\cdots\uparrow \;\triangleq\; E_{parallel} , \tag{2b}$$

$$\frac{z_\perp}{2} \cdot N \cdot x_A \cdot (1-x_A) \;\triangleq\; \uparrow\cdots\downarrow \;\triangleq\; E_{antiparallel} . \tag{2c}$$

This yields an $U_\perp$ of

$$U_\perp = z_\perp \cdot N \cdot \left[\left(x_A^2 - x_A \right) \cdot \Delta E_\perp + \tfrac{1}{2} E_p \right] \tag{3}$$

(where $z_\perp$: coordination number in 2D, N: total number of dipoles within the ad-layer).

The longitudinal contribution to U is

$$U_\parallel = N \cdot \left[E_{AA} - x_A \cdot \Delta E_\parallel \right] , \tag{4}$$

where $\Delta E_\parallel \equiv E_{AA} - E_{AD}$, (for a substrate providing e.g. only A groups for ∥ interaction). E_{AA}: (S)←$\cdots$→(L), E_{AD}: (S)←$\cdots$←(L) (S: substrate, L: ad-layer).
The configurational entropy, resulting from the possibilities in arranging the dipoles with respect to ↓ and ↑ is

$$S = -N \cdot k_B \cdot \left[x_A \cdot \ln x_A + (1-x_A) \cdot \ln(1-x_A) \right] \tag{5}$$

(for more details on S, see textbooks on physical chemistry or [13, 14, 15]; k_B: Boltzmann constant).
By summing up the terms given by Eqs. 3, 4, 5 we obtain $F(x_A, T)$. x_A results from the global minimum of $F(x_A, T)$, therefore we set $\dfrac{\partial F}{\partial x_A} = 0$ to find

$$z_\perp \cdot (2x_A - 1) \cdot \Delta E_\perp - \Delta E_\parallel + k_B T \cdot \ln \frac{x_A}{1-x_A} = 0 . \tag{6}$$

Fig. 4 demonstrates that polarity can occur already at $\Delta E_\perp = 0$, although not at its maximum value, where all dipoles would be oriented parallel. To the left side where $E_p - E_{ap} < 0$, lateral coupling finally aligns all dipoles to form an almost perfectly oriented layer (↑↑) with maximum polarity. To the right side where $E_p - E_{ap} > 0$ we would approach a state of antiparallel ordering (↓↑) which can not be properly described by a mean-field model according to Eq. 2. A more complete prediction was obtained by a

Monte Carlo simulation assuming the same type of nearest neighbour interactions ($z_\perp = 4$), but described by methods of statistical mechanics [16a]. Following this approach, a *non-polar* state is in our example approached at $\Delta E_\perp > 4.5$ kJmol^{-1}.

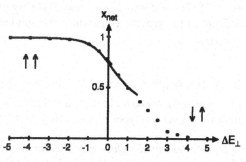

Fig. 4: Numerical solutions for the net polarity $x_{net} \equiv 2x_A - 1$ as a function of the lateral interaction energy $\Delta E_\perp$ [kJmol^{-1}]. Straight line: derived from Eq. 6; dots: results from a stochastic simulation [16a]. $T = 300$ K, $z_\perp = 4$ and $\Delta E_\parallel = 5$ kJmol^{-1}. $\uparrow\uparrow$, $\downarrow\uparrow$: state of parallel or antiparallel ordering, respectively.

So far, we have considered a surface providing a *homogeneous* interaction to dipoles. However, realistic crystals may contain surface sites yielding different energies with respect to both orientations of dipoles attached to them. In such cases we have at least *two* different $\Delta E_\parallel$ values ΔE_A, ΔE_D ($\Delta E_A \equiv E_{AA} - E_{AD}$, $\Delta E_D \equiv E_{DD} - E_{AD}$), which under the influence of $\Delta E_\perp$ control the resultant polarity of each layer attached during growth. Typical examples are the growth of topologically *centric* crystals, where at least two sites occupied by dipoles in the bulk are related by a center of symmetry. In this case the $\Delta E_{A,D} \equiv \Delta E_A - \Delta E_D$ can introduce some polarity through the influence of configurational entropy. The most frequently occuring space group in organic crystals P2$_1$/c (2/m) (Fig. 5) is hence affected to form defects (T$_{growth}$ >> 0), i.e. to undergo orientational disorder [17]. This can introduce some net polarity, i.e. *splitting crystals into macroscopic domains* which are related by a center of symmetry. An essential assumption made here is that during thermal relaxation of an ad-layer, the substrate layer is kept *invariant*. The reason is that for a site close to bulk conditions, the energy of activation for a re-orientation is much larger than for molecules within the ad-layer. Polarity acquired through the process of slow crystal growth represents therefore a *metastable state*, stabilised by a kinetic barrier.

Fig. 5: Representation of P2$_1$/c with a dipolar molecule in a general position (screw axes are not shown). Corresponding sites on (010) and (0$\bar{1}$0) faces showing an opposite dipole orientation are labeled by I and II. Preferred orientational disorder with respect to mis-aligned attachments at I or II along both directions of the b-axis produces non-vanishing dipole sums (projection onto b) parallel or antiparallel to either the b or $-b$ direction [17].

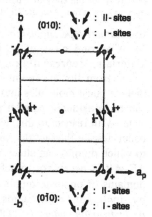

Our discussion above is leading to a quite universal statement about spontaneous polarity formation in molecular crystals: in case of $\Delta E_{i,hkl} = E(\leftarrow \cdots S_{i,hkl}) - E(\rightarrow \cdots S_{i,hkl}) \neq 0$, i.e. $\Delta E_{i,hkl}$ is different for at least two sites i and i' related by a center of symmetry in the bulk, polarity develops because of an unbalanced energy difference while the crystal is subject to orientational disorder [10].

354

The "causing event" (dipoles interacting with a surface) and the second law of thermodynamics require that dipolar molecules included into any growing crystal lattice, will always form growth sectors belonging to a pyroelectric symmetry group ($\theta > 0°$). The extent of any measurable pyroelectric effect will, however, primarily depend on the growth temperature, the solid state concentration of dipoles, the energy differences given by $\Delta E_{i,i',hkl}$, on the particular packing of dipoles and the dipole moment [10].

3. A classification of materials exhibiting spontaneous polarity formation

Phenomena of *symmetry* reduction and *twinning* related to polarity formation have been reported for various types of materials [18, 10]:

I: *One-dimensional diffusion of dipolar molecules into parallel channels of a zeolite* [19] represents an example which is conceptionally very close to our mechanical analogue discussed above: the dipoles enter a zeolite channel with different probabilities in regard to the terminal groups of A-π-D molecules. If channels at corresponding sides of crystals are open, molecules may enter from both sides and form a bipolar state of aligned dipole chains (Fig. 6). Experimental confirmation of this situation emerged from a new type of a scanning probe microscopy, based on the measurement of the *local* pyroelectric effect of crystals [19]. It is interesting that a mechanism of polarity formation by 1D diffusion does not need $\Delta E_\perp \neq 0$. Nevertheless, the dipole-channel interaction can influence the final state, because the probability to enter the channel may depend on the orientation of dipoles, and diffusion in one direction may be faster than in the opposite one. According to Fig. 1, the selectivity of interactions at the entry into a zeolite channel represents a sufficient criterium to obtain a bipolar state featuring finally two maroscopic domains of opposite polarity.

Fig. 6: Representation of a zeolite channel where open ends on either side can provide the same selectivity for the entry of dipoles. Thus, a bipolar state may result from a centric zeolite structure.

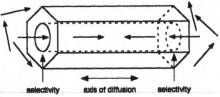

selectivity   axis of diffusion   selectivity

II: *Channel-type inclusion compounds* [3-4]. A second class of purely molecular materials, representing *supramolecular solids* from the viewpoint of self-assembly, is given by host compounds which can include guest molecules into parallel channels (Fig. 7a). Channel and polarity formation can be considered parallel processes forming (i) a new compound and (ii) a bipolar state. With respect to this, we will have to discuss (Chap. 4) why polarity formation can occur despite the fact that there are *no* preformed channels. However, there is a surface to which dipoles can attach with a tendency for a preferred orientation. As in case I, no $\Delta E_\perp \neq 0$ is necessary to achieve polarity, which again is split into two macroscopic domains of opposite dipole orientation (Fig. 7b). Experimental examples are provided by inclusion compounds of thiourea, tri-*o*-thymotide and in particular by *perhydrotriphenylene* (PHTP) [20].

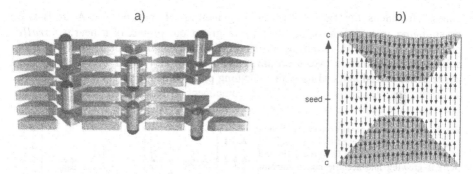

Fig. 7: a) Schematic representation of the growth of host-guest (triangles-pills) crystals: stabilising and destabilising interactions between A and D terminal groups (dark and light cappings of the pills) fix the orientation of an attaching dipole in longitudinal growth direction. Lateral growth of every new channel starts at a random location with random orientation of the first dipole. b) As a result, crystals are structured into two cone-like macro-domains of opposing polarity (dark regions). For details on the lateral growth mechanism, see [23, 6b, 6c].

III: *Organic solid solutions* ($H_{1-x}G_x$) [18]. We may consider another example where the general principle of polarity formation [10] is valid, namely *solid solutions*, if we add further degrees of freedom to the engineering of an architecture: we will define the components H (host) and G (guest) which can allow fairly different situations. If we specify molecules which form a polar lattice we again refer to the topological definition given above, differences are due to the *terminal functional groups*:

IIIa	H:	dipolar molecule A-π-D
	G:	molecule identical to H, but faulted in the dipolar orientation; dipolar molecule A'-π'-D', faulted or not faulted in orientation; X-π-X, X = A or D or R (R: less polarisable group)

IIIb	H:	X-π-X, X = A or D or R (R: less polarisable group)
	G:	dipolar molecule A-π-D

IIIa, IIIb differ with respect to $H_{1-x}G_x$ (x << 1, IIIa; x >> 0, IIIb), i.e. the role of *hosting* or *being hosted*.

Class IIIa takes up ideas put forward by A. I. Kitaigorodsky (†1985) a long time ago [13]. He considered *orientational defect formation as formally equivalent to solid solution formation by two different species*. By a solid solution we generally mean miscibility in the crystalline state which may cover only part or the entire range of the composition $H_{1-x}G_x$ (0 < x < 1). So, if x << 1, we may consider G a perturbation to a centric or acentric structure. General considerations take us to the conclusion that both symmetry types can undergo orientational defect formation (i) either by 180° faulted H or (ii) by inclusion of a different component G. *Centric* structures will adopt a certain extent of polarity by faulted H within particular domains related by symmetry, whereas *acentric* structures can exceed a certain amount of polarity reduction or increase (Fig. 8a). Examples of class III symmetry reduction for x << 1 have been investigated extensively at the Weizmann institute [18].

Class IIIb opens the field to crystal engineering. H may be A-π-A or D-π-D, crystallising in a centric or acentric space group but providing a nearly *parallel alignment* of bricks. If - from the structural, thermodynamic point of view - there is solid solution formation over a certain range of x, we may begin with a purely non-pyroelectric crystal and end up with something polar, although we do not need to have an acentric native structure for the dipolar component G! This sounds like a paradox but is nevertheless quite simple to retrieve from a summary given in Figs. 8, 14.

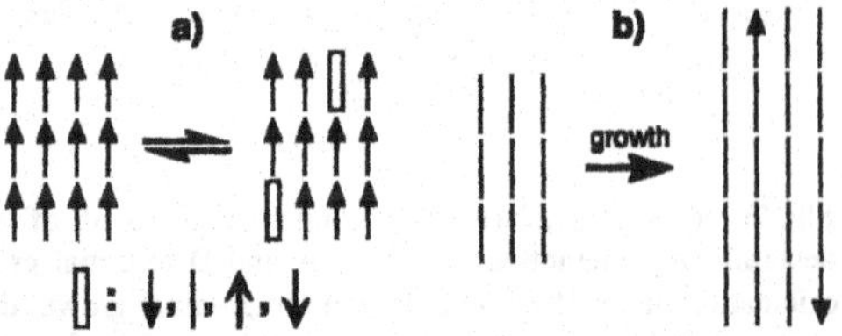

Fig. 8: a) Types of possible defects ☐ in a $H_{1-x}G_x$ crystal. ↓: 180° faulted H, |: non-polar G (e.g. A-π-A), ↑, ↓: different G. b) A centric seed (left) is growing into acentric macro-domains (right).

Up to here, we have outlined formal criteria to assemble molecules which can lead to spontaneous polarity formation in various existing materials. Recognition of a *general principle* behind all this was stated just recently [10]. Before we get more technical, let us recall these basic types of materials I, II, III, with respect to interactions relevant to polarity formation:

I   zeolites   : $\Delta E_{\parallel} \neq 0$; $\Delta E_{\perp}$ may be zero or not

II   inclusion compounds   : ΔE_A, $\Delta E_D \neq 0$, $\Delta E_{\perp} < 0$: can enhance polarity formation; $\Delta E_{\perp} > 0$: may cancel polarity formation

III   solid solutions   : ΔE_A, $\Delta E_D \neq 0$; $\Delta E_{\perp} \gtreqless 0$: can be the dominant term deciding whether a structure is acentric or centric. $\Delta E_{\perp} < 0$ is not a necessary condition to allow polarity fomation

4. Polarity evolution in channel type inclusion compounds

Particular views to be presented in this chapter have been elaborated elsewhere [5-7]. However, we will discuss here a generalised theoretical framework. Starting with the general principle for polarity formation we recognise easily that in inclusion compounds providing parallel and well seperated channels for the hosting of guest molecules, growth along the channel direction is associated with a surface. This surface is considered to be a heterogeneous substrate giving rise only to a small $\Delta E_{\perp}$. If we apply thermodynamic arguments for one ad-layer, as discussed in Chap. 2, we have to solve two coupled equations for $x_A(A)$ and $x_D(D)$ (for a definition of variables, see Fig. 9):

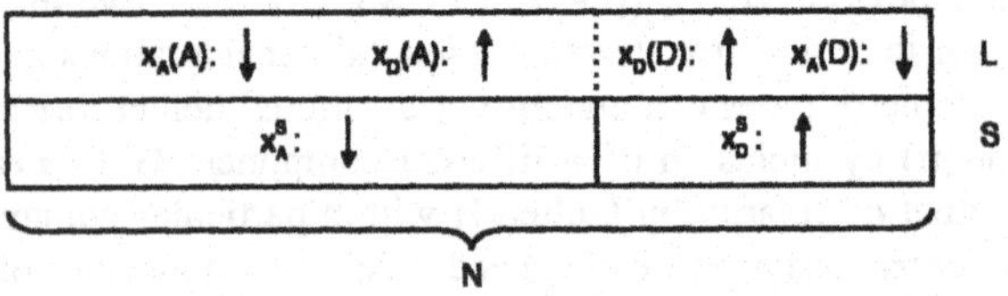

Fig. 9: Definition of variables (molar fractions x) for two adjacent layers representing the fixed substrate (S) and the ad-layer (L). N: total number of dipoles attached to S, L. x_A^S, x_D^S: molar fractions of either dipole orientation on S. $x_A(A)$, $x_D(A)$: corresponding fractions accommodating to the sub-ensemble x_A^S; $x_D(D)$, $x_A(D)$ accommodating to x_D^S, respectively.

$$0 = 2z_\perp \Delta E_\perp \left[x_A(A) - x_D(D) - \tfrac{1}{2}\left(x_A^s - x_D^s\right) \right] - \Delta E_A + k_B T \cdot \ln \frac{x_A(A)}{x_A^s - x_A(A)} \quad , \qquad (7)$$

$$0 = -2z_\perp \Delta E_\perp \left[x_A(A) - x_D(D) - \tfrac{1}{2}\left(x_A^s - x_D^s\right) \right] - \Delta E_D + k_B T \cdot \ln \frac{x_D(D)}{x_D^s - x_D(D)} \quad . \qquad (8)$$

Explicit expressions for $x_A(A)$, $x_D(D)$ can be found by an iterative solution starting from $\Delta E_\perp = 0$ and increase to final values. For the purpose of a discussion of channel-type inclusion crystals of perhyrotriphenylene (PHTP) we assume a neglectable $|\Delta E_\perp|$. Therefore, Eqs. 7, 8 separate to

$$x_A(A) = x_A^s \cdot \frac{f_A^{(0)}}{f_A^{(0)} + 1} \quad , \quad x_D(D) = x_D^s \cdot \frac{f_D^{(0)}}{f_D^{(0)} + 1} \quad ; \qquad (9, 10)$$

with
$$f_A^{(0)} \equiv e^{\frac{\Delta E_A}{k_B T}} \quad , \quad f_D^{(0)} \equiv e^{\frac{\Delta E_D}{k_B T}} \quad . \qquad (11, 12)$$

Taking into account that

$$x_A(A) + x_D(A) \equiv x_A^s \quad , \quad x_A(D) + x_D(D) \equiv x_D^s \quad , \qquad (13, 14)$$

and

$$x_A(A) + x_A(D) \equiv x_A^L \quad , \quad x_D(A) + x_D(D) \equiv x_D^L \quad , \qquad (15, 16)$$

the final result can be written in matrix notation (where x_A^L, x_D^L: resulting dipoles with $\downarrow(x_A^L)$ or $\uparrow(x_D^L)$ within the ad-layer on the substrate providing x_A^s, x_D^s):

$$\begin{bmatrix} x_A^L \\ x_D^L \end{bmatrix} = \begin{bmatrix} \dfrac{f_A^{(0)}}{f_A^{(0)} + 1} & \dfrac{1}{f_D^{(0)} + 1} \\ \dfrac{1}{f_A^{(0)} + 1} & \dfrac{f_D^{(0)}}{f_D^{(0)} + 1} \end{bmatrix} \cdot \begin{bmatrix} x_A^s \\ x_D^s \end{bmatrix} \quad . \qquad (17)$$

Starting from a substrate with $x_A^s = x_D^s$ (no polarity), the matrix $P(A, D)$ (Eq. 17) can generate x_A^L, x_D^L values where $x_A^L \neq x_D^L$. If we move from one layer to the next and so on, Eq. 17 can be applied as many times as necessary because every layer q grown on the q-1 layer will act as a substrate for the growth step q. Therefore, continuing polarity evolution can be described by

$$\begin{bmatrix} x_A^L(q) \\ x_D^L(q) \end{bmatrix} = \begin{bmatrix} p_{11} & p_{12} \\ p_{21} & p_{22} \end{bmatrix}^q \cdot \begin{bmatrix} x_A^s \\ x_D^s \end{bmatrix} \quad , \quad q = 1, 2, \ldots, \infty \quad . \qquad (18)$$

p_{ij} matrix elements of P (Eq. 17) are considered *transition probabilities*, according to a growth process shown by Fig. 1 in [6b, 6c]). Fig. 10 shows polarity evolution calculated from Eqs. 17, 18.

It is important to note that Eqs. 7, 8 describe only polarity development *along* the channel axis, relying on a pre-existing substrate of any possible composition with respect to x_A^s, x_D^s. Seen this way, our theory is identical to a general description of a

358

"growth" process stated many years ago by the Russian mathematician A. A. Markov (1856 - 1922). So called *homogeneous Markov chains* (see [21] for mathematical definitions) describe a new state i based on its previous state j assuming a constant probability p for the transition i→j. At large q (→ ∞) x_A^L, x_D^L reach stationary values which do *not* depend on x_A^S, x_D^S [5, 6]:

$$x_A^L(\infty) = \frac{p_{12}}{p_{12} + p_{21}} \quad , \quad x_D^L(\infty) = \frac{p_{21}}{p_{12} + p_{21}} \tag{19, 20}$$

If we want to calculate the polarity evolution as q is increased, diagonalisation of P yields [22]:

$$x_{net}(q) \equiv x_A^L(q) - x_D^L(q) = x_{net}(\infty) \cdot (1 - \lambda^q) \quad , \quad x_{net}(\infty) = \frac{p_{12} - p_{21}}{p_{12} + p_{21}} \quad ; \tag{21, 22}$$

where $x_{net}(\infty)$ is the final state of the net polarity after infinite attachment steps, and $\lambda = 1 - p_{21} - p_{12}$ is one of the two eigenvalues of the Markov matrix P (Eq. 18, $\lambda' = 1$). According to Fig. 10, x_{net} reaches a stationary value after a fairly low up to a very large number of attachment steps q, depending on corresponding energy differences.

Fig. 10: Influence of temperature on averaged polarity evolution. Interaction energies: $E_{AD} = -5.7$, $E_{DD} = -2.8$, $E_{AA} = 0$ [kJmol^{-1}]. a) $T = 50$ K, $x_{net}(\infty) = 0.99$; b) $T = 100$ K, $x_{net}(\infty) = 0.93$; c) $T = 200$ K, $x_{net}(\infty) = 0.65$; d) $T = 300$ K, $x_{net}(\infty) = 0.44$. Increasing the temperature leads to a down-scaling of the interaction energies.

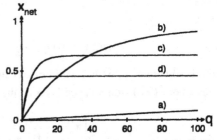

Summarising the main results of this chapter, we conclude that polarity formation in channel-type inclusion compounds is possible, although there is no $\Delta E_\perp < 0$ assumed. As outlined for other materials (I, II, III, Chap. 3) a bipolar state is obtained. For a mechanism accounting for effects of *lateral growth* (Fig. 7b), see [6b, 6c, 23, 24].

So far, we have discussed guest molecules featuring typical -A···D-, -A···A- and -D···D- interactions, well known from the supramolecular concept of synthons [25]. As stated in Fig. 10, (i) a rather *low* level of the maximum polarity may be obtained, or (ii) many growth steps may be necessary in order to yield a x_{net} close to 1. After having developed the basics of our model, we may ask: *is there a particular set of the aforementioned interactions which would favour (i) $x_{net} \approx 1$, (ii) at a low value of q?*

If we introduce the guest molecules A-π-D-R or D-π-A-R (R: non-polarisable group), the following interactions are of importance:

$$-A\cdots R\text{- or -}D\cdots R\text{- :} \quad \text{small, -}R\cdots R\text{- : small}$$
$$-A\cdots A\text{- or -}D\cdots D\text{- :} \quad \text{significant, e.g. } E_{DD} < 0 < E_{AA}$$

Keeping only interactions -A···A- or -D···D- different from zero, the solution to our design problem is easily available [9]:

Fig. 11: a) Development of the net polarity $|x_{net}|$ as a function of the longitudinal interaction energy $\Delta E_\parallel$ [kJmol^{-1}] at 300 K. For destabilising ($\Delta E_\parallel > 0$) interaction energies, the attained maximum polarity can be up to three times higher than for stabilising ($\Delta E_\parallel < 0$) interaction energies. b) Number of minimum attachment steps q until ~ 95% of the maximum possible net polarity x_{net} is attained, plotted as a function of $\Delta E_\parallel$.

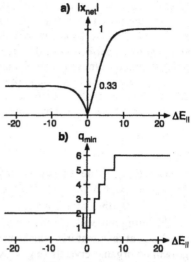

A nice example of the effect of *molecular recognition* in channel-type inclusion compounds results if A-π-D and D-π-D or A-π-A molecules are simultaneously present in the nutrient [6b]. In the case of A-π-D / D-π-D a relatively low concentration of D-π-D in the nutrient can cancel or invert x_{net}, i.e. the resultant polarity may change sign by adding a *non-polar* co-guest. Contrary, the addition of A-π-A to a system raises x_{net} above the value of a pure system. More details on this rather strange behaviour can be found in [6b, 6c] and on our poster contribution to this school.

To close this chapter, we conclude that channel-type inclusion compounds represent ideal systems to *achieve maximum polarity at low q*, if appropriate guest molecules are selected. Preferential guest molecules fortunately are members of a large class of very effective nonlinear optical molecules. According to Fig. 12, inclusion compounds of PHTP and many dipolar compounds can be obtained by a) a solid state reaction, b) crystal growth from (i) not co-included solvents, (ii) from the melt (if A-π-D(ℓ) is miscible with PHTP(ℓ)), and (c) from the vapour (if the vapour pressure of guest molecules is sufficiently high). Details on the crystal growth of PHTP-guest compounds can be found in [3, 24].

Fig. 12: Synthesis of a PHTP-guest inclusion compound: (i) the host-guest stoichiometry (n, m, resp.) is set; (ii) growth conditions are chosen to achieve a negative ΔG_R; (iii) a polar or non-polar seed may be formed at first, which undergoes polarity evolution.

5. The design of organic solid solutions featuring polar properties

The potential of solid solutions to exhibit polar properties has been recognised by L. Leiserowitz, M. Lahav and co-workers [18], although A. I. Kitaigorodsky put forward similiar concepts much earlier by his analysis of the structure and properties of *mixed organic crystals* [13].

Assuming complete miscibility (often limited because of ordered phase formation), we may distinguish two classes of $H_{1-x}G_x$ (H: host compound, G: guest compound) with respect to composition x: (i) x << 1; (ii) 0 < x < 1.

360

Case (i) can be considered to be a description of faulted crystals, faulted by inverted H or some impurity G. Centric crystal structures of H are of particular interest, because at $x \neq 0$, these structures can develop polarity independent of the structural and compositional perfection of the seed. An estimation of x for the growth of a crystal of point symmetry 2/m has been obtained recently [17] by application of a model which is formally equivalent to W. Schottky's (1886 - 1976) treatment of point defects [27]. In cases where the molar fraction x_i ($i = A, D$) of defects at temperature T is below e.g. ~ 0.01, x_i can be described by a most simple expression:

$$x_i \cong e^{\frac{-\Delta E_{defect,i}}{k_B T}} , \qquad (23)$$

where $\Delta E_{defect, i}$, the energy of defect formation, may in some cases be given by $\Delta E_{defect, i} \equiv \Delta E_i + \Delta E_\perp$ ($i = A, D$). We assume $E_{antiparallel} < 0$, $|E_{antiparallel}| > |E_{parallel}|$, therefore is $\Delta E_\perp > 0$ (the lateral energy difference comprises a particular coordination by neighbouring molecules in the centric lattice); and $\Delta E_i > 0$. In Fig. 13 we have plotted x vs ΔE_{defect} at temperatures which correspond to different experimental conditions for the growth of organic crystals (e.g. 4-cyano-4'-iodobiphenyl [17]). For more details and a discussion of the effect of impurities, see [10, 28].

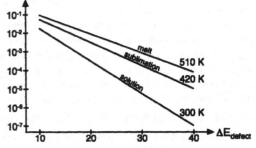

Fig. 13: Schematic representation of the concentration of defects x_{defect} plotted against the energy of defect formation ΔE_{defect} [kJmol⁻¹]. The three plots at different temperatures may be representative for different growth methods.

If methods of statistical mechanics are applied to dipoles in e.g. 2D, preliminary results confirm that in case of $\Delta E_\perp > 0$, a non-vanishing polarisation can exist near the surface. The number of layers where polarisation occurs, strongly decreases if $\Delta E_\perp$ is increased from positive values near zero to realistic energies of a few kJmol⁻¹ [16b.]

Having shown that centric crystals of dipolar compounds can - strictly speaking - not exist, we will add some comments to case (ii). In this area, crystal engineering will certainly produce examples of interest to fundamental experimental studies or practical applications.

Fig. 14: Schematic view of spontaneous polarity formation around a seed (dotted box) in a solid solution $H_{1-x}G_x$ ($x < 1$). Attachment of — at a seed composed of — is most likely if interactions -A···D- are stabilising. Therefore (at low x) almost all incoming → will attach as shown. In case the concentration of → in the nutrient is fairly high, lateral interactions between → may become important, driving polarity to either side depending on $\Delta E_\perp \gtrless 0$.

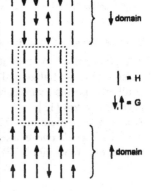

Let us consider a solid solution $H_{1-x}G_x$ of H ≡ symmetrical, but bifunctionalised at the ends of the π-frame, and of G ≡ asymmetrical, showing a similar van der Waals surface as H. In order to obtain a macroscopic crystal domain featuring polar properties it will *not* be necessary to have acentric native crystal structures of H and/or G. An appropriate choice of terminal functional groups will control the attachment to e.g. chain-type motives in $H_{1-x}G_x$ according to the schematic view of Fig. 14. $\Delta E_\perp$ between G molecules will, however, start to interfere with the simple process operating only at low x. At a composition where considerable G···G lateral contact occurs, polarity can start to drop, if $\Delta E_\perp$ (pair) is dominantly > 0. For a given set of ΔE_A, ΔE_D and $\Delta E_\perp$, stochastic simulations will have to explore possible maximum values of net polarity as the content of G in $H_{1-x}G_x$ is increased. Preliminary experiments using a number of different H-type materials (e.g. 4,4'-dinitro-bi (or tri)-phenyl) doped with asymmetrical molecules G (e.g. 4-amino-4'-nitro-bi (or tri)-phenyl) showed frequency doubling for crystallites which represent a solid solution of H and G. However, this is yet an unexplored field which at first needs syntheses of corresponding molecules H and G forming solid solutions over a wide range of composition. It is similarly of interest to investigate $H_{1-x}G_x$ in case of two dipolar components if one of them crystallises in a centric (4-iodo-4'-cyanobiphenyl) and the other in an acentric (4-iodo-4'-nitrobiphenyl) structure.

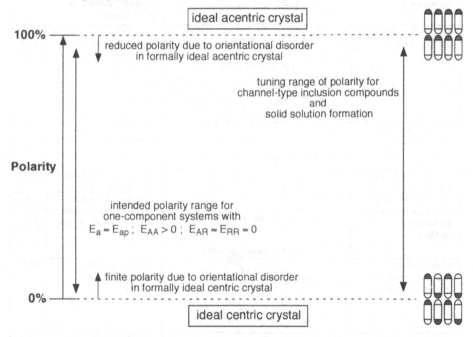

Fig. 15: Pictorial summary of the stochastic approach to polarity formation [9].

If we compare the efficiency (electro-optic effect) of such *mixed crystals* with *single component* and other *classical organic nonlinear optical materials*, we have to account for the effect of dilution, both with respect to polarity and optical absorption. Whereas classical materials provide the highest density of dipoles, a drawback often is optical absorption and reactivity due to the intermolecular π-π overlap. If dipolar compounds are diluted into a matrix (similar to the polymeric approach [1,2]), absorption bands

may exceed a shift towards shorter wavelengths, as well as the photochemical stability
may be increased.

6. Conclusions: polarity, a tunable property of molecular crystals formed by dipolar compounds

Possible phenomena of aggregation leading to polarity in various types of molecular
crystals are summarised in Fig. 15 [9]. This chart expresses the present ability of the
stochastic approach. Other routes leading to polar materials are shown in Tab. 1.

Design Principle	Examples	Comments	Refs.
probability (ca. 25% crystal structures in CSD[a] are acentric)	MNBA ($\theta_p = 19°$)[b]	high density of NLO-active component; parallel alignment relatively rare	[33] [34]
supramolecular synthons (strong, directional interactions)	($\theta_p \approx 0°$)	parallel alignment of polar molecular "strings"; antiparallel alignment possible → centric crystal	[25] [35]
co-crystallisation or salt formation	MC-PTS ($\theta_p = 0°$) Mero-DBA ($\theta_p = 0°$)	greater synthetic flexibility. dilution effect on NLO-active component	[36] [37]
minimum ground-state dipole moment		large β values possible; examples rare	[38]
inclusion formation with inorganic zeolites	H_2N—NO$_2$ in AlPO$_4$-5	small guests preferable; small crystals; dilution	[39]
inclusion formation with organic, channel-forming host molecules	[PHTP$_x$-(guest)$_y$] ($\theta_p \approx 0°$)	applicable to many rod-like NLO-active guests; dilution ~ 10 (w.r.t. pure guest)	[4] [26] [40]
solid-solution formation (reduction in crystal symmetry)	in	requires a parallel packing of the host strcuture	[18]
inclusion formation with α- or β-cyclodextrins	H_2N—NO$_2$	limited to certain guests; dilution effect	[41]

Tab. 1: Crystal design strategies for optimising the parallel alignment of dipolar molecules. [a] Cambridge Structural Database; [b] angle between the charge-transfer axis and the polar crystallographic axis.

Seemingly, the supramolecular route using channel-type inclusion compounds represents the most tunable solution. Crystals grown this way show only a tenth of the net dipole density compared to classical materials, though. The reason is that most of the molecules exhibiting largest β values crystallise in centric native structures. Inclusion formation of A-π-D-R with PHTP will lead to a polar material which, in theory, can adopt the maximum polarity given for a supramolecular system.

The following *reaction scheme* can be derived from Fig. 15 as a general concept:

I   crystal formation: H + G → mixed crystal
II property formation: H + G → spontaneous polarity formation into two separate macro-domains of opposing polarity

It is essential to note that in the case of inclusion compounds ($\Delta E_\perp = 0$) the final state of polarity in each macro-domain does not depend on the polarity of the seed. Any primary state of reaction II can finally yield polarity, as long as I occurs. This enunciates a fundamental difference to all other attempts in growing polar molecular crystals by self-aggregation (no external E-fields). In other cases known up to now, crystal engineering has to control the *primary step of polar ordering*, because further growth steps are nothing but a replicate of the primary code (no phase transitions assumed). Contrary, the stochastic approach is leading to an *evolution of polarity* (Fig. 10) en route to a macroscopic object. This rise of polarity in q-space can be as fast as 2 to 6 steps (Fig. 11b) or it may need up to ~ 10^6 steps to reach stationary values only at the macroscopic level [22]. What else can we learn from the stochastic analysis? Once we have demonstrated the feasibility of the growth of ad-layers on a frozen substrate (a most simple model, which could be extended to a system taking into account some relaxation over n layers, similar to the theory of surface melting [11, 12]) we can think of polarity formation in general:

Assuming a molecular crystal - built up by rod-shaped molecules as defined above - is growing by flat faces, we conclude that at each step (upon growth on a seed for which we do not need to have a theory for the polar state of its final topological structure; see remarks in [29]) polarity is due to

$$p(q+1) = P(q) \cdot p(q) \tag{24}$$

(p: vector representing the polarity of the ad-layer, P: transition matrix; q: number of attachments).

Presently developed theoretical models show [30] cases, where the matrix P can be regarded as a constant of the growth process when stationarity is reached. Therefore, the evolution or production of polarity may in special cases be described by

$$p(q) = P^q \cdot p(q = 0) \ . \tag{25}$$

An approximative analytical solution for stationary values of x_A^L has been found in the case of A-π-D-R molecules, provided that $E_{AA} \gg 0$ ($E_{AR} \cong E_{RR} \cong 0$):

$$x_A^L(\infty) \cong \frac{1}{1 + e^{-(E_{AA} - 2z_\perp \Delta E_\perp)/k_B T}} \tag{26}$$

(limited validity for $\Delta E_\perp / E_{AA} > 0.1$ if $\Delta E_\perp > 0$; see also comment to Fig. 4).

364

Eq. 26 nicely shows the influence of the lateral interaction which at $\Delta E_\perp > 0$ tends to align dipoles into the $\uparrow\downarrow$ configuration. In essence, there is only *one effective energy parameter of free choice* determining the polarity of channel-type inclusion crystals or some one component crystals [42].

Eq. 25 is a most condensed notation providing a criterion for the prediction of a solid state property such as *polarity*. Todays standard approach is a computer assisted structure prediction [31] in order to decide whether an aggregate of dipolar compounds is crystallising in a centric or acentric structure. The present state of our view seems to provide an answer based on a *minimum set of necessary parameters*. The "crunch question" arising at this point is, however: can we predict the effective parameters needed to give an explicit value for P without a precise structure prediction? Hopefully, this and other issues will stimulate partcipants of this course and readers later on to think of any mental link promoting progress in our learning and understanding of complex phenomena in supramolecular chemistry.

About ten years ago, when experimental work on polar inclusion compounds was initiated, D. F. Eaton and collaborators wrote: *"We consider this* [approach by inclusion formation] *to be an important achievement that points to new methodology, a paradigm of guest-host induced dipolar alignment that can be used by chemists to engineer solid-state materials with specific properties."* [32].

This review summarises specifically the polar properties of channel-type inclusion compounds. However, nanoporous materials show in general numerous other solid state properties of interest. For an overview covering a much broader area, see [43].

Acknowledgments

This work has received substantial support from the Swiss National Science Foundation (grants NFP 36 4036-0439932, NF 20-4316.95, NF 21-50828.97). At this place I take the opportunity to thank my colleagues around the world with whom I had many inspiring discussions on this subject over the last few years. In particular, my cordial thanks go to Prof H. Bebie (University of Berne) for the ongoing collaboration in the theoretical description of polarity formation in molecular solids.

365

References

[1] Chemla, D. S. and Zyss, J. (1987) *Nonlinear optical properties of organic molecules and crystals* 1, Academic Press, Inc., London; Zyss, J. (1994) *Molecular nonlinear optics*, Academic Press, Inc., London

[2] Bosshard, C., Sutter, K., Prêtre, P., Hulliger, J., Flörsheimer, M., Kaatz, P. and Günter, P. (1995) *Organic nonlinear optical materials*, Gordon and Breach, Basel

[3] Hulliger, J., König, O. and Hoss, R. (1995) Polar inclusion compounds of perhydrotriphenylene and efficient nonlinear optical molecules, *Adv. Mater.* 7, 719-721

[4] Hoss, R., König, O., Kramer-Hoss, V., Berger, U., Rogin, P. and Hulliger, J. (1996) Crystallization of supramolecular materials: perhydrotriphenylene (PHTP) inclusion compounds with nonlinear optical properties, *Angew. Chem. Int. Ed. Engl.* 35, 1664-1666

[5] Hulliger, J., Langley, P. J., König, O., Roth, S. W., Quintel, A. and Rechsteiner P. (1998) A supramolecular approach to the parallel alignment of nonlinear optical molecules, *Pure Appl. Opt.* 7, 221-227

[6] a) Hulliger, J., Rogin, P., Quintel, A., Rechsteiner, P., König, O. and Wübbenhorst, M. (1997) The crystallization of polar, channel-type inclusion compounds: property-directed supramolecular synthesis, *Adv. Mater.* 9, 677-680; b) Roth, S. W., Langley, P. J., Quintel, A., Wübbenhorst, M., Rechsteiner, P., Rogin, P., König, O. and Hulliger J. (1998) Statistically controlled self-assembly of polar molecular crystals, *Adv. Mater.* 10, 1543-1546; c) König, O., Bürgi, H.-B., Armbruster, T., Hulliger, J. and Weber, T (1997) A study in crystal engineering: structure, crystal growth, and physical properties of a polar perhydrotriphenylene inclusion compound, *J. Am. Chem. Soc.* 119, 10632-10640

[7] Hulliger, J., Roth, S. W., Quintel, A. and Rechsteiner, P. (1998) On a rational principle for designing polar organic crystals, in J. Schreuer (ed.), *Predictability of Physical Properties of Crystals*, Berichte aus Arbeitskreisen der deutschen Gesellschaft für Kristallographie 2, Deutsche Gesellschaft für Kristallographie, Kiel, 9-24

[8] Bürgi, H.-B. and Dunitz, J. D. (1994) *Structure Correlation*, Verlag Chemie, Weinheim

[9] Hulliger, J., Langley, P. J. and Roth, S. W. (1998) A new design strategy for efficient electro-optic single-component organic crystals, *Cryst. Eng. Suppl. Mater. Res. Bull.* 1, 177-189

[10] Hulliger, J. (1999) Orientational disorder at growing surfaces of molecular crystals: general comments on polarity formation and on secondary defects, *Z. Krist.* 214, 9-13

[11] Wilke, K.-Th. and Bohm, J. (1988) *Kristallzüchtung*, Verlag Harri Deutsch, Frankfurt; Bohm, J. (1995) *Realstruktur von Kristallen*, E. Schweizerbart'sche Verlagsbuchhandlung, Stuttgart

[12] Desjonquères, M. C. and Spanjaard, D. (1996) *Concepts in surface physics*, Springer-Verlag, Berlin

[13] Kitaigorodsky, A. I. (1984) Mixed crystals, *Solid State Sciences* 33, Springer-Verlag, Berlin

[14] Meyer, K. (1977) *Physikalisch-chemische Kristallographie*, VEB Verlag für Grundstoffindustrie, Leipzig

[15] Ragone, D. V. (1995) *Thermodynamics of materials* 1 and 2, John Wiley & Sons, Inc., New York

[16] a) Bebie, H. and Hulliger, J. (to be published), work on the ordering behaviour of a single layer grown on a homogeneous substrate; b) an analysis of the difference in the thermodynamic description of the channel-type system assuming equilibrium for i) the volume or (ii) the surface; c) stochastic treatment of the 3D growth of dipolar materials (layer-by-layer growth)

[17] Hulliger, J. (1998) On an intrinsic mechanism of surface defect formation producing polar, multidomain real-structures in molecular crystals, *Z. Kristallogr.* **213**, 441-444

[18] Weissbuch, I., Popovitz-Biro, R., Lahav, M. and Leiserowitz, L. (1995) Understanding and control of nucleation, growth, habit, dissolution and structure of two- and three-dimensional crystals using 'tailor-made' auxiliaries, *Acta Cryst.* **B51**, 115-148

[19] Marlow, F., Wübbenhorst, M. and Caro, J. (1994) Pyroelectric effects on molecular sieve crystals loaded with dipole molecules, *J. Phys. Chem.* **98**, 12315-12319; Klap, G. J., Van Klooster, S. M., Wübbenhorst, M., Jansen, J. C., Van Bekkum, H. and Van Turnhout, J. (1998) Polarization reversal in $AlPO_4$-5 crystals containing polar or nonpolar organic molecules: a scanning pyroelectric microscopy study, *J. Phys. Chem. B* **102**, 9518-9524

[20] Farina, M. (1984) Inclusion compounds of perhydrotriphenylene, in J. L. Atwood, J. E. D. Davis and D. D. MacNicol (eds.), *Inclusion Compounds* 2, Academic Press, Inc, London, 69-95; Allegra, G., Farina, M., Immirzi, A., Colombo, A., Rossi, U., Broggi, R. and Natta, G. (1967) Inclusion compounds in perhydrotriphenylene. Part 1. The crystal structure of perhydrotriphenylene and of some inclusion compounds, *J. Chem. Soc. B*, 1020-1028

[21] Kemeny, J. G. and Snell, J. L. (1983) *Finite Markov chains*, D. van Nostrand Company, Inc., London; Iosifescu, M. (1980) *Finite Markov processes and their applications*, John Wiley & Sons, Inc., New York; Bharucha-Reid, A. T. (1960) *Elements of the theory of Markov processes and their applications*, McGraw-Hill Book Company, Inc., New York

[22] Quintel, A., Roth, S. W. and Hulliger, J. (submitted) 3D-Imaging and simulation of the polarisation distribution in molecular crystals, *Mol. Cryst. Liq. Cryst. A*

[23] Hulliger, J., Quintel, A., Wübbenhorst, M., Langley, P. J., Roth, S. W. and Rechsteiner, P. (1998) Theory and pyroelectric characterization of polar inclusion compounds of perhydrotriphenylene, *Opt. Mater.* **9**, 259-264

[24] Hulliger, J., Langley, P. J., Quintel, A., Rechsteiner, P. and Roth, S. W. (1999) The prediction and production of polar molecular materials, in J. Veciana, C. Rovira, D. B. Amabilino (eds.), *Supramolecular engineering of synthetic metallic materials: conductors and magnets*, Kluwer Academic Publishers, Dordrecht, 67-81

[25] Desiraju, G. R. (1995) Supramolecular synthons in crystal engineering - a new organic synthesis, *Angew. Chem. Int. Ed. Engl.* **34**, 2311-2327

[26] König, O. and Hulliger, J. (1997) Channel-type inclusion lattices of perhydrotriphenylene: a new route to orientationally confined nonlinear optical molecules, *Mol. Cryst. Liq. Cryst. Sci. Technol. Sect B*, Nonlinear Optics, **17**, 127-139

[27] Wagner, C. and Schottky, W. (1931) Theorie der geordneten Mischphasen, *Z. physik. Chem.* B **11**, 163-210; Schottky, W. (1935) Über den Mechanismus der Ionenbewegung in festen Elektrolyten, *Z. physik. Chem* B **29**, 335-355

[28] Sarma, J. A. R. P., Allen, F. H., Hoy, V. J., Howard, J. A. K., Thaimattam, R., Biradha, K. and Desiraju, G. R. (1997) Design of an SHG-active crystal, 4-iodo-4'-nitrobiphenyl: the role of supramolecular synthons, *Chem. Commun.* 101-102; Hulliger, J. and Langley, P. J. (1998) On intrinsic and extrinsic defect-forming mechanisms determining the defect structure of 4-iodo-4'-nitrobiphenyl crystals, *Chem. Commun.*, 2557-2558

[29] A mean field description shows that in some cases the substrate state can have an influence on the final state.

[30] This point needs much more discussion than can be provided here. The general case of polarity formation in single component crystals is presently analysed by means of stochastic simulations and an analytical mean-field description [16c].

[31] See this book: contributions by M. Schmidt or A. Gavezzotti and including references.

[32] Tam, W., Eaton, D. F., Calabrese, J. C., Williams I. D., Wang, Y. and Anderson, A. G. (1989) Channel inclusion complexation of organometallics: dipolar alignment for second harmonic generation, *Chem. Mater.* **1**, 128-140

[33] Wong, M. S., Bosshard, C. and Günter, P. (1997) Crystal engineering of molecular NLO materials, *Adv. Mater.* **9**, 837-842

[34] Knöpfle, G., Bosshard, C., Schlesser R. and Günter, P. (1994) Optical, nonlinear optical, and electrooptical properties of 4'-nitrobenzylidene-3-acetamino-4-methoxyaniline (MNBA) crystals, *IEEE J. Quantum Electron.* **30**, 1303-1312

[35] Masciocchi, N., Bergamo, M. and Sironi, A. (1998) Comments on the elusive crystal structure of 4-iodo-4'-nitrobiphenyl, *Chem. Commun.* 1347-1348

[36] Duan, X. M., Okada, S., Nakanishi, H., Watanabe, A., Matsuda, M., Clays, K., Persoons, A. and Matsuda, H. (1994) Evaluation of β of stilbazolium p-toluenesulfonates by the hyper Raleigh scattering method, *Proc. SPIE* **2143**, 41-51; Okada, S., Masaki, A., Matsuda, H., Koike, T., Ohmi, T. and Yoshikawa, N. (1990) Merocyanine-p-toluenesulfonic acid complex with large second order optical nonlinearity, *Proc. SPIE* **1337**, 178-183

[37] Wong, M. S., Pan, F., Bösch, M., Spreiter, R., Bosshard, C., Günter, P. and Gramlich, V. (1998) Novel electro-optic molecular crystals with ideal chromophoric orientation and large second-order nonlinearities, *J. Opt. Soc. Am. B* **15**, 426-431

[38] Shiahuy Chen, G., Wilbur, J. K., Barnes, C. L. and Glaser, R. (1995) Push-pull substitution versus intrinsic or packing related N-N gauche preferences in azines. Synthesis, crystal structures and packing of asymmetrical acetophenone azines, *J. Chem. Soc. Perkin Trans.* **2**, 2311-2317

[39] Cox, S. D., Gier, T. E., Stucky, G. D. and Bierlein, J. D. (1988) Inclusion tuning of nonlinear optical materials: switching the SHG of p-nitroaniline and 2-methyl-p-

368

nitroaniline with molecular sieve hosts, *J. Am. Chem. Soc.* **110**, 2986-2987; Girnus, I., Pohl, M.-M., Richter-Mendau, J., Schneider, M., Noack, M., Venzke, D. and Caro, J. (1995) Synthesis of $AlPO_4$-5 aluminiumphosphate molecular sieve crystals for membrane applications by microwave heating, *Adv. Mater.* **7**, 711-714

[40] Ramamurthy, V. and Eaton, D. F. (1994) Perspectives on solid-state host-guest assemblies, *Chem. Mater.* **6**, 1128-1136

[41] Tomaru, S., Zembutsu, S., Kawachi, M. and Kobayashi, M. (1984) Second harmonic generation in inclusion complexes, *J. Chem. Soc., Chem. Commun.* 1207-1208; Eaton, D. F. Anderson, A. G., Tam, W. and Wang, Y. (1987) Control of bulk dipolar alignment using guest-host inclusion chemistry: new materials for second-harmonic generation, *J. Am. Chem. Soc.* **109**, 1886-1888

[42] Hulliger, J., Bebie, H. and Roth, S. W., preliminary results derived of a mean-field model (see chap. 2) using Eqs. 7, 8. These equations are not appropriate to describe the process of predominantly anti-parallel ordering.

[43] Langley, P. J. and Hulliger, J. (1999) Nanoporous and mesoporous organic structures: new openings for materials research, *Chem. Soc. Rev.*, submitted

MOLECULAR MAGNETIC CLUSTERS: A BRIDGE BETWEEN MOLECULES AND CLASSICAL MAGNETS

ANDREA CANESCHI[†], ANDREA CORNIA[¶], ANTONIO C. FABRETTI[¶] AND DANTE GATTESCHI[†]
Department of Chemistry, University of Florence[†], Via Maragliano 77 I-50144 Florence, Italy and Department of Chemistry, University of Modena[¶], via Campi 183 I-41100 Modena, Italy

Abstract. The synthetic aspects associated to the formation of large metal ion clusters are reviewed with the aim to show the relevance of these materials to develop new magnetic properties at the interface between the quantum and the classical world. Particular emphasis is given to the structural aspects of the clusters and to the relation between structure and magnetic properties.

1. Introduction

The magnetic properties of large metal ion clusters have recently attracted much attention [1-3]. On one side they provide unique opportunities to observe quantum-size effects in magnets [4-7], and on the other side they can be used as models for understanding the behaviour of clusters occurring in metallo-enzymes and metallo-proteins [8-11]. An important example of the latter type of application is represented by the manganese clusters currently under intense investigation as models to help understanding the water oxidising complex, WOC, of Photosystem II [12-14].

These materials certainly give rise to formidable synthetic problems, because one must find the way to assemble a large but finite number of metal ions. The natural tendency of metal ions to aggregate in infinite lattices must be effectively counterbalanced by a sufficient stabilisation of the cluster. An example may help to clarify what we mean. $[Mn_{12}O_{12}(RCOO)_{16}(H_2O)_4]$, Mn12, is a class of very well known clusters which comprise a $Mn_{12}O_{12}^{16+}$ core, formed by an external octagon of manganese(III) with an internal tetrahedron of manganese(IV) ions bridged by oxo groups [15-17]. This core is not too different from the structures seen in mixed valence manganese oxides, like in the perovskites, which are currently under intense investigation for the colossal magneto-resistance. In fact $Mn_{12}O_{12}^{16+}$ can be considered as a tiny piece of an oxide, which is stabilised by the presence of the organic ligands as shown in Figure 1.

D. Braga et al. (eds.), Crystal Engineering: From Molecules and Crystals to Materials, 369-388.
© *1999 Kluwer Academic Publishers. Printed in the Netherlands.*

370

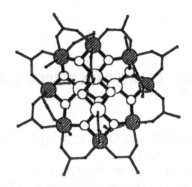

Figure 1. Structure of [Mn$_{12}$O$_{12}$(RCOO)$_{16}$(H$_2$O)$_4$]. The internal core Mn$_{12}$O$_{12}$$^{16+}$ and the external shell [(RCOO)$_{16}$(H$_2$O)$_4$], which is needed in order to ensure electroneutrality and octahedral coordination for the manganese ions, are clearly shown.

Similar structures were previously observed in metal clusters [18], like for instance Pd$_{38}$(CO)$_{28}$(PEt$_3$)$_{12}$ [19] or [H$_2$Ni$_{38}$Pt$_6$(CO)$_{48}$]$^{4-}$ [20], where the metal-metal bonded structures are stabilised by the CO and phosphine ligands which stretch out of the clusters, or like Pd$_{561}$phen$_{36}$O$_{190-200}$ [21], where the same role is played by the phenanthroline and oxygen ligands. To this same class of compounds belong the copper(I) chalchogenide clusters stabilised by the phosphine ligands [22], of which Cu$_{70}$Se$_{35}$(PtBu$_2$Me)$_{21}$ is a representative example [23].

Other beautiful examples of large clusters have been provided by the polyoxometalates [24-25]. In this case the stabilisation of the clusters is determined by the presence of M=O groups, which stretch out of the growing cluster and do not allow the growth of the bulk oxides. Recently clusters comprising 154 molybdenum ions were reported by Müller et al. [26] , which have now been extended to 248 ions! [27]. They are not so exciting from the magnetic point of view, because most of the ions are d^0, non-magnetic, but they clearly show that the goal of assembling a very large number of spins can be achieved. Similar clusters comprising also magnetic rare-earth ions were reported by Pope et al. [28]. Of course the final aim is that of learning to assemble very large clusters comprising up to ca. 4,000 ions as observed for iron(III) in ferritin [29]. In fact it would be extremely tempting to follow the example of nature which stepwise produces complex molecules which can at best perform the functions they have been designed for. The natural approach is that of either using genetic control or an epigenetic one through self-assembly under dissipative conditions [30]. Chemists are not so lucky for the moment, and in general they have to use techniques based on quasi-equilibrium conditions. However it is becoming increasingly clear how it is now possible to use pre-assembled structures in order to make larger clusters.

A detailed study of the growth technique of some polyoxometallates has been recently reported [31] for [H$_{14}$Mo$_{37}$O$_{112}$]$^{14-}$, 1, and

$[H_3Mo_{57}V_6(NO)_6O_{189}(H_2O)_{12}(MoO_6)]^{31-}$, **2**. The self-assembly of **1**, which takes place in aqueous molybdate solution was rationalised assuming the initial formation of a cluster of the α-Keggin type, $[H_xMo_{12}^{VI}O_{40}]^{(8-x)-}$, which by reduction is transformed into the highly nucleophilic ε-Keggin type cluster $[H_xMo_{12}^{V}O_{40}]^{(20-x)-}$. This species is stabilised by protonation and by the capture of four electrophilic $\{Mo^{VI}O_3\}$ groups, thus forming an anion of the type $[H_xMo_{12}^{V}O_{40}(Mo^{VI}O_3)]^{(20-x)-}$ which can be isolated as a methyl ammonium salt. By further reduction the four Mo^{VI} ions become nucleophilic, and work as templates, attracting electrophilic polyoxometallate fragments with 10 and 11 molybdenum atoms respectively ($\{Mo_{10}\} = \{H_3Mo_4^{V}Mo_6^{VI}O_{29}\}^+$, and $\{Mo_{11}\} = \{H_5Mo_6^{V}Mo_5^{VI}O_{31}\}^{3+}$). It is apparent that an unexpected symmetry breaking occurs in this case, opening interesting questions of a very general nature.

A symmetric cluster is on the other hand formed in the case of **2**. It contains three $\{Mo_{17}\}$ fragments linked by six magnetic V^{IV} centres and three $\{Mo_2^{V}\}$ species. There are cavities in the structure, which under reducing conditions can determine the step-by-step growth process whereby $\{Mo_{57+x}V_6\}$, x= 0-6 species can be formed. In fact the degree of occupation of the cavities is associated with the degree of reduction of the cluster which becomes nucleophilic. This cluster has also a magnetic interest, because it contains the d^1 ions V^{IV}, and indeed some interesting properties have been observed.

In this contribution we will focus on a particular class of clusters, which is based on bridging oxide type ligands. We will give particular emphasis to the alkoxides that appear to be very versatile for giving rise to large clusters. We will refer to systems developed both in our laboratory and in others. We will concentrate first on the synthetic aspects, trying to rationalise the synthetic routes leading to large rings containing up to eighteen metal ions. The discussion of the magnetic peculiarities of these clusters will follow. In particular we will try to stress the features that make them attractive not only from the fundamental point of view, but also for novel applications.

2. Synthetic Aspects of Large Iron Clusters

The investigation of oxygen-bridged iron clusters has a long standing tradition, starting from the pioneering work of Walter Schneider who aimed to understanding the mechanism of formation of iron hydroxides and oxides in solution [32]. This problem has also a large biological relevance, due to the fundamental role of iron in living organisms. In describing the clusters it is always important to define the bridging ligands, which are responsible for their growth, and the terminating ligands which are responsible for the stop of the growth. Actually it will be the result of the competition between these two forces that in the end will actually determine the size of the clusters that can be isolated. We want also to stress here that we will report only on clusters which can be isolated in the solid state and which give rise to crystals which are large enough to allow a X-ray crystal structure determination. We will in general be unable to conclude about the thermodynamic stability of the clusters, because other factors, like the kinetics of crystallisation may as well play a fundamental role on the type of cluster that is actually isolated.

The first class of compounds we want to report comprises alkoxides (MeO⁻, EtO⁻, PrO⁻) as bridging ligands, and β-diketonates, dik⁻, as limiting ligands. In all cases we will only take into consideration complexes of iron(III), which with this type of ligands is always high spin, i.e. the individual ion has five unpaired electrons and $S = 5/2$. The diketonates dik⁻ which have been taken into consideration can be classified according to the R and R' substituent groups of 1,3-propanedionate as shown below:

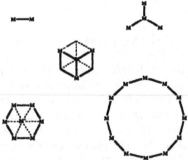

We have extensively investigated this class of compounds, and we have found 2-, 3-, 4-, 6-, 10-, and 12-nuclear species. The structures of the 2-, 3-, 4-, 6-, and 12-nuclear species are sketched below:

In the 3- and 6-nuclear species also additional alkali-metal ions are present, as will be shown below. The synthesis is carried out by reacting resublimed ferric chloride with 1 equiv. of Hdik and 3 equiv. of an alkali-metal alkoxide in the corresponding anhydrous alcohol:

$$FeCl_3 + Hdik + 3RO^- \xrightarrow{ROH} [Fe(OR)_2(dik)] + ROH + 3Cl^-$$

An additional equivalent of alkoxide, ensuring alkaline conditions, is used to favour aggregation. A compound of approximate formula $[Fe(OR)_2(dik)]_n$ is invariably obtained which is quite insoluble in the reaction medium, but can be easily dissolved in moderately polar organic solvents, like chloroform and dichloromethane. Recrystallisation with the addition of a slowly diffusing polar solvent, like methanol or ethanol, leads to the appropriate crystals. An exception is represented by the 3-nuclear species, which is directly obtained from the filtered reaction mixture. The most complete series of clusters so far synthesised by this technique is that containing methoxide ligands and dbm⁻ (1,3-diphenyl-1,3-propanedionate), which led to the isolation of 2-, 3-, 6-, 10-, and 12-nuclear species. The 3-, 6-, and 12-nuclear clusters have the most interesting features. In fact it was found that the 3-nuclear species can only be formed with potassium methoxide [33], while the 6-nuclear species can only be formed with the lithium and sodium methoxides [34,35]. Further it was discovered that with potassium and cesium methoxide it is also possible to obtain the 12-nuclear species [36]. In fact the six-membered rings, which have the general formula $[M'Fe_6(OR)_{12}(dik)_6]^+$, M'Fe6, and the structure shown in Figure 2 host in the centre of the ring an alkali ion (M'=Li, Na) even if charge balance considerations do not require it [34,35].

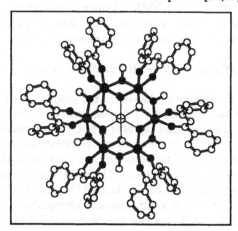

Figure 2. Structure of $[LiFe_6(OMe)_{12}(dbm)_6]^+$. After ref. [35].

At first one may think that the lithium and sodium ions act as templates to favour the formation of the six-membered rings, while potassium can either template the 3-nuclear clusters, or indirectly favour the formation of the 12-nuclear species. In that sense the formation of the clusters is certainly due to the development of host-guest interactions between the growing Fe(dik)(OR)$_2$ clusters and the alkali metal ions. In fact both in the 3- and in the 6-nuclear clusters the alkali metal ion occupies an empty cavity in a close-packed array of oxygen atoms. Therefore, although in these systems the presence of organic ligands may give rise to some distortions it is tempting to look at these clusters as small pieces of oxides. It is certainly surprising to see how the oxygen atoms order according to the close-packed requirements of the solid state. The structure of the hexanuclear species, with the central alkali ion, is the same as that of the Anderson anion. On the other hand, if we look at the structure of the cluster more from the organic point of view we may describe it as that of a crown ether, [Fe$_6$(OR)$_{12}$(dik)$_6$], which coordinates the alkali ion. In fact for this class of compounds the name metalla-crown ethers has been coined [37].

These clusters are soluble in organic solvents, therefore the relative stability can be investigated by NMR. In particular we have investigated the relative stability of the Li$^+$ and Na$^+$ derivatives of the hexanuclear clusters, and that of the dodecanuclear vs. the hexanuclear species [35, 36]. The ^{7}Li spectrum of LiFe6 in chloroform (Figure 3a) shows a broad downfield-shifted singlet at about 440 ppm, in agreement with a lithium ion in close contact with paramagnetic centres. The broad signal is not changed if a soluble lithium salt is added to the solution and a new sharp singlet is observed at ca. 0 ppm. This suggests that the exchange of the encapsulated lithium with the free lithium ion is slow on the NMR time scale. Analogous results are observed in the ^{23}Na NMR for NaFe6. If free lithium is added to a solution of NaFe6 no change is observed in the ^{23}Na NMR spectrum, and a sharp signal at ca. 0 ppm is observed in the ^{7}Li NMR. If free Na$^+$ is added to a solution of LiFe6 the ^{7}Li NMR spectrum shows after a few hours the presence of free lithium (Figure 3b), while the ^{23}Na NMR shows the presence of encapsulated sodium (Figure 3c) This means that the reaction:

$$[LiFe_6(OMe)_{12}(dbm)_6]^+ + Na^+ \leftrightarrow [NaFe_6(OMe)_{12}(dbm)_6]^+ + Li^+$$

is favoured towards the right. Apparently the size of the sodium ion (0.97 Å) is better suited than that of the lithium ion (0.68 Å) for the cavity of [Fe$_6$(OMe)$_{12}$(dbm)$_6$]. Addition of a soluble potassium salt does not alter NMR spectra, presumably because potassium is too large (1.33 Å) to fit into the cavity of [Fe$_6$(OMe)$_{12}$(dbm)$_6$].

This is confirmed by the fact that if the potassium or cesium methoxides are used in the synthesis, no hexanuclear species is obtained [35]. On the contrary, an appealing 12-nuclear species [Fe$_{12}$(OMe)$_{24}$(dbm)$_{12}$], Fe12, is obtained [36]. The twelve iron(III) ions are linked by bis(methoxo) units to give a rather unusual twisted ribbon (Figure 4). The smooth oscillations of the dodecairon backbone can be rationalised by noticing that the optimal Fe-Fe-Fe angle in edge-sharing octahedra is ca. 120°, whereas a planar ring requires Fe-Fe-Fe= 150°. All the Fe-Fe-Fe angles in the structure are indeed well below 150° (117.3-136.2°).

By using ^{1}H NMR spectroscopy it was possible to follow the reactivity of Fe12 with Li$^+$ and Na$^+$ ions. In fact the solution of the cluster in CD$_2$Cl$_2$:CD$_3$OD (3:1) shows paramagnetic effects on the protons of the aromatic rings.

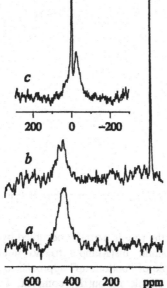

Figure 3. (a) ^{7}Li NMR spectrum of [LiFe$_6$(OMe)$_{12}$(dbm)$_6$]$^+$ in CDCl$_3$. (b) ^{7}Li NMR spectrum recorded 16 h after the addition of Na$^+$ to a solution of [LiFe$_6$(OMe)$_{12}$(dbm)$_6$]$^+$. (c) ^{23}Na NMR spectrum of the same solution. Chemical shifts are referenced to external LiCl (saturated) and NaCl (saturated), respectively. After ref. [36].

On addition of 3 equiv. of either NaBPh$_4$ or LiBPh$_4$ the proton signals show a time dependent shift, indicating that a reaction is occurring. After ca. 6 h the spectra become time independent and identical to those recorded for NaFe6 and LiFe6, respectively. Therefore the reaction:

$$[Fe_{12}(OMe)_{24}(dbm)_{12}] + 2M^+ \rightarrow 2\ [MFe_6(OMe)_{12}(dbm)_6]^+$$

is completely moved to the right. The conclusion is that indeed *the sodium and lithium ions act as templates for the synthesis of 6-nuclear clusters*. However, after all their presence prevents the isolation of larger clusters. The self-assembly of large cyclic structures like [Fe$_{12}$(OMe)$_{24}$(dbm)$_{12}$] *without* template effects may seem surprising. However, molecular rings with an empty cavity in the solid state are not uncommon. The most representative examples are the 10-nuclear clusters [Fe$_{10}$(OMe)$_{20}$(O$_2$CR)$_{10}$], (R = Me, CH$_2$Cl) called the "ferric wheels" [38], and the 18-membered ring[Fe(OH)(XDK)Fe$_2$(OMe)$_4$(OAc)$_2$]$_6$, Fe18, which represents the largest-known cyclic ferric cluster (Figure 5).

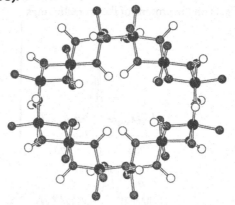

Figure 4. Structure of [Fe$_{12}$(OMe)$_{24}$(dbm)$_{12}$]. The carbon atoms of dbm ligands are omitted for clarity. After ref. [36].

In the latter, the middle of the ring is partially occupied by the bulky XDK ligands (the dianions of m-xylylenediamine bis(Kemp's triacid imide)). The synthesis of this compound is a nice example of "bottom-up" approach to nanoscale materials. In fact, the starting product is a simple diiron(III) complex, Fe$_2$O(XDK)(MeOH)$_5$(H$_2$O)]$^{2+}$, which already contains the basic unit Fe$_2$O(XDK) and is allowed to react with Et$_4$NOAc in weakly alkaline methanolic solution [39].

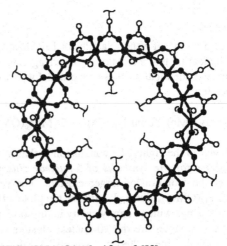

Figure 5. Structure of [Fe(OH)(XDK)Fe$_2$(OMe)$_4$(OAc)$_2$]$_6$. After ref. [39].

An empty six-membered ring, and its host-guest complexes with lithium and sodium ions, is also afforded by the ligand triethanolaminate(3-), tea. In this case the polydentate ligand provides both the terminal and the bridging oxygen ions [40].

The decanuclear species $[Fe_{10}O_4(OMe)_{16}(dbm)_6]$, Fe10, is formed under conditions which favour partial hydrolysis yielding oxide ions. From the structural point of view, it is perhaps the most interesting of the series. As apparent in Figure 6a, the core of the cluster comprises face-sharing, open-cubane fragments, Fe_3O_7, structurally analogous to the 3-nuclear species already described. In both compounds, the oxygen atoms are provided by dbm, μ_2-OMe, μ_3-OMe, and terminal OMe ligands. In addition, the iron-oxygen core of Fe10 comprises four μ_4-O ions of hydrolytic origin. The metal ions are arranged in two pentanuclear layers related by an inversion centre, each showing an Anderson-type structure, as shown in Figure 6b. The oxygen atoms are in three layers corresponding to a cubic closest packing of spheres as shown in Figure 6c, and the iron ions occupy octahedral cavities [41].

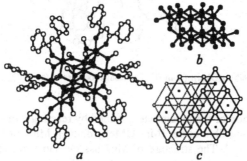

a   b   c

Figure 6. Structure of $[Fe_{10}O_4(OMe)_{16}(dbm)_6]$. After ref. [41].

3. Synthetic Aspects of Large Manganese Clusters

The corresponding manganese clusters are prepared from a manganese(II) salt by simultaneous oxidation and methoxide promoted aggregation. In the presence of sodium methoxide $[NaMn_6(OMe)_{12}(dbm)_6]BPh_4$, NaMn6, is obtained, whose structure is sketched in Figure 7. The structural differences from the corresponding iron(III) compound are striking. They can be justified by the tendency of octahedral manganese(III) ions to undergo axial elongation due to Jahn-Teller instability of the ground ^5E_g state. The elongation axes are painted black in Figure 7. The cluster is centrosymmetric and the six elongation axes are parallel to the x, y, and z axes, respectively, in order to minimise steric repulsion between neighbouring elongated octahedra [42]. It is certainly interesting to observe the three elongation axes present in the same compound. The three possible Jahn-Teller distortions, corresponding to three minima in the mexican-hat diagram of the potential energy, are all present. This has never been observed in bulk lattices. In both molecular and ionic lattices only ferro- or antiferro-distortive arrangements of elongated octahedra were observed. Ferrodistortive means that all the elongation axes are parallel to each other, while antiferrodistortive means that the elongation axes lie in the xy plane and neighbouring octahedra are elongated parallel to x and y, respectively. The microcosm of clusters is providing new spatial arrangements that are not observed in infinite lattices.

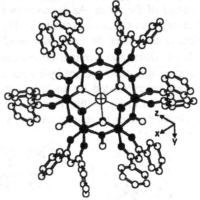

Figure 7. Structure of $[NaMn_6(OMe)_{12}(dbm)_6]^+$. The elongation axes of the MnO_6 chromophores are painted black. After ref. [42].

By recrystallisation of the precipitate obtained by reacting a manganese(II) salt with sodium methoxide and dbm either $[Mn_7(OMe)_{12}(dbm)_6]$, Mn7, or $[Mn_2(OMe)_2(dbm)_4]$, Mn2, are obtained [43]. The structure of Mn7 has the same overall appearance of the six-membered rings already described with the alkali ion in the middle, but this time manganese is present instead of sodium. Charge balance considerations indicate a mixed valent species. Assuming that the possible oxidation states are +II, +III, and +IV, there are three possible stoichiometries giving rise to an uncharged species, namely $Mn_3^{II}Mn_4^{III}$, $Mn_4^{II}Mn_2^{III}Mn^{IV}$, and $Mn_5^{II}Mn_2^{IV}$. The first hypothesis seems to be the most probable on the basis of charge separation considerations and by comparison with other existing clusters. The central ion is a manganese(II), because it is essentially isotropic and the bond distances correspond well to those typical of that oxidation state. The other ions in the ring are presumably disordered, as suggested by UV-Visible and NMR spectra [43]. A similar structure was reported for $[Mn_7(OH)_3Cl_3(hmp)_9]^{2+}$ which has charge distribution $Mn_4^{II}Mn_3^{III}$. In this compound the crystal structure clearly shows that the central ion is a manganese(II) and those on the ring are alternately manganese(II) and manganese(III) [44].

4. Magnetic Properties of Iron Clusters

A common feature of this class of materials is the absence of specific intermolecular interactions. The polar iron/oxygen cores of the clusters are in fact well isolated by the bulky organic groups of terminating ligands and the 3-D arrangement in the crystal is simply ruled by closest-packing requirements. An important consequence is that magnetic interactions between clusters in the lattice become vanishingly small and can be detected only in the mK region. In other words, the crystal provides a natural "amplification of the single-unit signal", unless perhaps at the lowest temperatures, so that molecular properties can be investigated in considerable detail.

Most of the compounds reported in Sect. 2 have a ring structure. This is a particularly appealing one, because rings have been long used by the theoreticians in order to obtain by extrapolation the thermodynamic properties of infinite chains. The availability of real rings now allows a fine check of the calculated properties. In passing we may just notice that it is somewhat disappointing that all the reported rings are even membered rings. Odd membered rings are expected to give rise to interesting spin frustration effects, and it would be very desirable to have real objects to play with. It would also be interesting to tackle the problem of the relative abundance of odd- vs. even-membered rings from a theoretical point of view in order to understand whether the dramatic difference is accidental or is bound to some fundamental requirement.

All the compounds are characterised by antiferromagnetic coupling between the iron(III) ions, as expected for oxo-bridged compounds. However the extent of the coupling, expressed through a spin-hamiltonian of the type: $H = J S_1 \cdot S_2$, varies throughout the series. In Table 1 we show the J coupling constants, obtained from the analysis of the temperature dependence of the magnetic susceptibility, for a series of 2-, 3-, 4-, and 6-nuclear species. We show also some structural parameters that are usually expected to influence the coupling constants, namely the Fe-O bond length for the bridge, the Fe···Fe distance and the Fe-O-Fe angle in the bridge. When different values are present we provide the average. The data of the dinuclear species provide evidence for a strong angular dependence of J [45]. Matters are less clear for the high nuclearity clusters. The magnetic properties of the 6-nuclear species are striking: by replacing the lithium ion with a sodium ion the coupling constant increases from 14 to 20 cm^{-1}. In other terms, the host-guest interactions are responsible of an increase in the coupling by 40%! It is not clear whether this is due to changes in the bond distances and angles induced by the alkali cation or by some indirect mechanism [35]. However host-guest interactions are a new powerful tool for tuning the magnetic properties of molecular magnets.

The antiferromagnetic rings are also characterised by important quantum-size effects, which make them particularly attractive in solid-state physics. The most obvious observation of these effects comes from the magnetisation at very low temperature. At 0.7 K the rings are in the ground $S = 0$ state, and the magnetisation is zero. On increasing the applied magnetic field the energies of the excited states decrease with a slope $-\mu_B g S$, where $g \cong 2$.

TABLE 1. Structural and magnetic parameters for some iron(III) clusters

Compound	Fe-O(br) (Å)	Fe···Fe (Å)	Fe-O-Fe (°)	J (cm^{-1})
[Fe(OMe)(dbm)$_2$]$_2$	1.987	3.087	102.0	15.4
[Fe(OMe)(dpm)$_2$]$_2$	1.974	3.105	103.7	19.0
[Fe(OEt)(bpm)$_2$]$_2$	1.965	3.049	101.8	14.8
[Fe(OPr)(npm)$_2$]$_2$	1.974	3.093	103.1	18.0
[KFe$_3$(OMe)$_7$(dbm)$_3$]	2.077	3.241	102.8	9.7
	2.062	3.243	103.2	12.9
[Fe$_4$(OMe)$_6$(dpm)$_6$]	1.986	3.137	104.3	21.1
[NaFe$_6$(OMe)$_{12}$(dbm)$_6$]Cl	2.015	3.195	104.8	20.4
[NaFe$_6$(OMe)$_{12}$(pmdbm)$_6$]ClO$_4$	2.020	3.215	105.6	19.9
[LiFe$_6$(OMe)$_{12}$(dbm)$_6$]PF$_6$	2.014	3.140	102.6	14.7

It is apparent that for sufficiently high magnetic fields a crossover in the ground state will occur, and the states with large S become the ground states. It has been found that the energies of the lowest S states are, to a good approximation, given by:

$$E(S) = J_{eff} \, S(S+1)/2 \qquad (1)$$

where $J_{eff} = 4J/N$, and N is the number of iron ions in the ring [34,38]. With this spectrum of energy levels the cross-over from S=0 to S=1, from S=1 to S=2, etc. will regularly occur for fields $B(S) = S J_{eff}/g\mu_B$. This behaviour is similar to that observed in bulk antiferromagnets, in which the increasing field flips some of the spins until eventually at extremely high field the magnetisation limit for the ferromagnetic state, i.e. the one with all the spins parallel to each other, is achieved. However in bulk magnets the increase of the magnetisation is gradual, while in the antiferromagnetic rings it has a stepped behaviour as shown in Figure 8. The steps are an indication of the quantum nature of the levels of the clusters. Eq. 1 indicates that for sufficiently large N it should be possible to reduce the quantum step almost to zero. It would be extremely desirable to synthesise very large rings in order to reach the region where quantum and classical effects coexist. The 18-membered ring is already very promising, but up to now no low temperature magnetisation data are available for this compound.

It is also possible to analyse the magnetic properties in a more sophisticated way, and discover that host-guest interactions can tune not only the magnetisation, by affecting the isotropic exchange interactions, but also the magnetic anisotropy. In order to determine the magnetic anisotropy it is necessary to perform magnetisation measurements on single crystals. This is always a difficult task, especially because with molecular materials small crystals seem to be the rule, with very few exceptions. Recently new experimental techniques have been introduced which open new and exciting perspectives for measuring magnetic susceptibility and magnetisation of small single crystals. For antiferromagnetic clusters, the so-called cantilever torque magnetometry is certainly the best suited. In this technique, whose applications in the field of molecular magnetism are only quite recent, a single-crystal sample is glued on the surface of a flexible metallic platelet (a Cu-Be alloy). When a magnetic field is applied, paramagnetic anisotropy tends to align the crystal along one of the easy directions.

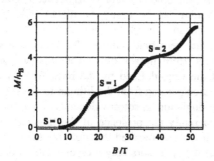

Figure 8. Stepped magnetisation of [NaFe$_6$(OMe)$_{12}$(dbm)$_6$]Cl at 1.5 K.

This causes a deflection of the cantilever, which is usually detected by a capacitive method. With this technique, it is possible to measure the magnetisation steps of 10-μg single crystals and from these to obtain the anisotropy [46]. In fact assuming that the anisotropy of the S states is axial the inflection points of the first step in the magnetisation is given by:

$$B_c = \left(g\mu_B\right)^{-1}\left(\Delta_1 + \tfrac{1}{3}D_1\right)\left(\Delta_1 - \tfrac{2}{3}D_1\right)^{\frac{1}{2}}\left[\Delta_1 + \tfrac{1}{3}D_1\left(1 - 3\cos^2\theta\right)\right]^{-\frac{1}{2}} \quad (2)$$

where Δ_1 is the singlet-triplet separation, D_1 is the axial zero-field splitting parameter for the triplet, and θ is the angle of the external magnetic field with the perpendicular to the ring plane. The results of the experimental determination on the two rings with lithium and sodium in the middle, respectively, are given in Table 2.

TABLE 2. Anisotropy parameters of the excited triplets of MFe$_6$ rings (M=Li, Na)

compound	J (cm^{-1})	Δ_1 (cm^{-1})	D_1 (cm^{-1})
[LiFe$_6$(OMe)$_{12}$(dbm)$_6$]PF$_6$	14.68(4)	9.673(8)	1.11(4)
[NaFe$_6$(OMe)$_{12}$(pmdbm)$_6$]ClO$_4$	19.89(3)	15.28(1)	4.32(3)

The experimental singlet-triplet separations are slightly different from the values calculated from the isotropic coupling constants, which were obtained by fitting the temperature dependence of the magnetic susceptibility [34,35]. Both sets of values however agree on a large difference between the Li and the Na derivative. The differences in the zero-field splitting of the excited triplets are even more striking: the value found in the sodium-containing ring is ca. 4 times larger than in the lithium derivative. The zero-field splitting parameter in iron(III) clusters is given by the sum of two contributions. The first is the projection of the single-ion anisotropy on the total spin S, and the second is due to the dipolar interactions between pairs of iron(III) ions of the ring. The latter contribution can be easily calculated if it is assumed that the individual iron centres can be considered as point dipoles. In this hypothesis it is calculated that the experimental zero-field splitting for $LiFe_6$ is essentially dipolar in origin, while for $NaFe_6$ it is 26% dipolar only. The dipolar contribution depends essentially on the shape of the ring, defined by the iron(III) ions only, therefore the changes in the dipolar values on passing from the lithium to the sodium derivatives are small. On the other hand the single ion anisotropy depends on the coordination sphere of the individual ions, and apparently the presence of the larger sodium ion introduces some significant distortion. It must be recalled that not only the magnitude, but also the orientation of single-ion anisotropy tensors play a crucial role in determining the zero-field splitting of the total spin states, as will be more extensively illustrated in the next Section.

The crossing of levels is an important manifestation of quantum effects. Quantum phenomena in mesoscopic magnets are currently under intense investigation, because they provide fundamental information on the transition from the quantum world to the classical one. Further it is expected that magnetic materials showing quantum-size effects may be used for novel applications, including quantum computing. In fact we want to show that the stepped magnetisation of antiferromagnetic rings can provide relatively simple ways of measuring quantum-tunnelling phenomena. If we extend the above-described rings to infinite chains the ground state is described by the Néel vector, which has two degenerate states. Tunnelling between these two states is possible, and recently a semiclassical theory, based on instantons, has been developed for the interpretation of the tunnel splitting in molecular clusters [47]. No direct measurement of the tunnelling has been performed so far on molecular clusters, but the static magnetisation data reported above were used to test the results of the semiclassical calculations, and the agreement was found to be excellent.

A possibility of directly measuring tunnel frequencies may come from NMR, where the nuclei are used as probes of the local spins. In fact preliminary T_1 measurements on the cluster $[Fe_{10}(OMe)_{20}(O_2CCH_2Cl)_{10}]$, Fe10ring, at very low temperature have shown distinct peaks corresponding to the crossing of S levels [48]. Since NMR techniques are currently actively investigated as candidates for developing quantum computers it is extremely appealing to discover that molecular antiferromagnetic rings can be in principle ideal materials. In fact it is very easy to tune the tunnelling conditions by varying the metal ions, the number of ions in the ring, the exchange coupling constants, the magnetic anisotropy, and so on.

Solid-state NMR experiments on LiFe6, NaFe6 and Fe10ring were reported in which protons have been used to probe the spectrum of electron spin fluctuations as the system evolves from the high-temperature limit of uncorrelated spins to the S = 0 ground state.

In all cases, a peak in the spin-lattice relaxation rate, T_1^{-1}, is observed at temperatures close to J/k_B [49]. Since protons are expected to probe the low-frequency region of electron spin-spin correlations, the sharp increase in T_1^{-1} points to a critical slowing-down of spin fluctuations and recalls the behaviour of classical spin-5/2 chains. However, the subsequent drop-to-zero of T_1^{-1} reflects the discreteness of the low-lying magnetic states. The exponential decrease of T_1^{-1} at the lowest temperatures can in fact be used to evaluate the singlet-triplet gap. These values are in acceptable agreement with the values obtained through magnetisation data.

[$Fe_4(OMe)_6(dpm)_6$], Fe4, is an example of a single molecule magnet, i.e. of a cluster whose magnetisation relaxes slowly at low temperature [50]. The advantage of this cluster compared to others like Mn12 and Fe8 lies in the simplicity of its structure. The small number of ions present allows detailed calculations of the magnetic properties. In fact a detailed analysis of its properties provided important information on the origin of the magnetic anisotropy in iron(III) clusters.

The conditions in order to have slow relaxation of the magnetisation of a cluster at low temperature are the following: large ground spin state and Ising-type magnetic anisotropy. At the minimum level of complication the anisotropy of the ground state, characterised by a spin S, can be described by the Hamiltonian:

$$H = D\,[S_z^2 - 1/3\,S(S+1)] + E\,(S_x^2 - S_y^2) + \text{higher order terms} \qquad (3)$$

D and E are parameters depending on the corresponding terms of the individual ions (single-ion anisotropy) and on contributions generated by the spin-spin interaction. For iron(III) it is a good approximation to assume that the latter contributions can be described in the point-dipole approximation. Ising-type anisotropy is achieved if $D<0$ and under these conditions the $M = \pm S$ levels lie lowest. If the system is magnetised at low temperature, the $M=-S$ state will be selectively populated. On switching the field off it will revert to thermal equilibrium, i.e. it will equalise the populations of the $M=-S$ and $M=+S$ states, by climbing all the levels, one at a time, up to $M=0$, and then descending. Under this simplifying approximation the barrier for the re-orientation of the magnetisation is given by:

$$\Delta = |D|\,S^2 \qquad (4)$$

and the relaxation of the magnetisation is expected to follow the Arrhenius law:

$$\tau = \tau_0 \exp(\Delta/kT) \qquad (5)$$

where τ_0 is expected to be proportional to S^6/Δ^3.

The structure of Fe4 is shown in Figure 9.

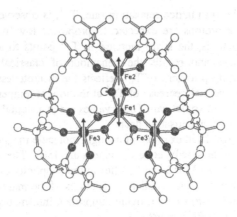

Figure 9. Crystal structure of $[Fe_4(OMe)_6(dpm)_6]$. After ref. [50].

The cluster has two-fold symmetry, with a binary axis passing through Fe1 and Fe2. The three external iron(III) ions almost define an equilateral triangle. The antiferromagnetic coupling between the individual centres determines a ground S=5 state, with the external spins up and the internal one down. Below 1 K the cluster shows slow relaxation of the magnetisation. The relaxation time follows the Arrhenius law with $\tau_0 = 1.1 \times 10^{-6}$ s, $\Delta/k=$ 3.5 K. This means that at 0.2 K, the lowest measurement temperature, the relaxation time is of the order of 1 minute. This relaxation is much faster than that of Mn12 or Fe8, as it should be expected given the smaller S and the lower barrier.

The zero-field splitting parameters were determined through HF-EPR spectra, yielding D= -0.20 cm^{-1}, E~ 0. The negative sign ensures that the magnetic anisotropy is of the Ising type. The height of the barrier, calculated with eq. (4), is larger than the experimental one. This has been found to be the case in all the system showing slow relaxation of the magnetisation investigated so far. It must be concluded that (4) is only a loose approximation.

In a system as simple as Fe4 it is possible to calculate both the single ion and the dipolar contributions to the zero-field splitting. The latter are easily calculated in the point-dipole approximation, while for the former a ligand-field approach, namely the Angular Overlap Model, was used. The conclusion, which could be reached, is that the dipolar contribution is far from negligible in iron(III) clusters, and in this sense the shape of the cluster can play a significant role in determining the magnetic anisotropy. As for single-ion contributions, large angular distortions from octahedral symmetry do not necessarily lead to large zero-field splitting since the effects of different types of distortion can cancel each other.

5. Magnetic Properties of Manganese Clusters

The magnetic properties of the manganese clusters of Section 3 have so far been less investigated than those of the corresponding iron clusters. However they also have provided some important hint to better understand the general properties of large spin clusters. In fact the magnetic properties of NaMn6 are very useful to clarify how structural features affect the value of S of the ground state and its anisotropy. The structure shows that the unpaired electrons of the manganese(III) ions are in the xz, yz, xy, and z^2 orbitals. It is the electron in the magnetic orbital z^2, which dominates the exchange pathway. The z^2 orbital of one ion points to the oxygen atom interacting with the empty x^2-y^2 orbital of the neighbouring manganese(III) ion. The unpaired electron is partially transferred to the empty x^2-y^2 orbital with up-spin. According to Hund's rule, this fraction of unpaired electron will polarise the spins in the xz, yz, xy, and z^2 orbitals of the neighbouring manganese(III) ions to be parallel to its spin; i.e. the two ions are ferro-magnetically coupled. The temperature dependence of the magnetic susceptibility of NaMn6 confirms this prediction, indicating a ferromagnetic coupling of the order of 10 cm^{-1}, and a ground S= 12 state [42]. Unfortunately the same structural features, which determine the high-spin ground state, namely the different orientations of the elongation axes in neighbouring octahedra are also responsible of the small magnetic anisotropy of the cluster. In fact the magnetic anisotropy in manganese(III) clusters is expected to be dominated by the single-ion anisotropy. Consequently, the zero-field splitting of the ground S state is given by the weighed sum of the contributions of the individual ions. Since two ions are elongated along x, two along y, and two along z, there will be large cancellation and the zero-field splitting of the cluster must be small. In fact the analysis of the magnetisation data at low temperature suggests D= -0.12 cm^{-1}.

6. Conclusions

Molecular clusters are a class of compounds which have provided in the recent years many new opportunities for testing and developing sophisticated theories of mesoscopic matter. They are an ideal meeting place for chemists and physicists and provide unique interdisciplinary opportunities. For chemists they represent a challenge to learn how to organise larger and larger number of metal ions in complex structures. In this sense the synthesis of large magnetic clusters is just another aspect of the development of supramolecular techniques. It must be further stressed that it is also possible to pursue similar goals also using organic radicals. For instance polycarbenes containing up to nine centres have already been reported [51]. The difficulties with organic radicals are larger due to the instability of the building blocks, and the difficulty to obtain magnetically active ground states. On the other hand for these reasons they are more challenging, also because the building blocks have usually small S values, either ½ or 1, and reaching the semiclassical limit is certainly hard. An intermediate approach may consist in assembling metal ions and organic radicals. A ring comprising six manganese(II) and six nitronyl nitroxide radicals was found to have a ground S= 12 state [52].

Another important point which must be stressed is that the control that is needed to obtain suitable clusters is not only limited to the choice of the ground state, but also to its magnetic anisotropy, and/or to that of the low lying excited states. This is even more demanding and still much theoretical advance is needed.

Finally in our opinion these materials have a large potential for future applications. In some of the clusters the magnetisation relaxes slowly at low temperature [53] and in principle they can be used to store information in one molecule. Further with their quantum nature they may be employed for quantum computing. Whether this is just a dream depends on the efforts of many ingenuous synthetic chemists who may wish to accept this hard challenge and are ready to collaborate with physicists.

7. Acknowledgements

This work has been made possible by the interactions with many colleagues, whose names are indicated in the references. Certainly the role of Roberta Sessoli, Claudio Sangregorio, Ferdinando Borsa, Alessandro Lascialfari, GianLuca Abbati has been a very important one for the development of the whole field (RS) or for important contributions to some specific applications.

The financial support of MURST and CNR is gratefully acknowledged.

8. References

1. Gatteschi, D.; Caneschi, A.; Pardi, L.; Sessoli, R., *Science*, **1994**, *265*, 1054.
2. Eppley, H.J.; Tsai, H.L.; Folting, K.; Christou, G.; Hendrickson, D.N. *J. Am. Chem. Soc.* **1995**, *117*, 301.
3. Murray, K.S. *Adv. Inorg. Chem.* **1995**, *43*, 261.
4. Awschalom, D.D.; DiVincenzo, D.P.; Smyth, J.F. *Physics Today* **1995**, *48*, 43.
5. Stamp, P.C.E. *Nature*, **1996**, *383*, 125.
6. Friedman, J.; Sarachik, M.; Tejada, J., Ziolo, R. *Phys. Rev. Lett.* **1996**, *76*, 3830.
7. Thomas, L.; Lionti, F.; Ballou, R.; Gatteschi, D.; Sessoli, R.; Barbara, B. *Nature* **1996**, *383*, 145
8. Solomon, E.I.; Sundaram, U.M.; Machonkin, T.E. *Chem. Rev.*, **1996**, *96*, 2563.
9. Dismukes, G.C. *Chem. Rev.* **1996**, *96*, 2909
10. Law, N.A.; Caudle, M.T.; Pecoraro, V.L. *Adv. Inorg. Chem.* **1999**, *46*, 305
11. Caneschi, A.; Gatteschi, D.; Sessoli, R. *J. Chem. Soc., Dalton Trans.*, **1997**, 3963
12. Aromi, G.; Wemple, M. W.; Aubin, S. J.; Folting, K.; Hendrickson, D. N.; Christou, G. *J. Am. Chem. Soc.* **1998**, *120*, 5850-5851.
13. Horner, O.; Riviere, E.; Blondin, G.; Un, S.; Rutherford, A. W.; Girerd, J. J.; Boussac, A. *J. Am. Chem. Soc.* **1998**, *120*, 7924.
14. Yachandra, V.K.; Sauer, K.; Klein, M.P. *Chem. Rev.* **1996**, *96*, 2927.
15. Lis, T. *Acta Cryst.* **1980**, *B36*, 2042
16. Sun, M.Z.; Ruiz, D.; Rumberger, E.; Incarvito, C.D.; Folting, K.; Rheingold, A.L.; Christou, G.; Hendrickson, D.N. *Inorg. Chem.* **1998**, *37*, 4758

17. Aubin, S.M.J.; Sun, Z.M.; Guzei, I.A.; Rheingold, A.L.; Christou, G.; Hendrickson, D.N. *J. Chem. Soc., Chem. Commun.* **1997**, 2239

18. de Jongh, L.J.(Ed.) *Physics and Chemistry of Metal Cluster Compounds*, Kluwer Academic Publisher, Dordrecht, **1994**.

19. Mednikov, E.G.; Eremenko, N.K.; Slovokhotov, Y.L.; Stuchkov, Y.T. *J. Chem. Soc. Chem. Commun.* **1987**, 210.

20. Ceriotti, A.; Demartin, F.; Longoni, G.; Manassero, M.; Marchionna, M.; Piva, G.; Sansoni, M. *Angew. Chem. Int. Ed. Engl.* **1985**, *24*, 697.

21. Schmid, G. *Polyhedron* **1988**, *7*, 2321.

22. Dehmer, S.; Fenske, D. *Chem. Eur. J.* **1996**, *2*, 1407.

23. Dance, I.; Fisher, K. *Progr. Inorg. Chem.* **1994**, *41*, 637.

24. Pope, M.T.; Müller, A. *Angew. Chem. Int. Ed. Engl.* **1991**, *30*, 34.

25. Müller, A. ; Peters, F. ; Pope, M.T. ; Gatteschi, D. *Chem. Rev.* **1998**, *98*, 239-272.

26. Müller, A.; Krickemeyer, E.; Meyer, J.; Bögge, H.; Peters, F.; Plass, W.; Diemann, E.; Dillinger, S.; Nonnebruch, F.; Randerath, M.; Menke, C. *Angew. Chem. Int. Ed. Engl.* **1995**, *34*, 2122

27. Müller, A.; Shah, S.Q.N.; Bögge, H.; Schmidtmann, M. *Nature* **1999**, *397*, 48.

28. Wassermann, K.; Dickman, M.; Pope, M.T. *Angew. Chem. Int. Ed. Engl.* **1997**, *36*, 1445

29. Mann, S.; Webb, J. W. R. J. P. *Biomineralization: Chemical and Biological Perspective;* VCH: New York, 1988.

30. Cramer, F. *Chaos and Order*, VCH, Weinheim, **1993**, p.37.

31. Müller, A.; Meyer, J.; Krickemeyer, E.; Bengholt, C.; Bögge, H.; Peters, F.; Schmidtmann, M.; Körgeler, P.; Koop M.J. *Chem. Eur. J.* **1998**, *4*, 1000.

32. Schneider, W. *Chimia* **1988**, *42*, 9; *Comments Inorg. Chem.* **1984**, *3*, 205.

33. Caneschi, A.; Cornia, A.; Fabretti, A.C.; Gatteschi, D.; Malavasi, W. *Inorg. Chem.* **1995**, *34*, 4660.

34. Caneschi, A.; Cornia, A.; Lippard, S.J. *Angew. Chem. Int. Ed. Engl.* **1995**, *34*, 467; Caneschi, A.; Cornia, A.; Fabretti, A.C.; Foner, S.; Gatteschi, D.; Grandi, R.; Schenetti, L. *Chem. Eur. J.* **1996**, *2*, 1379.

35. Abbati, G.L.; Caneschi, A.; Cornia, A.; Fabretti, A.C.; Gatteschi, D.; Malavasi, W.; Schenetti, L. *Inorg. Chem.* **1997**, *36*, 6443.

36. Caneschi, A.; Cornia, A.; Fabretti, A.C.; Gatteschi, D. *Angew. Chem. Int. Ed. Engl.* **1999**, in press.

37. Pecoraro, V.L.; Stemmler, A.J.; Gibney, B.R.; Bodwin, J.J.; Wang, H.; Kampf, J.W.; Barninski, A. *Progr. Inorg. Chem.* **1997**, *45*, 83.

38. K. L. Taft, C. D. Delfs, G. C. Papaefthymiou, S. Foner, D. Gatteschi, S. J. Lippard, *J. Am. Chem. Soc.* **1994**, *116*, 821; Benelli, C.; Parsons, S.; Solan, G.A.; Winpenny, R.E.P. *Angew. Chem. Int. Ed. Engl.* **1996**, *35*, 1825.

39. Watton, S.P.; Fuhrmann, P.; Pence, L.E.; Caneschi, A.; Cornia, A.; Abbati, G.L.; Lippard, S.J. *Angew. Chem. Int. Ed. Engl.* **1997**, *36*, 2774.

40. Saalfrank, R.W.; Bernt, I.; Uller, E.; Hampel, F. *Angew. Chem. Int. Ed. Engl.* **1997**, *36*, 2482.

41. Caneschi, A.; Cornia, A.; Fabretti, A.C.; Gatteschi, D. *Angew. Chem. Int. Ed. Engl.* **1995**, *34*, 2716.

42. Abbati, G.L.; Cornia, A.; Fabretti, A.C.; Caneschi, A.; Gatteschi, D. *Inorg. Chem.* **1998**, *37*, 1430.

43. Abbati, G.L.; Cornia, A.; Fabretti, A.C.; Caneschi, A.; Gatteschi, D. *Inorg. Chem.* **1998**, *37*, 3759.

44. Bolcar, M.A.; Aubin, M.J.; Folting, K.; Hendrickson, D.N.; Christou, G. *Chem. Commun.* **1997**, 1485.

45. Le Gall, F.; Fabrizi de Biani, F.; Caneschi, A.; Cinelli, P.; Cornia, A.; Fabretti, A.C.; Gatteschi, D. *Inorg. Chim. Acta* **1997**, *262*, 123.

46. Cornia, A.; Jansen, A.G.M.; Affronte, M. *Phys. Rev. B* submitted.

47. Chiolero, A.; Loss, D. *Phys. Rev. Lett.* **1998**, *80*, 169.

48. Julien, M.H.; Jang, Z.M.; Lascialfari, A.; Borsa, F.; Horvatic, M.; Caneschi, A.; Gatteschi, D. submitted for publication

49. Lascialfari, A.; Gatteschi, D.; Borsa, F.; Cornia, A. *Phys. Rev. B* **1997**, *55*, 14341.

50. Barra, A.-L.; Caneschi, A.; Cornia, A.; Fabrizi de Biani, F.; Gatteschi, D.; Sangregorio, C.; Sessoli, R.; Sorace L. *J. Am. Chem. Soc.* submitted.

51. Nakamura, M.; Inoue, K.; Iwamura, H. *J. Am. Chem. Soc.* **1992**, *114*, 1484.

388

52. Caneschi, A., Gatteschi, D., Laugier, J., Rey, P., Sessoli, R., Zanchini, C. *J. Am. Chem. Soc.*, **1988**,. *110*, 2795.
53. Sessoli, R.; Gatteschi, D.; Caneschi, A.; Novak, M. A. *Nature*, **1993**, *365*, 141-143.

MODELLING HYDROGEN-BONDED STRUCTURES AT THERMODYNAMICAL TRANSFORMATIONS

ANDRZEJ KATRUSIAK
Faculty of Chemistry, Adam Mickiewicz University,
Grunwaldzka 6, 60-780 Poznań, Poland

1. Introduction

Any substance is an assembly of interacting atoms, molecules or ions. Types of the interactions are usually associated with the chemical composition of a substance. Thus atoms of noble gases or many organic compounds interact with van der Waals forces in molecular crystals, structures of salts are dominated by electrostatic interactions, and silicon or diamond crystals are in fact huge covalently bonded molecules. Both the interactions and properties can change at varying thermodynamic conditions. Such changes of properties can be subtle and monotonous, but also abrupt and drastic [1,2], for example an insulator can become a semi- or a superconductor, a paraelectric can turn into a ferroelectric, a paramagnet into a ferromagnet, a dielectric into a metal, a fluid into superfluid, a gas into plasma, a crystal can suddenly become longer by nearly 50% [3]. Most of materials sciences and technologies nowadays are soundly based on the knowledge of the interactions between atoms, ions or molecules. It is also important to understand the role of the interactions for transformations of the substances and their properties. This knowledge is essential for verifying theories on solid-state chemistry and physics, for predicting properties of a substance at varied thermodynamic conditions, to model structural changes at the transition point, and also to identify and to synthesise a substance of requested properties. In the following chapter the role of hydrogen bonds for properties of substances, transformations of their structures, as well as the detailed analysis of the hydrogen bond geometry at phase transitions are discussed. It will be shown that hydrogen bonds are convenient objects for investigating structural transformations, for describing these transformations analytically, and for understanding their origins and mechanisms. Owing to relatively simple structure of hydrogen bonds, their geometrical transformations can be explicitly analysed as a set of trigonometric equations, provided that the electronic structures of the donor and

D. Braga et al. (eds.), Crystal Engineering: From Molecules and Crystals to Materials, 389–406.
© 1999 *Kluwer Academic Publishers. Printed in the Netherlands.*

390

acceptor atoms do not change. The results can be applied to many other substances, irrespective to the nature of interactions in their structures.

2. Why hydrogen bonds?

Research on hydrogen bonds has boomed in recent decades [4] and one may wonder if this interest is justified. Is it right that all these resources of thousands of scientists and immense means are dedicated to the hydrogen bond? What spectacular results or breakthroughs, published on hydrogen bonds made the "hydrogen bond" the most frequent key word in scientific papers in recent years. What is the reason for this interest?

Undoubtedly hydrogen bonds are abundant in Nature and they are particularly characteristic for many biological substances. However they always exist aside other types of interactions: covalent bonds, electrostatic and van der Waals forces. None of these interactions can be ignored when crystal structures are analysed. If the molecules of a substance contain H-donor and H-acceptor groups, they are very likely to form hydrogen bonds in the crystal structure. However, it does not mean that the hydrogen bonds would dominate the molecular arrangement. On contrary, in most cases there are also other types of contacts and interactions and some of them are stronger than hydrogen bonds. Thus electrostatic forces clearly dominate the arrangements in ionic structures, particularly of inorganic substances where small inorganic ions generate strong Coulomb forces. The long known Goldschmidt's or Pauling's rules of ionic coordination can be easily applied to hydrogen-bonded ionic crystals, and easily extended to the crystals built of bigger organic ions [5]. Most molecular crystals observe Kitaigorodsky's close-packing rule [6] maximising van der Waals contacts and minimising volume of interacting molecules, but considerable electrostatic cohesion forces are also present when molecules have net atomic charges associated with their atoms. Even if hydrogen bonds bind the molecules into dimers, trimers or other supramolecules, these aggregates are close-packed to minimise all other interactions [7]. What clearly distinguishes the hydrogen bond from the other intermolecular interactions is its *directionality*. In this respect hydrogen bonds resemble covalent bonds. Electrostatic or van der Waals interactions are *central*, which means that an atom or ion can be rotated, and the energy of interactions depend only on the distance*. The directional hydrogen bond can be watched effectively at varied thermodynamic conditions, which may be very helpful for understanding the mechanisms leading to structural transformations or the structure-property relations for a given substance. The directionality also allows that the patterns of molecular association can be predicted, which makes hydrogen bonds ideal tools for designing and engineering crystal

---

* This is true in the approximation of spherical atoms or ions, *e.g. see* Ref. [8].

structures of required properties. It will be further shown that due to the directionality of the hydrogen bond, and its several other features, one can reduce complex structural problems to the region of the hydrogen bond and its vicinity and give simple microscopic explanations to thermodynamic properties of hydrogen-bonded substances. Naturally, the thermodynamic transformations are not specific for the hydrogen-bonded crystals – in this way the hydrogen bonds may contribute to understanding general rules governing the properties of solids.

3. Diversity of hydrogen bonds

There are many types of hydrogen bonds [9], which thus provide a wealth of possible building blocks for engineering crystal structures. However, for effective study of hydrogen-bond transformations it is convenient to minimise the interference of other interactions in crystal structure. Thus, the most suitable for studies are strong hydrogen bonds, which are likely to transform, and which are possibly weakly affected by other interactions [10,11]. Several classifications of hydrogen bonds can be considered. There are intermolecular and intramolecular hydrogen bonds. The intramolecular hydrogen bonds are usually strongly strained by the rigid molecular skeletons, so the further discussion will be focused on intermolecular hydrogen bonds only. In another classification two-centre hydrogen bonds are usually much stronger than 'bifurcated' three-centre or 'trifurcated' four-centre hydrogen bonds [10], where additionally one of the hydrogen bonds interfere with the others. The two-centre hydrogen bond can be described by formula DH- -A, where D is the H-donor and A the H-acceptor. The hydrogen bonds can be also classified according to the types of the donor and acceptor groups, or in the first approximation according to the elements of atoms D and A, for example OH--O, NH--O, OH--N, NH--N, OH--Cl, ClH--Cl, NH--S etc. Each of these types can be further subdivided into classes depending on the chemical groups involved in the bonding, such as O(hydroxyl)H--O(carbonyl), O(H_2O)H--O(lactam). Another classification can be concerned with the ionicity of the hydrogen bonded groups, which may be neutral, cationic, or anionic. Hence conjugated anions DH--A$^-$ and cations DH$^+$- -A, which are termed homoconjugated when the donor and acceptor groups are the same, for example DH=RCOOH and A$^-$=RCOO$^-$, or D and A are both pyridines, or both R_3N; and heteroconjugated, when the donor and acceptor groups are chemically different [12]. Finally, monostable and bistable hydrogen bonds can be distinguished. In the bistable bonds the H-atom can assume one of two possible sites. So, hydrogen bond DH--A can transform to A--HD where the donor and acceptor groups exchange their functions. Such an H-transition is usually connected with a transformation of hydrogen-bonded molecules, for example alternative double and single bonds change theirs sequence from HO–C=C–C=O to O=C–C=C–OH, or it may transform an OH--N bonded molecular crystal into an O$^-$--H$^+$N bonded ionic one. Alternative single and double bonds usually strengthen the hydrogen bond, as the charges can move along such conjugated systems to increase electronegativity of the oxygen atoms. Other possible classifications of hydrogen bonds are not relevant to the further discussion and will not be mentioned [13]. Bistable hydrogen bonds OH--O appear most suitable for studying their transformations and interactions with the crystal lattice, simply because they are likely to transform. Hence details of a hydrogen-bond transformation can be followed for the same compound. Subtle effects of intermolecular interactions can be easily blurred when even slightly different molecules are considered, which can be avoided by investigating the same transformable structure at varied thermodynamic conditions. Although the presented discussion concerns the hydrogen bonded structures, identical or similar rules apply to transformations of any other types of crystals.

4. Composition of interactions

In most substances several types of interactions between molecules or ions coexist and sum up into cohesion forces of a crystal. To simplify the analysis of the role of a hydrogen bond in the crystal lattice, we will further separate the potential energy of the hydrogen bond from all other interactions approximated by one potential-energy function of the crystal field. Let us consider an $-OH--O=$ hydrogen bond in a crystal; for the sake of simplicity let's assume its two dimensional structure, as shown in Fig. 1 (and as it is often encountered in Nature!). Kroon et al. [14] showed that such H-bonds are preferably linear, i.e. the O-H-O angle equals 180^0. This angle will be further denoted η_H. Thus the potential energy of the isolated H-bond will have its minimum at 180^0. It is extremely unlikely that the interactions of the hydrogen-bonded aggregate with crystal lattice will favour the same angle η_H equal 180^0: the sum of the other forces

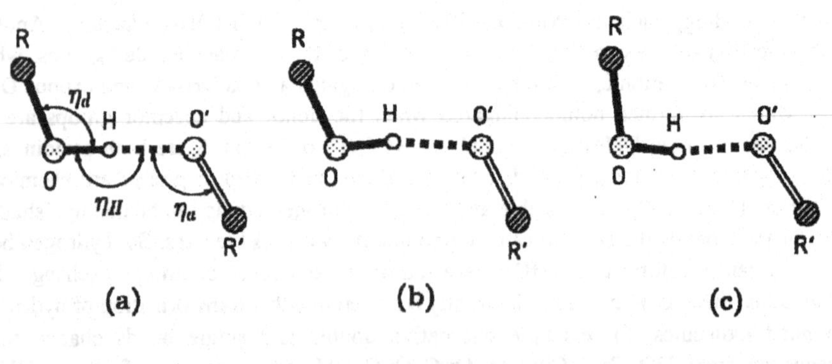

Fig. 1. An isolated $-OH--O=$ hydrogen bond (a), and its possible deformations decreasing (b) and increasing (c) angle η_H. Angles R-O-H, denoted η_d, and H--O'$=R$, η_a, are constant in these drawings.

will either decrease or increase the η_H angle. These possible potential-energy functions are illustrated in Fig. 2, and the resultant deformations of the hydrogen bond are shown in Figures 1b and 1c. Angles R-O-H, denoted η_d, and H--O'=R', η_a, are not equal, as indicated in Figure 1. Due to hybridisation sp^3 of oxygen O and sp^2 of oxygen O', angles η_d is closer to the ideal value of 109^0 and angle η_a is closer to ideal 120^0, respectively. The difference between η_d and η_a is significant for the orientation of the hydrogen-bonded molecules. When η_H equals 180^0 the R-O bond is inclined by η_d to the O···O' line, and the O'=R' bond is inclined by angle η_a. Intermolecular forces are weak compared to intramolecular interactions, and in the first approximation we may assume that the electronic structure of the molecule does not change when small changes occur to the crystal lattice in which it is embedded. Thus we will assume that angles η_d and η_a, and distances O-H, denoted d_d, and H--O', d_a, do not change. The crystal field can easiest affect the η_H angle and the mutual orientation of the molecules, as shown in Figure 1. In fact these are the only parameters allowed to change after constraining d_d, d_a, η_d and η_a (for a planar system η_H is the sole variable). The orientation of hydrogen-bonded molecules can be conveniently described by angles R-O···O', denoted $\eta_d{'}$, and O···O'=R', denoted $\eta_a{'}$, which both refer to the opposite oxygen atom in the H-atom stead, and so avoid problems with often poorly located hydrogens from X-ray structural studies. Owing to the premise of energetically favoured angle η_H equal $180°$, a similar relation as for angles η_d and η_a, applies also to angles $\eta_d{'}$ and $\eta_a{'}$:

$$\eta_d{'} < \eta_a{'} \tag{1}$$

As can be seen from Figure 1, the crystal field can decrease or increase the difference between $\eta_d{'}$ and $\eta_a{'}$. It is apparent from Figures 1 and 2 that the crystal field will either make the arrangement of the molecules more similar (E_{sym}) or less similar (E_{asym}) with respect to the O···O' direction.

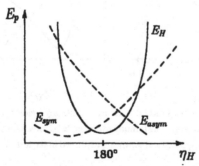

Fig. 2. Schematic representation of the hydrogen bond potential energy (solid line) in the function of angle O-H--O, denoted η_H (compare Fig. 1a), and possible contributions to this energy from the crystal field (broken lines). If the sum of crystal-field contributions makes the arrangement of the hydrogen-bonded molecules more symmetric it is termed E_{sym} (see Fig. 1b), and if less symmetric, E_{asym} (Fig. 1c).

394

5. Pseudosymmetry and symmetry

In certain types of structural transformations the atoms, molecules or ions move between sites which are symmetry-related or very similar. Owing to the symmetry or pseudosymmetry between these sites, the energy of interactions with their crystal environments is similar, and in principle the problem can be reduced to a potential function with two or more energy wells separated by barriers. Such a transformation is often illustrated by potential-energy functions of the proton moving along the hydrogen bond. Figure 3 shows two possible types of such potential: a double-well Morse function for the fast H-hopping (a); and two separate one-minimum functions (b) when slow H-jumps are coupled to rearrangements of the hydrogen-bonded molecules or ions. The transformations like this can occur for the molecules or ions which allow that the H-donor and H-acceptor can interchange their functions. It is reasonable to expect that the H-atoms are likely to transfer in these hydrogen bonds where two H-sites are alike. The main reason for this is that the energies of the system before and after the H-transfer between similar sites are similar, thus the process requires mainly an energy to surpass the energy barrier, or minute energy is used for rotating the molecules and reducing the strains in the H-bond after transformation.

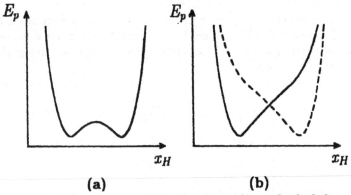

(a)   **(b)**

Fig. 3. Two-well potential energy in the function of H-site in a quickly transforming hydrogen bond (a), and two energy functions (b) for a slowly transforming hydrogen bond, when the H-transfer modifies the crystal environment. Relative terms *quickly* and *slowly* are explained in the text of Sections 5 and 6.

When the H-atom arrives in the other site, the energy is retained and can be used for transforming the same hydrogen bond again or for transforming another system. For this energetical requirement the pseudosymmetrical structures are more likely to transform than the asymmetrical ones. The pseudosymmetric and asymmetric hydrogen bonds are exemplified in Figure 4.

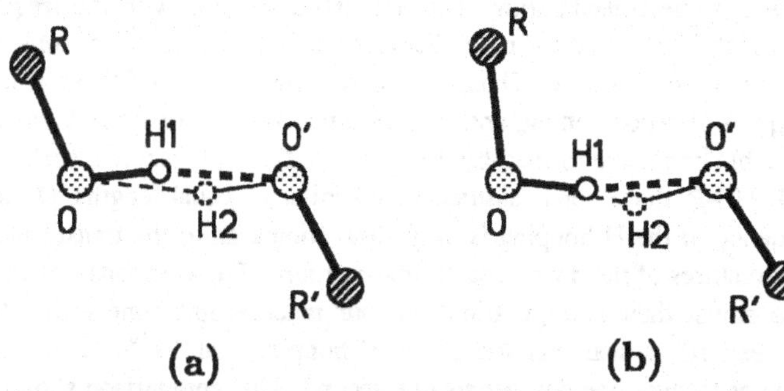

Fig. 4. Two possible arrangements of hydrogen bonded moieties, illustrating the role of pseudosymmetry for transformations of the hydrogen bonds. The alternative H-site, indicated by the dashed circle and its bonds by thin lines, does not significantly change the H-bond dimensions in the pseudosymmetric hydrogen bond (a), and does so in the asymmetric one (b).

6. Hydrogen-bond transformations

Several types of transformations of hydrogen bonds can be clearly distinguished. One classification divides the transformations into those breaking the hydrogen bonds, and those without breaking hydrogen bonds. Both these transformations can proceed dynamically, or as single occurrences transforming one stable structure into another stable one. Dynamical breaking and formation of hydrogen bonds is observed when hydrogen-bonded molecules, ions or groups rotate with high frequency, for example in high-temperature phases of guanidinium nitrate [3] and is characteristic for aggregates of water molecules [15,16]; the static transformations can be exemplified by breaking of O-H--O and formation of O-H--Cl hydrogen bonds in 3,6-dichloro-2,5-dihydroxytelphtalate [17]. The transformations without breaking hydrogen bonds can be exemplified by the H-transfer between donor and acceptor groups. Analogously, a single H-transfer can occur, as at the phase transition of crystals 1,3-cyclohexanedione [18,19]: =O--H-O- changes to -O-H--O= below 286K, or above 8MPa. As already stated, such H-transfers are coupled to atomic displacements. In this respect it appears extremely unlikely that there are any hydrogen-bonded crystal structures, where an H-transfer from one stable site to the other stable site would not induce any changes to the molecular arrangement. Even for highly pseudosymmetrical hydrogen bonds small differences in the H-sites can be clearly described. Several examples of such differences are given below.

Similar hydrogen bonds can transform dynamically when the H-hopping become coupled with lattice-mode vibrations. The time scale of this process is a characteristic feature and merits some consideration. The light H-atom hops with the frequency of about 10^{11}Hz, which is very fast for much heavier atoms of molecules or ions involved in forming the hydrogen bonding. Thus, due to the high inertia of these atoms, they cannot follow quick H-repositioning, and they assume average positions. Consequently, if single and double bonds are formed by the donor and acceptor atoms, such as C=O--HO-C, or P=O--HO-P, these bonds assume very similar or equal lengths. On the other hand, the frequency of the H-hopping is very slow compared to the transformations of the electronic structures of the donor and acceptor groups. The electrons usually require about 10^{-16}s to adjust their configurations. In our macroscopic time-scale, this huge difference between 10^{-11}'s characteristic of the H-hopping and 10^{-16}s of the electronic structure compares to over one day *versus* one second. This comparison shows that the quick H-hopping cannot average the electronic structures of the donor and acceptor oxygen atoms involved in an -OH--O= hydrogen bond. Even though the H-atom disorder (two half-maxima each about 1 Å from the oxygen atoms) is observed in the diffractometric experiments due to short correlation length, and though the oxygen atoms assume averaged positions, the electronic structure of the donor and acceptor groups change in the rhythm of the H-hopping. Thus in an disordered -OH--O=/=O-HO- hydrogen bond the oxygen atoms are either in sp^3 or sp^2 hybridisation, depending on the H-site. A significant consequence of this is that there are no reasons why the O-H covalent bond or H--O contact should change. Indeed, it can be observed in accurate neutron-diffraction experiments that the O-H bond length, H--O contact as well as the R-O-H valency angle and angle H--O'-R' do not change significantly between the structures of one compound with frozen or disordered H-atoms in the -OH--O= hydrogen bonds. This observation will have a significant bearing for understanding the properties of substances with dynamically disordered -OH--O= bonds.

In the context of the above discussion it is worthwhile to address terms *adiabatic* and *nonadiabatic* often used for describing transformations of hydrogen bonds. The thermodynamic term *adiabatic* means that a process is impassable to energy, or in other words that when a transformation of a system occurs there are no interactions with the environment. It was shown above that when the H-atom moves from its one site to the other, the mutual arrangement of the hydrogen-bonded molecules change: the molecules or ions are rotated by a certain angle and this perturbation further propagate in the crystal structure. Thus this transformation of the hydrogen bond is clearly nonadiabatic, and the H-atom potential energy has the form of two separate functions shown in Figure 3b. In fact one can also consider a converse situation, when the crystal environment enforces an H-transfer. When the molecules rotate, the potential energy functions interchange, the H-atom finds itself 'lifted up' on the slope of the other potential function, and 'slides' down to its minimum corresponding to the opposite site. Such transformations occur as single transformations, stochastically, or as solitary waves. When the H-transfers become so fast that the heavy atoms assume average positions, the H-transfers proceed independently of the molecular arrangement around, and in this respect a transformation like this can be termed *adiabatic*. However, as already stated, any H-transfer modifies the electronic structure of molecules or ions, not to mention its electrostatic interactions with other charges around, and in this respect neither of the H-transfers is adiabatic. Terminology sometimes used by quantum chemists adds some more confusion. The well known Born-Oppenheimer approximation separating the movement of nuclei from electrons is also termed *adiabatic*. This approximation is used in most *ab initio* and other calculations and the term *adiabatic* is then used for describing the so obtained model, irrespective of any time-scale considerations presented above. For all the above reasons terms *adiabatic* and *nonadiabatic* have not been used in the further discussion. Instead terms *slow* and *quick* have been used, which are referred to the systems where the heavy atoms adjust their positions following the H-transfers, and where the positions of heavy atoms become averaged, respectively.

For our present symmetry considerations it is not important how the H-bond transforms: in certain systems the H-atom may 'jump' over the energy barrier to the other site; in others the donor and acceptor groups may rotate, so the H-bond is broken for a short instant before it is formed again by another H-atom located on the opposite site of the bond. Type of the transformation would naturally have profound structural and thermodynamic consequences, and would be easily discriminated by the isotope effect of H/D substitution for the critical temperature, T_c, or by the $T_c(p)$ dependence. However, for assessing the likelihood of an H-bond transformation it may be of primary importance to understand how the energy of the structure would change if the H-atom is moved, in this way or another, to the other side of the H-bond.

6.1. REARRANGEMENT -O-H--O= BONDED AGGREGATES

Transformations involving H-transfers in hydrogen bonds --OH--O= binding transformable molecules or ions into aggregates are most likely when the initial and resultant states are symmetry related. Hydrogen bonds in such structures are pseudosymmetric. If the pseudosymmetry of a hydrogen bond is common for all the crystal lattice, then the dimensions of the hydrogen bond before and after the H- transfer do not change; if this symmetry is only local, than small changes in the dimensions can occur. As already noted, the H-site is combined with angular dimensions of a hydrogen bond, which would reverse or change when the H-atom transfers to the other site. One can easily analyse the pattern formed by hydrogen-bonded molecules to see if such an rearrangement is possible [18]. Because the η_d' angle increases, and η_a' decreases following the H-transfer, one may mark these angles or their complements which would

398

increase their opening, as shown in Figure 5. If the marks are distributed evenly on both sides of the chain, than the transfers are likely to occur. However all marks on one side indicate that the aggregate would have to significantly change its dimensions, or bend a chain, and thus the H-transfer is highly unlikely.

It is remarkable that slow H-transfers connected with stochastic process or solitary waves proceed in many structures, but they are difficult to detect by diffractometric measurements. When looking at such a structure determined by X-ray or neutron elastic diffraction, one gets impression that the atomic positions are very stable. Naturally one can see the thermal vibrations of atoms, which can be very small, but often all the atoms are located within thousandths of an Ångström, and reliability factors are so low. Meanwhile the diffraction methods are not best suited for observing all types of dynamics in crystal lattice. The slow H-transfers often transform a structure into its C_i-related image. The diffraction of such coexisting domains, which are larger than the coherence length of the scattered X-rays, differs by mere anomalous correction and only for noncentrosymmetric structures can be detected. On the other hand stochastic movements can be easily detected by second moment ^{1}H-NMR measurements. It was shown indeed, that in the structures of CHD1 and CHD2 such H-transfers accompanied by molecular re orientations exist down to 200K. Moreover, it can be seen from Figure 5 that the tilts of molecules in CHD1 are smaller than in CHD2. These tilts are equal $3.38(2)^\circ$ for CHD1 at 290K, and $7.20(1)^\circ$ for CHD2 at 273K [18]. Indeed, the M_2 ^{1}H-NMR showed that the amplitudes of molecular rearrangements increase when the crystals are cooled [23].

6.2. CARBOXYLIC ACID DIMERS

The role of crystal field for H-stability in hydrogen bonds can be convincingly illustrated by the structures of carboxylic dimers. They were intensively investigated by various methods in the gas, fluid and solid states [24-26]. The activation energy for the H-transfers in isolated dimers is of few kcal/mol only [27-29]. It is well known that at the same thermodynamic conditions (most crystal data were measured at room temperature) the dimers of chemically very similar molecules of carboxylic acids behave differently in their crystal structures in this respect that in some of them the H-atoms are ordered, and in the others disordered in the hydrogen bonds. It was shown [30,31] that the rate of H-disordering can be correlated to the skewness angle, s, defined in Figure 6, reflecting the strain induced to the pair of hydrogen bonds by the crystal field. In the structures where the crystal field retains or even increases the skewness angle to 5^0 or more, the H-atoms are ordered and the double and single bonds are well located. For the s values of 2^0 or less the H-atoms are disordered and the single and double bonds delocalised. As equally numerous carboxylic acids structures with ordered

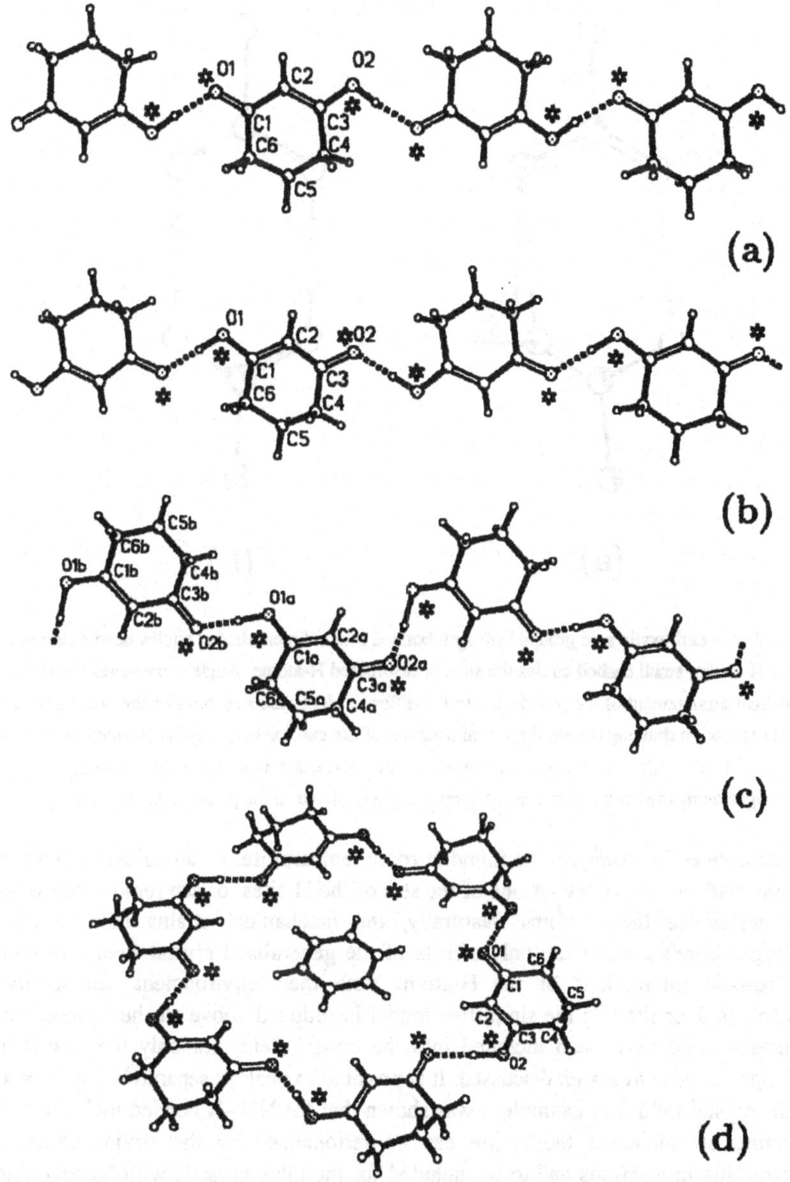

Fig. 5. Hydrogen-bonded aggregates of 1,3-cyclohexanedione in polymorphs CHD1 (a) [19]; CHD2 stable below 286K [20] or above 8Mpa (b) [21]; CHD3 (c) [22]; and in the inclusion compound with benzene (d) [19]. The asterisks mark the η′ angles or their complements which would increase if H-atoms transfer to the other sides of the hydrogen bonds, and their distributions indicate that the H-transfers are more likely for polymorphs CHD1 and CHD2.

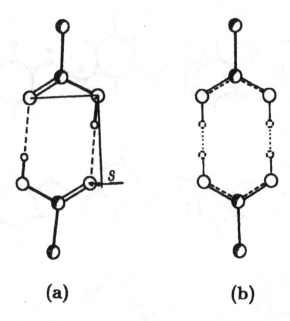

$$\text{(a)} \qquad \qquad \text{(b)}$$

Fig. 6. Pair of carboxylic acid groups hydrogen bonded into a dimer: shaded circles denote carbons, small circles H-atoms, small dashed circles the sites of disordered H-atoms. Angle s, measures the skewness in the mutual arrangement of the groups due to the different electronic structures of the donor and acceptor oxygen atoms. In drawing (a) the skew conformation of the carboxylic groups is retained or increased by the crystal field, while in drawing (b) it is made parallel and destabilises the H-sites. Owing to the strains in the eight-membered rings, they are only approximately planar in the plane of the drawing.

and disordered H-atoms can be found at room temperature, it can be concluded that the crystal field can either favour one of the sets of the H-sites, or can remove the skewness and destabilise the H-atoms. Naturally, the mechanical strains induced into the hydrogen bonds are not the only effects of the generalised crystal field. For example electrostatic interactions of the H-atoms with their environment can stabilise or destabilise their sites. In the simplified model introduced above all the interactions of a hydrogen bond have been included into the crystal field, but only the strains in the hydrogen bonds have been discussed. It is possible to analyse separately any component of the crystal field. For example it was shown that for NH--N bonded molecular crystal structure of imidazole the H-site can be rationalised by the strains alone, while electrostatic interactions had to be included for the ionic crystals with homoconjugated NH^+--N bonds [32].

6.3. -OH--O= BONDS IN FERROELECTRICS

Hydrogen bonds –OH--O= play an important role in the mechanism of phase transitions between ferroelectric and paraelectric phases of a large group of crystals. The best known representative of this group is potassium dihydrogen phosphate, KH_2PO_4, abbreviated KDP. Hence the substances of this type are termed KDP ferroelectrics, even though their structures are very diverse. All they contain in the ferroelectric phase an – OH--O= hydrogen bond which above a critical temperature, T_c, when the crystal transforms to the paraelectric phase, becomes disordered and the H-atom hops between two sites. Due to these quick H-transfers, the heavy atoms involved in the hydrogen bond as well as the whole structure acquire a higher symmetry, which is approximated by the ferroelectric structure below T_c. This same paraelectric symmetry concerns the sites of the disordered H-atoms, and all the crystal structure when looking at it for a time longer than several H-hop cycles. Thus for a paraelectric phase the relation describing the mutual orientation of the hydrogen-bonded groups, molecules or ions assumes the form:

$$\eta_d' = \eta_a' \qquad (2)$$

differently than Equation (1) for the ordered –OH--O= hydrogen bonds. However, as already explained in Section 6, an instantaneous snapshot would show the H-atoms on

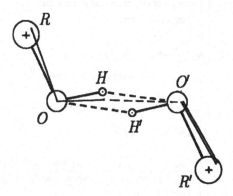

Fig. 7. Hydrogen bond –OH--O= in the ordered (point atoms thin lines, double line represent double bond O'=R) and disordered (big circles, thicker lines; the crosses indicate the centres of atoms R and R') structures. The illustrated structural changes, rotations of bonds R–O and O'–R', shifts of atoms R and R', and shortening of the O···O' distance, have been somewhat enhanced for clarity.

402

Table 1. Possible symmetries relating both sides of a disordered hydrogen bond, and the formula for calculating the hydrogen-bond dimensions from bond lengths d_d and d_a and angles η_d and η_a assumed constant and characteristic for a given compound.

H-bond symmetry and schematic drawing	Trigonometric relations between dimensions of disordered H-bond	Unrestricted angular dimensions
C_{2h}	$\eta' = \eta_d + \mu_d = \eta_a - \mu_a$, where $\mu_d = tan^{-1}\left(\dfrac{d_a \sin\mu}{d_d \cos\mu + d_a}\right)$, $\mu_a = tan\left(\dfrac{d_a \sin\mu}{d_d \cos\mu + d_d}\right)$ $\mu = \eta_a - \eta_d \,(= \mu_d + \mu_a)$ $\eta_H = 180° - \mu$ $r = \sqrt{d_d^2 + d_a^2 - 2d_d d_a \cos\eta_H}$ $\delta = \sqrt{d_d^2 + d_a^2 - 2d_d d_a \cos\mu}$ $\psi = \sin^{-1}\left(\dfrac{2d_d \sin\mu_d}{\delta}\right)$	—
C_i	$\begin{cases} \cos\eta_d = \cos\mu_d \cos\eta + \sin\mu_d \sin\eta \cos\upsilon \\ \cos\eta_a = \cos\mu_a \cos\eta - \sin\mu_a \sin\eta \cos\upsilon \\ d_d \sin\mu_d = d_a \sin\mu_a \end{cases}$ $\mu = \mu_d + \mu_a$ η_H, r, δ and ψ as for C_{2h}	υ
C_2	$\begin{cases} \cos\eta_d = \cos\mu_d \cos\eta + \sin\mu_d \sin\eta \cos\left(\dfrac{\theta - \tau}{2}\right) \\ \cos\eta_a = \cos\mu_a \cos\eta + \sin\mu_a \sin\eta \cos\left(\dfrac{\theta + \tau}{2}\right) \\ d_d \sin\mu_d = d_a \sin\mu_a \end{cases}$ $\mu = \mu_d + \mu_a$, η_H, r and ψ* as for C_{2h}	θ, τ
C_{2v}	$\eta' = \eta_d + \mu_d = \eta_a + \mu_a$, where μ_d and μ_a as for C_{2h} $\mu = 2\eta' - \eta_d - \eta_a$ η_H and r as for C_{2h} $\delta = r - 2d_d \cos\mu_d = 2d_a \cos\mu_a - r$ $\psi = 0°$	—
C_s	$\begin{cases} \cos\eta_d = \cos\mu_d \cos\eta + \sin\mu_d \sin\eta \cos\upsilon \\ \cos\eta_a = \cos\mu_a \cos\eta + \sin\mu_a \sin\eta \cos\upsilon \\ d_d \sin\mu_d = d_a \sin\mu_a \end{cases}$ $\mu = \mu_d + \mu_a$ η_H, r and δ as for C_{2v} $\psi = 0°$	υ

* see Ref. [33] for two possible definitions of angle ψ in C_2-symmetric H-bonds.

one side or the other of the bonds, with a short correlation length between structural units, and accordingly modified electronic structure of the donor and acceptor groups, even though the heavy atoms would have averaged positions (we ignore their thermal vibrations). The main difference between the donor and acceptor sites is the orientation of the H-bonded molecules or ions, as explained in Section 4.

Figure 7 illustrates structural changes in an –OH--O= hydrogen bond when it transforms between the ferroelectric and paraelectric phases. The H-disordering is coupled with displacement of heavy atoms, which assume the positions for which both H-sites preserve the dimensions d_d, d_a, η_d and η_a of the ferroelectric phase (see Section 4). Owing to this condition the geometry of the hydrogen bond in ferroelectric and paraelectric phases can be related by mathematical formulae, listed in Table 1. The formulae depend on the symmetry element of the hydrogen bond in the paraelectric phase. For symmetries C_{2h} and C_{2v} all the dimensions of the disordered H-bond in paraelectric phase can be explicitly determined [33]. Symmetry C_{2v} is very unlikely, and is not encountered even in carboxylic acid dimers, where due to strains the carboxylic groups are not coplanar. Thus displacement of molecules or ions changing angles η_d' and η_a' can be calculated along with the decrease of angle η_H, shortening* of r, the distance between the sites of the disordered H-atom, δ, the inclination of the trajectory of the H-hopping with respect to the O···O' line, ψ. The non-zero ψ angle, usually of few degrees, was a matter of controversies since it was first observed in 1953 [34]. For the other possible symmetries, listed in Table 1, there are parameters which are not constrained by symmetry: one parameter for symmetries C_i and C_s, and two for symmetry C_2. However the values of these parameters can be also assessed from the dimensions in the ferroelectric phase. The atoms involved in C_i, C_s and C_2-symmetric hydrogen bonds are not confined to one plane, as in C_{2h} and C_{2v}-symmetric bonds. The solutions of the equations of Table 1 well agree with experimentally determined dimensions of the disordered hydrogen bonds [33]. The detailed information obtained in this way allow to model transformations in whole of the crystal structure of a ferroelectric crystal. For example it was established that the H-disordering is coupled with angular displacements, equalising the η_d' and η_a' angles by small rotations of molecules. Thus these phase transitions have a clear displacive contribution to the order-disorder character, further enhanced by R-O shortening and O'=R' lengthening, which are not discussed here. The displacements of the heavy molecules or ions can be easily measured by determining the structure of the ferroelectric phase, and they were shown to well correlate with T_c, which provides a means of assessing the critical temperature from structural data only [35,36]. It was also possible to explain for these materials the structural origin of the so called tricritical point at which the character of

---

* Distance O···O shrinks when the H-atom becomes disordered, because angle η_H decreases. This is based on the observation, that angle η_H is preferably close to 180° when H is ordered. However, if it happened that η_H would be smaller for the ordered hydrogen bond, then the O···O would increase on H-disordering. No such an exceptional case was till now observed.

the phase transition changes from first-order non-continuous to second-order continuous one [37], and to obtain the values of critical pressure in excellent agreement with experimental data. The model presented above also explains the anomalous thermal expansion of hydrogen-bonded crystals at T_c.

7. Conclusions

The presented model of transformations of hydrogen-bonded crystals is not limited to the region of the hydrogen bond only. Although concentrated on the hydrogen bond, it involves also the angular parameters η which combine the H-transfers with arrangement of the hydrogen-bonded moieties. Thus the model describes both the hydrogen bond itself and the hydrogen bonded molecules and ions, which often constitute the whole or most of the crystal structure. It also includes the other interactions, as a general crystal field. The model can be applied for any hydrogen bonds, although for weak hydrogen bonds the effects of transformations can be difficult to observe among stronger interactions, also undergoing transformations. The transformations may equally concern complex living tissues [38] and simple pure substances [39]. Most recently similar considerations were applied for analysing the structural factors responsible for H stability in NH- -N hydrogen bonds [32], and for finding NH- -N hydrogen bonded ferroelectric crystals [40]. A similar model was applied for investigating the negative thermal expansion of ice at 80K [41]. As stated in *Introduction*, the considerations of crystal symmetry and interactions between the structural units can be applied to any crystals, irrespective of the nature of the interactions.

Acknowledgement

This project was partly supported by the Polish Committee of Scientific Research, project 3T09A 01511.

8. References

1. Cowley, R. A. (1980) Structural phase transitions I. Landau theory, *Advances in Physics* **29**, 1-110.
2. Klamut, J., Durczewski, K. and Sznajd, J. (1979) *Wstęp do fizyki przejść fazowych* [*Eng. Introduction to the Physics of Phase Transitions*] Zakład Narodowy im. Ossolińskich, Wrocław.
3. Katrusiak, A. and Szafrański, M. (1996) Structural phase transitions in guanidinium nitrate, *J. Mol. Struct.* **378**, 205-223.
4. Jeffrey, G. A. And Saenger, W. (1991) *Hydrogen Bonding in Biological Structures*, Springer-Verlag Berlin Heidelberg New York.

405

5. Pauling, L. (1960) *Nature of the Chemical Bond*, Cornell Univ. Press.

6. Kitaigorodskii, A. I. (1973) *Molecular Crystals and Molecules*, Academic Press, New York-London, Academic Press [Polish edition: *Kryształy molekularne*, PWN Warszawa (1976)].

7. Laing, M. (1975) The Packing of Molecules in Crystals, *South African Journal of Science* **71**, 171-175.

8. Nyburg, S. C. and Faerman, C. H. (1985) A Revision of van der Waals Atomic Radii for Molecular Crystals: N, O, F, S, Cl, Se, Br and I Bonded to Carbon, *Acta Cryst.* **B41**, 274-279.

9. Novak, A. (1974) Hydrogen Bonding in Solids. Correlation of Spectroscopic and Crystallographic Data, *Structure and Bonding* **18**, 177-215.

10. Taylor, R., Kennard, O. and Versichel, W. (1984) Geometry of the N-H···O=C Hydrogen Bond. 2. Three-Center ("Bifurcated") and Four Center ("Trifurcated") Bonds, *J. Am. Chem. Soc.* **106**, 244-248.

11. Emsley, J. (1980) Very Strong Hydrogen Bonding, *Chem. Soc. Rev.*, 9_1, 91-124.

12. Joesten, M. D. (1982) Hydrogen Bonding and Proton Transfer, *J. Chem. Educ.* **59**, 362-366.

13. Yongping Pan and McAllister, M. A. (1997) Characterisation of Low-Barrier Hydrogen Bonds. 5. Microsolvation of Enol-Enolate. An ab Initio and DFT Investigation, *J. Org. Chem.* **62**, 8171-8176.

14. Kroon, J., Kanters, J. A., van Duijneveldt-van de Rijdt, J. G. C. M., van Duijneveldt, F. D. and Vliegenhart, J. A. (1975) O-H...O hydrogen bonds in molecular crystals. A statistical and quantum-chemical analysis, *J. Mol. Struct.* **24**, 109-129.

15. Saykally, R. J. and Blake, G. A. (1993) Molecular Interactions and Hydrogen Bond Tunneling Dynamics: Some New Perspectives, *Science* **259**, 1570-1575.

16. Franks, F. (1972) The Properties of Ice, in F. Franks (ed.) *Water. A Comprehensive Treatise*, Vol. 1. Plenum Press, New York-London, pp. 115-149.

17. Byrn, S. R., Curtin, D. Y. And Paul I. C. (1972) The X-ray crystal structures of the yellow and white forms of dimethyl 3,6-dichloro-2,5-dihydroxyterephthalate and a study of the conversion of the yellow form to the white form in the solid state, *J. Am. Chem. Soc.* **94**, 890-898.

18. Katrusiak, A. (1994) Molecular motion and hydrogen-bond transformation in crystals of 1,3-cyclohexanedione, in D. W. Jones and A. Katrusiak (eds.), *Correlations, Transformations and Interactions in Organic Crystal Chemistry*, Oxford University Press, pp. 93-113.

19. Etter, M. C., Urbańczyk-Lipkowska, Z., Jahn, D. A. and Frye, J. S. (1986) Solid-state structural characterisation of 1,3-cyclohexanedione and of a 6:1 cyclohexanedione:benzene cyclomer, a novel host-guest species, *J. Am. Chem. Soc.* **108**, 5871-5876.

20. Katrusiak, A. (1991) The Structure and Phase Transition of 1,3-Cyclohexanedione Crystals as a Function of Temperature, *Acta Cryst.* **B47**, 398-404.

21. Katrusiak, A. (1990) High-Pressure X-ray Diffraction Study on the Structure and phase Transition of 1,3-Cyclohexanedione Crystals, *Acta Cryst.* **B46**, 246-256.

22. Katrusiak, A. (1992) Stereochemistry and transformation of -OH-O= hydrogen bonds. I. Polymorphism and phase transition of 1,3-cyclohexanedione crystals, *J. Mol. Struct.* **269**, 329-354.

23. Pająk, Z., Latanowicz, L. and Katrusiak, A. (1992) NMR Study of Molecular Motions in 1,3-Cyclohexanedione, *Phys. Stat. Solidi (a)* **130**, 421-428.

406

24. Pauling, L. and Brockway, L. O. (1934) Structure of the carboxyl group. I. Investigation of formic acid by the diffraction of electrons, *Proc. Natl. Acad. Sci. US* **20**, 336-340.

25. Karle, J. and Brockway, L. O. (1944) An electron-diffraction investigation of the monomers and dimers of formic, acetic and trifluoroacetic acids and the dimers of deuterium acetate, *J. Am. Chem. Soc.* **66**, 574-584.

26. Martinache, L., Kresa, W., Wegener, M., Vonmont, U. And Bauder, A. (1990) Microwave spectra and partial substitution structure of carboxylic acid bimolecules, *Chem. Phys.* **148**, 129-140.

27. Nagaoka, S., Terao, T., Imashiro, S., Saika, A., Hirota, N. and Hayashi, S. (1981) A study on the proton transfer in benzoic acid dimer by carbon-13 high-resolution solid-state NMR and proton T_1 measurements., *Chem. Phys. Lett.* **80**, 580-584.

28. Meier, B. H., Graf, F. and Ernst, R. R. (1982) Structure and dynamics of intermolecular hydrogen bonds in carboxylic acid dimers, *J. Chem. Phys.* **76**, 767-774.

29. Hayashi, S., Umemura, J., Kato, S. and Morokuma, K. (1984) Ab Initio Molecular Orbital Study on the Formic Acid Dimer, *J. Phys. Chem.* **88**, 1330-1334.

30. Katrusiak, A. (1996) Macroscopic and structural effects of hydrogen-bond transformations, *Crystallogr. Rev.* **5**, 133-180.

31. Katrusiak, A. (1996) Stereopopulation control in 3-(2,4-dimethyl-6-methoxyphenyl)-3-methylbutyric acid and proton stability in hydrogen-bonded carboxylic groups, *J. Mol. Struct.* **385**, 71-80.

32. Katrusiak, A. (1999) Stereochemistry and transformations of NH- -N hydrogen bonds. Part I. Structural preferences for the H-site, *J. Mol. Struct.* **474**, 125-133.

33. Katrusiak, A. (1993) Geometric effects of H-atom disordering in hydrogen-bonded ferroelectrics, *Phys. Rev. B* **48**, 2992-3002.

34. Bacon, G. E. and Pease, R. S (1953) A neutron diffraction study of potassium dihydrogen phosphate by Fourier synthesis, *Proc. Royal Society of London A* **220**, 397-421.

35. Katrusiak, A. (1995) Coupling of displacive and order-disorder transformations in hydrogen-bonded ferroelectrics, *Phys. Rev. B* **51**, 589-592.

36. Katrusiak, A. (1996) Stereochemistry and transformation of -OH- -O= hydrogen bonds. II. Evaluation of T_c in hydrogen-bonded ferroelectrics from structural data, *J. Mol. Struct.* **374**, 177-189.

37. Katrusiak, A. (1996) Structural Origin of Tricritical Point in KDP-Type Ferroelectrics, *Ferroelectrics* **188**, 5-10.

38. Nagle, J. F., Mille, M. and Morovitz, H. J. (1980) Theory of hydrogen bonded chains in bioenergetics, *J. Chem. Phys.* **72**, 3952-3971.

39. Katrusiak, A. (1998) Modelling Hydrogen-bonded Crystal Structures beyond Resolution of Diffraction Methods, *Pol. J. Chem.* **72**, 449-459.

40. Katrusiak, A. and Szafrański, M. (1999) Ferroelctricity in NH- -N hydrogen-bonded crystals, *Phys. Rev. Lett.* **82**, 576-579.

41. Katrusiak, A. (1996) Rigid H_2O Molecule Model of Anomalous Thermal Expansion of Ices, *Phys. Rev. Lett.* **77**, 4366-4369.

DISCRETE AND INFINITE HOST FRAMEWORKS BASED UPON RESORCIN[4]ARENES BY DESIGN

LEONARD R. MACGILLIVRAY[†] and JERRY L. ATWOOD[‡]

[†] *Steacie Institute for Molecular Sciences, National Research Council of Canada*

Ottawa, Ontario, Canada K1A 0R6

[‡] *Department of Chemistry, University of Missouri-Columbia*

Columbia, Missouri, USA 65211

1. Introduction

Chemistry has witnessed the emergence of an approach to chemical synthesis that focuses on the exploitation of non-covalent forces (*e.g.* hydrogen bonds, □-□ interactions) for the design of multi-component supramolecular frameworks.[1] Fascination with Nature's ability to direct the self-assembly of small subunits into large superstructures (*e.g.* viruses, fullerenes), coupled with a desire to construct materials possessing unique bulk physical properties (*e.g.* optical, magnetic), has undoubtedly provided a major impetus for their design.[2]

2. Multi-Component Hosts

Along these lines, there is much interest in utilizing non-covalent forces, particularly in the form of hydrogen bonds, for the construction of multi-component hosts (*e.g.* molecular capsules 1) that display recognition properties analogous to their monomolecular predecessors

407

D. Braga et al. (eds.), Crystal Engineering: From Molecules and Crystals to Materials, 407–419.

(*e.g.* carcerands **2**).[3-4] Such frameworks typically involve replacing covalent bonds with supramolecular synthons that retain the structural integrity of the parent host molecule.

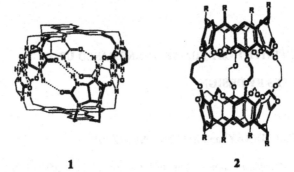

1   **2**

Notably, in addition to providing access to systems which are difficult to obtain using conventional covalent approaches to molecular synthesis,[5] such frameworks can display properties not found in the molecular analog (*e.g.* reversible formation) which, in some instances, can bear relevance in understanding related biological phenomena (*e.g.* virus formation).[6]

3. Resorcin[4]arenes for Multi-Component Host Design

With this in mind, we recently initiated a program of study aimed at extending the cavities of resorin[4]arenes supramolecularly.[7] As a starting point, we chose the readily available *C*-methylcalix[4]resorcinarene **3** as a platform for the assembly process.[8] Indeed, solid state studies had revealed the ability of **3** to adopt a bowl-like conformation with C_{2v} symmetry in which four upper rim hydroxyl hydrogen atoms of **3** are pointed upward above its cavity which, in turn, effectively make **3** a quadruple hydrogen bond donor.[9] Using a resorcinol-based supramolecular synthon[10] **4** for host design, we reasoned that co-crystallization of **3** with hydrogen bond acceptors such as pyridines would result in formation of four O-H···N hydrogen

bonds between the upper rim of **3** and four pyridine units which would extend the cavity of **3** and

yield a discrete, multi-component host, **3**·4(pyridine) **5** (where pyridine = pyridine and

derivatives), capable of entrapping a guest, **3**·4(pyridine)·guest. Furthermore, in addition to a

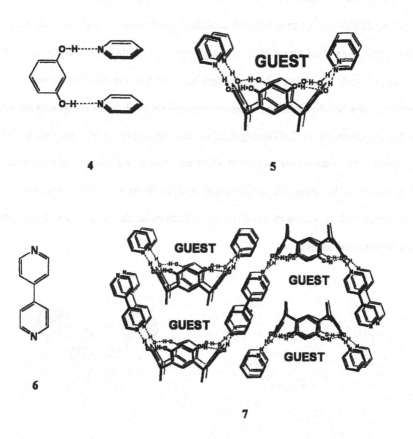

discrete system, we reasoned that by using a linear *exo*-bidentate spacer unit such as 4,4'-

bipyridine **6**, it should be possible to link together **3** in the solid state[11] for the formation of an

infinite, 1D array, **3**·2(4,4'-bipyridine)·guest **7**.

4. Discrete Systems

4.1 Deep Cavities Based Upon Rigid Extenders

The product of the co-crystallization of 3 with pyridine from boiling pyridine is shown in Fig. 1.[7] The assembly is bisected by a crystallographic mirror plane and consists of 3 and five molecules of pyridine, four of which form four O-H···N hydrogen bonds, as two face-to-face dimers, such that they adopt an orthogonal orientation, in a similar way to 4, with respect to the upper rim of 3. As a consequence of the assembly process, a cavity has formed inside which a disordered molecule of pyridine is located, interacting with 3 by way of C-H···π-arene interactions. Notably, the remaining hydroxyl hydrogen atoms of the six-component assembly form four intramolecular O-H···O hydrogen bonds along the upper rim of 3 resulting in a total of eight structure-determining O-H···X (X = N, O) forces. Indeed, the inclusion of an aromatic such as pyridine within 3·4(pyridine) is reminiscent of the ability of covalently modified calix[4]arenes, such as p-tert-butylcalix[4]arene, to form molecular complexes with aromatics such as benzene and toluene.[12]

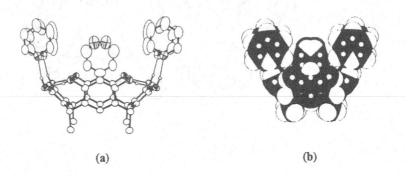

(a)   (b)

Figure 1. X-ray crystal structure of 3·4(pyridine)·pyridine: (a) ORTEP perspective and (b) space-filling view.

To determine whether it is possible to isolate a guest within **5** which, unlike

3·4(pyridine), is different than that the 'substituents' hydrogen bonded to the upper rim of **3**, we

next turned to pyridine derivatives, namely, 4-picoline (monopyridine) and 1,10-phenanthroline

(bipyridine).[13] In a similar way to pyridine, both molecules possess hydrogen bond acceptors

along their surfaces and π-rich exteriors which we anticipated would allow these units to

assemble along the upper rim of **3** as stacked dimers. As shown in Fig. 2, co-crystallization

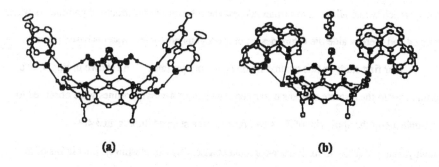

(a)   (b)

Figure 2. ORTEP perspective of: (a) **3·4**(4-picoline)·MeNO$_2$ and (b) **3·4**(1,10-phenanthroline)·
MeCN.

of **3** with either 4-picoline or 1,10-phenanthroline from MeNO$_2$ and MeCN, respectively, yielded

six-component complexes, **3·4**(4-picoline)·MeNO$_2$ and **3·4**(1,10-phenanthroline)·MeCN, which

are topologically equivalent to the parent pyridine system.[7] Unlike the parent assembly,

however, the cavities created by the five molecules were occupied by guests different than the

walls of the host. Indeed, this observation illustrated that this approach to discrete, extended

frameworks based upon **3** is not limited to two different components. Notably, whereas the 4-

picoline moiety was observed to interact with **3** by way of conventional O-H···N hydrogen bonds, the 1,10-phenanthroline extender was also observed to interact with **3** by way of a

bifurcated O-H···N force.

4.2 Deep Cavities Based Upon Flexible Extenders

With the realization that **5** may be exploited for the inclusion of guests different than

the supramolecular extenders of **3** achieved, we next shifted our focus to pyridines that possess

flexible substituents. In addition to introducing issues of stereochemistry, we anticipated that

this approach would allow us to further address the robustness and structural parameters which

define **4** and thereby aide the future design of analogous host-guest systems based upon **3**.

Our first study in this context involved 4-vinylpyridine.[14] As shown in Fig. 3, in a

similar way to the discrete systems described above, four 4-vinylpyridines were observed to

assemble along the upper rim of **3**, as two face-to-face stacked dimers, in 3·4(4-

vinylpyridine)·MeNO$_2$, to form a six-component assembly. Interestingly, the olefins of this

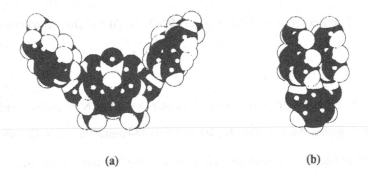

(a)   (b)

Figure 3. Space-filling view of the X-ray crystal structure of: (a) 3·4(4-vinylpyridine)·MeNO$_2$
and (b) resorcinol·4-vinylpyridine. Note the parallel alignment of the double bonds in the former
which contrasts the anti-parallel arrangement adopted by the latter.

system, in contrast to resorcinol·4-vinylpyridine, adopted a parallel orientation in the crystalline state, the bonds being separated by a distance of 4.18 Å. Indeed, approaches that utilize host frameworks to promote alignment of olefin bonds in the solid state, for conducting [2+2] photochemical reactions, for example, are rare[15] and these observations suggest that similar complexes based upon 3 may provide a route to achieving this goal.

From similar studies aimed at determining the ability of a flexible pyridine to elaborate 3, we also discovered the ability of the 'boat' conformation[16] of 3 to self-assemble in the solid state, as a 'T-shaped' building block,[17-22] to form a linear 1D hydrogen bond array 8.[23]

8

As shown in Fig. 4, co-crystallization of 4-benzylpyridine with 3 yielded 3·4(4-benzylpyridine) in which four pyridines were observed to assemble along the upper rim of 3 such that they participate in four O-H···N hydrogen bonds with two opposite resorcinol units of the macrocycle. Unlike the discrete systems, however, the pyridines formed edge-to-face, rather than face-to-face, π-π interactions in which the benzyl substituents of the ligands are effectively 'wrapped' around each other. As a consequence of these forces, the resorcinol moieties of 3 that interact with the pyridines were 'pulled' together which, in turn, induced the cavity of 3 to close. The remaining resorcinol units of 1, which adopt a co-planar orientation and lie approximately perpendicular to the resorcinol moieties that participate in the O-H···N forces, were then observed to form four intermolecular O-H···O hydrogen bonds with two neighboring molecules of 3, making 3 an overall eightfold hydrogen bond donor. As a result, a 1D hydrogen bonded polymer, in which the 4-benzylpyridines alternate along each side of the array, formed.

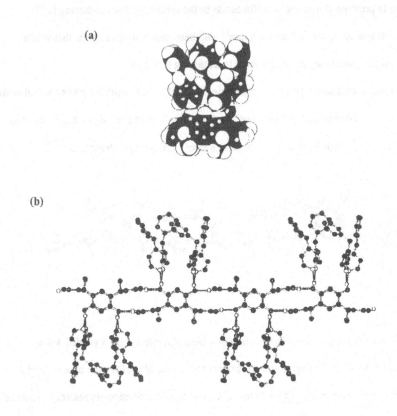

(a)

(b)

Figure 4. X-ray crystal structure of 3·4(4-benzylpyridine): (a) space-filling view and (b) the 1D polymeric array.

It is of interest to note that the ability of 3 to self-assemble in the solid state as a 'T-shaped' building block is reminiscent of the ability of metal centers in a number of Cd, Ag, and, Co coordination polymers,[17-21] as well as ions of guanidinium-based lattices,[22] to serve a similar structural role in the formation of 1D ladder and 2D brick host frameworks. In such cases, the 'T-shaped' units define corners of cavities. Although the components of 3·4(4-benzylpyridine) do not assemble to form any one of the four possible structure types based upon a 'T-shaped' moiety,[18] such structures could, in principle, be generated from an assembly process, as in 7, involving 3 and a linear bipyridine (e.g. 4,4'-bipyridine).

5. Infinite Systems

An attractive feature of this design strategy lies in its modularity. In addition to permitting the design of discrete multi-component frameworks, it is possible, in principle, to link together molecules of 3, in the bowl conformation, to give rise to infinite assemblies. Owing to the molecular recognition properties of 3, such materials would be expected to display inclusion properties based upon the ability of 3 to selectively include guests within its cavity.

5.1 Inclusion of Single Guests

As shown in Fig. 5, co-crystallization of 3 with 4,4'-bipyridine from MeCN yielded 3·(4,4'-bipyridine)·MeCN in which four O-H···N hydrogen bonds form between four hydroxyl groups of 3 and two stacked pyridines.[7] Owing to the ability of 4,4'-bipyridine, and its stacked dimer, to serve as a linear bifunctional hydrogen bond acceptor, the components were observed to assemble such that they form a 1D wave-like polymer 7 in which the MeCN solvent molecule is located within the cavity of 3. Notably, that this assembly may be exploited to host a different guest (e.g. MeNO$_2$, Me$_2$CO), as well as be produced using a different linear bridging unit (e.g. 4,4'-dipyridyl butanedioate), was also realized.[24] Thermal studies of these host-guest materials also revealed that the stability of the framework is, in general, independent of the nature of the guest, the guest dissociating from the solid at approximately 190 °C.[24]

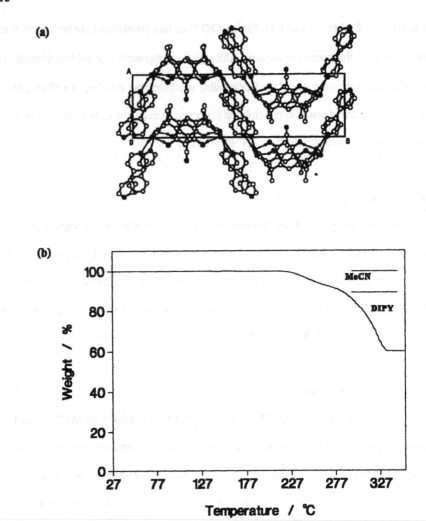

Figure 5. (a) X-ray crystal structure and (b) TGA trace of 3·2(4,4'-bipyridine)·MeCN.

5.2 Inclusion of Multiple Guests

In an effort to further develop the host-guest chemistry of 7, the ability of 7 to include two guests, which assemble as a van der Waals type complex within 3, was also discovered.[25] In particular, as shown in Fig. 6, co-crystallization of 3 with 4,4'-bipyridine from THF and THF/MeCN (8:1) yielded 1D wave-like arrays, 3·2(4,4'-bipyridine)·guest (where guest = 2(THF), THF·MeCN, respectively), in which the dipole moments of the included guests are aligned in an approximate anti-parallel fashion, an orientation which presumably maximizes attractive electrostatic forces between constituent molecules.

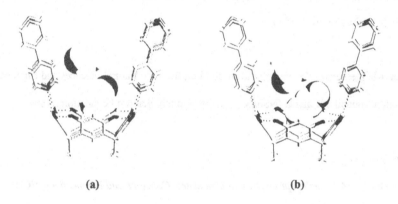

(a)   (b)

Figure 6. X-ray crystal structure of (a) 3·2(4,4'-bipyridine)·2(THF) and (b) 3·2(4,4'-bipyridine)·THF·MeCN.

Indeed, these multi-guest systems represent the first examples in which two guests have been isolated within a resorcin[4]arene in the solid state and suggest a possibility of using this wave-like array as a host for guests capable of undergoing reactions in the solid-state.

6. Conclusion

The studies presented here illustrate an approach for the rational design of discrete and infinite multi-component host-guest assemblies based upon 3. By recognizing 3 as a quadruple hydrogen bond donor and employing a crystal engineering strategy, the ability of pyridine[7] and, its derivatives,[13-14,23-25] to elaborate the cavity of 3 has been illustrated which, in turn, has given rise to multi-component hosts that exhibit recognition properties analogous to their mono-molecular predecessors. With these observations realized, we anticipate that, in addition to expanding the library of components that may be used to construct these frameworks (*i.e.* spacer units, guests), focus will be placed upon incorporating higher levels of function within these solids (*e.g.* chirality, reactivity).

Acknowledgements. We are grateful for funding from the Natural Sciences and Engineering Research Council of Canada (NSERC) and the National Science Foundation (NSF).

7. References

1. Lehn, J. M. (1995) *Supramolecular Chemistry, Concepts and Perspectives*, VCH, Weinheim.

2. Philip, D. and Stoddart, J. F. (1996) *Angew. Chem., Int. Ed. Engl.* **35**, 1155.

3. Heinz, T. H., Rudkewich, D. M. and Rebek, J., Jr. (1998) *Nature* **394**, 764.

4. Jetti, R. K. R., Kuduva, S. S., Reddy, D. S., Xue, F., Mak, T. C. W., Nangia, A. and Desiraju, G. R. (1998) *Tetrahedron Lett.* **39**, 913.

5. de Mendoza, J. (1998) *Chem. Eur. J.* **4**, 1373.

6. MacGillivray, L. R. and Atwood, J. L. (1997) *Nature* **389**, 469.

7. MacGillivray, L. R. and Atwood, J. L. (1997) *J. Am. Chem. Soc.* **119**, 6931.

8. Högberg, A. G. S. (1980) *J. Am. Chem. Soc.* **102**, 6046.

9. Murayama, L. and Aoki, K. (1997) *Chem. Commun.*, 119, and references therein.

10. Desiraju, G. R. (1995) *Angew. Chem., Int. Ed. Engl.* **34**, 2311.

11. Krishnamohan, S. and Zaworotko, M. J. (1996) *Chem. Commun.*, 2655.

12. Andretti, G. D., Ugozzoli, F., Ungaro, R. and Pochini, A. in *Inclusion Compounds*, Oxford University Press, Oxford: 1991; Vol. 4, pp. 64.

13. MacGillivray, L. R. and Atwood, J. L. (1999) *Chem. Commun.*, 181.

14. MacGillivray, L. R., Reid, J. L., Atwood, J. L. and Ripmeester, J. A. (1999) *Cryst. Eng.*, in press.

15. Toda, F. (1988) *Top. Curr. Chem.* **149**, 211.

16. Shivanyuk, A., Schmidt, C., Böhmer, V., Paulus, E. F., Lukin, O. and Vogt, W. (1998) *J. Am. Chem. Soc.* **120**, 4319.

17. Fujita, M., Kwon, Y. J., Sasaki, O., Yamaguchi K. and Ogura, K. (1995) *J. Am. Chem. Soc.* **117**, 7287.

18. Robinson, F. and Zaworotko, M. J. (1995) *J. Chem. Commun., Chem. Commun.*, 2413.

19. Yaghi, O. M. and Li, H. (1996) *J. Am. Chem. Soc.* **118**, 295.

20. Losier, P. and Zaworotko, M. J. (1996) *Angew. Chem., Int. Ed. Engl.* **35**, 2779.

21. Hennigar, T. L., MacQuarrie, D. C., Losier, P., Rogers, R. D. and Zaworotko, M. J. (1997) *Angew. Chem., Int. Ed. Engl.* **36**, 972.

22. Swift, J. A., Pivovar, A. M., Reynolds, A. M. and Ward, M. D. (1998) *J. Am. Chem. Soc.* **120**, 5887.

23. MacGillivray, L. R. and Atwood, J. L., submitted.

24. MacGillivray, L. R., Holman, K. T. and Atwood, J. L. (1998) *Cryst. Eng.* **1**, 87.

25. MacGillivray, L. R., Holman, K. T. and Atwood, J. L. *Trans. Amer. Cryst. Assoc.*, in press.

CRYSTAL ENGINEERING: FROM MOLECULES AND CRYSTALS TO MATERIALS

DARIO BRAGA
Dipartimento di Chimica G. Ciamician
Università di Bologna,
Via Selmi 2, 40126 Bologna, Italy
dbraga@ciam.unibo.it, http://catullo.ciam.unibo.it

FABRIZIA GREPIONI
Dipartimento di Chimica
Università di Sassari,
Via Vienna 2, 07100 Sassari, Italy
grepioni@ssmain.uniss.it, http://catullo.ciam.unibo.it

ABSTRACT: Modern crystal engineering is the planning and utilisation of crystal-oriented syntheses and the evaluation of the physical and chemical properties of the resulting crystalline materials. Crystal engineering is an interdisciplinary area of chemistry research that cuts horizontally across all traditional subdivisions of the discipline, e.g. organic, inorganic, organometallic, and co-ordination chemistry.

1. Introduction

"WHAT IS A CRYSTAL ?"
Were they asked, most people (including many chemists) would answer this question by probably saying that crystals are regularly shaped, often colourful, shining objects which are used mainly as ornament or in devices of popular use, e.g. piezoelectric devices or liquid crystal displays. Chemistry text books describe crystals as kind of ('low entropy') containers in which molecules or ions are neatly packed in three-dimensional order. A distinction is usually made between extended covalent solids, e.g. diamond and quartz, ionic crystals, where coulombic interactions are predominant, and molecular crystals held together by weak intermolecular forces. Synthetic chemists and crystallographers are, on the other hand, used to thinking of crystals as fundamental and unbeatable tools to obtain hyperfine information on molecular or ionic structures, and structural parameters (bonds, angles, torsions, contacts etc.). This is a static perception. Dunitz opened an important feature article by recalling the perception, expressed by some chemist, of crystals as *chemical cemeteries* where molecules are buried, lifeless, after they have done their job in the real chemistry life (gas phase, solution etc.).[1] This view (see Figure 1) is shared by many crystallographers, those who are supposed to be more than others culturally

D. Braga et al. (eds.), Crystal Engineering: From Molecules and Crystals to Materials, 421–441.
© 1999 Kluwer Academic Publishers. Printed in the Netherlands.

equipped to appreciate crystals as chemical entities whose properties are the result of the collective behaviour of billions of billions of atoms and molecules in continuous motion (at least at the temperatures of routine diffraction experiments).

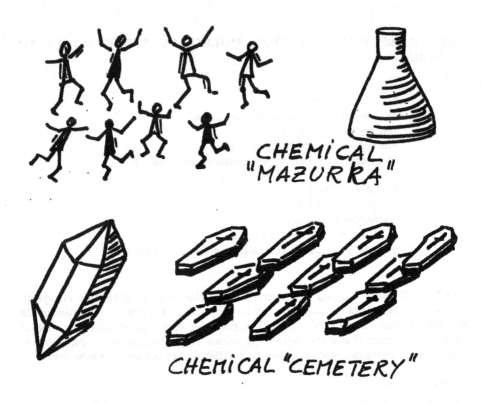

CHEMICAL "MAZURKA"

CHEMICAL "CEMETERY"

Figure 1. A crystal is seen as a chemical cemetery, while "real chemistry" occurs in solution.

Although molecules are still being synthesised by chemists (and so they will for a long time to come), the recent past has witnessed a true explosion of interest in synthetic strategies that have taken from molecules to molecular aggregates. This shift of interest has led to the development of synthetic strategies that abandon the focus on covalent bond chemistry to begin *making use* of the weaker and less directional bonds that are responsible for the organisation of molecules into superstructures. This is known as supramolecular chemistry.[2]

Jean-Marie Lehn defined supramolecular chemistry as *"chemistry beyond the molecule bearing on the organised entities of higher complexity that result from the association of two or more chemical species held together by intermolecular forces"*.

Since nucleation and growth of a molecular crystal implies self-recognition and self-organisation, and since these processes generate *collective* physical and chemical properties, molecular crystals can be viewed as supermolecules, or even as hypermolecules, in which numbers of Avogadro of molecules interact *via* a plethora of non-covalent interactions. In the following we will use *molecular* as a synonym of *non-covalent*; the epithet will therefore encompass all sort of interactions (from electrostatic, to hydrogen bonds, to van der Waals ones) that do not imply rupture or formation of two-electron σ-bonds. Indeed, as sophisticated procedures have been developed to sequentially assemble molecules *via* covalent synthesis, crystal engineers are now concentrating their efforts on developing crystal-oriented synthetic strategies. This means a substantial shift of interest: from the one focused on atoms and bonds between atoms to the one focused on molecules and *bonds between molecules*. This logical process is depicted in Figure 2 for the simple case of $Co_2(CO)_8$

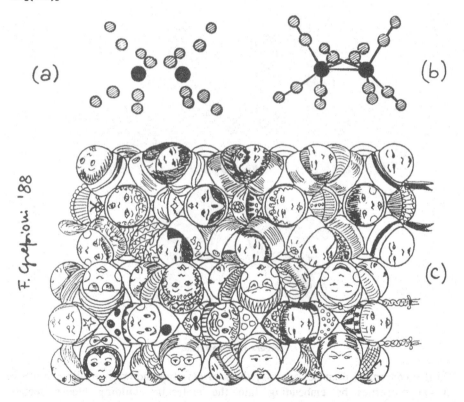

Figure 2. The logical process that takes from molecular engineering to crystal engineering: (a) isolated atoms, (b) the isolated atoms are joined by bonds to form a molecule of $Co_2(CO)_8$, (c) A view of the crystal formed by $Co_2(CO)_8$ molecules ... in the international spirit of the Erice School.

424

Covalent synthesis requires a large amount of energy. Text books again tell students that ΔH values to break or form covalent bonds are of the order of 10^2 kJ.mol^{-1}, an energy equivalent to blasting molecules apart if compared with the tiny energies required to break non-covalent interactions, including hydrogen bonds. However, the key point, which more adequately describes the supramolecular nature of a molecular crystal, is the recognition that collective crystal properties (*viz.* solid state properties) are different from, though intimately related to, the molecular properties of the building blocks: the properties of ice, resulting from hydrogen bond aggregation of water molecules, are different from those of an isolated water molecules in the vapour phase. The snow flake in Figure 3 represents, in all its elegance and fragility, a (naive) example of a supramolecular aggregate of water molecules with properties well distinct from those of the isolated gas phase water molecule.

Figure 3. A snow flake.

On this conceptual premise, one may think of devising and/or controlling collective crystal properties by embedding into the molecular building blocks specific recognition and binding sites that may lead to the target supramolecular aggregation, hence to the target chemical or physical property. Once the extra-molecular functionality has been *written* into the building block by means of common laboratory chemical synthetic procedures, the crystallisation process, typically a process of molecular self-organisation - *i.e.* molecules or ions pack up adjusting their relative positions to reach a thermodynamic minimum - becomes a process of self-assembly, *i.e.* molecules are not *left alone* to self-organise but are directed by the extra-molecular *tug-boat* interactions. The stability and cohesion, phase transitional behaviour and other more utilitarian properties, such as those exploited in optoelectronic, magnetic, conductivity applications of the material will depend on the strength and number of non-covalent interactions in use (see Table 1). This is, in very simplistic terms, the basic idea sustaining molecular crystal engineering. This is also the idea Guy Orpen and I (D.B.) have followed to plan and organise the 28th Course of the International School of Crystallography (Erice 1999) on *Crystal Engineering: from Molecules and Crystals to Materials* which has generated this Book and this contribution.

TABLE 1. A (non-exhaustive) chart of target properties

TARGET MATERIALS	TARGET PROPERTIES
chiral frameworks	NLO properties
mixed metal / valence / spin materials	molecular magnets conductivity
charge transfer - π stacks	charge transfer - conductivity superconductivity
honeycomb - channels - host/guest - zeotypes	nanoporosity - sieves - catalysis solid-state sensors

The use of the term 'molecular' implies some kind of boundaries: while it would be artificial to force vertical subdivisions (organic, organometallic, inorganic *etc.*) into crystal engineering, there is a more fundamental issue that needs to be appreciated when *making crystals with a purpose*, namely the energy involved in the process.

Covalent solids (quartz, diamond, graphite, BN *etc.*) are probably amongst the most popular *materials* but all functional modification of covalent networks of these solids require high energies. Often new properties can only be obtained by utilising the non-covalent features. Graphite intercalation chemistry, for example, dates back almost 150 years and has constituted a major issue of research for many years (see Figure 4). Similar reasoning applies to zeolites, the other large class of inorganic systems whose solid state properties have been know for centuries. The word 'zeolite' means 'boiling stone' in ancient Greek and refers to the ability of releasing large quantities of water from the channelled structure. The catalytic activity, the molecular sieve properties *etc.* are all due to the non-covalent interactions that groups of atoms and ions may establish with the zeolite framework. The preparation of *zeotypes*, *i.e.* of molecular-based porous systems, is one of the targets of materials chemistry.

426

Graphite

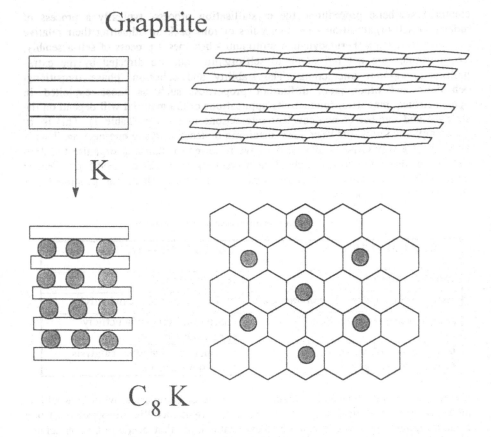

Figure 4. (a) The structure of graphite, an example of mixed, covalent/van der Waals solid. (b) The weak interaction between graphite plane allows intercalation, for example of alkali metal atoms.

These examples serve the scope of defining an energetic basis to discriminate between covalent and non-covalent crystal engineering. It is almost unnecessary to remind the reader that *Natura non facit saltus*, thus there is a continuum of intermediate situations, and the energy scale depicted in the cartoon below (Figure 5) can only be taken as a broad reference. It is worth noting, however, how covalent and non-covalent crystal engineering admit an intermediate situation, that is coordination crystal engineering. In co-ordination crystal engineering the link between building blocks is provided by polydentate ligands that can join together co-ordination complexes in extended networks ('co-ordination polymers'). Molecular solids formed by molecules are at the bottom of the energy scale. The difference between a crystal building process involving only non-covalent interactions and one in which covalent bonds are broken and formed has important methodological consequences. In the building up of non-covalent crystals use is made prevalently of molecules or ions held together by bonds weaker than those between atoms forming the building blocks (see below). In covalent crystal engineering, on the contrary, use is made of covalent bonds between components that might not exist as independent entities.

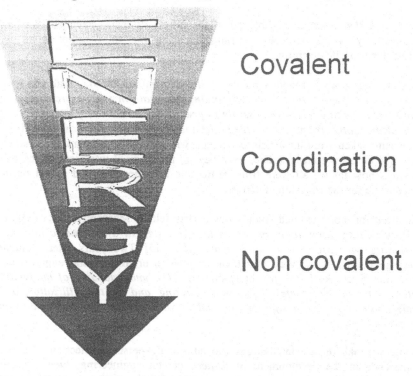

Crystal engineering

Covalent

Coordination

Non covalent

Figure 5. An energy scale for crystal engineering.

2. Crystal Engineering

The use of the qualifier "engineering" associated to crystals predates its use to describe materials design strategies. It was put forward by Schmidt in the late sixties to explain the photodimerization of olefins in the solid state.[3] The idea was beautifully simple: the double bonds were locked in place by the crystal packing at an appropriate distance for photochemical activated cyclization reactions, and topochemical control on crystal reactivity was thus achieved.

Schmidt wrote: *"The systematic development of our subject will be difficult if not impossible until we understand the intermolecular forces responsible for the stability of the crystalline lattice of organic compounds: a theory of the organic solid state is a requirement for the eventual control of molecular packing arrangement. Once such a theory exists we shall, in the present context of synthetic and mechanistic photochemistry, be able to 'engineer' crystal structures having intermolecular contact geometries appropriate for chemical reaction, much as, in other context, we shall construct organic conductors, catalysts, etc."*

In spite of the initial optimism and of the efforts of many scientists, the lack of predictability of the molecular arrangements in the solid inhibited an earlier development of the discipline.

In September 1988 J. Maddox published an editorial, in *Nature*, that was going to make a big impact: *"One of the continuing scandals in the physical sciences is that it remains in general impossible to predict the structure of even the simplest crystalline solid from a knowledge of their chemical compositions"*.[4] This statement, taken from an article concerned with *ab*-initio calculations of silica has been quoted many times (perhaps even beyond the original intentions of the Author) to stress how far we all were from being able to understand and model the forces responsible for the cohesion of solids.

The need for more detailed studies was clearly felt by Margaret Etter: *"Organizing molecules into predictable arrays is the first step in a systematic approach to designing (organic) solid- state materials"*. [5] Two years later, Gautam R. Desiraju published the very first book devoted to organic crystal engineering; the term was given the following interpretation: *"The understanding of intermolecular interactions in the context of crystal packing and in the utilization of such understanding in the design of new solids with desired physical and chemical properties"*.[6]

While we take these contributions (an admittedly personal selection !) as sort of milestones in the development of *modern* crystal engineering, there are, in our opinion, three main facts that are responsible for the birth (or rebirth) of crystal engineering as a 'stand-alone' science:

i) the appropriate cultural environment provided by the success of supramolecular chemistry and the consequent shift of interest from a molecular based chemistry to the chemistry of molecular aggregates;

ii) the urge for more utilitarian objectives for the chemical sciences, such as those provided by materials chemistry, as a consequence of diffuse funding restrictions for fundamental studies;

iii) last but not least, the progress in computing and diffraction tools that allow to tackle on a reasonable time scale theoretical and experimental problems of great complexity, such as those associated with complex molecular solids, interdigidated networks and supramolecular aggregates.

The ambitious goal is that of constructing intelligent materials, *smart materials*, that may have useful properties, possibly selected and predefined by an adequate choice of the component building blocks. Of course 'crystals with a purpose' were prepared long before the modern definition of crystal engineering. The chemistry of molecular materials, for example, has been explored extensively in the areas of charge transfer, conductivity and superconductivity, and magnetic materials (see contributions, and literature entries, in this Book). It is, however, the supramolecular awareness that has given impetus to the field by generating useful interdisciplinary connections.

As pointed out before, the common denominator between molecular crystal engineering and supramolecular chemistry is the utilisation of non-covalent bonding. Indeed, as non-covalent bonds control the behaviour of supermolecules, the periodical distribution of such interactions is responsible for the properties of the molecular crystalline material. Scientists do often resolve to paradigms in the effort of simplifying complex problems. If the utilisation of non-covalent interactions is taken as the paradigm of supramolecular chemistry, and periodicity as paradigmatic of crystal structure, then the preparation and utilisation of periodical supermolecules may well be the paradigm of molecular crystal engineering. This relationship is shown in Table 2.

TABLE 2. A Molecular Crystal is a Periodical Supermolecule

DISCIPLINE	PARADIGM
Supramolecular Chemistry	Non-Covalent Interactions
Crystal Chemistry	Periodicity
Molecular Crystal Engineering	Periodical distribution of non covalent interactions (periodical supermolecules)

Therefore, molecular crystal engineering can be seen as the logical intersection between the route leading from supramolecular chemistry to nanochemistry and molecular machines and the route leading from solid state chemistry to a molecular-based chemistry of materials (see the cartoon in Figure 6).

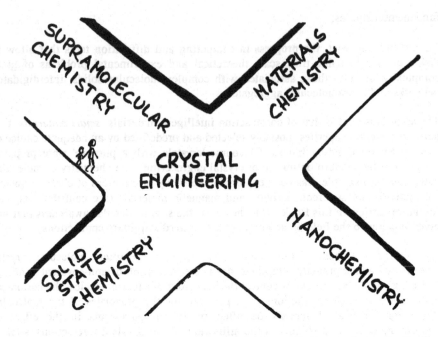

Figure 6. Molecular crystal engineering can be seen as the logical intersection between the route leading from supramolecular chemistry to nanochemistry and molecular machines and the route leading from solid state chemistry to a molecular-based chemistry of materials.

On these conceptual premises it should be clear that crystal engineering does not differ from classical chemical experiments in which molecules are *imagined*, synthetic strategies *devised*, products *characterised* and their properties *measured*. Crystal engineering, in a sense, follows these processes *twice* taking chemical synthesis products (the molecular building blocks) into the synthesis of crystals as shown in Table 3.

The discussion of all these aspects is well beyond the scope of this contribution. Moreover, most of them are dealt with in this Book. In the following we will confine our considerations to some aspects of the modeling and synthesis stages, in which we have been more directly involved.

TABLE 3. Crystal Engineering

MODELING	⇒ Design of molecular building blocks (knowledge of functional groups) ⇒ Database analysis ("data mining", occurrence, distribution, clustering, graph-sets) ⇒ Graphical representation (van der Waals interlocking, crystal symmetry, hydrogen bonds) ⇒ Theoretical calculations of building blocks (EH, ab-initio, dft) ⇒ Theoretical crystal structure generation
SYNTHESIS	⇒ Molecular synthesis of building blocks ⇒ Crystal synthesis of aggregates ⇒ Co-ordination chemistry ⇒ Nucleation and growth ⇒ Biomineralization
CHARACTERISATION (solid-state techniques)	⇒ Powder and single crystal X-ray diffraction ⇒ Ab-initio determination from X-ray powder diffractograms ⇒ SS-NMR - IR / Raman ⇒ Calorimetric techniques DSC / TGA ⇒ Electron microscopy, AFTM, STM
EVALUATION (of solid state properties)	⇒ Second and third harmonic generation ⇒ Magnetism / conductivity / superconductivity ⇒ Charge transfer ⇒ Spectroscopical / photochemical ⇒ Molecular and ionic sieves / nanoporous materials

2.1. MODELING

Modeling is the examination of available information on intermolecular interactions and molecular recognition, and the theoretical study of nucleation and crystallisation processes. The cartoon in Figure 7 shows the fundamental aspects of modeling.

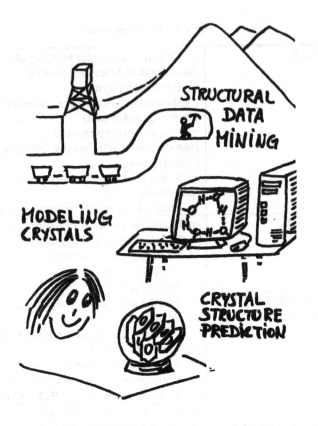

Figure 7. The fundamental aspects of modeling.

We are only beginning to perceive the complex relationship between a given molecular (or ionic) structure and the structure of possible corresponding periodical supermolecules. To take Gavezzotti's words: *"The physical nature of the forces acting between organic molecules in crystals is rather well understood; what is difficult to grasp is the complex spatial pattern of such forces"*.[7]

The phenomenon of polymorphism, for example, is a manifestation of the intriguing interplay between molecules and crystals. Different crystalline forms containing molecules of identical composition and geometry are called polymorphs.[8] It is well known that alternative experimental packing modes, that are still chemically and crystallographically acceptable, differ very little in terms of energy. The relationship between polymorphs depends on whether the solid-solid transition occurs below the solid-liquid transition (enantiotropic systems) or the transition is preceded by melting of the polymorphs (monotropic systems). The chemical and physical properties of a crystalline material can change dramatically with the solid state transformation. In a supramolecular approach to crystal polymorphism, one may say that molecular crystal polymorphs represent *supramolecular* isomers, and that the change in crystal structure associated with a phase transition, in which intermolecular bonds are broken and formed, is the *crystalline* equivalent of an isomerization at the molecular level.

Data mining is also a fundamental component of the modeling stage. It is hard to deny that, without the forsight of scientists such as O. Kennard and others, crystal engineers would today be deprived of one of their most efficacious tools: the possibility of cross analysis of numerical data on more than 180.000 organic and organometallic molecular crystal structures deposited in the Cambridge Structural Database.[10]

Scientists 'learn from Nature' and the thousands of molecular crystals obtained and characterised in thousands of crystallographic laboratories, mostly in uncorrelated ways, contain collective information which have relevance on a statistical basis. Analysis of the CSD depositories provides an immediate perception on the occurrence and transferability of given interaction motifs. Of course a word of caution is in order. Databases are investigated by queries, and queries are developed *via* computer softwares. A golden rule of datamining is that 'the computer will always answer your question'. However, whether the answer will be the correct one, or have any decent chemical meaning at all, will depend on how the question is asked. In other words, datamining never gives wrong answers, but the scientist often asks wrong questions.

An educative example is provided by a CSD analysis of the metrics of O(H)---O hydrogen bonded interactions below 2.8 Å in polycarboxylic or polycarboxylate systems. Figure 8 shows what the histograms look like if a distinction between O-H---O and O-H$^{(-)}$---O$^{(-)}$ is made: two distinct populations are observed corresponding to completely different energetic situations. It has been recently argued that, while the neutral O-H---O bond is a cohesive interaction which *holds* particles together in the solid state, the O-H$^{(-)}$---O$^{(-)}$ interaction is more controversial and, at least in the solid state, plays the role of a supramolecular organiser providing a means to attain the least repulsive assembly without actually *binding* the anions together.[11]

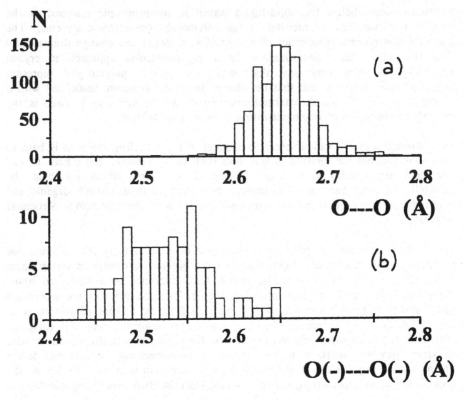

Figure 8. Comparison of intermolecular (a) O(H)---O and (b) O(H)$^{(-)}$----O$^{(-)}$ distances in polycarboxylic and polycarboxylate systems.

Apart from the obvious need to 'ask the correct question', on approaching CSD datamining one should keep in mind that the Database contains a heavily biased sample. Since the CSD is a 'real-life' database (as adversative to theoretical), the taste of the experimentalist, which often depends on the scientific fashion of the moment, and other utilitarian reasons, such as fund-rising and the necessity of tackling projects that guarantee results in relatively short times have all great impact on the population of compounds that is stored in the CSD. In addition, all those crystalline materials that fail to satisfy the numerous pre-conditions that precedes data collections (e.g. size, shape and habit, unit cell dimensions, chemical and physical stability, diffracting power etc.) are not represented in the database. Since they are not there, all speculations are possible, including the fact that these materials contain extremely important information on the *why* they do not meet the crystallographer wishes.

2.2. STRUCTURE GENERATION AND PREDICTION, THEORETICAL EVALUATION

Talking about ab-initio determination of crystal structures Leiserowitz and Hagler wrote in 1983: *"The interest* (in ab-initio determination of crystal structures) *covers various fields, including the basic question as to why a molecule adopts its particular packing arrangement,... ".*[12] More than eleven years later, Whitesell *et al.* wrote: *"There is as yet no generally successful approach to predicting, let alone controlling, molecular orientations in crystals. Thus the rational design and preparation of crystalline and other supra-molecular materials is hampered by insufficient knowledge of those factors that control packing."* [13]

Since the work *prediction* carries too much 'magics' (see the cartoon above), it would be advisable to speak about theoretical generation of crystal structures. The design of crystals for magnetic, conductivity, superconductivity or non-linear optical applications would become much easier and efficient if one could model crystal structures computationally from molecular structures. The molecular or ionic structure needs not necessarily to be determined by means of crystallography tools. Spectroscopic techniques or *ab initio* methods may be just as satisfactory. Indeed, the generation of hypothetical crystal structures is logically connected to *ab-initio* calculations of molecular structures. The prerequisite for the construction of a molecular crystal is the ability to model intermolecular interactions correctly and this can only be done with a good knowledge of the factors that control molecular recognition and crystal cohesion. [14]

The problem of computational crystal structure generation has been tackled in different ways, but essentially there are two main approaches, namely 'number crunching' and *aufbau*.[15] In this latter approach the molecular crystal is constructed by translation in the three-dimensional space of a nucleus formed by molecules which are related by symmetry operators such as a screw axis, an inversion centre or a glide plane. Both the nucleus energy and the crystal energy are optimised under the action of adequate atom-atom potential functions. Theoretical crystal structures have also been generated by using Monte Carlo simulated annealing. This is the so-called dynamical approach and it has the advantage that no particular crystallographic assumption is necessary. The method relies solely on a brute force computation in which a collection of randomly oriented molecules are moved in space until the energy minimum is reached. Other methods are based on a combination of molecular dynamics and packing analysis.

It should be pointed out that there is need to explore only a very limited subset of space group arrangements since the vast majority of molecules pack in very few space groups. It may be anticipated that computational experiments that attempt to rationalise, model and predict crystal structures and properties will almost routinely accompany laboratory experiments of crystal engineering in the future. These aspects

436

are dealt with in details in the contributions of Perlstein, Schmidt and Williams in this Book.

2.3. SYNTHESIS AND CHARACTERISATION

Crystal engineering means crystal synthesis. Molecular crystal engineering is made with molecules (or ions) - the building blocks in the construction process. What is, at times, difficult to grasp is the complex relationship between molecular structure, supramolecular features, and intermolecular bonding. Synthesis is where the chemist works on his/her more congenial ground.

The *chemistry of covalent bond* is so much developed that there is no practical limitation to the preparation of adequate building blocks. A more serious problem is, instead, that of the characterisation of the reaction products, which, in crystal engineering, are solids. This is to point out that on moving from a molecular based chemistry to a solid-state supramolecular chemistry - most of the well established - routine - tools for the isolation and characterisation of *molecular* species (mainly in solution, sometimes in the gas phase) become useless.

Since a crystal synthesis product is ultimately a solid materials, methods and techniques for solid state are required. Solid state analytical techniques are more complicate and, in general, less widely accessible in departmental laboratories than routine techniques for solution chemistry. Diffraction methods are, of course, the methods of choice for the characterisation of crystalline products. With the advent and diffusion of CCD diffractometers, the measurements of a large number of single crystal data sets within a single research project is becoming fairly common. But crystallisation from solvents usually leads to a solid material composed of single crystals and of bulk material. The correspondence between the structural characteristics of the bulk and those of the single crystals needs to be checked. This ought to be routinely done by comparing observed powder spectra with those calculated on the basis of the structures determined by single crystal X-ray diffraction.

We have devoted our efforts, also in collaboration with others, to apply crystal engineering strategies within the organometallic chemistry field, with the purpose of bringing in the solid the electronic, magnetic, and structural properties of transition metal atoms. Although the role of metal atoms in crystal engineering has been addressed in review articles,[16] it may be useful to recall the distinct *functions* of metal atoms in crystal engineering:

i) Ligand structure and co-ordination geometry can be used *to preorganise in space* non-covalent links with neighbouring molecules.[17]

ii) The electronics of metal-ligand bonding interactions, such as donation and backdonation, permit tuning of ligand polarity and acid/base behaviour.[18]

iii) Metal atom variable oxidation states and/or the utilisation of non-neutral ligands permit 'charge assistance' to weak bonds.[17]

iv) Electron deficient metal atoms may accept electron density intermolecularly from suitable Lewis bases, while electron rich metal atoms may have sterically unhindered lone pairs that accept hydrogen bonds.[19]

v) Size and shape of the complexes can be used to control the self-assembly and the host-guest systems.[20]

These studies have constituted the basis of the organometallic-based crystal engineering we have developed in our laboratory to produce mixed organic, inorganic and organometallic crystals. The relatively simple example provided by the product of the reaction of *squaric acid* (3,4-dihydroxy-3-cyclobutene-1,2-dione, H_2SQA) with cobalticinium hydroxide will be used as an entry point into this chemistry: if a non-co-ordinating and non-hydrogen bonding base, such as $[(\eta^5\text{-}C_5H_5)_2Co)]^+[OH]^-$ is reacted with polycarboxylic acids such as H_2SQA, self-aggregation of the partially deprotonated acid molecules is attained. The $[(\eta^5\text{-}C_5H_5)_2Co)]^+$: $[H_2SQA]^-$ 1:1 system is constituted of infinite $\{[(HSQA)]^-\}_n$ ribbons and of ribbons of cobalticinium cations, as shown in Figure 9. The good matching in size and shape between the cyclopentadienyl ligands and the $[(HSQA)]^-$ ions, leads to a superstructure in which the squarate ribbons *intercalate* between cobalticinium cations. The π-π distance is *ca.* 3.35 Å. The oxygen atoms from the rims of the $\{[(HSQA)]^-\}_n$ ribbons interact with the $[(\eta^5\text{-}C_5H_5)_2Co)]^+$ cations *via* charge assisted $C\text{-}H^{\delta+}\text{---}O^{\delta-}$ hydrogen bonds. An almost isomorphous compound can be obtained if the paramagnetic cation $[(\eta^6\text{-}C_6H_6)_2Cr]^+$ is used.

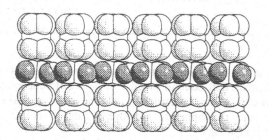

Figure 9. Ribbons of $[(HSQA)]^-$ monoanions joined *via* $O\text{-}H^{(-)}\text{---}O^{(-)}$ interactions and ribbons of cobalticinium cations form the crystal of the 1:1 system.

438

The self-assembly strategy is controlled by two factors: (i) the absence of hydrogen bond acceptors that could compete with the acid itself and (ii) the complementary role of strong and weak hydrogen bonds reinforced by Coulombic contributions. The combination of redox processes, acid-base and solubility equilibria that is behind the synthesis of $[(\eta^5\text{-}C_5H_5)_2Co)]^+[H_2SQA]^-$ is summarised in Table 4.

TABLE 4. A chart for the preparation of mixed organic/organometallic supersalts

Oxidation of neutral sandwich organometallic molecules to the corresponding cations	Reduction of O_2 to strongly basic O_2^-, deprotonation of the acid and generation of OR/OM anions
⇒ formation of stable organometallic cations with metal centres unavailable for co-ordination ⇒ neutral species are insoluble in water or polar solvents and soluble in thf	⇒ the water solutions are those of hydroxides ⇒ the supramolecular salts are insoluble in low polarity solvents while they are soluble in water
The anions form O-H---O, O-H$^{(-)}$---O$^{(-)}$ and O-H---O$^{(-)}$ interactions and self-assemble into Supramolecular anions	The cations mould the O-H---O hydrogen bonded frameworks via charge assisted C-H---O bonds
⇒ The strongly basic -COO$^{(-)}$ groups seek strong donors which can be found only on the partially deprotonated acid	⇒ The aggregates precipitate immediately in apolar solvents and can be recrystallised from water or nitromethane

The utilisation and combination of organometallic acids or bases allow the preparation of crystals that contain metal atoms in different oxidation, charge and spin states as well as the construction of non-centrosymmetric crystals when enantiomerically pure reactants are employed. Table 5 collects examples of the different types of compounds.

TABLE 5. Some supersalts and their target structres

Target structures	Species involved
chiral frameworks	L-tartaric, L-*trans*-acotinic acid $[(\eta^5\text{-}C_5H_5)_2Co]^+$ or $[(\eta^5\text{-}C_5Me_5)_2Co]^+$ (diamagnetic) $[(\eta^6\text{-}C_6H_6)_2Cr]^+$ or $[(\eta^6\text{-}C_6H_5Me)_2Cr]^+$ (paramagnetic)
mixed metal / *valence / spin*	$[(\eta^5\text{-}C_5H_4COOH)_2Fe]$ $[(\eta^5\text{-}C_5H_4COOH)_2Co]^+$
charge transfer *π stacks*	squaric acid $[(\eta^5\text{-}C_5H_5)_2Co]^+$ or $[(\eta^5\text{-}C_5Me_5)_2Co]^+$ $[(\eta^6\text{-}C_6H_6)_2Cr]^+$ or $[(\eta^6\text{-}C_6H_5Me)_2Cr]^+$
honeycomb	D,L-tartaric, phthalic, trimesic acid $[(\eta^5\text{-}C_5H_5)_2Co]^+$ or $[(\eta^5\text{-}C_5Me_5)_2Co]^+$ $[(\eta^6\text{-}C_6H_6)_2Cr]^+$ or $[(\eta^6\text{-}C_6H_5Me)_2Cr]^+$
channels *zeotypes*	$[(\eta^5\text{-}C_5H_4COOH)_2Co]^+$
host / guest	1,3-cyclohexanedione $[(\eta^6\text{-}C_6H_6)_2Cr]^+$ or $[(\eta^6\text{-}C_6H_5Me)_2Cr]^+$

This very brief description and the chart above serve, obviously, the primary scope of providing the key reference entries to our small contribution to crystal engineering. Our *organometallic super-salts* are essentially ionic systems that own their cohesion and stability mainly to electrostatic forces, but whose structure is controlled by the directionality, predictability and reproducibility of complementary strong and weak hydrogen bonds. Similar to "conventional" salts, these supersalts are very stable and highly soluble in polar solvents, such as water or nitromethane, while the presence of extended networks introduces anisotropy in the ion arrangements (*e.g.* the squarate salts above) and characteristics that are typical of hydrogen bonded molecular crystals. These are all useful features in the quest for reproducible and transferable strategy to build crystalline materials.

440

3. Conclusions

Crystal engineering is the modelling, synthesis and exploitation of crystalline materials with predefined aggregation of molecules or ions to obtain solids with desired chemical and physical properties. The interest in the potentials of this research field is increasing rapidly. This originates from various, often uncorrelated, research areas such as solid state reactivity, and dynamics, interest in new molecular and ionic materials, the packing of organic, and organometallic molecules in crystals, and the need to understand the popularity of space groups and to predict how molecules recognise each other and aggregate to form three-dimensional arrays. The spectacular growth in the chemistry and properties of complex supramolecular systems has fueled further interest in controlling and exploiting the aggregation of molecules in crystals.

Crystal engineering is at the intersection of supramolecular chemistry and materials chemistry. There is a synergistic relationship between design and synthetic methods of supramolecular chemistry and those exploited to build crystalline aggregates. In both cases the *collective properties* of the aggregate depend on the choice of intermolecular and interion interactions between components. Materials for applications in optoelectronics, conductivity, superconductivity, magnetism as well as in catalysis, molecular sieves, solid state reactivity, and mechanics are being sought.

In this contribution, we have discussed some aspects of this booming field of reaserch. The use of cartoons drawn by us is a way to stress that our views are very personal (hence obviously opinionable), but we also hope that they show how much we enjoy working in crystal engineering.

The brief historical outline presented in the first part should have demonstrated how, over a short period of time, crystal engineering has grown away from its cradle, which was essentially organic in nature, and is now spanning all areas of chemistry, with relevant interdisciplinary interactions with biology, informatics and physics. The high quality contributions collected in this Book demonstrates that crystal engineering is now a 'stand-alone' discipline.

4. References

1. Dunitz, J. D.; Schomaker, V.; Trueblood, K. N. (1988) *J. Phys. Chem.* **82**, 856.
2. (a) Lehn, J. M. (1990) *Angew. Chem., Int. Ed. Engl.*, **29**, 1304. (b) Lehn, J. M. (1995) *Supramolecular Chemistry: Concepts and Perspectives*, VCH, Weinheim.
3. Schmidt, G.M.J. Photodimerizations in the solid state (1971) *Pure Appl. Chem.* **27**, 647-657.
4. Maddox, J. (1988) *Nature*, **335**, 201.
5. M.C. Etter (1987) *J. Am. Chem. Soc.* **109**, 7786
6. Desiraju, G. R. (1989) *Crystal Engineering: The Design of Organic Solids*; Elsevier, Amsterdam.
7. Gavezzotti, A. (1991) *J. Am. Chem. Soc.* **113**, 4622.
8. Dunitz, J.; Bernstein, J. (1995) *Acc. Chem. Res.* **28**, 193.

9.   Gavezzotti, A.; Filippini, G. (1995) *J. Am. Chem. Soc.* **117**, 12299.
10. Allen, F.H.; Kennard, O. (1993) *Chemical Design Automation News* **8**, 31.
11. (a) Braga, D.; Grepioni, F.; Tagliavini, E.; Novoa, J. J.; Mota, F. (1998) *New J. Chem.* 755. (b) Braga, D. Grepioni, F.; Novoa, J. J. (1998) *Chem. Commun.* 1959.
12. Leiserowitz, L.; Hagler A. T. (1983) *Proc. R. Soc. London A* **388**,133
13. Whitesell, J. K.; Davis, R. E.; Wong, M-S, Chang, N-L. (1994) *J. Am. Chem. Soc.* **116**, 523.
14. Gavezzotti, A. (1994) *Acc. Chem. Res.* **27**, 309.
15. (a) Williams, D.E. (1996) *Acta Cryst.* A52, 326. (b) Whitten, D.G., Chen, L., Geiger, H.C.; Perlstein, J.; Song, X. (1998) *J. Phys. Chem. B* **102**, 10098. (c) Karfunkel, H. R.; Gdanitz, R.J. (1992) *J. Comput. Chem.* **13**, 1171. (d) Gdanitz, R.J. (1992) *Chem. Phys. Lett.* **190**, 391.
16. (a) Braga, D.; Grepioni, F. (1996) *Chem. Commun.* 571. (b) Braga, D.; Grepioni, F.; Desiraju, G. R. (1998) *Chem. Rev.* **98**, 1375.
17. Braga, D.; Grepioni, F. *Coord. Chem. Rev.* (1999) in press
18. Braga, D.; Grepioni, F. (1997) *Acc. Chem. Res.* **30**, 81.
19. See for example: Braga, D.; Grepioni, F.; Tedesco, E.; Biradha, K.; Desiraju, G. R. (1997) *Organometallics* **16**, 1846.
20. Braga, D.; Grepioni, F. (1999) *J. Chem. Soc. Dalton Trans.* 1.
21. A. Gavezzotti, (1998) *Cryst. Rev.* **7**, 5.

NMR AND CRYSTALLINITY OF NANOSTRUCTURED MATERIALS

P. SOZZANI, A. COMOTTI, R. SIMONUTTI
Dept. of Materials Science, Milan University "Bicocca",
Via R. Cozzi 53, I-20125 Milan, Italy

1. Introduction

Nuclear magnetic resonance can provide valuable information about organic and inorganic solids. Nanocomposites and nanostructured materials contain groups of atoms or molecules arranged in nanophases showing intermediate degrees of order and a variety of motional states. In addition polimorphic behavior is frequently observed. In this context magnetic resonance is able to measure some parameters which can be exploited for an exhaustive description of the materials. The aim of this chapter is tracing a method for the correlation of XRD and NMR data of these materials. After a brief outline of the main useful parameters, a supramolecular system will be presented as a rich example of application of the technique.

Most NMR parameters of interest are listed below and related to a conventional crystallographic description. When not explicitly indicated we are dealing with decoupled ½ spin nuclei.

1.1. CHEMICAL SHIFT (C.S.) AND CHEMICAL SHIFT ANISOTROPY (C.S.A.)

The frequency at which the nuclei resonate in a specific environment is called the Chemical Shift (C.S.) [1]. The chemical shift of a nucleus showing anisotropic interactions, as in most cases, is dependent on the orientation of its coordinates in the main magnetic field H_0 (laboratory frame). The condition for the absence of anisotropy requires that the nucleus experiences a cubic or higher symmetry (tetrahedral, octahedral, icosahedral, or spherical); in the remaining cases, even nuclei placed on special positions in the crystal cell may have anisotropy, although limited e.g. to axial anisotropy for an atom placed on a symmetry axis, with either prolate or oblate interactions.

Isotropic C.S. (δ_i) is measured when short correlation times ($\tau < 10^{-8}$ sec) are allowed in the solid. A reduced anisotropy is observed, e.g. for methyls, at room temperature and in rubbery or gel phases. Isotropic C.S. can be artificially obtained by fast spinning (>5 kHz) the same sample about a tilt angle to the main magnetic field vector (magic angle spinning MAS technique) [2]. Microcrystalline solids and fragmented solids spun at the magic angle show a single resonance for each independent carbon atom. Single crystals

D. Braga et al. (eds.), Crystal Engineering: From Molecules and Crystals to Materials, 443–458.
© 1999 *Kluwer Academic Publishers. Printed in the Netherlands.*

placed in the magnetic field also show a single peak for each independent carbon atom, but the resonance value is dependent on the tilting angle to the magnetic field. The use of single crystals is not widespread due to the time required for acquiring the angle dependent data and the low response of such crystals. Mats of aligned crystals or fibers are sometimes studied. The convolution of the signals due to the different orientations can be calculated by the C.S. dependence on tilt angle and the probability of distribution of the angles, being the orientation perpendicular to the main magnetic field more frequent than the other ones and producing the dominant singularities as indicated in Figure 1.

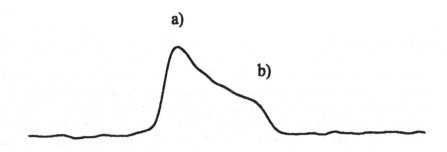

Figure 1. CSA of an axial symmetric nucleus of an isotropic powder: a) crystallites perpendicular to the main magnetic field; b) crystallites parallel to the main magnetic field.

The carbonyl carbon in a series of homologue solids retains the C.S.A. consistent with the symmetry elements of the molecule; the electronic structure dominates over the space interactions by a large anisotropy [3]. Gas absorbed into porous matrices can imprint the symmetry of the pores onto their anisotropic line shape. A few examples of gases absorbed on zeolites show this anisotropy. The axial symmetry due to a cylindrical channel must be retained by the C.S.A. of the included gas atoms, as described by xenon gas confined into channels.

Sharp isotropic signals obtained by MAS (sometimes a few Hz full width at half-height) are extremely sensitive to symmetry conditions within the crystal cell [4]. The symmetry properties are transmitted to the C.S. by the electron cloud surrounding the observed nucleus. The symmetry conditions are thus indirectly transferred to the effective clusters of atoms in the cell, unlike XRD in which the symmetry operation of translation is the essential condition itself of the signal. In any case through space interactions are to be taken into account. In molecular systems, for example, one must always consider more than the single molecule. The cause of the shift is not generally a through space magnetic interaction, but a change of bond length and angles. *Ab initio* calculation of the chemical shift by the electronic distribution were performed, although generally not precise. Some very minor changes of the structure seem to determine the chemical shift in silicates [5,6]. A small fraction of Å can be estimated.

1.2. MULTIPLICITY OF SIGNALS AND SIGNAL INTENSITY

NMR spectra cannot contain more signals than the number of nuclei contained in the crystal cell. In addition, symmetry elements in the cell can lower this number, because a subunit in the cell can be subjected to symmetry operations (operations of rotoreflection) that can reproduce the entire cell. The environment must remain identical as seen from the viewpoint of the nucleus. It is obvious that the internal symmetry of single molecules or cluster of atom identified by a certain connectivity can be disregarded. The comparison of the single molecule multiplicity (like in solution or in the gas phase) to the observed multiplicity due to the symmetry in the cell can locate the molecule in the cell. The intensity of a signal due to a particular atom is reduced if it takes part into more than one subcell. If n is the number of the shared subcells 1/n is the reduction factor. For example, an atom on a face of a subcell shows a factor of ½. The predicted intensity distribution can be eventually normalized to entire numbers. The calculations for approximating the signal intensity to entire numbers is relevant for the correct building up of a compatible structure hypothesis. Degeneration of signals due to lack of resolution makes this task sometimes harder.

Motion reduces both anisotropy and multiplicity of signals. It can partly or totally average out the differences of the atoms arranged in independent positions within the crystal cell and lowers the symmetry. The symmetry becomes dependent on the observation time scale. The response to motional disorder does not appear as blurring or broadening of the signals, but as an averaging, forming signals even sharper than the signals due to ordered and static arrangements.

Anisotropy of a static sample can be partly averaged by internal motions and the powder pattern can be calculated. Quadrupolar nuclei, as deuterium, are the nuclei of choice for describing motions by the anisotropy analysis, because the line shape is dependent on the reorientations of selectively labeled groups in the sample [7].

1.3. NUCLEAR RELAXATION

Several relaxation times can be measured in a solid material. The use of relaxation times for describing a crystal can be a rich source of information. The rate of relaxation can be enhanced by paramagnetic species. Their location can be thus determined, but many times they can quench the signal. If the relaxation due to motion prevails, the spin-lattice relaxation times (T_1) in a solid material are inversely proportional to the rate of reciprocal motions and of motions with respect to the main magnetic field. The shortest values (less than 1 s) indicate the most efficient relaxation process is occurring. If the motional regime matches the frequencies of the order of 10^8 Hz, a particular atom or group of atoms can provide an efficient relaxation path for the surrounding and for themselves. Very slow relaxation, of the order of several minutes or hours, are measured by very rigid structures like ^{13}C in diamond or ^{29}Si in quartz. Vibrations are effective for relaxation. A molecular rigid crystal generally needs a few hundred seconds for being mostly relaxed.

Slow diffusive motions in the kHz regime are addressed by the measurements of T_1 relaxation times in the rotating frame, conventionally called $T_1\rho$.

Surprisingly several relaxation times are detected in the same system. Included molecules, diffusing species, dangling chains and reorienting groups are especially short relaxing. The measurement of T_1 relaxation times is of help for deducing if the high symmetry of a group of atoms, thus showing a single signal, is realized by localized motions [8].

Spin-spin relaxation times (T_2) appear in the expression indicating the homogeneous broadening of the line in the frequency domain ($1/\pi T_2$). A high spin density of the observed nucleus (abundant spin active nuclei) allows an efficient spin-spin relaxation, also described as a spin propagation or spin diffusion. When abundant spins like ^1H are observed, the spin diffusion phenomenon is considerably broadening the line-width. On the other hand, spin propagation can travel a large space in rigid solids during a few milliseconds or seconds of observation time, collecting information about microphases or phase interactions. The magnetization can be than detected under high resolution conditions as collected by the rare nuclei dispersed in the phase [9]. A material domain or a nanophase composed by nuclei interacting together relax at the same time. A sort of microscopy is thus realized.

Mixed crystals are identified by this mean, showing a single relaxation in the hydrogen domain.

1.4 . CROSS POLARIZATION DYNAMICS

The cross polarization process allows the rare nuclei X (^{13}C, ^{29}Si ...) to come into thermal contact with the neighboring system of abundant nuclei I (^1H, ^{19}F ...) by means of a radio frequency pulse sequence. When the two nuclei have the same nutation frequency ($\gamma_x H_{1x} = \gamma_I H_{1I}$), magnetization can flow from the cold abundant nuclei to the hot rare nuclei. The rate of building-up of magnetization of a rare nucleus, during magnetization transfer by cross polarization (CP), follows a time constant which is informative of distances of the observed nuclei by the irradiated abundant nuclei. The fastest is the rate, the more close are the communicating spins by a function of $1/r^6$. This is a powerful and precise tool for determining distances in the 0.3-1.5 nm range [10]. The approach is valuable in crystal engineering: an example will be highlighted below.

2. Supramolecular Chemistry of Spirocyclophosfazenes

The tris(2,3-naphthalenedioxy)cyclotriphosphazene (TNP) and tris(o-phenylenedioxy)cyclotriphosphazene (TPP) are good candidates for exploiting the above NMR parameters in order to obtain a rich crystal description.

They show the following features:

a) they can form channel inclusion compounds (ICs) with macromolecules, but they can include even volatile molecules and gas atoms either in cages and in channels.

b) they show a polymorphic behavior, both in the pure state and in the inclusion compounds.

c) a crystalline phase was recently discovered containing large empty channels, like a "molecular zeolite", and gases can easily diffuse into them.

2.1. MULTIPLICITY OF SIGNALS AND CRYSTAL SYMMETRY IN TNP ICs

The TNP molecules form inclusion compounds (ICs) with several small molecules such as benzene, xylene, cyclohexane. In the past the ICs were mainly characterized by X-ray diffraction technique: most of the crystalline TNP inclusion compounds present a hexagonal unit cell (Figure 2): the structure contains channels with 10 Å diameter parallel to the z axis [11]. In some cases a cage-type structure was found, e.g., in TNP/p-xylene IC [12].

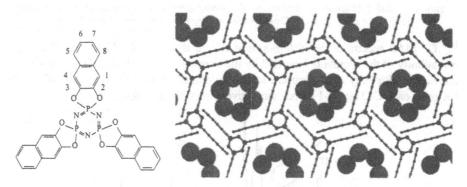

Figure 2. Schematic representation of the TNP molecule (on the right side). Crystal cell of the TNP channel-type structure as reported by Allcock et al. (on the left side).

In this report we present the ^{13}C and ^{31}P magic angle spinning MAS NMR spectra of the TNP ICs [13]. The channel-type structure of TNP ICs with benzene, and the cage-type structure of TNP ICs with p-xylene give rise to ^{13}C signal multiplicities due to non equivalent carbon atoms in the unit cell. ^{31}P CP MAS spectra reflect the symmetry of the crystal cell. A residual dipolar coupling with ^{14}N is also observed. The ^{13}C CP MAS spectrum of guest free TNP structure is interpreted in accordance with the formation of the monoclinic cell [14].

The ^{13}C CP MAS spectrum of TNP/benzene IC shows five peaks in the aromatic chemical shift region (Figure 3b), thus, there is a single peak for each type of aromatic carbon atom, in accordance with the high symmetry of the hexagonal crystalline phase [15]. The single paddles are perpendicular to the phosphazene ring. The benzene chemical shift overlaps the peak at 128.1 ppm, as selectively recorded in the ^{13}C MAS NMR spectrum obtained with a short recycle time (Figure 3a).

448

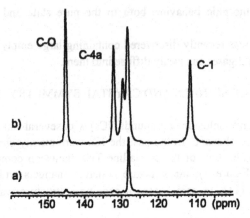

Figure 3. *¹³C MAS NMR spectra of TNP/benzene IC, spinning speed of 5530 Hz: a) Single Pulse Sequence (SPE) with a delay of 10 s; b) Cross Polarization Sequence with a contact time of 8 ms.*

¹³C CP MAS spectra of the TNP/*p*-xylene IC are shown in the Figure 4a. The high crystallinity of the sample is indicated by the sharpness of the signals. Quite striking is the high multiplicity of the carbon atoms, compared to the spectra of the ICs with benzene; each C-O and C-1 carbon atom gives rise to three peaks.

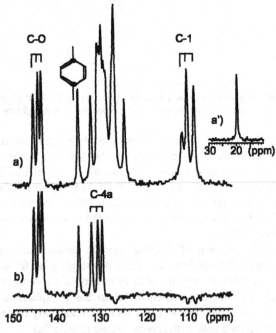

Figure4. *a) ¹³C CP MAS NMR spectra with a contact time of 4 ms, spinning speed of 5476 Hz of the TNP/p-xylene, a') ¹³C MAS Single Pulse Sequence with a recycle delay of 10 s; b) protonated carbon suppression pulse sequence with a delay of 50 µs.*

The quaternary carbon atoms can also be observed separately by the non quaternary carbons suppression pulse sequence as shown in Figure 4b. With this specific pulse sequence we can selectively detect the 135 ppm resonance, corresponding to the quaternary carbons of the p-xylene guest and the three peaks assigned to C-4a carbon atoms which, in the conventional CP MAS spectrum, overlap other signals. The intensity ratio of 1 for the p-xylene peak, compared with the C-4a carbon signals (three peaks with intensity 1:1:1), leads to a guest to host ratio of 1:1. This result is consistent with the crystal structure suggested by Kubono. The authors propose a cage-type structure with the p-xylene and the TNP molecule in the 1:1 ratio.

Three peak multiplicity of C-4a atoms (and of C-O and C-1 atoms) suggest, at first sight, a different environment for each paddle (with the maintenance of the symmetry plane within the molecule). This is not the case: in fact, one of the three signals (downfield signal in each group) is assigned to naphthalenedioxyphosphole unit lying close to the plane **a-b** (paddle A, Figure 5), this unit bisects the N-P-N angle as for the paddles of TNP inclusion compounds with THF and benzene. The intensity of each downfield signal is thus due to two carbon atoms.

This is consistent with the proposed crystal structure, containing a paddle (A) lying on a mirror plane of the TNP molecule. The remaining two paddles (B) are equivalent and form an angle of 140°, one with respect to the other. Therefore, the two remaining signals of each group (hereafter called "doublet") cannot be explained by a different geometry of the two paddles. However, in this conformation, the mirror plane of the TNP molecule lying on the phosphazene ring is retained, but the mirror plane is lost in the crystal cell.

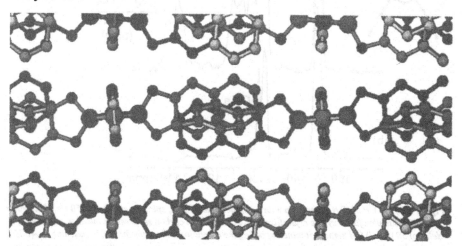

Figure 5. *Crystal structure of TNP/p-xylene IC. The projection of the unit-cell along the b-axis is presented (paddle A lies in the plane a-b).The crystal structure derives from the atomic coordinates reported by Kubono et al.*

Because of the tilt of the mean plane of the phosphazene ring around the **b** axis, the carbon atoms of the paddles B are tilted (Figure 5). Each doublet is due to two carbon atoms in the same paddles. The tilt produces a large displacement of these paddles, whereas on the single one which is perpendicular to c axis, it is much less.

2.1.1. Aromatic Paddles of Guest Free TNP

The ^{13}C CP MAS spectrum of guest free TNP shows more signals for each type of carbon atom (Figure 6a). The lower symmetry of the monoclinic guest free cell is reflected in the signal multiplicity. By applying the protonated carbon suppression pulse sequence with a delay of 50 μs (Figure 6b) only the quaternary carbon peaks survive, corresponding to the C-O and C-4a carbon atoms. This pulse sequence is particularly useful as the chemical shift overlaps some carbon atoms in the region, at about 130 ppm. It is possible to detect the multiplicity of the quaternary carbon atoms and to distinguish two peaks for C-4a atoms (intensity ratio of 2:1).

Also visible is a splitting in the C-O carbon region, with an intensity ratio of 1:2. The multiplicity of the C-O and C-4a carbon peaks reflects the symmetry of the TNP molecule and, in particular, the bending of the five-member hexocyclic rings.

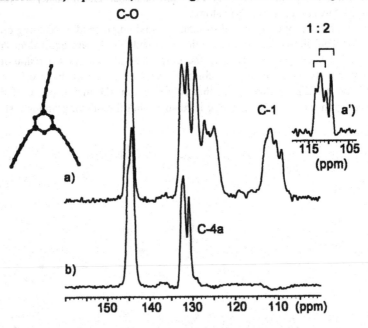

Figure 6. a) ^{13}C CP MAS NMR spectra with a contact time of 8 ms, spinning speed of 5550 Hz of the guest free TNP, sample after o-xylene crystallization; b) protonated carbon suppression pulse sequence with a delay of 50 μs; a') C-1 signals obtained applying the resolution enhancement: line broadening = -40 Hz, gaussian broadening = 50 %. On the left side: geometry of the TNP molecule in the monoclinic crystal cell. The geometry is calculated by the atomic coordinates reported by Kubono et al.

In the TNP molecule, there are two naphathalenedioxy residues bonded to the phosphorous atoms that show a considerable deviation from the mean plane and give rise to peaks with double intensity. The ^{13}C signals of C-O and C-4a atoms of intensity equal to 1 are associated with the third residue which deviates from the mean plane slightly. The signal of intensity 1 resonates at the same chemical shift as ICs with benzene and THF, consistent with the symmetry of the paddles (hereafter described as perpendicular).

The interpretation of the C-1 carbon atom is more complex as it is particularly sensitive to the dihedral angle C(1)-C(2)-P-O. In the TNP ICs, this dihedral angle is close 180° and therefore the C-1 atoms resonate upfield like the *ortho* carbon of 1,4-dimethoxybenzene in the *trans* conformation to the methoxy side group. In the TNP guest free matrix, the C-1 carbon atom presents a high multiplicity, but it is possible to distinguish two groups of signals of intensity 1:2. Each group is split further, giving rise to a doublet which presents an intensity ratio of 1:1 (Figure 6a'). The distribution of the 1:2 intensity ratio reflects the bending distortion of the exocyclic rings, as shown before for the quaternary carbon atoms. The reason for the further multiplicity might be the loss of symmetry on the cyclophosphazene ring, justifying the 1:1 splitting.

2.1.2. ^{31}P MAS NMR of TNP polymorphs

^{31}P CP MAS of TNP/benzene IC is shown in Figure 7a. We can observe an asymmetric triplet at 35.1, 32.8 (main peak), 30.7 ppm which resonates in the chemical shift region of spirocyclophosphazene and therefore is diagnostic of the formation of the cyclophosphazene molecules.

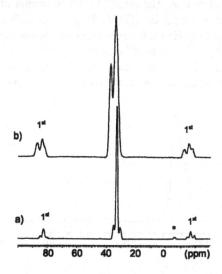

Figure 7. ^{31}P *CP MAS of TNP/benzene IC (a) and monoclinic TNP structure (b), contact time of 2 ms, spinning speed of 6000 Hz, recycle delay of 10 s.*

The ^{31}P CP MAS of the guest free TNP monoclinic structure presents two broader signals than the asymmetric triplet of TNP ICs with an intensity ratio of 1:2 that resonate at 37.2, 33.7 ppm (Figure 7b).

452

The symmetry loss of TNP molecule generates non equivalent phosphorous atoms and, at first sight, we can expect that the phosphorous atom, bonded to naphthalenedioxy residue that slightly deviates from the mean plane, has a different resonance with respect to resonances of the two remaining phosphorous atoms. The presence of the doublet with an intensity ratio agrees with the geometry of the TNP molecule in the monoclinic cell.

2.1.3. Dynamics of TNP Crystal Structure

The dynamic behavior of the TNP/benzene IC can be determined by a few complementary perspectives with the application of specific pulse sequences: [13]C Single Pulse Experiment with proton decoupling (SPE). The selective detection of the benzene signals (Figure 3a) in the ^{13}C SPE spectrum with short recycle delay is indicative of the higher mobility of the guest molecules than those of the TNP matrix. The short [13]C T_1 relaxation times of the benzene molecules (12 s) confirms the presence of relatively fast motions at room temperature. The matrix signals for spin-lattice relaxation times (T_1) of the inclusion compound, show, respectively, 69s, 66s, 56s, 57s, 54s for carbons C-4a, C-O, C-1, C-5 and C-6. The measurements suggest a crystalline rigid system, but not so rigid as the pure host monoclinic crystal structure where values of more than 1000 s are estimated for each carbon atom. The drastic reduction of the matrix relaxation times in the IC derives from the high mobility of the guest molecules and by a looser crystal packing that must be taken into. The static ^{13}C SPE spectrum of the TNP/benzene IC, performed with a short recycle delay (Figure 8), presents a Chemical Shift Anisotropy (CSA) of about one thousand Hz for benzene carbons. Taking into account the principal components of the carbon CSA in solid benzene (σ_{11}=-88.2, σ_{22}=-12.0, σ_{33}=127.7 ppm) [16] the small residual chemical shift anisotropy indicates the fast range of motions experienced by the guest molecules, even though an isotropic dynamics has not yet been reached

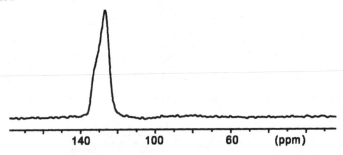

Figure 8. [13]C SPE static spectrum of TNP/benzene IC, delay time = 5s.

In the TNP/p-xylene IC the long relaxation times of both the matrix (600÷1000s) and the guest molecule (about 500s) confirm that, at room temperature, the degrees of freedom explored by the p-xylene molecules are not so many as found by benzene in the channel-type structure.

2.2. Measurement of Distances by Cross Polarization Dynamics in TPP ICs

The channel-type structure of TPP ICs (about 5 Å diameter) provides an aromatic environment to trap some molecules such as benzene, tetrahydrofurane and p-xylene (Figure 9).[17].

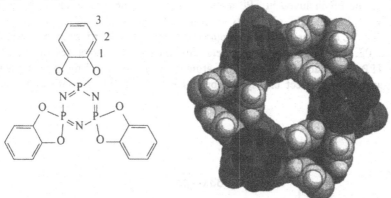

Figure 9. Schematic representation of the TPP molecule (on the left side) and of the channel-like structure (on the right side).

Novel channel-like inclusion compounds formed by TPP with linear hydrocarbons and polyethylene are presented. These compounds provide the opportunity to observe confined hydrocarbons in the crystal state. Furthermore, the aliphatic chains in TPP ICs are in the unusual state of being surrounded by aromatic rings displaced parallel to the channel. New channel-like ICs with either perdeutero-guests or perdeutero-TPP matrices have been prepared for the first time [19].

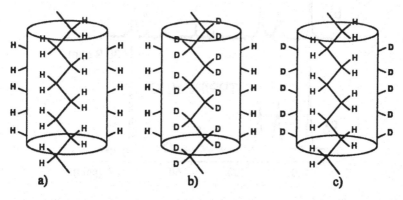

Figure 10. Schematic representation of the channel-like TPP ICs: a) hydrogenated host and hydrogenated guest; b) hydrogenated host and deuteroguest-labelled;c) deuterohost-labelled and hydrogenated guest.

454

2.2.1. Supramolecular Structure of TPP/PE-d₄ IC

The ^{13}C MAS spectrum of TPP/PE IC exhibits three peaks in the aromatic region that demonstrates the formation of the channel-like structure and a signal at 32.1 ppm in the aliphatic region that is associated with the PE included in TPP nanochannels (Figure 11a). The upfield shift (more than 1 ppm) of deuterated methylene units (30.9 ppm) with respect to the PE included in TPP matrix (Figure 11b) derives from the isotope effect. In Figure 11a the ^{13}C CP MAS spectrum of the TPP/PE IC is shown for comparison. The ^{13}C chemical shift at 33.6 ppm corresponds to the orthorhombic crystal form of bulk PE. The large 1.5 ppm upfield shift on the ^{13}C resonances of the PE included in TPP nanochannels is due to the aromatic ring current of the o-phenylene units. The current ring effect clearly demonstrates that the PE is confined in the channels.

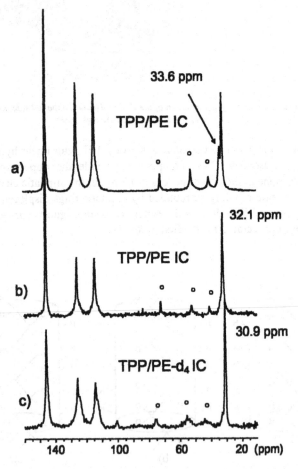

Figure 11. a) ^{13}*C CP MAS spectrum of TPP/PE IC, spinning speed of 5550 Hz, contact time of 3 ms.* ^{13}*C SPE MAS spectra: b) TPP/PE IC sample, spinning speed 5550 Hz; c) TPP/PE-d₄ sample, spinning speed of 5440 Hz. A recycle delay of 100 s is applied. The circles indicate the spinning side bands.*

2.2.2 $^1H \rightarrow {}^{13}C$ Cross Polarization Dynamics

The cross polarization process allows the rare nuclei (^{13}C) to come into thermal contact with the neighboring system of abundant spins (protons) by means of a radio frequency pulse sequence. When the Hartmann-Hahn condition is fulfilled (Figure 7) the inverse of the cross polarization time (T_{CH}^{-1}) is proportional to the second moment of the dipolar coupling between the ^{13}C and ^1H nuclei, and considering that the second moment of the heteronuclear dipolar coupling varies, at a first approximation, as the inverse sixth power of the proton carbon distance, *the cross polarization time (T_{CH}) is proportional to the sixth power of the proton carbon distance ($1/r^6$).*

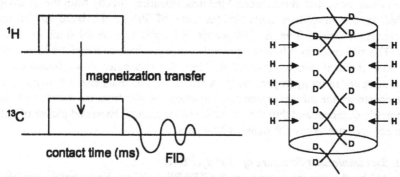

Figure 12. *Schematic representation of the NMR Cross Polarization Pulse Sequence. Schematic representation of the TPP/PE-d$_4$ IC. The hydrogen magnetization in the CP NMR experiment comes only from the TPP host molecules.*

To demonstrate the formation of the inclusion compound of PE-d$_4$ and TPP molecules we exploited the cross polarization phenomenon by applying ^1H-^{13}C CP pulse sequence (Figure 12) at varying contact times on this supramolecular structure that contains deuterons in the guest molecules and protons only in aromatic rings.

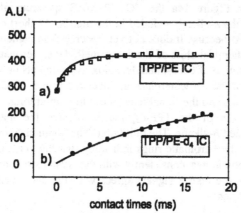

Figure 13. *Intensities of the polymethylene ^{13}C chemical shifts vs contact times of a) TPP/PE and b) TPP/PE-d$_4$ ICs.*

This experiment is made possible in spite of the mobility of the guest molecules because the cross polarization pulse sequence efficiently reveals the carbon nuclei of the methylene units even for a short contact time of 3 ms (Figure 11a). The efficient cross polarization phenomenon is due to the presence of spectral density components that belong to slow motion regime. The CP experiments are also carried out on TPP/PE IC at varying contact times, for comparison.

The ^{13}C signal intensities of the inner methylene of PE-d_4 and PE included in the TPP channels are reported in Figure 13.

The signal intensity ratio of PE builds up faster than those of PE-d_4 because, in the former case, the carbon nuclei receive the magnetization directly from the ^1H covalently bonded to the methylene units. In the case of PE-d_4, the build up of the ^{13}C magnetization derives from the TPP matrix hydrogens that are far from the methylene units. The difference in behavior represents the magnetization transferred from the TPP protons to the methylene units of PE-d_4. Since this phenomenon is efficient for carbon nuclei surrounded by hydrogens within a few Å, the ^{13}C magnetization build up of PE-d_4 demonstrates the through-space magnetization transfer from the matrix hydrogens at nanometric distances and thus the formation of the adduct. Extended phases of bulk PE-d_4 do not give rise to any CP signal at 32.4 ppm.

2.2.3. Supramolecular Structure of TPP-d_{12}/PE IC

Prompted by the previous results on the TPP/PE-d_4 IC we have applied the $^1H \rightarrow {}^{13}C$ Cross Polarization Pulse Sequence on ICs selectively deuterated on the TPP matrix. By transferring the magnetization from the hydrogens of molecules confined in the channels to the carbon nuclei in the surrounding the ^{13}C CP NMR spectra give an insight into the proximity of the TPP-d_{12} host to the guests.

An interesting case is provided by the TPP-d_{12}/C_{36} IC, being the deuterated molecules of this adduct complementary to the TPP/PE-d_4 IC previously reported. By ^{13}C CP Pulse Sequence (Figure 14a) we can discriminate between hydrogenated guests against deuterated host. In Figure 14a the ^{13}C CP MAS spectrum at short contact times selectively records the n-alkane included in the matrix. At short contact times the TPP-d_{12} matrix is masked because it does not yet receive the magnetization from the guest protons. The ^{13}C chemical shift of the inner methylenes resonates upfield due to the current ring contribution of the o-phenylenedioxy substituents in the TPP matrix.

The carbon peaks of the perdeuterohost are detected in the ^{13}C CP MAS spectrum at a contact time of 1.4 ms and the signal intensities build up at longer contact times (Figure 14). The three peaks of the TPP-d_{12} matrix receive the magnetization from the hydrogens of included n-alkane. The build up of the magnetization is the same for each carbon atom of the TPP aromatic rings indicating that the guest is at the same distance with respect to carbon atoms, consistently with the crystal structure already presented.

The fast growth of the matrix signals agrees with the short distance (less than 3 Å) between guests and host.

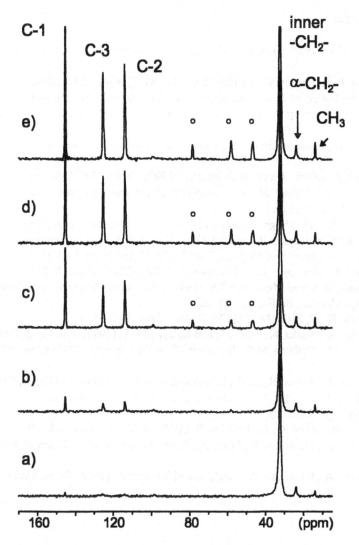

Figure 14. *¹³C CP MAS spectra of TPP-d₁₂/C₃₆ IC at different contact times. A recycle delay of 20 s is applied; the spinning speed is of 5066 Hz. The contact times applied are the followings: a) 100 μs; b) 1.4 ms; c) 7 ms; d) 16 ms and e) 22 ms.*

It is remarkable to note that the detected signals are only due to the matrix in close contact with the guest molecules. The close contact can just occur in the crystals of the inclusion compound. Defective structure and impurity not belonging to the inclusion compound cannot be in close contact with the guest accordingly to the observation of narrow signals.

458

3. References

1. Slichter, C.P. (1990) *Principles of Magnetic Resonance*, Chapter 4, Springer-Verlag, Berlin.

2. Andrew, E.R., Bradbury, A. and Eades, R.G. (1959) *Nature* **183**, 1802.

3. Pines, A., Gibby, M.G. and Waugh, J.S. (1973) *J. Chem. Phys.* **59**, 669.

4. Fyfe, C.A., Grondey H., Feng, Y. and Kokotailo, G.T. (1990) *J. Am. Chem. Soc.* **112**, 8812.

5. Comotti, A., Castaldi, Gilioli, C., Torri, G. and Sozzani, P. (1994) *J. Mater. Sci.* **29**, 6427.

6. Skibsted, J., Hjorth, J. and Jakobsen, H.J. (1990) *Chem. Phys. Lett.* **106**, 262.

7. Bovey, F.A. (1988) *Nuclear Magnetic Resonance Spectroscopy*, Chapter 8.8, Academic Press Inc., London.

8. Laupetre, F. (1993) High Resolution NMR Investigations of Local Dynamics in Bulk Polymers at Temperatures Below and Above the Glass Transition Temperature, p. 63 in *NMR Basic Principles and Progress*, Vol. 30, Springer-Verlag, Berlin.

9. Comotti, A., Simonutti, R. and Sozzani, P. (1996) *Chem. Mater.* **8**, 2341.

10. Veeman, W.S. and Mass, W.E.J.R. (1994) *NMR Basic Principles and Progress* Vol. 32, Springer-Verlag; Berlin,; pp 127-162.

11. Allcock, H. R., Stein, M. T (1974) *J. Am. Chem. Soc.*, **96**, 49.

12. Kubono, K., Asaka, N., Isoda, S. and Kobayashi, T. (1993) *Acta Cryst.* **C49**, 404.

13. Comotti, A., Gallazzi, M.C., Simonutti, R. and Sozzani P. (1998) *Chem. Mater.* **11**, 3589.

14. Kubono, K., Asaka, N., Isoda, S. and Kobayashi, T. (1994) *Acta Cryst.*, **C50**, 324.

15. Allcock, H. R., Allen, R. W., Bissell, E. C., Smeltz, L. A. and Teeter, M. (1976) *J. Am. Chem. Soc.*, **98**, 5120.

16. Linder, M., Höhener, A., Ernst, R. R. (1979) *J. Magn. Reson.*, **35**, 379.

17. Comotti, A., Simonutti, R., Stramare, S. and Sozzani P. (1999) *Nanotechnology* **10**, 70.

18. Comotti, A., Simonutti, R., Catel, G. and Sozzani P., (1999) *Chem. Mater.* **12**, in press.

MOLECULAR-IONIC HALOGENOANTIMONATES AND BISMUTHATES - A RICH FAMILY OF CRYSTALS SHOWING ATTRACTIVE PROPERTIES

G.BATOR, R.JAKUBAS AND L.SOBCZYK*
*Faculty of Chemistry, University of Wrocław, Joliot-Curie 14
50-383 Wrocław, Poland*

1. Introduction.

The halogenoantimonate and bismuthate crystals containing organic cations seem to be interesting from a point of view of the crystals engineering for several reasons. The most important feature consists in that the halocoordinated Sb and Bi octahedra (Sb and Bi on the 3rd oxidation stage) can be easily joined via corners, edges or faces. These octaedra form varieties of polyanionic sublattices, which the organic cations of varying size and symmetry can be incorporated to. The cations are able to form weak hydrogen bonds. Simultaneously in many cases these cations show a freedom of reorientations. Their dynamics is a main factor ruling transitions to interesting phases such like pyroelectric, ferroelectric, ferrielectric or ferroelastic. Some of crystals from this family seem to be interesting from a point of view of the non-linear optical properties.

The family of crystals under consideration was the subject of a few reviews [1-4]. 23 different types of assembling of halocoordinated MX_6 depending on the cations used were reported so far. Up-to-date the properties of 46 crystals grown on the base of halocoordinated Sb(III) and Bi(III) octahedra were recognised. Recently the attempts of the growth of crystals containing Sb on the 5th oxidation stage were undertaken. In this case exclusively isolated SbX_6^- octahedra appear. However, some of these crystals show interesting properties from the point of view of material sciences [5, 6].

In the present contribution we would like to focus our attention to the following aspects: polyanionic structures in crystals of some importance in material sciences, crystals showing ferroelectric and ferroelastic phase transitions, electro-optic properties, antimonates on 5th oxidation stage and conclusions.

D. Braga et al. (eds.), Crystal Engineering: From Molecules and Crystals to Materials, 459–468.
© 1999 *Kluwer Academic Publishers. Printed in the Netherlands.*

460

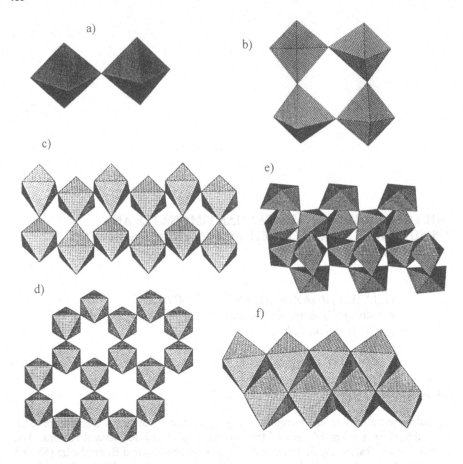

Fig. 1 Examples of polyanionic structures: a) isolated corner-sharing bioctahedra $Bi_2X_{11}^{5-}$, b) squaric structures containing four corner-sharing octahedra of MX_5^{2-} stochiometry, c) one-dimensional pleated ribbon structure of $M_2X_9^{3-}$ polyanions, d) two-dimensional corrugated sheets of $M_2X_9^{3-}$ polyanions, e) infinite tapes of differently joined octahedra (MX_4^- stoichiometry), f) isolated $M_8X_{28}^{4-}$ units.

2. Polyanionic structures.

To illustrate the possibilities of polyanionic assembling of halogenoantimonates(III) and bismuthates(III) on the base of MX_6 (M = Sb, Bi; X = Cl, Br, I) octahedra we present in Fig. 1 a few of such polyanions. We selected these, which appear in crystals of particularly interesting properties. In addition some structures are shown which exemplify variety of possibilities to promote further searches. The formation of concrete architecture depends on several factors but stoichiometry of a salt is certainly most important. Also the size and symmetry of organic cations as well as of the halogen used i.e. Cl, Br or I are of importance [2].

The studies in the field of crystal growth, performed till now, were concentrated mainly on salts containing simple methylammonium (primary, secondary, tertiary and quaternary) cations, but in many papers the possibility of crystal growth with exotic cations including bulky crown ether complexes has been shown. Besides of alkylammonium some other cations like R_4P^+, R_3S^+ and many others can be involved to the crystalline lattice. It seems that there exist further potential possibilities in searches for new crystals. For instance one could suggest growing of crystals containing cations as well as anions of high electric polarizability.

3. Ferroelectric and ferroelastic properties.

The crystals recognised so far to show ferroelectric phase transitions are collected in Table 1.

As shown the ferroelectric and ferroelastic properties discovered up-to-now are limited to the three types of anionic sublattices, namely to the two-dimensional honey-comb like structures, face-sharing and corner-sharing bioctahedra [7-16]. In all cases except MABB (and maybe MABA) the ferroelectric transitions are of an order-disorder type related to the reorientations of alkylammonium cations. So far we dealt with methyl, dimethyl- or trimethylammonium cations.

The simplest mechanism of ferroelectric phase transition was revealed and well evidenced in the DMACA crystal [17]. In the crystalline lattice of this compound two types of the dimethylammonium cations appear: one type is seized in the hexagonal vacancies inside corrugated layers and second one is located between layers. The cations placed inside the layers are responsible for the ferroelectric properties. In the paraelectric phase these cations can occupy two equivalent positions as shown in Fig. 2. The reorientations proceed around the axis parallel to the C...C direction and the barrier to rotation forms two N-H...Cl hydrogen bonds. Such a picture is consistent both with the temperature X-ray diffraction [18] and dielectric relaxation [19] as well as NMR studies [20].

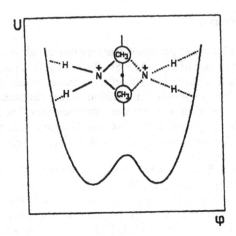

Fig. 2. Flip-flap mechanism of ferroelectric phase transition
in [NH$_2$(CH$_3$)$_2$]$_3$Sb$_2$Cl$_9$ (DMACA).

The most attractive appeared to be the MAPCB and MAPBB crystals [21]. They are ferroelectrics at room temperature and undergo the transitions to paraelectric phases at 307 and 312 K, respectively. Both the $\varepsilon'(T)$ peak and the spontaneous polarizations shown in Fig. 3 make these materials comparable to the TGS family.

The mechanisms of the phase transitions and spontaneous polarizations in both crystals are very complicated. There are three types of CH$_3$NH$_3^+$ cations of different orientation and dynamics that was quite well evidenced by X-ray diffraction [15], dielectric relaxation [22], NMR [23] and spontaneous polarization studies [24]. Particularly interesting appeared to be the dielectric relaxation studies over broad frequency range which show that there are three independent relaxators of quite different relaxation times. What is something unusual one observes at lower temperatures the dependence of spontaneous polarization on temperature [25] that can be related to the involvement of different cations to the polarization.

The mechanism of the ferroelectric phase transition in MABB and similarly in MABA is presumably quite different. This transition takes place at low temperatures when the reorientations of cations are frozen. The detailed spectroscopic and NQR studies [26, 27] seem to show that we are dealing in this case with displacive type of transition connected with a deformation of halocoordinated octahedra. Similar phase transition is observed in the isomorphous caesium salt Cs$_3$Bi$_2$Br$_9$ [28].

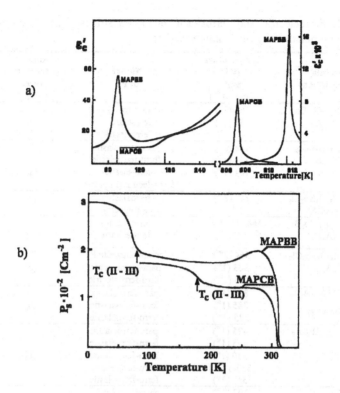

Fig. 3. ε'(T) (a) and P_s(T) (b) plots for $(CH_3NH_3)_5Bi_2Cl_{11}$ (MAPCB) and $(CH_3NH_3)_5Bi_2Cl_{11}$ (MAPBB).

A few words one should devote to the possibility of growth of mixed crystals by exchange of Sb and Bi as well as halogen atoms tending towards searching for the dipole glasses. The studies performed up-to-date [29] revealed sharp frustration phenomena and substantial changes of transition temperatures and dielectric properties but the glassy state was not discovered so far.

The layer ferroelectrics of the $R_3M_2X_9$ composition reveal additionally ferroelastic properties already in the high temperature phases, i.e. in the paraelectric ones. The ferroelasticity in such systems is presumabely due to the deformation of the anionic

TABLE 1. Ferroelectric and ferroelastic phase transitions (PT) in alkylammonium halogenoantimonates(III) and bismuthates(III).

Formula (abbreviation)	Temperature [K] and order of PT	Nature of PT	Anionic structure (acc. to Fig. 1)	P_s $(10^3 Cm^{-2})$
$(CH_3NH_3)_3Sb_2Br_9$ (MABA)	168 (1st) 131 (1st)	para-ferroelastic ferroelastic-ferroelectric	(d)	1.3
$(CH_3NH_3)Bi_2Br_9$ (MABB)	188 (1st) 140 (1st) 104 (1st)	para-ferroelastic ferro-ferroelastic ferroelastic-ferroelectric (improper)	(d)	3.0
$[NH_2(CH_3)_2]_3Sb_2Cl_9$ (DMACA)	242 (2nd)	ferroelastic-ferroelectric	(d)	6.9
$[NH_2(CH_3)_2]_3Sb_2Br_9$ (DMABA)	164 (1st or 2nd)	ferroelastic-ferroelectric	(d)	7.0
$[NH(CH_3)_3]_3Sb_2Cl_9$ (TMACA)	364 (2nd) 363 (1st) 125 (2nd)	incommensurate PT para-ferroelectric ferro-ferroelectric	(d)	30.0
$[NH_2(CH_3)_2]_3Bi_2I_9$ (DMAIB)	305 (1st) 285 (1st) 235 (2nd)	para-ferroelastic ferro-ferroelastic ferro-ferroelastic	(b)	
$[C(NH_2)_3]_3Bi_2Br_9$	425 (1st) 415 (1st) 350 (1st) 333.5 (1st) 311 (1st)	para-ferroelastic ferro-ferroelastic ferro-ferroelastic ferro-ferroelastic ferro-ferroelastic	(b)	
$(CH_3NH_3)_5Bi_2Cl_{11}$ (MAPCB)	307 (2nd) ≈170	para-ferroelectric ferro-ferroelectric	(a)	12 (T>250K) 18 (T<100K)
$(CH_3NH_3)_5Bi_2Br_{11}$ (MAPBB)	312 (2nd) 77 (1st)	para-ferroelectric ferro-ferroelectric	(a)	20.0 (T>250K) 30.0 (T<70K)

sublattice. The alkylammonium cations during inducing the ferroelastic phases show a considerable dynamic disorder and sometimes even isotropic reorientational disorder. This suggests that the contribution of cations in the mechanism of ferroelastic transition is of minor importance. According to the Sapriel classification [30] one can propose, based on optical observation and birefringence, the following types of para-ferroelastic transitions:

MABA – $\bar{3}$mF2/m at 168 K, MABB - $\bar{3}$mF2/m at 188 K, DMACA- 2/mF$\bar{3}$m , DMABA – 2/mF$\bar{3}$m, TMACA - mF$\bar{3}$m [3]. The two further ferroelastics of the $R_3M_2X_9$ composition are characterised by isolated face-sharing bioctahedra $Bi_2X_9^{-3}$. Thus DMAIB undergoes at 305 K transition of the 6/mmmFmmm type, while the ferroelastic transition at 425 K in $(Gu)_3Bi_2Br_9$ (Gu – guanidinium cation) [14] is connected with the symmetry change from orthorhombic to tetragonal. Also in these crystals the proton magnetic resonance studies show the crucial role of the anionic sublattice in the spontaneous deformation.

4. Electro-optic properties.

Up-to-now the linear electro-optic effect was studied for three ferroelectrics: $[NH(CH_3)_3]Sb_2Cl_9$ (TMACA) [31], $(CH_3NH_3)_5Bi_2Cl_{11}$ (MAPCB) [32] and $(CH_3NH_3)_5Bi_2Br_{11}$ (MAPBB) [33]. The all crystals exhibit strong electro-optic activity. Their electro-optic parameters are summarised in Table 2.

The values of v_π (the voltage which when applied across the crystal causes a phase retardation of π) obtained for MAPCB and MAPBB are comparable to those established in the KDP family crystals. It is interesting to notice that electro-optic properties of TMACA are comparable to those quoted for materials used commercially in an electro-optic shutter.

TABLE 2. Electro-optic crystals and their properties at
room temperature (λ = 632.8 nm).

Crystal	$v_\pi^{*)}$ (kV)	$r_{eff}^{**)}$ (10^{12}m/V)
TMACA	1.2	122.0
MAPBB	a-axis 14.4	a-axis 5.8
	b-axis 23.8	b-axis 3.5
MAPCB	a-axis 7.5	a-axis 18.8
	b-axis 17.1	b-axis 8.3

$^{*)}$ half-wave voltage
$^{**)}$ effective electro-optic coefficient

5. Halocoordinated antimonates(V) RSbCl$_6$.

The structure of these crystals is distinguished by the presence of isolated $SbCl_6^-$ octahedra. Among the investigated $RSbCl_6$ compounds new ferroics appeared to be these which contained small in size of high symmetry organic cation such like tetramethyl-phosphonium [6], guanidinium [5] and piperidynium [34]. They show ferroelastic properties at room temperature and variety of the phase transition sequences (Table 3).

TABLE 3. Ferroelastic RSbCl$_6$ crystals.

Compound	Temperature [K] and order of PT	Symmetry	Character of PT
[(NH$_2$)$_3$C]SbCl$_6$	351 (1st) 265 (2nd)	hexagonal→C2/m C2/m→P2$_1$/a	para-ferroelastic ferro-ferroelastic
[P(CH$_3$)$_4$]SbCl$_6$	405 (1st) 350 (2nd)	C2/m→ P$\bar{1}$	para-ferroelastic
[C$_5$H$_{12}$N]SbCl$_6$	368.6 (1st) 307.3 (2nd) 294.6 (1st)	orthorhombic→C2/m C2/m→P$\bar{1}$ P$\bar{1}$→P1	para-ferroelastic ferro-ferroelastic ferro-ferroelastic

An example of the ferroelastic domains of [C$_5$H$_{12}$N]SbCl$_6$ (in the *ab* plane) at room temperature is presented in Fig. 4.

The organic cations in high temperature phases show isotropic reorientation. The freezing of reorientational motion can (not necessarily) contribute to the mechanism of phase transitions. The X-ray diffraction and ^{1}H NMR studies show that the para-ferroelastic phase transitions are accompanied by the deformation of anionic sublattice and this deformation is presumably the factor ruling transitions. Majority of phase transitions, however, is of „order-disorder" type. Usually they are accompanied by weak dielectric anomalies. The dielectric response resembles in this case those encountered in the systems with so called „rotation" transitions. The dipole-dipole interactions are relatively weak and one can not expect a long-range electrical ordering.

The low electric polarizability in the case of RSbCl$_6$ crystals is connected, most probably, with a lack of layer structures and the lone electron pair 5s^2 which is characteristic of the Sb(III) compounds.

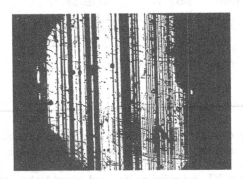

Fig. 4. A micrograph of ferroelastic domains of [C$_5$H$_{12}$N]SbCl$_6$ between crossed polarizers taken along the c-axis at room temperature.

6. Conclusions

Based on the critical review of the succeeded crystal growths, the phase transitions and the structural as well as dynamical studies, by using the X-ray diffraction and a number of other techniques, we tried to show that the halogenoantimonates and bismuthates containing variety of the organic cations represent an attractive group of crystals from the point of view of the materials sciences. The antimonates and bismuthates on the 3rd oxidation stage seem to be particularly promising. The halocoordinated octahedra can join in this case forming polyanionic units of various architectures adapted to the size and symmetry of the organic cations. In the case of haloantimonates(V) only isolated MX_6 are formed. The nature of rich phase transition sequences and physical properties of crystals under consideration were discussed. It follows that many of crystals reveal several interesting properties such like ferroelectric, ferroelastic and non-linear optical.

7. References

1. Jakubas, R. (1990) *Structure and Phase Transitions in alkylammonium halo-genoantimonates(III) and bismuthates(III)*, Acta Univ. Wrat. No 1250, Wrocław.
2. Zaleski, J. (1995) *Structure, phase transitions and molecular motions in chloro-antimonates(III) and bismuthates(III)*, Opole University Press, Opole.
3. Sobczyk, L., Jakubas, R. and Zaleski, J. (1997) Self-Assembly of Sb(III) and Bi(III) Halo-Coordinated Octahedra in Salts of Organic Cations, Structure, Properties and Phase Transitions, *Polish Jour. Chem.* 71, 265-300.
4. Bator, G., Baran, J., Jakubas, R. and Sobczyk, L. (1998) The structure and vibrational spectra of some ferroelectric and ferroelastic alkylammonium halogeno-antimonates(III) and bismuthates(III), *J. Mol. Struct.* 450, 89-100.
5. Jakubas, R., Ciapała, P., Pietraszko, A., Zaleski, J. and Kusz, J. (1998) Structure and Phase Transitions in the [C(NH$_2$)$_3$]SbCl$_6$ crystals, *J. Phys. Chem. Solids* 59, 1309-1319.
6. Ciapała, P., Jakubas, R., Bator, G., Pietraszko A. and Kosturek B. (1998) Phase transitions in ferroelastic [P(CH$_3$)$_4$]SbCl$_6$ crystal, *J. Phys. Condensed Matter* 24, 5439-5447.
7. Jakubas, R., Bator, G., Sobczyk, L. and Mróz, J. (1994) Dielectric and pyroelectric properties of (CH$_3$NH$_3$)$_3$Me$_2$Br$_9$ (Me = Sb, Bi) crystals in the ferroelectric phase transition regions, *Ferroelectrics* 158 , 43-48.
8. Jakubas, R., Krzewska, U., Bator, G. and Sobczyk, L. (1988) Structure and phase transition in (CH$_3$NH$_3$)$_3$Bi$_2$Br$_9$. A novel improper ferroelectric. *Ferroelectrics* 77, 129-135.
9. Iwata, M. and Ishibashi, Y. (1992) Ferroelastic phase transition in (CH$_3$NH$_3$)$_3$Bi$_2$Br$_9$ single crystal, *Ferroelectrics* 135, 283-289.
10. Jakubas, R. (1986) Ferroelectric phase transition in tris(dimethylammonium) nona-chlorodiantimonate(III), [NH$_2$(CH$_3$)$_2$]$_3$Sb$_2$Cl$_9$, *Solid State Commun.* 60, 389-391.
11. Jakubas, R., Sobczyk, L. and Matuszewski, J. (1987) Ferroelectricity and phase tran-sitions in tris (dimethylammonium) nonabromodiantimonate (III), *Ferroelectrics* 74 , 339-345.
12. Jakubas, R., Miniewicz, A., Bertault, M., Sworakowski, J. and Ecolivet, C. (1989) Phase Transitions in ferroelectric nonachlorodiantimonate [(CH$_3$)$_3$NH]$_3$Sb$_2$Cl$_9$ studied by calorimetric and dielectric methods, *J. Phys. France* 50, 1483-1491.
13. Zaleski, J., Jakubas, R. and Sobczyk, L. (1990) Successive phase transitions and ferroelasticity of [(CH$_3$)$_2$NH$_2$]$_3$Bi$_2$I$_9$, *Phase Transitions* 27, 25-36.

468

14. Jakubas, R., Zaleski, J., Kosturek, B. and Bator, G. (1999) Structure and phase transition in the ferroelastic [C(NH₂)₃]₃Bi₂Br₉ crystal, $J.Phys.:Condens. Matter$, submitted for publication.

15. Carpentier, P., Lefebvre, J. and Jakubas, R. (1995) Structure of pentakis (methyl-ammonium)undecachlorodibismuthate (III) (CH₃NH₃)₅Bi₂Cl₁₁ at 130 K and mecha-nism of the phase transition, $Acta Cryst.$ **B51**, 167-174.

16. Jakubas, R. (1989) A new ferroelectric compound: (CH₃NH₃)₅Bi₂Br₁₁, $Solid State Commun.$ **69**, 267-269.

17. Gdaniec, M., Kosturkiewicz, Z., Jakubas, R. and Sobczyk, L. (1988) Structure and mechanism of ferroelectric phase transition in tris(dimethylammonium) nona- chlorodiantimonate (III). $Ferroelectrics$ **77**, 31-37.

18. Zaleski, J. and Pietraszko, A. (1996) Structure at 200 K and 298 K and X-ray investigations of the phase transition at 242 K of [(CH₃)₂NH₂]₃Sb₂Cl₉, $Acta Cryst.$ **B52**, 287-295.

19. Bator, G. and Jakubas, R. (1988) Dielectric dispersion in ferroelectrics [NH₂(CH₃)₂]₃Sb₂Cl₉ and [NH₂(CH₃)₂]₃Sb₂Br₉, $Phys. Stat. Sol. (a)$ **147**, 591-600.

20. Jagadeesh, B., Rajan, P.K., Venu, K. and Sastry, V.S.S. (1994) Order - disorder phase transition and dimethylammonium group dynamics in [NH₂(CH₃)₂]₃Sb₂Cl₉ - proton NMR study, $Solid State Commun.$ **91**, 843-847.

21. Jakubas, R and Sobczyk, L. (1990) Phase transitions in alkylammonium halogenoantimonates and bismuthates, $Phase Transitions$ **20**, 163-193.

22. Pawlaczyk, Cz., Jakubas, R., Planta, K., Bruch, Ch. and Unruh, H-G. (1992) Dielectric dispersion in pentakis methylammonium bismuthates single crystals; II (CH₃NH₃)₅Bi₂Cl₁₁, $J. Phys.: Condens. Matter$ **4**, 2695-2700.

23. Medycki, W., Piślewski, N. and Jakubas, R. (1993) Molecular Dynamics of the methylammonium cation in [CH₃NH₃]₅Bi₂Cl₁₁. $Sol. State Nucl. Mag. Reson.$ **2**, 197-200.

24. Mróz, J. and Jakubas, R. (1990) Ferroelectric phase transitions of (CH₃NH₃)₅Bi₂Cl₁₁, $Ferroelectrics Lett.$ **11**, 53-56.

25. Carpentier, P., Zielinski, P., Lefebvre, J. and Jakubas, R. (1997) Phenomenological analysis of the phase transitions sequence in the ferroelectric crystal (CH₃NH₃)₅Bi₂Cl₁₁ (PMACB), $Z.Phys. B Condens. Matter$ **102**, 403-414.

26. Miniewicz, A., Jakubas, R., Ecolivet, C. and Girard, A. (1994) Raman scattering in ferroelectric [CH₃NH₃]₅Bi₂Br₉ single crystal., $Jour. Raman Spectroscopy$ **25**, 371-375.

27. Ishihara, H., Watanabe, K., Iwata, A., Yamada, K., Kinoshita, Y., Okuda, T., Krishnan, V. G., Dou, S. and Weiss, A. (1992) NQR and X-ray studies of [N(CH₃)₄]₃M₂X₉ and (CH₃NH₃)₃M₂X₉ (M = Sb, Bi ; X = Cl, Br), $Z.Naturforsch.$ **47a**, 65.

28. Bator, G., Baran, J., Jakubas, R. and Karbowiak, M. (1998) Raman studies of structural phase transition in Cs₃Bi₂Br₉ , $Vibrational Spectroscopy$ **16**, 11-20.

29. Bator, G., Mróz, J. and Jakubas, R. (1997) Dielectric and ferroelectric properties of the mixed crystals system (CH₃NH₃)₅Bi₂(1-x)Sb₂ₓCl₁₁, $Physica B$ **240**, 362-371.

30. Sapriel, J. (1975) Domain wall orientation in ferroelastics, $Phys. Rev.$ **B12**, 5128-5140.

31. Miniewicz, A. (1990) $Search for molecularionic and molecular crystals exhibiting ferroelectric and electro-optic properties$, Wydawnictwo Politechniki Wrocławskiej, Monografie 37(21), Wrocław.

32. Miniewicz, A. and Jakubas, R. (1991) Linear electro-optic effect in a ferroelectric crystal of (CH₃NH₃)₅Bi₂Cl₉ (PMACB), $J. Mol. Electronics$ **7**, 55-62.

33. Miniewicz, A. and Jakubas, R. (1990) Linear electro-optic effect in a ferroelectric crystal (CH₃NH₃)₅Bi₂Br₉ , $J. Mol. Electronics$ **6**, 113-117.

34. Jakubas, R. et al, results unpublished.

THERMOCHROMISM OF ORGANIC CRYSTALS.

A CRYSTALLOGRAPHIC STUDY ON *N*-SALICYLIDENEANILINES.

KEIICHIRO OGAWA

The University of Tokyo, Komaba,

Meguro, Tokyo 153-8902, Japan

E-mail: ogawa@ramie.c.u-tokyo.ac.jp

1. Introduction

Change in color of substances induced by variation of the temperature is known as thermochromism.[1] Thermochromism occurs in wide range of substances from inorganic to organic compounds, and from solution to solid state.

Thermochromism in organic compounds has various origins: cis-trans isomerism, tautomerism, ionic or radical cleavage of chemical bonds, or conformational change, etc. These processes can be regarded as a change from species A to B, where the longest wavelength of the absorption bands is considerably different between two species. For reversible thermochromism, species A and B are in an equilibrium. Usually a colorless or faint-color species A is more stable than a colored species B. In such a case, the population of species B increases with raising the temperature, resulting color deepening. If the colorless species A is less stable than the colored species B, the population of species A increases with raising the temperature, resulting color fading. Such a color change is called the negative thermochromism.

For the crystals in which species A and B are in equilibrium, the observed

469

D. Braga et al. (eds.), Crystal Engineering: From Molecules and Crystals to Materials, 469–479.

structure would be a superposition of A and B. If the populations of two species are large enough to be detected and their molecular structures are considerably different, a disorder will appear in the observed structure. If the molecular structures are too similar to be resolved, the observed structure will become a weighted average of two species. In any case, the observed structure will be changed with variation of the temperature. It is therefore expected that the crystal structure change associated with thermochromism of organic crystals can be detected by variable temperature X-ray analysis. We recently succeeded in this for the first time for N-salicylideneaniline crystals.[2]

2.   Thermochromism of N-Salicylideneanilines

N-Salicylideneanilines, which are condensation products of salicylaldehydes with anilines, belong to a class of the most popular compounds which show thermochromism as well as photochromism, a light-induced color change, in the crystalline state.[3] Before entering the story of a specific derivative of N-salicylideneanilines, let us survey briefly on the chromic behavior of this class of compounds. This will give a reason why not the parent compound but a specific derivative was selected to study.

1

N-Salicylideneanilines usually exist as crystalline materials with colors ranging from pale yellow through orange to deep red at room temperature. For example, the stable form of the parent compound (1) is pale yellow at room temperature. This form does not change in color with variation of the temperature, whereas it becomes red when it is irradiated with ultraviolet light.[4] In contrast, the crystals of a chloro derivative N-(5-choloro-2-hydroxybenzylidene)aniline (2) are thermochromic. The crystals of 2 are orange-red at room temperature and become pale yellow when they are cooled to 77 K. The color change is reversible.[4] Thus, the chromic behavior of the crystals of N-salicylideneanilines is different from compound to compound. Furthermore, the behavior strongly depends on the crystal structure. For example, in the case of N-salicylidene-2-methylaniline, the crystals obtained from methanol are photochromic, whereas the crystals from petroleum-ether are thermochromic. [4]

The thermochromism of crystalline N-salicylideneanilines is ascribed to the tautomerism between the OH and NH forms (eq 1). In most cases, the OH form is more stable than the NH form. With raising the temperature, the population of the less stable NH form is increased. For example, in the electronic spectrum of crystalline 2, the absorption band at ca. 470 nm, which is assigned to the NH form, increases in intensity with raising the temperature. Although X-ray crystallographic analyses of 2 at different temperature were carried out, no significant change was observed for the crystal structure with variation of the temperature. A change in the crystal structure associated with the thermochromism of N-salicylideneanilines has never been observed.[5] Such an observation was achieved by the use of variable temperature X-ray analysis of N-(5-chloro-2-hydroxybenzylidene)-4-hydroxyaniline (3).[2]

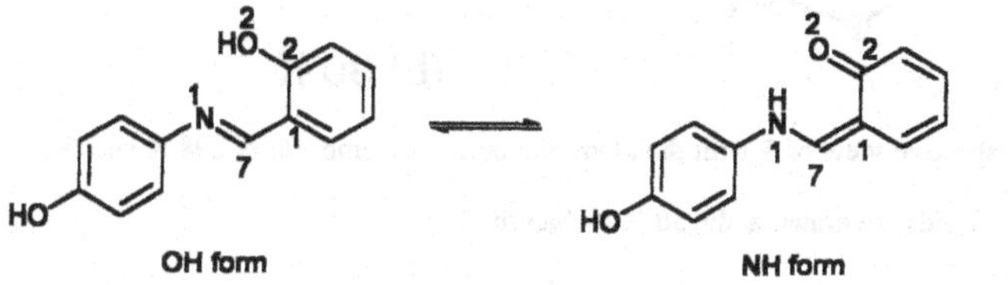

$$(1)$$

OH form   NH form

2: X=H   3: X=OH

3. Variable Temperature X-ray Analysis

Perspective views of the molecule of 3 are shown in Figure 1. Selected bond lengths obtained from the X-ray crystallographic analyses of 3[6] are listed in Table 1. Bond lengths of 2[7], which are regarded as of the pure OH form,[8] are also listed in Table 1 for comparison.

Table 1 shows that the each length of the bonds which could change in bond order by the tautomerism is significantly different between 2 and 3. Thus, O2–C2 bond of 3 is shorter than that of 2 and C1–C7 bond of 3 is also shorter than that of 2; C2–C1 bond of 3 is longer than that of 2 and C7–N1 bond of 3 is also longer than that of 2. The results suggest that the NH form might coexist in the crystals of 3.

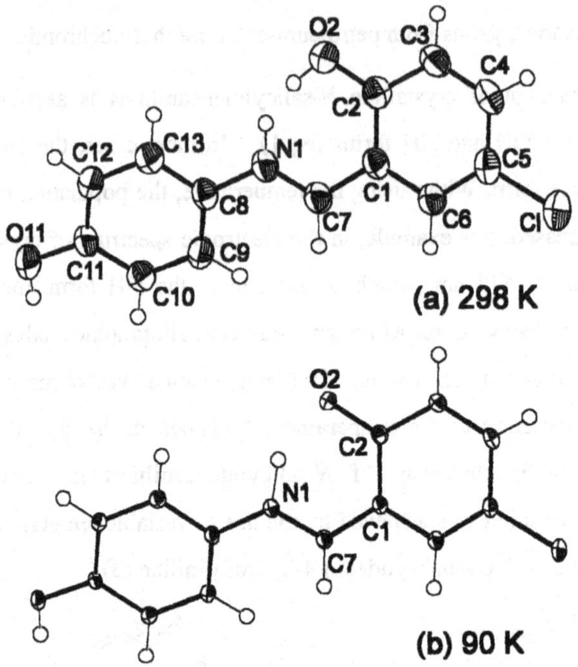

(a) 298 K

(b) 90 K

Figure 1. Perspective view of 3 with the atom numbering scheme: (a) at 298 K and (b) at 90 K. The ellipsoids are drawn at the 50 % probability level.

TABLE 1. Selected bond lengths of 2 and 3 (Å)

compd	T (K)	O2–C2	C2–C1	C1–C7	C7–N1
2	90	1.350(2)	1.412(3)	1.457(3)	1.291(2)
3	375	1.320(2)	1.414(3)	1.434(3)	1.288(3)
	298	1.321(2)	1.422(3)	1.433(2)	1.293(2)
	220	1.318(2)	1.427(2)	1.428(2)	1.297(2)
	160	1.313(2)	1.427(2)	1.429(2)	1.303(2)
	90	1.310(1)	1.433(2)	1.425(1)	1.308(1)

The most important point is that lengths of these bonds of **3** systematically vary with the temperature. Thus, the lengths of C2–C1 and C7–N1 bonds increase and the lengths of O2–C2 and C1–C7 bonds decrease with lowering the temperature. The results are interpreted as follows: (i) The observed structure is the superposition of the OH and NH forms, which remain unresolved. (ii) Each of the observed bond lengths is the weighted average of the corresponding length of the OH and NH forms according to the population of two forms. (iii) Their populations vary with the temperature. As a result, the observed structure changes with variation of the temperature. It is therefore concluded that there is an equilibrium between the OH and NH forms in crystals and that the population of the NH form increases with lowering the temperature.

This conclusion becomes definitive from the difference Fourier synthesis using the refined structure from which only the tautomeric hydrogen atom is removed. The difference synthesis for the structure at 160 K or higher locates two peaks assigned to two hydrogen atoms, one connected to O2 and the other to N1. In contrast, the difference synthesis for the structure at 90 K locates only a single peak assigned to the hydrogen atom connected to N1. Accordingly, the two hydrogen atoms were treated as disordered in the refinement of structures at 160 K or higher [Figure 1 (a)] whereas the structure at 90 K was refined as the pure NH form [Figure 1 (b)]. Thus, the X-ray diffraction unambiguously displayed the occurrence of the tautomerism that favors the NH form in crystals.

4. Electronic Absorption Spectra in the Solid State

Electronic spectra show that the crystals of **3** are thermochromic. The absorption band at 485 nm, which is assigned to the NH form, appears at room temperature and increases in intensity with the lowering the temperature. The results are consistent with those from X-ray diffraction: the NH form exists appreciably in the solid state at room temperature and increases in population with lowering the temperature. The change in the X-ray structure of **3** with variation of the temperature is, therefore, certainly for the thermochromy.

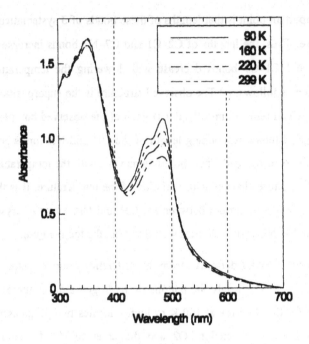

Figure 2. Electronic spectra of 3 in a transparent KBr disk.

5. Energy Difference between the OH and NH Forms

The populations of the NH and OH forms at different temperatures were estimated from the temperature dependence of the intensity of the absorption band (Table 2), by the use of the method reported by Theilacker et al.[9] Almost the same results were also obtained by the application of a similar method to the temperature dependence of the bond lengths which were determined by X-ray crystallographic analyses (Table 2). The results reveal that ca. 90 % of 3 exists as the NH form in crystals at 90 K and that the energy difference between the NH and OH forms is ca. 0.4 kcal mol^{-1} in crystals.

TABLE 2. Populations of the NH and OH forms in the crystals of 3

	populations (%)			
	from electronic spectra		from X-ray analysis [a]	
T (K)	OH form	NH form	OH form	NH form
90	11	89	10	90
160	19	81	17	83
220	27	73	26	74
299	37	63	31	69

[a] Estimated from the length of O2–C2 bond determined by X-ray crystallographic analysis. Estimations from the other lengths gave essentially the same results.

6. Zwitter Ionic Character of the NH Form

The geometry of 3 at 90 K is significantly different from that expected for a typical "keto" form 3k. The length of O2–C2 bond and that of C1–C7 bond are considerably longer than the standard length of the C=O bond [1.222 Å] and that of C=C bond [1.340 Å] in conjugated enones,[10] respectively, and the length of C2–C1 and that of C7–N1 bond are considerably shorter than the standard length of the C–C bond [1.464 Å] in conjugated enones and that of $C(sp^2)$–N bond [1.355 Å] in enamines, respectively. The results suggest that the NH form in the crystals of 3 has the character of the zwitter ion 3ki considerably.

(2)

3k   **3ki**

7. Behavior in Solution

In contrast to the crystalline state, **3** favors the OH form in solution. This was evidenced by electronic spectra. The spectra of the EPA solution[11] show no absorption band for the NH form at ca. 485 nm in the temperature range from 298 to 77 K. The results show that **3** exists exclusively as the OH form in the solution. The stabilization of the NH form in crystals is, therefore, ascribed to intermolecular interactions.

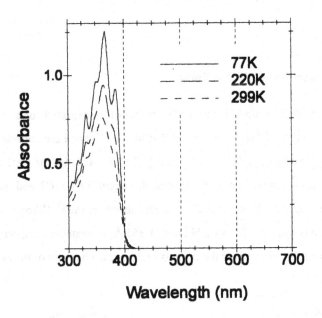

Figure 3. Electronic absorption spectra of **3** in the EPA solution.

8. Intermolecular Molecular Hydrogen Bonding in Crystals

Examinations of the molecular packing of the crystals of **3** reveal the occurrence of the intermolecular hydrogen bonding. The distance between O2 and H11 of the adjacent molecule was 2.622(1) Å and the angle of O2$\cdots$H11–O11 was 178(2) ° at 90 K.[12,13] It is therefore concluded that the stabilization of the NH form in the crystals of **3** results primarily from the intermolecular hydrogen bonding in crystals.

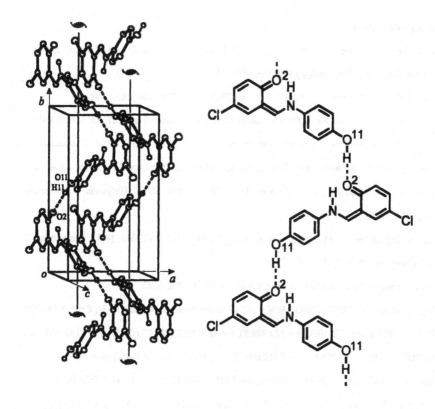

Figure 4. Packing diagram of **3** (left) and intermolecular hydrogen bonding for **3** (right).

9. Conclusion

This study demonstrates that the variable temperature X-ray crystallographic analysis is a powerful tool for the study on the thermochromism of organic crystals. The study evidences the following points: (i) The thermochromism of **3** in crystals is originated from the tautomerism between the OH and NH forms. (ii) The NH form has a character of zwitter ion considerably in crystals. (iii) The NH form is stabilized due to intermolecular hydrogen bondings in crystals.

478

References and Notes

[1] Review: (a) Day, J. H. *Chem. Rev.* **1963**, *65*, 65–80. (b) Nassau, K. *The Physics and Chemistry of Color*; John Wiley and Sons: New York, **1983**; pp 132–133, 347–348.

[2] Ogawa, K.; Kasahara, Y.; Ohtani, Y.; Harada, J. *J. Am. Chem. Soc.*, **1998**, *120*, 7107–7108.

[3] Review: (a) Hadjoudis, E.; Vittorakis, M.; Moustakali-Mavridis, I. *Tetrahedron* **1987**, *43*, 1345–1360. (b) Hadjoudis, E. Tautomerism by Hydrogen Transfer in Anil, Aci-Nitro and Related Compounds. In *Photochromism*; Dürr, H., Bouas-Laurent, H., Eds.; Studies in Organic Chemistry 40; Elsevier, Amsterdam, **1990**; pp 685–712. (c) Inabe, T. *New J. Chem.* **1991**, *15*, 129–136. (d) E. Hadjoudis, E. *Molecular Engineering* **1995**, *5*, 301–337.

[4] (a) Cohen, M. D.; Schmidt, G. M. *J. Phys. Chem.* **1962**, *66*, 2442–2445. (b) Cohen, M. D.; Schmidt, M. J.; Flavin, S. *J. Chem. Soc.* **1964**, 2041–2051.

[5] Bregman, J.; Leriserowitz, L.; Schmidt, G. M. *J. Chem. Soc.* **1964**, 2068–2085.

[6] Crystallographic data for 3: $C_{13}H_{10}ClNO_2$, MW = 247.67, monoclinic, space group $P2_1/a$, $Z = 4$, λ(Mo, Kα) = 0.71073 Å. $T = 375$ K, $a = 7.2240(3)$, $b = 12.6908(8)$, $c = 12.5188(7)$, $\beta = 93.606(5)°$, $V = 1145.4(1)$ Å^3, $R = 0.0486$, GOF = 0.991. $T = 298$ K, $a = 7.1805(4)$, $b = 12.638(1)$, $c = 12.514(1)$, $\beta = 93.184(7)°$, $V = 1133.9(2)$ Å^3, $R = 0.0430$, GOF = 0.991. $T = 220$ K, $a = 7.119(2)$, $b = 12.563(2)$, $c = 12.490(2)$, $\beta = 92.71(2)°$, $V = 1115.8(4)$ Å^3, $R = 0.0469$, GOF = 0.996. $T = 160$ K, $a = 7.086(2)$, $b = 12.518(2)$, $c = 12.485(2)$, $\beta = 92.42(2)°$, $V = 1106.5(4)$ Å^3, $R = 0.0458$, GOF = 1.033. $T = 90$ K, $a = 7.043(2)$, $b = 12.459(3)$, $c = 12.4827(19)$, $\beta = 92.031(18)°$, $V = 1094.7(4)$ Å^3, $R = 0.0388$, GOF = 1.057.

[7] The X-ray structure of 2 was redetermined with higher accuracy in this study. Our structure was essentially identical with that determined by Bregman et al.[5] The crystallographic data: $C_{13}H_{10}ClNO$, MW = 231.67, orthorhombic, space group $Pca2_1$, $Z = 4$, λ(Mo, Kα) = 0.71073 Å. $T = 298$ K, $a = 12.177(3)$, $b = 4.483(3)$, $c = 19.271(3)$, $V = 1051.9(7)$ Å^3, $R = 0.0350$, GOF = 1.042.

[8] It was proved by Schmidt and co-workers that 2 exists exclusively as the OH form at 90 K in crystals.[5]

[9] Theilacker, W.; Kortüm, G.; Friedheim, G. *Chem. Ber.* **1950**, *83*, 508–519.

[10] Allen, F. H.; Kennard, O.; Watson, D. G.; Brammer, L.; Orpen, A. G.; Taylor, R. Typical interatomic distances: organic compounds. In *International Tables for Crystallography*; Vol. C. Wilson, A. J. C. Ed.; The International Union of Crystallography; Kluwer Academic Publishers: Dordrecht, The Netherlands, **1995**; pp 685–706.

[11] The EPA solution is the mixed solution of ether, isopentane, and ethanol in the volume ratio of 5:5:2.

[12] The distance and angle shows that the intermolecular hydrogen bonding is of medium strength, see: Jeffrey, G. A. *An Introduction to Hydrogen Bonding*, Oxford University Press: New York, 1997; pp 11–16.

[13] A similar intermolecular O⋯H–O hydrogen bonding occurs in the crystals of 2-hydroxy-*N*-(2-hydroxybenzylidene)aniline,[14] *N*-(2,3-dihydroxybenzylidene)-2-hydroxymethylaniline,[15] and *N*-(2,3-dihydroxybenzylidene)isopropylamine.[16] For these compounds the NH form was detected in crystals.

[14] (a) Lindeman, S. V.; Antipin, M. Yu.; Struchkov, Y. T. *Kristallografiya*, **1988**, *33*, 365–369. (b) Zheglova, D. K.; Gindin, V.; Kol'tsov, A. I. *J. Chem. Res., Synop.* **1995**, 32–33.

[15] Puranik, V. G.; Tavale, S. S.; Kumbhar, A. S.; Yerande, R. G.; Padhye, S. B.; Butcher, R. J. *J. Crystallogr. Spectrosc. Res.* **1992**, *22*, 725–730.

[16] Mansilla-Koblavi, F.; Tenon, J. A.; Ebby, T.N.; Lapasset, J.; Carles, M. *Acta Crystallogr. Sect. C* **1995**, *51*, 1595–1602.

CRYSTAL ENGINEERING: FUNCTIONALITY AND AESTHETICS

C. V. KRISHNAMOHAN SHARMA
Department of Chemistry, Texas A&M University,
College Station, Texas, 77843-3255, USA

Abstract: Functionality and aesthetics are two major driving forces for the advancement of science and technology and of course crystal engineering is no exception. In this article we discuss our rational designing strategies for controlling the functional properties of materials using both organic and inorganic molecular building blocks. The targeted functional properties include light-harvesting, ion exchange, porosity and intercalation. Our efforts to construct porous solids by exploiting directional forces in conjunction with molecular symmetry unraveled beautifully interwoven and threaded supramolecular structures that resemble carpets and Chinese blinds of the real world. Efforts have been made to discuss our results in light of the narrowing gap between 'discrete' and 'infinitely large' supermolecules.

1. Introduction

The emergence of supramolecular chemistry revolutionized the concepts and thinking process of chemists in the 21st century.[1] Now researchers are successfully finding solutions to complex problems such as the origin of life and biological processes simply by looking *beyond the molecule*, i.e. by understanding how molecules recognize (or organize with) other molecules to execute certain preprogrammed functions.[2,3] As we slowly unravel the vast range of intricate and beautiful structures with complexity and wonderfully efficient functions in nature, our urge to mimic nature structurally and functionally is constantly growing.[4,5] In fact, the discovery of many novel artificial systems with functionality (e.g., light harvesting, porosity) and aesthetics (e.g., entangled molecular structures such as helices, catenanes and rotaxanes) was made possible only through the inspiration drawn from mother nature.[2-9]

Crystals have always been the source of enchantment, elegance and beauty to laymen, emperors and scientists alike. While the external beauty of crystals delight every one of us, the internal structure (i.e., atomic and molecular) of the crystals has been the subject of both fascination and serious debate among scientific community.[10,11] After all, the chemical and physical properties of the materials are governed by the geometrical arrangement of molecules or ions in the solid state.[12] Interestingly, molecules/ions in the solid state bind together with a high degree of internal order (i.e., periodicity or symmetry) and yet balance the intermolecular interactions with an amazing level of precision! This

D. Braga et al. (eds.), Crystal Engineering: From Molecules and Crystals to Materials, 481–500.

482

fundamental characteristic property of the crystals won them the title of "supermolecule *par excellence*" or an "infinitely large supermolecule". [13,14]

The crystal is a supermolecule concept integrating the subject of crystal engineering with mainstream supramolecular chemistry and led to the free flow of ideas from conventional "molecular recognition" to crystal engineering and *vice versa*. [14] So, whether we construct a discrete supermolecule just by binding two molecular species or construct an infinitely large supermolecule by linking millions of molecules, the primary designing tools, *supramolecular synthons*, will be identical. [15] The only fundamental difference in these two instances is that formation of discrete supramolecular species requires convergent binding, while infinitely large supramolecular assemblies require divergent binding.

Understanding the intriguing correlation between shape, symmetry and intermolecular forces/coordination bonds is key to the successful design of nanoscale/crystalline architectures. [15,16] The ingenuity of chemists plays a major role in identifying the complementary molecular building blocks for a proposed purpose/architecture. For example, we can achieve the target of constructing a molecular square using several different ways, i.e. by assembling, organic, inorganic or hybrid organic-inorganic molecular building blocks with appropriate symmetry as shown in Scheme 1. Further, depending upon the choice of the building block, the resultant square could be a charged or a neutral or a chiral molecular square for recognizing certain guest molecular species or for templating a beautifully entangled molecular system. [8,16] Similar ideas can be extended to crystalline architectures, where we target certain 1D, 2D or 3D networks and then work backwards to figure out appropriate molecular building blocks required to achieve the objective. [15]

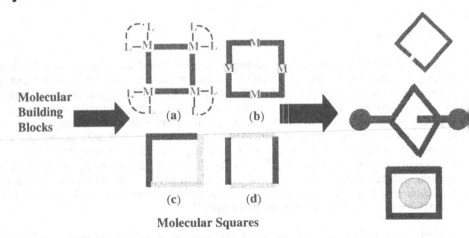

Molecular Building Blocks

(a) (b)

(c) (d)

Molecular Squares

Scheme 1

The underlying principle of current crystal engineering strategies is propagation of molecular symmetry into crystalline symmetry (or more appropriately crystalline networks) with the aid of robust and directional interactions (e.g., strong hydrogen bonds, coordination bonds) for the realization of predictable crystalline architectures. [17-22] For this reason, molecular complexes bearing robust hydrogen bonded functional groups and coordination polymers dominate most of the designer materials reported in recent years. However, there are also significant number of examples, where the directional nature of weak intermolecular interactions has been successfully exploited for the design of new materials.[23,24]

While there is no doubt that a wide variety of crystalline architectures of organic and inorganic materials are designed by exploiting molecular symmetry and robust directional bonds, such requirements greatly restricted our choice of molecular building blocks and thereby the scope of crystal engineering![25,26] Unfortunately, in the absence of powerful and reliable theoretical models for predicting the crystal structures for a given molecular structure, we will have to continuously rely upon current empirical, trial and error methods for widening the purview of crystal engineering.

1.1 SCOPE OF THIS ARTICLE

Functionality and aesthetics are the two fundamental aspects for the development of science and technology. Interestingly, these two aspects go hand in hand in nature and it is hard to view one without the other (including in this article). Although, aesthetics is a subjective terminology (beauty lies in the eyes of the beholder!), symmetrical shapes and analogous objects have been traditionally fascinating human minds and the design of such objects at macroscopic and nanoscopic level has been a challenging task for scientists.[3,8,15,16] In this article we discuss our deliberate designing attempts and serendipitous discoveries of crystalline architectures formed by *organic, inorganic and organic-inorganic building blocks*. The functional properties addressed here include light harvesting, ion exchange, porosity and guest inclusion. While the beautifully interwoven structures that resemble carpets and Chinese blinds add on to the aesthetic dimension of this article.

2. Light Harvesting Model Complexes

2.1 INTRODUCTION

Nature uses well-organized supramolecular arrays of chromophores (pigments) to trap solar energy and convert it into chemical potential that drives the chemistry of photosynthetic organisms.[6,27] We could mimic the functional properties of natural systems by developing strategies for the construction of similar but well-defined artificial chromophoric assemblies. In recent years there has been a tremendous interest in the synthesis of various multichromophoric complex systems for converting sunlight energy into spectroscopic energy of molecular components for light harvesting purposes.[28-30] However, tedious experimental procedures associated with the synthesis of highly organized multichromophoric assemblies made it very difficult to design structural models suitable for light harvesting. It may be mentioned here that green plants contain as many as 360 chromophoric groups or antenna units for efficiently absorbing the incident sun light energy! Here we report a novel crystal engineering strategy to imbed virtually unlimited numbers of porphyrin chromophores in desirable, controlled metallation states (including mixed-metalloporphyrins and freebase-porphyrins) into highly organized, rigid one-dimensional and/or two-dimensional crystalline, supramolecular arrays.

Indeed, freebase porphyrins and metallated porphyrins exhibit distinct absorption and emission spectra. Freebase porphyrins show four absorption bands in the visible region, but a corresponding metalloporphyrin reveals an essentially two-banded one in the visible region, furthermore, the insertion of the metal cations may cause new bands in some cases.[31] For this reason, it is possible to overlap absorption and emission spectra of freebase and metalloporphyrins for attaining efficient electron/energy transfer properties simply by mixing and matching the composition of freebase and metallo porphyrins in the highly organized solid state supramolecular arrays. In this context, the design of rigid coordination polymers with *freebase-porphyrins* forms an essential first step toward the design of multichromophoric light harvesting model complexes. Because construction of the supramolecular polymer or sheet with the freebase-porphyrin utilizing functionality external to the cavity, may allow selective metallation and demetallation of the cavity without disturbing the overall crystalline architecture.

2.2 FREEBASE 1D AND 2D POLYMERS

The coordination complexes of metal halides, MX_2 (M = Cd, Hg, Pb; X = Br, I) with freebase-tetrapyridylporphyrin (TPyP) isolated from 1,1,2,2-tetrachloroethane (TCE) form predictable 1D, $[(HgI_2)_2(TPyP)] \cdot 2TCE$, 1, or 2D, $[(MI_2)TPyP] \cdot 4TCE$, (M= Pb, 2, Cd, 3) coordination networks depending upon the coordination geometry of the metal (Scheme 2).[32-34]

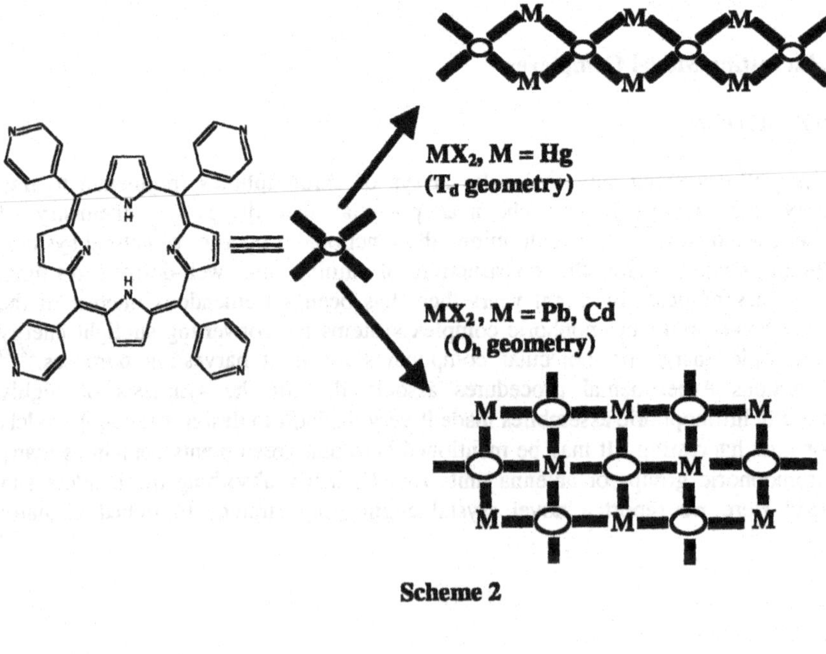

MX_2, M = Hg
(T_d geometry)

MX_2, M = Pb, Cd
(O_h geometry)

Scheme 2

The crystal structure of complex 1 has a nanoporous 1D polymeric architecture with each HgI$_2$ tetrahedrally coordinated with a pyridyl moiety of two TPyP molecules. A supramolecular cavity is formed between the linked porphyrins with an effective cavity size of 2.5 Å x 7.7 Å (Scheme 2, Figure 1). The 1D coordination polymers are arranged in layers inducing another supramolecular cavity (effective cavity size, 2.4 Å x 3.0 Å, Figure 1b). The 2D layers are offset stacked at 5.5 Å and form a partially open porous network and the TCE molecules are sandwiched between the layers.

The Pb and Cd centers in 2 and 3 adopt octahedral geometry with *trans*-halides and coordination of four pyridyl moieties to form isostructural nanoporous 2D coordination polymers (Scheme 2, Figure 2). The solvated TCE molecules are sandwiched between the layers and the 2D polymers are off-set stacked at *ca.* 6.9 Å such that a portion of the porphyrin cavity resides over the supramolecular pore and avoids the formation of continuous open channels.

2.2.1 Partial Metallation of Layered Structures

The porphyrin cavities in the supramolecular arrays of freebase complexes, 1-3 can be selectively populated either by crystallizing the metal halides and TPyP in the presence of suitable metal salts or by reacting metal halides with a mixture of freebase and metalloporphyrins in specific stoichiometric ratios.[35] However, we found that the latter method provides us a greater control over the formation of multichromophoric assemblies predictably for fine-tuning the electron/energy transfer properties of these systems. We have successfully introduced varying compositions of Zn^{2+}, Cu^{2+}, Ni^{2+}, and mixed metalloporphyrins into 1D or 2D networks of 1-3. The UV/Vis absorption and solid state fluorescence spectra of 1 metallated with 50% and 75% Zn are shown in Figures 3 and 4. The percentage of ZnTPyP was determined by spectroscopic and crystallographic methods and the metal contents determined match the ratio of ZnTPyP utilized in the complexation reactions. Similarly we have introduced Ni^{2+} and Cu^{2+} together into the crystalline lattices of 1-3 in controlled ratios.

The fluorescence spectrum of 1 is similar to TPyP and the spectra of 1•50%Zn and 1•75%Zn show an enhanced intensity around λ 660 nm due to the partial overlap of ZnTPyP and TPyP emission bands (Figure 4). Although, the fluorescence intensities for 1•50%Zn, and 1•75%Zn are very weak compared to their parent chromophores for approximately the same amount of sample, we are not yet able to quantify the energy transfer in the solid state because of limitations in preparing solid samples with accurate thickness of the surfaces.

The designing strategies reported here clearly demonstrate that we can virtually imbed all doubly charged metal cations in the periodic table into the highly organized polymeric porphyrin arrays of 1-3 and fine tune the electronic properties by adjusting relative percentage and composition of chromophores. At this stage we do not know about the relative positioning of metallo and freebase porphyrins in the solid state. Although, it is very difficult to determine such relative positioning of the chromophores, it is reasonable to assume that if the networks have two different chromophores (50% metallated) the probability of every alternative chromophore being freebase and metallo porphyrins is far greater. Similarly if three chromophores are imbedded into these networks equally (*ca.*

486

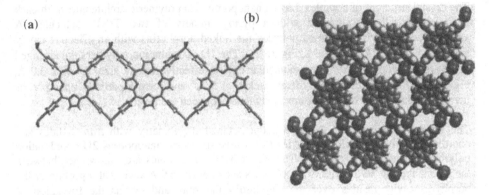

Figure 1. (a) The 1D coordination polymers of freebase-porphyrin **1** produce a supramolecular cavity. The two NH protons of the porphyrin cavity are not shown to highlight the porphyrin cavity. (b) Arrangement of 1D polymers into a layer induces a second cavity. The porphyrin cavities are partially metallated by Zn^{2+} (metal occupancy is 30%, 50%, or 75% in the complexes characterized).

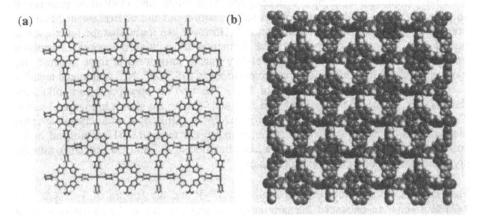

Figure 2. The 2D coordination polymers of **2** and **3**. (a) The iodides are perpendicular to the plane of the paper. The NH groups are omitted to highlight the porphyrin cavity. (b) The partial metallation of the porphyrin cavities is represented with random positioning of filled circles in the space-filling model.

33.3%) the possibility of every chromophore being adjacent to two other different porphyrinic chromophores is far greater.

3. Molecular Squares

3.1 INTRODUCTION

Discrete macrocyclic metal complexes constitute a new family of inorganic host materials with conformational rigidity and readily provide an alternative to organic macrocycles

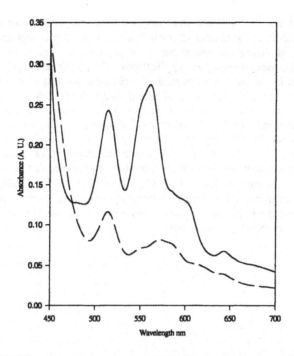

Figure 3. The UV/Vis absorption spectra of 75% Zn- (solid line) and 50% Zn- (broken line) metallated-1 single crystals dissolved in TCE. The percentage of metal content was estimated by comparing these spectra with the corresponding titration curves.

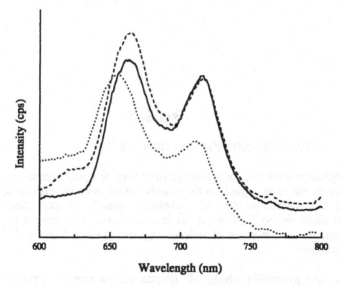

Figure 4. The solid state fluorescence spectra of freebase-porphyrin crystals 1 (—), 1•50%Zn (---), 1•75%Zn (...). The spectra of metallo-porphyrins were arbitrarily multiplied to compare with emission spectra of 1. Notice that the enhanced intensity around 660 nm is due to partial overlap of ZnTPyP and TPyP emission bands.

such as, crown ethers, cyclophanes, and calixarenes for selectively binding or exchanging anions.[16,36-38] Further, the synthesis of such discrete nanoscopic supramolecular species with predetermined shapes and geometries is of great importance in the development of novel materials for nanotechnological applications.[39] As mentioned in the introduction, the subtle differences in the binding properties of molecular building blocks (i.e., convergent versus divergent) determine the nature of supramolecular networks (i.e., discrete versus infinite). In general, linear bifunctional ligands form 1D coordination polymers when reacted with transition metal in 1:1 molar ratio.[40] However, analogous metal complexes of angular bifunctional ligands (e.g. pyrimidine) may form a variety of supramolecular isomers depending upon the metal coordination environment as shown in Scheme 3. Indeed, we have isolated all these three possible supramolecular isomers by reacting pyrimidine with different metal salts (e.g. Cu(SiF$_6$)•xH$_2$O, Co(NO$_3$)$_2$•xH$_2$O, Ag(NO$_3$)) in 1:1 molar ratio. Of these, the discrete molecular squares of Ag$^+$:pyrimidine complexes are of special interest here owing to their relevance in the design of novel inorganic macrocycles and ion exchange properties.

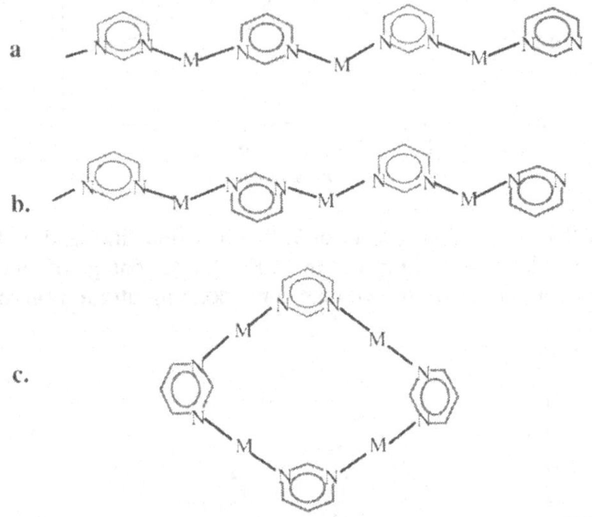

a

b.

c.

Scheme 3

3.2 METAL SPACERS AND LIGAND CORNERS

Current synthetic strategies for the design of molecular squares focus on using metal ions at the corners and linear ligands as the spacers, which often requires occupying several of the metal's coordination sites with chelating ligands to force discrete macrocycle formation (Scheme 1a).[16] However, an inverse approach to these strategies provides a simple route for the generation of molecular squares as exemplified by 1:1 Ag$^+$ complexes with pyrimidines.

Complex, [Ag(pyrimidine)][NO$_3$], 4, reveals an interesting cyclic self-assembly of pyrimidine and Ag$^+$ cations with internal cavity dimensions, ca. 7.2 x 7.2 Å (Scheme 1c,

Figure 5a).[41] Ag⁺ cations in the tetranuclear cationic square exhibit bent coordination geometries (Ag-N-Ag angles: 150.4° and 155.7°). Each of these squares are stacked by six other squares at 3.45 Å (three above and three below the square plane) forming continuous open channels that are occupied by nitrate anions (Figure 5b). The smaller size and weaker coordination bonds of NO_3^- anions with Ag⁺ cations in this open channel structure can be exploited for exchanging NO_3^- anions with ReO_4^- or TcO_4^- from solution.

(a)   (b)

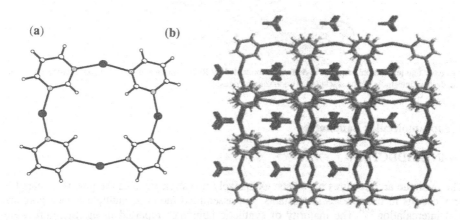

Figure 5. The 1:1 Ag⁺:Pyrimidine metal complexes form tetracationic molecular squares. (a) The angularity of the ligand and bent coordination geometry of the Ag⁺ promotes the formation of such a discrete supermolecule. (b) The molecular squares form continuous channels and nitrates anions reside over the channels and are exchangeable with pertechnetate.

3.3 ION EXCHANGE

The nitrate anions of **4** readily exchange with larger softer ReO_4^- and TcO_4^- anions. When complex **4** crystals were suspended in MeOH containing $NH_4^{99}TcO_4$, and stirred for 2 h, a 12% drop in the radioactivity of the solution was noticed. A significant reduction in the activity, *ca.* 40% was observed after 24 h and the distribution ratio for TcO_4^- was calculated to be 95. We also examined the radioactivity of the anion-exchanged crystals after washing them thoroughly with fresh MeOH. A 0.008 g of crystalline sample showed an activity of 885 counts per minute confirming the presence of TcO_4^-. Interestingly, dissolution of complex **4** in MeOH-Water containing $NaReO_4$ and recrystallization at room temperature yielded crystals of [Ag(Pyrimidine)][ReO_4], **5**. The crystal structure of **5** also consists supramolecular squares as observed in **4** with ReO_4^- anions both in the open channels and forming a 1D coordination polymer (Figure 6).[41]

These results suggest that usage of angular ligands in association with divalent coordinating metal complexes provide a simple route for the synthesis of supramolecular inorganic macrocylces. Further, with a suitable design strategy it is possible to engineer macrocycles that can efficiently remove hazardous pertechnetate anions from the solution. Indeed, removal of TcO_4^- anions from the Hanford radioactive waste tanks in the US has been a current technological challenge.

490

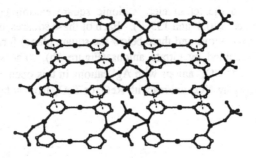

Figure 6. The tetracationic squares in **5** are interconnected by ReO_4^- anions and form 1D coordination polymers. The dashed lines indicate the stacking between square edges.

4. Supramolecular Laminates

4.1 INTRODUCTION

The laminate architectures have been extensively investigated since the past few years for applications in membranes, surfactants, self-assembled mono or multiple layers, porosity and intercalation.[42-44] The majority of synthetic laminates reported in the literature were designed by considering inorganic building blocks, the α- and γ- zirconium phosphonate complex series represent prime examples of such designer layered materials.[45] However, the rational design of organic laminates has never been attempted before because of the fact that the nature of intermolecular forces in organic materials is not well understood until recently. In principle, the design of layered structures with the aid of robust intermolecular forces effectively reduces the complex problem of controlling the three-dimensional structure of the solids into a one-dimensional problem. In this regard, our efforts to control the crystalline architectures of trimesic acid (1,3,5-benzenetricarboxylic acid) led to the discovery of a novel class of organic laminates that can readily intercalate guest molecules like natural clays.

4.2 INTERDIGITATION AND INTERCALATION

Trimesic acid (H_3TMA) has been known since at least 1867 and is a prototypal molecule in current crystal engineering studies for the fact that it predictably self-assembles through the well-known carboxylic acid dimer motif into a honeycomb grid with 11.0 Å cavities.[46,47] Further, the acid groups of trimesic acid readily ionize to form salts with inorganic and organic cations. Our modular self-assembly approach for expanding the honeycomb grids of trimesic acid through complementary hydrogen bonding interactions between carboxylate anions and secondary ammonium cations yielded fruitful results. For example, the crystal structure of triply deprotonated trimesic acid salt with N,N-dicyclohexylamine, [H_2N(cyclohexyl)$_2$]$_3$[TMA], **6** forms a honeycomb grid with effective cavity size, 12.7 Å (Scheme 4).[48] The cyclohexyl groups in this complex project above and below the plane of the honeycomb sheets and form a lamellar structure. This interesting observation led us to focus our attention on the rational design of densely packed robust hydrogen bonded sheet structures by exploiting the hydrogen bonding properties of secondary alkyl/aryl ammonium cations.

When trimesic acid is reacted with two equivalents of secondary amines, closely packed robust ionic hydrogen bonded lamellar structures result as depicted in Scheme 5.[49] The series of doubly deprotonated trimesic acid salts, $[H_2N(propyl)_2]_2[HTMA]$, **7**, $[H_2N(hexyl)_2]_2[HTMA]$, **8**, $[H_2N(octyl)_2]_2[HTMA]$, **9**, $[H_2N(decyl)_2]_2$ [HTMA], **10**, form identical two dimensional hydrogen bonding networks stabilized by N-H$^+$•••O$^-$ and

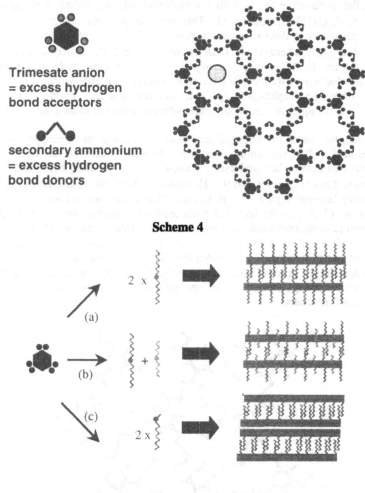

Trimesate anion
= excess hydrogen
bond acceptors

secondary ammonium
= excess hydrogen
bond donors

Scheme 4

Scheme 5

O-H•••O$^-$ ionic hydrogen bonds as shown in Figure 7. The hydrogen bonded sheet structure of **7-10** indicate that relevant hydrogen bond lengths and the repeating units in the sheets (i.e., 16.9 Å x 21.6 Å) are almost identical. The hydrophobic alkyl substituents of the ammonium cations project above and below the sheet and are interdigitated (Figure 8). The inter-layer separation distances in **7-10** are 7.0 Å, 10.3 Å, 12.4 Å, 14.56 Å respectively and are directly related to the alkyl chain lengths and the interdigitation capability.[49]

The laminate architecture of **7-10** did not form any intercalated complexes as we anticipated. Nevertheless, when we attempted to form a triply deprotonated complex of N, N-dibenzylamine with trimellitic acid, H_3TML, it rather formed a doubly deprotonated salt with lamellar architecture like in **7-10**, but clathrated with the neutral N,N-dibenzylamine, $[H_2N(dibenzyl)_2]_2[HTML]]$•Guest, **11**! This interesting result induced us to carry out further experiments to determine the clathration behavior of N, N-dibenzyl amine salts with H_3TMA, $[[H_2N(dibenzyl)_2]_2[HTMA]]$•Guest, **12**. Indeed, dibenzylammonium salts of both H_3TMA and H_3TML incorporate numerous aromatic guest molecules irrespective of the size and electronic nature of the guest molecules (e.g., nitrobenzene, anisole, veratrole, pentamethylbenzene, naphthalene, pyrene, and ferrocene).[50] Figure 9 shows a few representative examples of these complexes with and without aromatic guest molecules.

In retrospect, the reason for dibenzylammonium salt's ability to intercalate aromatic guest molecules unlike alkyl ammonium salts is not difficult to understand. In the absence of guest molecules both n-alkyl and benzyl groups interdigitate using hydrophobic and/or C-H•••π interactions (Figures 8 and 9). However, in the event of aromatic guest inclusion into the inter-lamellae region, benzyl amines (**11** and **12**) have a distinct advantage over alkyl groups (**7-10**), as the included guest molecules provide the extra stability to the system through additional stacking interactions (i.e., between guest and benzyl units). On the other hand complexes **7-10** do not have any such advantage, instead the guest inclusion requires the destabilization of robust hydrophobic interactions that already exist between interdigitated alkyl groups. More importantly, the benzylammonium salts **11** and **12** exhibit guest exchange properties and are poised to be a true new class of synthetic "organic clay" materials.

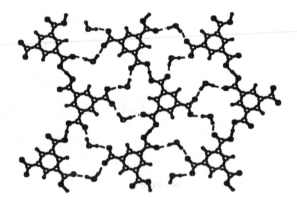

Figure 7. The two-dimensional hydrogen bonding network observed in **7-10**. The alkyl substituents of the ammonium cations are omitted for the sake of clarity. N and O atoms are shown as hatched and filled circles respectively. Hydrogen bonds are shown as dashed lines.

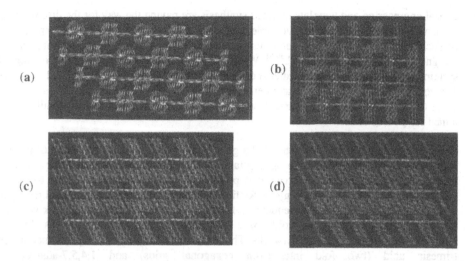

Figure 8. Supramolecular laminates of **7-10 (a-d)**. Note that interdigitation and tilt of the alkyl groups occurs as the chain length increases.

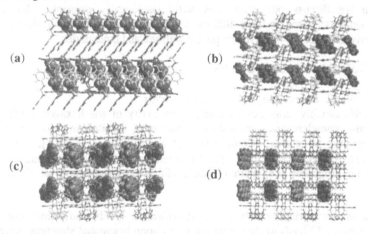

Figure 9. Supramolecular laminates of **11** and **12**. (a) Interdigitation of benzyl groups in **12**, solvated with MeOH. Inclusion of pentamethylbenzene (b), N,N-dibenzylamine (c) and pyrene (d) into the inter lamellae regions of **11** and **12**. All the guest molecules are shown in space-filling model.

5. The Art of Interweaving –Nature's Way

The molecular entanglement is a common phenomenon in biology; DNA forms helices, closed circles, catenanes, knots and extremely complex branched junctions.[51-53] The undisputed beauty of these systems and accompanying precious information about biological processes is invoking ever-increasing fascination toward the intertwined molecular complexes. Interestingly, in recent years research progress in the areas of crystal engineering and template directed synthesis are paving the way for the discovery of numerous synthetically intertwined molecular systems. The motivation for the exploration of such artificial entangled systems also comes from their possible applications in the design of "molecular machines".[39,54] While many of the discrete molecular entangled systems reported recently (e.g., catenanes and rotaxanes) are a result of targeted template directed synthesis, the solid state interwoven architectures (1D, 2D and 3D networks) seem to be a consequence of current crystal engineering strategies and especially those aimed for porous solids.[55]

Crystal engineering principles generated a great deal of interest in the synthesis of porous solids due to the fact that, with appropriate choice of symmetrical molecular building blocks and robust directional forces, it is possible to generate open crystalline networks (see scheme 4). While this strategy works fine in theory and in practice, the hard reality is that *nature abhors vacuum*. Consequently, if we do not engineer an open network with the aid of a template to fill the void space generated, nature will outsmart us through interweaving of identical open networks. The text book examples of this phenomenon are trimesic acid (two fold interwoven hexagonal grids) and 1,4,5,7-adamantane tetracarboxylic acid (five fold interwoven diamondoid networks).[46,56]

In this section we discuss our encounters with aesthetically pleasing solid state interpenetrated networks based on three different kinds of molecular building blocks, i.e., trimesic acid derivatives, bis(4,4'-dihydroxyphenyl)sulfone and lanthanides.

5.1 MOLECULAR CARPETS AND BOXES

5.1.1 Triply and Doubly Interwoven "Super Trimesic Acids"
Our modular self-assembly strategies to enlarge the cavity of the hexagonal grids of trimesic acid with complementary secondary amine units resulted in two interesting interwoven network structures. Complexes, $[4,4'\text{-dipyridine}]_{1.5}[H_3TMA]$, **13** and $[H_2N(\text{propyl})_2]_3[TMA]\bullet 2H_2O$, **14**, form open distorted hexagonal networks and exhibit distinct catenation properties (i.e., parallel and inclined interpenetration) as described below.[57,58]

The robustness of the pyridine-carboxylic acid hydrogen bonds and complementary donor:acceptor ratio in **13** leads to the formation of an open hexagonal structure (internal dimensions, *ca.* 35 Å x 26 Å, Figure 10a). Self-inclusion of these networks is facilitated by the chair conformation of the distorted hexagon and stacking interaction with two other independent networks (Figure 10b). Nevertheless, the resultant molecular carpet has partially open cavities within the layer and this void space is effectively filled in a cog-like manner because of the way adjacent layers stack.

Interestingly, minor variations in the ionic hydrogen bonding patterns of secondary ammonium cations-carboxylate anions in **14** cause the formation of a rectangular grid or a squashed hexagon with internal dimension 10 Å x 19 Å. The propyl groups of the ammonium cations project above and below the sheet like in supramolecular laminates discussed before and form a walled structure around the cavities. Therefore, unlike **13**, complex **14** undergoes an inclined interpenetration and forms an interlocked three dimensional laminate structure as shown in Figure 11b. The solvated guest molecules occupy the void space and are highly disordered.

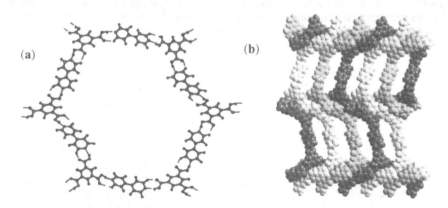

Figure 10. (a) The hexagon formed with chair conformation through pyridine-carboxyl hydrogen bonds in **13**. The hatched lines indicate O-H•••O and C-H•••O hydrogen bonds. (b) The parallel interpenetration of open hexagons result in a molecular carpet structure. Notice that void space remains even after interweaving.

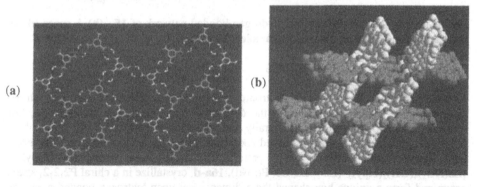

Figure 11. (a) The subtle differences in ammonium-carboxylate hydrogen bonding motifs in **14** lead to squashed hexagonal open network (compare with corresponding idealized hexagonal network shown in scheme 4). (b) Inclined two-fold interpenetration of distorted hexagonal sheets at 78°. The propyl substituents of the ammonium cations projecting above and below the sheet are not shown here for the sake of clarity.

5.1.2 Interwoven 2D Layer Instead of a 3D Diamondoid Network

In general tetrahedral molecules with rigid complementary hydrogen bonding sites adapt diamondoid architecture with levels of interpenetration based on the relative size of superdiamondoid cage and the volume of the tecton.[59] However, bis(4,4'-dihydroxyphenyl) sulfone, **15**, exploits its tetrahedrally disposed complementary hydrogen bonding sites to generate a unique doubly interwoven molecular carpet architecture in the solid state (Figure 12).[60] The two-dimensional hydrogen bonding in **15** is sustained by interaction between flexible hydroxy groups and rigid sulfone groups and creates cavities with dimensions *ca.* 8.0 Å x 10.0 Å. These cavities are large enough to facilitate a two-fold interwoven molecular carpet structure. Entangled 2D grids in **15** are essentially stabilized by van der Waals and herringbone interactions. Indeed, a 3D diamondoid network is also feasible for **15**, but presumably packs less efficiently.

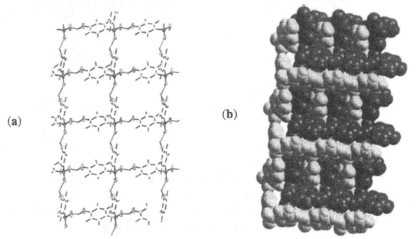

(a)   (b)

Figure 12. (a) The two-dimensional hydrogen bonded network of **15**. (b) A space filling diagram of the doubly interwoven supramolecualr carpet.

5.2 MOLECULAR CHINESE BLINDS

Lanthanides, with their unique electronic and magnetic properties and rich stereo chemistry offer a novel approach for the design of exotic crystalline architectures. For example, $Ln(NO_3)_3$ forms three structurally distinct series of complexes with simple 4,4'-bipyridine, i.e., neutral hydrogen bonded cocrystals, metal-ligand coordination complexes, and organic cation-inorganic anion pairs.[61] The isostructural complexes [BipyH+] $[Ln(NO_3)_4(OH)_2(bipy)]^-$ (Ln = La, Ce, Pr, Nd), **16a-d**, crystallize in a chiral $P2_12_12_1$ space group and form a unique box shaped three dimensional open hydrogen bonded networks that spontaneously undergo self-inclusion. Therefore, we have decided to explore the lanthanide's structural patterns further in order to bring them into the realm of crystal engineering.

When lanthanum nitrate is reacted with 1,2-bis(4-pyridyl)ethane (bpe), a charged complex, $[bpeH]^+[La(NO_3)_4(OH)_2(bpe)]^-$, **17**, was formed in a chiral space group $C222_1$.[62] The lanthanum cations in **17** are 11-coordinate, complexed with four bidentate nitrate anions, two bridging bpe ligands and one water molecule resulting in an anionic 1D coordination polymer (Figure 13a). The two protons of the water molecule along with oxygen atoms of adjacent nitrate anions constitute complementary divergent hydrogen bonding recognition sites on the backbone of the 1D coordination polymer. The hydrogen bonding nature of the 1D polymers leads to the formation of a negatively charged pleated sheet structure with unusual cavities. Interestingly, such cavities are threaded by two linear hydrogen bonded chains of the monoprotonated bpeH⁺ molecules resulting an unprecedented polypseudorotaxane type architecture or molecular Chinese blinds structure as shown in Figures 13b and 13c.

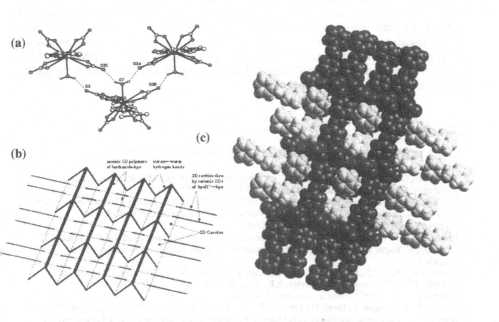

(a)

(b)

(c)

Figure 13. (a) Hydrogen bonded self-assembly of [La(NO₃)₄(H₂O)(bpe)]⁻ anions in **17**. The 1D polymers run perpendicular to the plane of the paper. (b) A schematic representation of "Chinese blinds" structure in **17**. (b) A space-filling representation of a single pleated sheet of **17** threaded by 1D linear chains of bpeH⁺•••bpeH⁺.

6. Conclusions

Crystal engineering would be three decades young as it enters into the new millenium. The scope of the subject has been constantly growing since its inception in the context of solid state chemical reactivity.[63] Our realization of the importance of ubiquitous intermolecular interactions in biological processes, materials and drug design elevated crystal engineering into a truly multidisciplinary field and self-sustainable discipline. In this article we tried to provide a sampling of diversified aspects of crystal engineering together. Our efforts to synthesize freebase and mixed metalloporphyrins for light-harvesting purposes and discrete macrocycles as inorganic host materials make use of principles for forming coordination polymers. In contrast, our ability to steer hydrogen bonds in trimesic acid salts for synthesizing open networks and clay like materials predictably comes from our knowledge base on hierarchical strength of hydrogen bonds provided by organic crystal engineering. The molecular blinds structure culminates this review article not only for its aesthetic appeal, but also for a variety of reasons. The blinds structure is formed as a result of highly synchronized self-assembly of coordination polymers, hydrogen bonds and ionic interactions and suggest that we can design exotic materials by combining organic and inorganic branches of crystal engineering. Another important crystal engineering aspect hidden behind the Chinese blinds is its chirality. In fact, the rational design of chiral networks has been an elusive goal for crystal engineers, probably for the reason that symmetry driven crystalline networks tend to adopt centrosymmetric space groups. However, if we use high coordinating lanthanides as building blocks, we can easily induce dissymmetry at metal centers simply by coordination with suitable complementary ligand(s), which may thus potentially result in controllable 1D, 2D or 3D chiral networks.

7. References

1. Lehn, J.-M. (1995) *Supramolecular Chemistry: Concepts and Perspectives*; VCH, Weinheim, Germany.
2. Whitesides, G.M., Mathias, J.P. and Seto, C.T. (1991) Molecular self-assembly and nanochemistry –a chemical strategy for the synthesis of nanostructure, *Science*, **254**, 1312-1319.
3. Sauvage, J.-P. and Hosseini, M.W. (volume eds.) (1996) Templating, self-assembly and self-organization, in J.L. Atwood, J.E.D. Davies, D.D. MacNicol, F. Vögtle and J.-M. Lehn (eds.), *Comprehensive Supramolecular Chemistry*, Vol. 9, Pergamon, Oxford.
4. Reinhoudt, D. N. (volume ed.) (1996) Supramolecular Technology, in J.L. Atwood, J.E.D. Davies, D.D. MacNicol, F. Vögtle and J.-M. Lehn (eds.), *Comprehensive Supramolecular Chemistry*, Vol. 10, Pergamon, Oxford.
5. Heuer, A.H., Fink, D.J., Laraia, V.J., Arias, J.I., Calvert, P.D., Kendall, K., Messing, G.I., Blackwell, J., Rieke, P.C., Thompson, D.H., Wheeler, A.P., Veis, A. and Caplan, A.I. (1992) Innovative materials processing strategies: a biomimetic approach, *Science*, **255**, 1098-1105.
6. Balzani, V., Campagna, S., Denti, G., Juris, A., Serroni, S. and Venturi, M. (1998) Designing dendrimers based on transition-metal complexes. Light-harvesting properties and predetermined redox patterns, *Acc. Chem. Res.* **31**, 26-34.
7. Clearfield, A. (1998) Organically pillared micro- and mesoporous materials, *Chem.Mater.* **10**, 2801-2810.
8. Amabilino, D.B. and Stoddart, J.F. (1995) Interlocked and intertwined structures and superstructures, *Chem. Rev.* **95**, 2725-2828.
9. Kaes, C., Hosseini, M.W., Rickard, C.E.F., Skelton, B.W. and White, A.H. (1998) Molecular tectonics, part 7 -synthesis and structural analysis of a helical coordination polymer formed by the self-assembly of a 2,2 '- bipyridine-based exo-ditopic macrocyclic ligand and silver cations, *Angew.Chem.Int.Ed.Engl.*, **37**, 920-922.
10. Pauling, L. (1960) *The nature of chemical bond and the structures of molecules and crystals: An introduction to modern structural chemistry*, 3rd Ed., Cornell University Press, Ithaca, New York.
11. Kitaigorodskii, A.I. (1973) *Moleccular crystals and molecules*, Academic Press, New York.
12. Desiraju, G.R. (1989) *Crystal engineering: the design of organic solids*, Elsevier, Amsterdam.
13. Dunitz, J.D. (1996) *Thoughts on crystals as supermolecules*, in G. R. Desiraju, (ed.) The Crystal as a Supramolecular Entity, pp. 1-30.
14. Desiraju, G. R. and Sharma, C.V.K. (1996) *Crystal engineering and molecular recognition- twin facets of supramolecular chemistry*, in G. R. Desiraju, (ed.) The Crystal as a Supramolecular Entity, pp. 31-61.
15. Desiraju, G. R. (1995) Supramolecular synthons in crystal engineering - a new organic synthesis, *Angew. Chem.Int.Ed.Engl.* **34**, 2311-2327.
16. Stang, P.J. and Olenyuk, B. (1997) Self-assembly, symmetry, and molecular architecture: coordination as the motif in the rational design of supramolecular metallacyclic polygons and polyhedra, *Acc.Chem.Res.* **30**, 502-518.
17. Braga, D., Grepioni, F. and Desiraju, G.R. (1998), Crystal engineering and organometallic architecture, *Chem. Rev.* **4**, 1375-1405.
18. Yaghi, O.M., Li, G. and Li, H. (1995) Selective binding and removal of guests in a microporous metal-organic framework, *Nature*, **378**, 703-706.

19. Russell, V.A., Evans, C.C., Li, W. and Ward, M.D. (1997) Nanoporous molecular sandwiches: pillared two-dimensional hydrogen-bonded networks with adjustable porosity, *Science*, **276**, 575-579.

20. MacGillivray L.R. and Atwood J.L. (1997) Rational design of multicomponent calix[4]arenes and control of their alignment in the solid state, *J.Am.Chem.Soc.*, **119**, 6931-6932.

21. Aakeroy, C.B. (1997) Crystal engineering: strategies and architectures, *Acta Crystallogr.* **B53**, 569-586.

22. Robson, R. (1996) Infinite Networks, in J.L. Atwood, J.E.D. Davies, D.D. MacNicol, F. Vögtle and J.-M. Lehn (eds.), *Comprehensive Supramolecular Chemistry*, Vol. **6**, Pergamon, Oxford, pp. 733-755.

23. Sharma, C.V.K. and Desiraju, G.R. (1994) C-H•••O Hydrogen bond patterns in crystalline nitro compounds: studies in solid state molecular recognition. *J.Chem.Soc., Perkin Trans.* 2, 2345.

24. Desiraju, G.R. (1996) The C-H•••O hydrogen bond: structural implications and supramolecular design, *Acc.Chem.Res.* **29**, 441-449.

25. Sharma, C.V.K. and Rogers, R.D. (1998) Perspectives of Crystal Engineering, *Materials Today*, **1**, issue, 3, 27-30.

26. Desiraju, G.R. (1997) Crystal gazing: structure prediction and polymorphism, *Science*, **278**, 404-405.

27. Wasielewski, M. R. (1992) Photoinduced electron transfer in supramolecular systems for artificial photosynthesis, *Chem.Rev.* **92**, 435-461.

28. Drain, C. M., Nifiatis, F., Vasenko, A. and Batteas, J. D. (1998) Porphyrin tessellation by design: metal-mediated self-assembly of large arrays and tapes, *Angew.Chem.Int.Ed.Engl.* **37**, 2344-2347.

29. Li, F.R., Yang, S.I., Ciringh, Y.Z., Seth, J., Martin III, C. H., Singh, D. L., Kim, D.H., Birge, R. R., Bocian, D. F., Holten, D. and Lindsey, J. S. (1998) Design, synthesis, and photodynamics of light-harvesting arrays comprised of a porphyrin and one, two, or eight boron-dipyrrin accessory pigments, *J.Am.Chem.Soc.* **120**, 10001-10017.

30. Hunter, C. A. and Hyde, R. K. (1996) Photoinduced energy and electron transfer in supramolecular porphyrin assemblies, *Angew.Chem.Int.Ed.Engl.* **35**, 1936-1939.

31. Buchler, J. W. (1978) Synthesis and properties of metalloporphyrins in D. Dolphin (ed.), *The Porphyrins*, Vol. 1, Academic Press, New York, pp. 389-483.

32. Abrahams, B. F., Hoskins, B. F., Michail, D. M. and Robson, R. (1994) Assembly of porphyrin building-blocks into network structures with large channels, *Nature*, **369**, 727-729.

33. Kumar, R. K., Balasubramanian, S., Goldberg, I. (1998) Supramolecular multiporphyrin architecture. coordination polymers and open networks in crystals of tetrakis(4-cyanophenyl) and tetrakis(4-nitrophenyl)metalloporphyrin, *Inorg.Chem.* **37**, 541-552.

34. Bhyrappa, P., Wilson, S. R. and Suslick, K. S. (1997) Hydrogen-bonded porphyrinic solids: supramolecular networks of octahydroxy porphyrins, *J.Am.Chem.Soc.* **119**, 8492-8502.

35. Sharma C.V.K., Broker, G.A., Huddlestone, J.G., Baldwin, J.W., Metzger, R.M. and Rogers, R.D. (1999) Design strategies for solid state supramolecular arrays containing both mixed-metallated and freebase porphyrins", *J.Am.Chem.Soc.* **121**, 1137-1144.

36. Kusukawa T. and Fujita M (1998), Encapsulation of large, neutral molecules in a self-assembled nanocage incorporating six palladium(II) ions, *Angew.Chem.Int.Ed.Engl.* **37**, 3142-3144.

37. Masciocchi, N., Ardizzoia, G.A., LaMonica, G., Maspero, A. and Sironi, A. (1998) Unique formation of a crystal phase containing cyclic oligomers and helical polymers of the same monomeric fragment, *Angew.Chem.Int.Ed.Engl.* **37**, 3366—3368.

38. Hunter, C.A. (1995) Self-assembly of molecular-sized boxes, *Angew.Chem.Int.Ed.Engl.* **34**, 1079-1081

39. Balzani, V., Gomez-Lopez, M. and Stoddart, J. F. (1998) Molecular machines, *Acc.Chem.Res.* **31**, 405-414.

40. Zaworotko, M. J. (1998) Coordination polymers in K. R. Seddon, and M. J. Zaworotko (eds.) *Crystal Engineering: The Design and Application of Functional Solids*, NATO, ASI series, Kluwer, Dordecht, Netherlands.

41. Sharma C.V.K., Griffin, S.T. and Rogers, R.D. (1998) Simple routes to supramolecualr squares with ligand corners: 1:1 Ag(I):pyrimidine cationic tetranuclear assemblies. *Chem.Commun.* 215-216.

42. Mallouk, T. E. and Gavin, J.A. (1998) Molecular recognition in lamellar solids and thin films, *Acc.Chem.Res.* **31**, 209-217.

43. Crooks, R.M. and Ricco, A.J. (1998) New organic materials suitable for use in chemical sensor arrays, *Acc.Chem.Res.* **31**, 219-227.

44. Menger, F.M., Lee J. and Hagen, K.S. (1991) Molecular laminates. Three distinct crystal packing modes, *J.Am.Chem.Soc.* **113**, 4017-4019.

45. Clearfield, A. (1998) Metal phosphonate chemistry, in K.D. Karlin (ed.) *Progress in Inorganic Chemistry*; John Wiley & Sons, New York, Vol. **47**, pp. 371-510.

500

46. Herbstein, F. H. (1996) 1,3,5-Benzenetricarboxylic acid (trimesic acid) and some analogues, in J.L. Atwood, J.E.D. Davies, D.D. MacNicol, F. Vogtle and J.-M. Lehn (eds.), *Comprehensive Supramolecular Chemistry*, Pergamon, Oxford, Vol. **6**, pp. 61-83.

47. Kolotuchin, S.V., Fenlon, E. E., Wilson, S. R., Loweth, C. J. and Zimmerman, S.C. (1995) Self-assembly of 1,3,5-benzenetricarboxylic acids (trimesic acids) and several analogues in the solid state, *Angew.Chem.Int.Ed.Engl.* **34**, 2654-2657.

48. Melendez, R. E., Sharma, C.V.K., Zaworotko, M. J., Bauer, C. and Rogers, R.D. (1996) Toward the design of porous solids: modular honeycomb grids sustained by anions of trimesic acid, *Angew.Chem.Int.Ed.Engl.* **35**, 2213.

49. Sharma, C.V.K., Bauer, C., Rogers, R.D. and Zaworotko, M.J. (1997) Interdigitated supramolecular laminates, *Chem.Commun.* 1559-1560.

50. Biradha, K., Dennis, D., MacKinnon, V.A., Sharma, C.V.K. and Zaworotko, M. J. (1998) Supramolecular synthesis of laminates with affinity for aromatic guests: a new class of clay mimics, *J.Am.Chem.Soc.* **120**, 11894-11903.

51. Radloff, R., Bauer, W. and Vinograd, J. (1967) A dye-buoyant-density method for the detection and isolation of closed circular duplex DNA: the closed circular DNA in Hela Cells, *Proc.Natl.Acad.Sci.* USA, **57**, 1514-1521.

52. Hudson, B. and Vinograd, J. (1967) Catenated circular DNA molecules in Hela Cell Mitchondira, *Nature*, **216**, 647-652.

53. Seeman, N.C. (1998) Nucleic acid nanostructures and topology, *Angew.Chem.Int. Ed. Engl.* **37**, 3220-3238.

54. Sauvage, J.-P. (1998) Transition metal-containing rotaxanes and catenanes in motion: toward molecular machines and motors, *Acc.Chem.Res.* **31**, 611-619.

55. Batten, S.R. and Robson, R. (1998) Interpenetrating nets: ordered, periodic entanglement, *Angew.Chem.Int.Ed.Engl.* **37**, 1460-1494.

56. Ermer, O. (1988) Fivefold-diamond structure of adamantane-1,3,5,7-tetracarboxylic acid, *J.Am.Chem.Soc.* **110**, 3747-3754.

57. Sharma, C.V.K. and Zaworotko, M.J. (1996) X-ray crystal structure of $C_6H_3(CO_2H)$-1,3,5• (4,4'-Bipyridine): 'a super trimesic acid' chicken-wire grid". *Chem.Commun.* 2655-2656.

58. Sharma, C.V.K. and Zaworotko, M.J. Unpublished results.

59. Brunet, P., Simard, M. and Wuest, J.D. (1997) Molecular tectonics. porous hydrogen-bonded networks with unprecedented structural integrity, *J.Am.Chem.Soc.* **119**, 2737-2738.

60. Davies, C., Langler, R. F., Sharma, C.V.K. and Zaworotko, M. J. (1997) A supramolecular carpet formed *via* self-assembly of bis(4,4'-dihydroxyphenyl)sulfone, *Chem.Commun.*, 567-568.

61. Al-Rasoul, K. and Weakley, T.J.R. (1982), The crystal structure of 4,4'-bipyridinium(+1) tetranitratodiaqua (4,4'-bipyridyl)-neodymiate(III) and trinitratotetraaquaytterbium(III)-4,4'-bipyridinium(+1)nitrate(1/2), *Inorg.Chim.Acta.* **60**, 191-196.

62. Sharma, C.V.K. and Rogers, R.D. (1999) Molecular Chinese blinds: spontaneous self-organization of tetranitrato lanthanides into open chiral hydrogen bonding networks, *Chem. Commun.* 83-84.

63. Schmidt, G.M.J. (1971) Photodimerization in the solid state, *Pure Appl. Chem.* **27**, 647-678.